A herd of beisa oryx in East Africa. (Photograph by John Dominis, LIFE Magazine © Time Inc.)

Mammalogy

Terry A. Vaughan

Department of Biological Sciences,
Northern Arizona University

1972
W. B. SAUNDERS COMPANY
PHILADELPHIA   LONDON   TORONTO

W. B. Saunders Company: West Washington Square
                           Philadelphia, Pa. 19105

                           12 Dyott Street
                           London, WC1A   1DB

                           833 Oxford St.
                           Toronto 18, Ontario

Mammalogy   ISBN 0-7216-9011-4

Print No.:   9   8   7   6   5   4   3   2   1

PREFACE

This text is intended for use in upper division college or university courses on mammals to be taken by students who have a basic background in zoology. The approach has been to discuss structure and function together whenever possible, for students are far more interested in the functioning mammal than in lists of diagnostic morphological characters. Probably no two instructors cover the same material in presenting a course in mammalogy, and with this in mind I have tried to treat enough aspects of the biology of mammals to make this book useful to different instructors with contrasting approaches to the subject.

I have made liberal use of anatomical drawings because many schools lacking adequate teaching collections of mammals offer courses in mammalogy, and students at these schools can become familiar with the structures of certain groups of mammals only by referring to drawings. Some of the drawings may also be useful as reference material during laboratory work.

In mammalogy, as in many other fields of science, new knowledge is being gained at such a rate that a textbook becomes to some extent a progress report. At best, a book on any aspect of science is somewhat out of date by the time it is published. The major diagnostic structural features of most families and genera of mammals have been well known for some time, but new studies yielding information bearing on the relationships between mammals and new fossil material are constantly forcing revisions of taxonomic schemes and of our ideas as to patterns of evolution. Within the broad fields of mammalian behavior, ecology, physiology, paleontology, and functional morphology, much remains to be learned. Even basic life history information is lacking for many species of mammals. But the job of assembling information on mammals has been eased within the last few years by the publication of books with collections of articles on such subjects as primate behavior, hibernation, the biology of bats, and aquatic mammals.

I have had considerable help on this book from many people. The help of the staff of W. B. Saunders Company has been invaluable. Carl W. May, former Biology Editor, shepherded this book through all but its final stages of preparation and had a hand in planning the organization and coverage. Richard H. Lampert, present Biology Editor, gave encouragement, advice, and editorial assistance in the final stages of writing and during production of the book.

For critically reviewing chapters and offering important comments and advice I would like to thank Sydney Anderson, Russell P. Balda, Gary C. Bateman, Tyler Buchenau, William H. Burt, William A. Clemens, Mary R. Dawson, James S. Findley, Edwin Gould, E. Raymond Hall, Milton Hildebrand, C. D. Johnson, Karl F. Koopman, William Z. Lidicker, Richard E. MacMillen, Robert T. Orr, James L. Patton, Oliver P. Pearson, Gilbert C. Pogany, Frank Richardson, Alfred S. Romer, Constantine Slobodchikoff, Hobart M. Van Deusen, B. J. Verts, Warren F. Walker, Jr., Olwen Williams, and William O. Wirtz II.

Many students assisted with various aspects of this project. I would especially like to thank Celeste Haren, Thomas R. Huels, Jo Ann Mendolia, Larry G. Marshall, O. J. Reichman, Samuel Semoff, Jan Smith, Roger B. Smith, Kent M. Van De Graaff, Kenneth Weber, and Gary J. Weisenberger.

For allowing me to use their photographs, I am indebted to many friends and colleagues whose names appear in the legends for their photographs. Special thanks are due to Hobart M. Van Deusen, who supplied photographs of New Guinea mammals; Fritz C. Eloff, who sent photographs of the Kalahari Desert and of gemsbok; Diana Harrison and Larry G. Marshall, who arranged for my use of photographs of marsupials and of Australian terrestrial communities; and W. Leslie Robinette, who kindly heeded my pleas for photographs of African mammals. For expert help in selecting appropriate photographs from the files of the San Diego Zoo, I extend my appreciation to Edalee Harwell.

Throughout much of the course of this project Mrs. Elouise Weisenberger did the typing, assisted with figures, and worked on a variety of laborious jobs that one finds it expedient to give to a capable assistant. In the latter stages of the work Miss Susan M. Beeston very ably took over this work, and Mrs. Karen Van De Graaff assisted with typing.

Finally and most importantly, I am indebted to my wife, Hazel A. Vaughan, who contributed optimism, tolerance, and a critical and artistic eye.

TERRY A. VAUGHAN

CONTENTS

viii   CONTENTS

CHAPTER 1

INTRODUCTION

THE DOMAIN OF MAMMALOGY

Mammalogy—the division of zoology dealing with mammals—has occupied the efforts of scientists of many kinds. Vertebrate zoologists have studied such aspects as the structure, taxonomy, distribution, and life histories of mammals; physiologists have considered mammalian hibernation and water metabolism; physicists and engineers have studied mammalian echolocation and locomotion; geologists and vertebrate paleontologists have outlined the patterns of mammalian evolution; and anthropologists and psychologists have considered mammalian behavior. Indeed, many significant observations of mammals under natural conditions have been made by keen observers who lacked formal training in zoology. In mammalogy, as in many other fields of zoology, there is room for the non-professional. An individual willing to devote time to careful observations of the activities of mammals, and effort to the accurate recording of these observations, may make worthwhile contributions. Wide use of the observations of trappers and outdoorsmen has added importantly to the published works of well-known zoologists. As an example, many observations of carnivores by the late W. H. Parkinson, a trapper who worked on the western slope of the Sierra Nevadas of California, were included in Grinnell, Dixon, and Linsdale's *Fur-bearing Mammals of California* (1937). Our present knowledge of mammals, then, has been contributed by scientists trained in zoology and in diverse non-zoological fields, and by perceptive but untrained observers.

Mammals have been regarded as worthy of study by this wide variety of workers for many reasons. The practical aspects have attracted some. Much has been learned of mammalian histology and of the effects of diseases and drugs by the use of various kinds of laboratory mammals; work on domestic breeds of mammals has improved meat production; and research on game species has shown how sustained yields of these animals can be achieved through appropriate management techniques. But to most students and researchers, practicality is not foremost. To this group, mammals are simply fascinating creatures with physiological, structural, and behavioral adaptations to many different

1

modes of life; living mammals in their natural settings are the focal point of interest. The adaptations themselves, how they enable mammals to efficiently exploit demanding environmental conditions, and how the adaptations evolved are all fascinating lines of inquiry. Studies of interactions between different species of mammals in their natural habitats, of population cycles and migrations, and of predator-prey relationships are ecological investigations frequently begun primarily because of a researcher's intense interest in a biological relationship rather than his preoccupation with solving a practical problem. The impressive literature on mammals has resulted largely from such basic research. In this book I deal primarily with the basic information about mammals; "practical mammalogy" is a secondary concern.

During the last 25 years our knowledge of mammalian biology has expanded tremendously: echolocation (natural "sonar") has been studied intensively in both bats and marine mammals; the remarkable ability of some mammals to live under conditions of extreme aridity with no drinking water has been explained; insulation and circulatory and metabolic adaptations in relation to temperature regulation and metabolic economy have been studied; adaptations to deep diving in marine mammals have been investigated; hibernation and migration and the mechanisms influencing them have received attention; important contributions have been made to our knowledge of population cycles of mammals and the factors that may control them; and studies of functional morphology have enlarged our understanding of mammalian terrestrial, aquatic, and aerial locomotion. Probably no field has been more tardy in developing than that of animal behavior, but within the last few years this has been a productive discipline. Considerable attention has been concentrated on the social behavior of anthropoid primates, and these and other mammals have been found to have remarkably complex patterns of social behavior. Explanations for the shapes of the horns and antlers of various deer and antelope and their kin have resulted from careful studies of breeding behavior of these animals (see p. 340), and predation by mammals has been put in reasonable perspective partially by recent behavioral research. There are still great and vital gaps in our knowledge, but these gaps have been narrowed markedly in roughly the last two decades, and work continues.

CLASSIFICATION

In any careful study, one of the vital early steps is the organization and naming of objects. As stated by Simpson (1945:1), "It is impossible to examine their relationships to each other and their places among the vast, incredibly complex phenomena of the universe, in short to treat them scientifically, without putting them into some sort of formal arrangement." The arrangement of organisms is the substance of taxonomy. But the modern taxonomist, perhaps better termed a systematist, is less interested in identifying and classifying animals than in studying their evolution. He brings information from such fields as genetics, ecology, behavior, and paleontology to bear on the subjects of his research. He attempts to include in each taxonomic category only animals that evolved from a common ancestor. Excellent discussions of the importance of systematics to our knowledge of animal evolution are given by Simpson (1945) and Mayr (1963).

Because of the difficulties arising from a single kind of animal or plant being recognized by different common names by people in different areas, or by many common names by people in one area, scientists more than 200 years ago adopted a system of naming organisms that would be recognized by biologists throughout the world. Each known kind of organism has been given a binomial (two-part) scientific name. The first, the *generic name*, may be applied to a number of kinds; but the second name refers to a specific kind, a *species*. As an example, the blacktailed jackrabbit of the western United States is *Lepus californicus*. To the genus *Lepus* belong a number of similar, but distinct, long-legged species of hares, such as *L. othos* of Alaska, *L. europaeus* of Europe, and *L. capensis* of Africa. Because considerable geographic variation frequently occurs within a species, a third name is often added; this designates a particular *subspecies*. Thus, the large-eared and pale-colored subspecies of *L. californicus* that occurs in the deserts of the western United States is *L. c. deserticola;* the smaller-eared and dark-colored subspecies from coastal California is *L. c. californicus.*

The species is the basic unit of classification. A modern and widely accepted definition of a species is given by Mayr (1942:120): "Species are groups of actually or potentially interbreeding natural populations, which are reproductively isolated from other such groups." Each species is generally separated from all other species by a "reproductive gap," but within each species there is the possibility for gene exchange; as put by Dobzhansky (1950), all members of a species "share a common gene pool."

Clearly, however, not all species resemble each other to the same degree or are equally closely related. The hierarchy of classification, based on the starting point of the species, has been developed to express degrees of structural similarity and, ideally, phylogenetic relationships between species and groups of species. The taxonomic scheme includes a series of categories, each higher category more inclusive than the one below. Using our example of the hares, a number of long-legged species are included in the genus *Lepus;* this genus, and other genera containing "rabbit-like" mammals, form the family Leporidae; this family, and the family Ochotonidae (the pikas), share certain structural features not possessed by the other mammals, and belong to the *order* Lagomorpha; this order, and all other mammalian orders, form the *class* Mammalia, members of which differ from all other animals in the possession of hair, mammary glands, and many other features. Mammals, birds, reptiles, amphibians, and fish all possess an endoskeleton, and these groups (in addition to some others) form the *phylum* Chordata. All of the phyla of animals (Protozoa, Porifera, Coelenterata, etc.) are united in the Animal Kingdom. The classification of our jackrabbit can be outlined as follows:

Kingdom Animal
  Phylum Chordata
    Class Mammalia
      Order Lagomorpha
        Family Leporidae
          Genus *Lepus*
            Species *Lepus californicus*

Further subdivision of this scheme of classification may result from the recognition of subgroups such as subclass, superorder, or subfamily.

Most ordinal names end in -*a*, as in Carnivora; all family names end in -*idae,* and all subfamily names end in -*inae.* In the following discussions, contractions of the names of orders, families, or subfamilies will often be used as adjectives for the sake of convenience: leporid will refer to Leporidae, leporine to Leporinae, and lagomorph to Lagomorpha.

Some similarities between different kinds of animals are due to parallelism or to convergence. *Parallelism* occurs when two closely related kinds of animals pursue similar modes of life for which similar structural adaptations have evolved. The similar specializations of the skull and dentition (elongate snouts and reduced teeth) that occur in a number of genera of nectar-feeding Neotropical bats are examples of parallelism. *Convergence* involves the development of similar adaptations to similar (or occasionally nearly identical) styles of life by distantly related species. The golden moles of Africa (see p. 76) and the marsupial "moles" of Australia (see p. 52) are examples of convergence. These animals belong to different mammalian infraclasses (Eutheria and Metatheria, respectively; see p. 39), and their lineages have been separate for over 70 million years; but their habits are much the same and structurally they resemble each other in many ways.

An outline of the classification of mammals used in this book is given in Chapter 4 (p. 36). It is not based on any single published classification, but in some ways it reflects current taxonomic thought. Although it departs from his system in many minor ways, the classification used here is based on that of Simpson (1945). It should be stressed that no universal agreement has been reached on the classification of mammals. Our knowledge of many groups of mammals is incomplete, and future study may demonstrate that some of the families listed here can be discarded because they contain animals best included in another family; perhaps a family or two are yet to be described. The present classification, then, is not used by all mammalogists, and it is by no means immutable.

DESCRIBING SPECIALIZATION

The terms "primitive," "specialized," and "advanced" are used repeatedly in the chapters on the orders and families. A primitive mammal is one that has not departed far from the ancestral type, or at least has retained many structural characters typical of the ancestral type. A monotreme (duck-billed platypus or spiny anteater) is more primitive than a house cat because it lays eggs, has a small brain, and has bones in the pectoral girdle that are found in reptiles but not in other mammals. In these ways the monotreme resembles reptiles, the ancestral group of mammals, whereas the cat does not.

A specialized mammal is one that, in becoming adapted to a particular mode of life, has departed strongly from the ancestral structural plan. A horse is specialized because, in becoming adapted to a life in which speed afoot and the ability to feed on grasses are important, the limbs became elongate, all digits but the third were essentially lost, and the cheek teeth developed complicated occlusal surfaces. In these ways (and many others) the specialized horse has departed strongly from the structure of the primitive mammal.

The term *advanced* is frequently used in comparisons of different members of an evolutionary line or taxonomic group. An advanced species is one in which the particular structural features that characterize the group are highly developed. As an example, the present-day big brown bat is more advanced than the early Eocene bat *Icaronycteris index*. Both are specialized for flight, but because of many refinements in the flight apparatus of the big brown bat it is the far more perfectly adapted flier.

PLAN OF THE BOOK

The first part of this book covers preliminary material: Chapter 2 deals with the characteristics of mammals, the structural features that characterize the group; Chapter 3 briefly covers the evolution of mammals from reptilian ancestry; and Chapter 4 is an introduction to the classification of mammals. In roughly the first half of the remainder of the book (Chapters 5 through 14) the orders and families of mammals are discussed; the second half (Chapters 15 through 21) treats selected aspects of the biology of mammals. My main hope is that from this coverage of the subject the student can gain a general understanding of, and an appreciation for, the form and function of mammals.

The anatomical drawings, which I regard as essential to the ordinal chapters, should help students understand the discussions of structure and function. I have made liberal use of these drawings, most of which illustrate skulls, teeth, or feet. The profiles of skulls are from the left side, and all occlusal views of teeth show the right upper or the left lower tooth row.

The bibliography in the appendix includes not only papers that are cited in the text, but also some additional publications that have material useful to students by authors whose papers are cited. Preceding this bibliography is a list of general references and publications dealing with mammals of various regions. In an attempt to compensate to some extent for the lack of emphasis on applied or practical mammalogy, this brief general bibliography includes some references on that topic.

CHAPTER 2

MAMMALIAN CHARACTERISTICS

Mammals owe their spectacular success to many features. Many of the most important and most diagnostic mammalian characteristics serve to further intelligence and sensory ability, to promote endothermy, or to increase the efficiency of reproduction or of securing and processing food. The senses of sight and smell are highly developed, and the sense of hearing has undergone greater specialization in mammals than in any other vertebrates. Efficient utilization of food is aided by specializations of the dentition and the digestive system. The perfection of endothermy has allowed mammals to remain active under a wide array of environmental conditions, and specializations of the postcranial anatomy, particularly the limbs and feet, have enabled them to make effective use of this activity. Extended periods of association between the parents and young of some species have facilitated the use of demanding types of foraging and the development of complex social behavior.

The basic structural plan of mammals was inherited from a relatively unspectacular reptilian group, the mammal-like reptiles of the synapsid order Therapsida (see p. 27). Members of this ancient order followed an evolutionary path that diverged strongly from that of the vastly more spectacular and more successful dinosaurs and other ruling reptiles (subclass Archosauria). The key to the marginal persistence of the therapsids through the Triassic was perhaps their ability to move and to think more quickly than their dull-witted archosaurian contemporaries. These same abilities probably enabled the descendents of the therapsids, the mammals, to survive through the Jurassic and Cretaceous, periods when the dinosaurs completely dominated the terrestrial scene.

An important morphological trend in the therapsid-mammalian line was toward skeletal simplification. An engineer may redesign a machine and increase its efficiency by reducing the number of parts to the minimum consistent with the effective performance of a particular function; a similar type of simplification occurred in the therapsid skeleton. In the skull and lower jaw, which in primitive reptiles consisted of many bones, a number of bones were lost or reduced in size.

The limbs and limb girdles were also simplified to some extent, and their massiveness was reduced. As a result, the advanced therapsid skeleton roughly resembles that of the egg-laying mammals (order Monotremata), but the limbs of some therapsids were less laterally splayed than are those of today's specialized monotremes.

When mammals first appeared in the Triassic, then, they represented no radical structural departure from the therapsid plan, but had simply attained a level of development (involving a dentary/squamosal jaw articulation; see p. 29) that is interpreted by most vertebrate paleontologists as indicating that the animals had crossed the mammalian-reptilian boundary. Many of the mammalian characters discussed in this chapter resulted from evolutionary trends clearly characteristic of therapsid reptiles. Unfortunately, the fossil record cannot indicate when various important features of the soft anatomy became established, and one can only speculate as to whether or not advanced therapsid reptiles had such features as mammae, hair, or a four-chambered heart.

The following sources have discussions of the characteristics of mammals: Romer, 1970; Smith, 1960; Weichert, 1965; Young, 1957. Especially useful is the treatment by Romer in *Vertebrate Paleontology* (1966:187–196).

SOFT ANATOMY

Skin Glands. The skin of mammals contains several kinds of glands not found among other vertebrates; the most important of these are the *mammary glands*. The nutritious milk secreted from these glands provides the sole nourishment for young during their initial period of rapid postnatal growth. In most mammals the openings of the glands are in projecting nipples (mammae), from which milk is sucked by the young. In monotremes nipples are lacking, and young suck milk from tufts of hair on the mammary areas (Burrell, 1927; Ewer, 1968:236). Cetaceans have muscles that force milk into the mouth of the young, a seemingly necessary adaptation in animals that have no lips and are therefore unable to suck. The number of nipples varies from two in a number of mammals to about 19 in the mouse-opossum (*Marmosa;* Tate, 1933:36).

Other types of skin glands are of importance in mammals. The watery secretion of the *sweat glands* functions primarily to promote evaporative cooling, but also eliminates some waste materials. In man and some ungulates sweat glands are broadly distributed over the body surface, but in most mammals they are more restricted. In some insectivores, rodents, and carnivores, sweat glands occur only on the feet or on the venter, and they are completely lacking in the Cetacea and in some bats and rodents. Hair follicles are supplied with *sebaceous glands* (Fig. 2–1); their oily secretion lubricates the hair and the skin. A variety of *scent glands* and *musk glands* occur among mammals. These glands are variously used as attractants, for marking territories, for communication during social interactions, or for protection. "Skunk smell" is familiar to all but the most city-bound, and has caused the temporary banishment of many a farm dog. A musk gland marked by a patch of dark hairs occurs on the top of the tail of wolves and coyotes, as well as on the tail of many domestic dogs. The functions of some of the mammalian scent glands in connection with social behavior are discussed in Chapter 17 (p. 330).

Hair. The bodies of mammals are typically covered with hair, a unique mammalian feature that has no structural homologue among other vertebrates. Hair was perhaps developed by therapsid reptiles before a scaly covering was lost. Hairs arise from between scales on the tails of many small mammals and from between the bony plates of armadillos; they retain the same distribution on the skin of some scaleless mammals that they would have if scales were present.

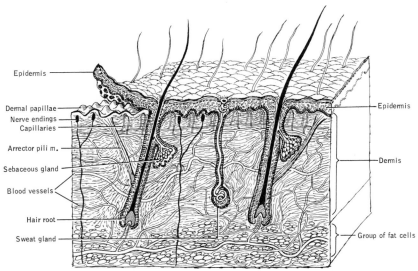

Epidermis

Dermal papillae
Nerve endings
Capillaries

Arrector pili m.

Sebaceous gland

Blood vessels

Hair root

Sweat gland

Epidermis

Dermis

Group of fat cells

FIGURE 2–1. Section of mammalian skin. (From Romer, A. S.: *The Vertebrate Body,* W. B. Saunders Company, 1970.)

A hair consists of dead epidermal cells that are strengthened by keratin, a tough horny tissue made of proteins. A hair grows from living cells in the hair root (Fig. 2–1). Each hair consists of an outer layer of cells arranged in a scale-like pattern, the *cuticle,* a deeper layer of highly packed cells, the *cortical layer,* and in some cases a central core of cuboidal cells, the *medulla* (Fig. 2–2). The color of hair depends on pigment in either the medulla or the cortical cells; the cuticle is usually transparent.

The coat of hair, termed the *pelage,* functions primarily as insulation. The dissipation of heat from the skin sur-

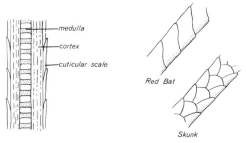

medulla

cortex

cuticular scale

Red Bat

Skunk

FIGURE 2–2. Structure of a hair and two types of reticular scale patterns. (After Storer, T. I., and Usinger, R. L.: *General Zoology,* 4th ed. Copyright 1965 by McGraw-Hill Book Company. Used with permission of McGraw-Hill Book Company.)

face to the environment, and the absorption of heat from the environment, are retarded by pelage. Pinnipeds, many of which live in extremely cold water, are insulated by both hair and subcutaneous blubber. Some mammals are hairless, or nearly so. These either live in warm areas or have specialized means of insulation other than hair. The essentially hairless whales and porpoises have thick layers of blubber that provide insulation. Hair is sparse on elephants, rhinos, and hippopotami; these animals occupy warm areas, have thick skins that offer some insulation, and have such favorable mass/surface ratios because of their large size (see p. 368) that retention of body heat is no problem.

Hair, being non-living material, is subject to considerable wear and bleaching of pigments. During periodic molts, usually once or twice a year, old hairs are lost and new ones replace them. This often occurs in a regular pattern of replacement (Fig. 2–3). In many north-temperate species the molts are in the spring and fall, and the summer pelage is generally shorter and has less insulating ability than the winter pelage. In some species that occupy areas with continuous snow-cover in the winter, the summer pelage is brown and the winter coat is

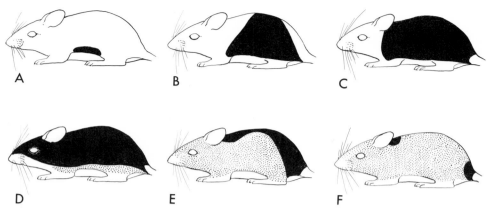

FIGURE 2–3. The pattern of post-juvenile molt in the golden mouse (*Ochrotomys nuttalli*). Black areas indicate portions where replacement of juvenile hair by adult hair is occurring. Stippled areas indicate new adult pelage. (After Linzey and Linzey, 1967.)

white. The arctic fox, several species of hares, and some weasels follow this pattern.

The color of most small, terrestrial mammals closely resembles the color of the soil on which they live. In his careful study of concealing coloration in desert rodents of the Tularosa Basin of New Mexico, Benson (1933) found that white sands were inhabited by nearly white rodents, whereas on adjoining stretches of black lava lived black rodents. Broadly speaking, mammals that occur in forests or beneath a thick canopy of vegetation are dark colored, whereas those that occur in more open situations are relatively pale.

Counter-shading is a color pattern common to mammals and many other vertebrates. Under most lighting conditions the back of an animal is more brightly illuminated than is the shaded underside. If a mammal were all of a single color, the underside would appear very dark relative to the back, and the form of the animal would be obvious. But when the back and sides are darkly colored and the underside and insides of the legs are white—an almost universal color pattern among terrestrial mammals—the well lighted back reflects little light and the shaded white venter tends to strongly reflect light. The result is that the form of the animal becomes obscured to some extent, and the animal becomes less conspicuous.

The color patterns of mammals serve a variety of purposes. The pelages of some ungulates and some rodents are marked by white stripes that tend to obliterate the shapes of the animals when they are against broken patterns of light and shade. If these markings only occasionally allow an animal to go unnoticed by a predator, or if they cause a predator to be indecisive in his attack for but a fraction of a second, they have adaptive value. But what about the glaringly white rump patch of the pronghorn, and other conspicuous white markings belonging to cursorial grassland or open-country species? These markings probably function as warning signals when an animal begins to run from danger, and the gaits that some ungulates use at these times serve to show off the markings. The black and white coloration of skunks, on the other hand, makes these defensively well-endowed animals conspicuous and unmistakable to their enemies.

Circulatory System. In keeping with their active lives and their endothermic ability, mammals have highly efficient circulatory systems. A complete separation of the systemic circulation and the pulmonary circulation has been achieved in mammals. The four-chambered heart functions as a double pump: the right side of the heart receives venous blood from the body and pumps it to the lungs; the left side receives oxygenated blood from the lungs and pumps it to the body.

Table 2–1. Heart Rates of Selected Mammals.
(Data from Altman and Dittmer, 1964:235.)

SPECIES	COMMON NAME	WEIGHT	HEART RATE BEATS/MIN.
Erinaceus europaeus	European hedgehog	500–900 g.	246 (234–264)
Sorex cinereus	Gray shrew	3–4 g.	782 (588–1320)
Eutamias minimus	Least chipmunk	40 g.	684 (660–702)
Sciurus carolinensis	Gray squirrel	500–600 g.	390
Phocaena phocaena	Harbor porpoise	170 kg.	40–110
Mustela vison	Mink	0.7–1.4 kg.	272–414
Phoca vitulina	Harbor seal	20–25 kg.	18–25
Elephas maximus	Asiatic elephant	2,000–3,000 kg.	25–50
Equus caballus	Horse	380–450 kg.	34–55
Sus scrofa	Swine	100 kg.	60–80
Ovis aries	Sheep	50 kg.	70–80

The fascinating evolution of the mammalian heart and circulatory pattern is described in detail by Romer (1970:429–432).

As might be expected because of the great size differential between the smallest and the largest mammal (the weights of the 2 g. shrew and the 160,000 kg. whale differ by a factor of 80,000,000) the heart rate is highly variable between species. The rate in non-hibernating mammals varies from under 20 beats/minute in whales to over 1,300 in a shrew (Table 2–1). Especially remarkable is the ability of some mammals to alter the heart rate rapidly. As an extreme example, a resting big brown bat (*Eptesicus*) has a rate of about 400 beats/minute; this rate increases almost instantly to about 1000 when the bat takes flight, and generally returns to the resting rate within one second after the flight (Fig. 2–4).

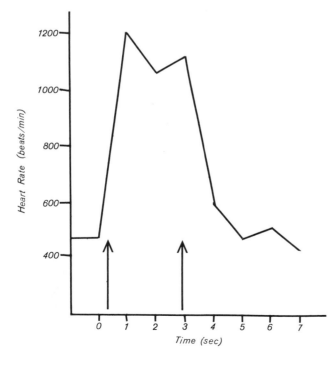

FIGURE 2–4. Heart rate of the big brown bat (*Eptesicus fuscus*) at rest and during flight. The arrows indicate the beginning and end of flight. (After Studier and Howell, 1969.)

The erythrocytes (red blood cells) of mammals are biconcave discs. They extrude their nuclei when they mature, apparently as a means of increasing oxygen carrying capacity.

Respiratory System. In mammals the lungs are large and, together with the heart, virtually fill the thoracic cavity. Air passes down the trachea, into the bronchi, and through a series of branches of diminishing size into the bronchioles, from which branch alveolar ducts. Clustered around each alveolar duct is a series of tiny terminal chambers, the alveoli. Exchange of gases between inspired air and the blood stream occurs in the alveoli; the thin alveolar membranes are surrounded by dense capillary beds. In man the lungs contain about 300 million alveoli, which provide a total respiratory surface of about 70 square meters (well over 600 sq. ft. — some 40 times the surface area of the body).

Air is forced into the lungs by muscular action that increases the volume of the thoracic cavity and decreases the pressure within the cavity. Some increase is gained by the forward and outward movement of the ribs under the control of intercostal muscles, but of greater importance is the muscular diaphragm (a structure unique to mammals). When relaxed the diaphragm is bowed forward, but when contracted its central part moves backward toward the celomic cavity, thus increasing the volume of the thoracic cavity.

Reproductive System. In mammals, both ovaries are functional and the ova are fertilized in the oviducts. The embryo develops in the uterus and lies within a fluid-filled amniotic sac. Nourishment for the embryo comes from the maternal blood stream by way of the placenta. (The female reproductive cycle and the establishment of the placenta is discussed in Chapter 18.) The structure of the uterus is variable (Fig. 2–5).

The male copulatory organ, the penis, contains erectile tissue and is surrounded by a sheath of skin, the prepuce. In many species the penis contains a bone, the *os penis* or baculum, which may differ markedly even between closely related species (Fig. 2–6) and may therefore be of considerable use in taxonomic studies. The tip of the penis has an extremely complicated form in some species (Fig. 2–7). The testes of mammals, instead of lying in the celomic cavity as in other vertebrates, are typically contained in the scrotum, a sac-like structure that lies outside of the body cavity but is an extension of the celomic cavity. The testes either descend permanently from the celomic cavity into the scrotum when the male reaches reproductive maturity or are withdrawn into the body cavity between breeding seasons

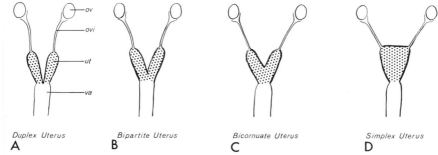

Duplex Uterus   Bipartite Uterus   Bicornuate Uterus   Simplex Uterus
A   **B**   **C**   **D**

FIGURE 2–5. Several types of uteri (stippled) found in placental mammals, showing degrees of fusion of the two "horns" of the uterus. A duplex uterus (A) occurs in the orders Lagomorpha, Rodentia, Tubulidentata, and Hyracoidea; and bipartite uterus (B) is known in the order Cetacea; a bicornate uterus (C) is found in the order Insecivora, in some members of the orders Chiroptera and Primates, and in the orders Pholidota, Carnivora, Proboscidea, Sirenia, Perissodactyla, and Artiodactyla; a simplex uterus (D) is typical of some members of the orders Chiroptera and Primates, and of the order Edentata. (From *Evolution of Chordate Structure: An Introduction to Comparative Anatomy* by Hobart M. Smith. Copyright © 1960 by Holt, Rinehart and Winston, Inc. Reproduced by permission of Holt, Rinehart and Winston, Inc.)

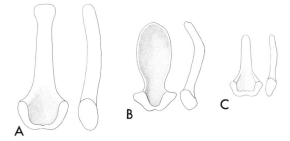

FIGURE 2-6. The bacula of several species of New Guinean murid rodents, showing differences in the structure of this bone in closely related mammals. (From Lidicker, 1968.)

and descend when the animal again becomes fertile. In most mammals the maturation of sperm cannot proceed normally at the usual deep body temperatures, and the scrotum functions as a "cooler" for the testes and developing sperm.

Brain. Compared to the brains of other vertebrates, that of the mammal is unusually large. This greater size is due largely to a tremendous increase in the size of the cerebral hemispheres. These structures were ultimately derived from a part of the brain important in "lower" vertebrates in re-ceiving and relaying olfactory stimuli. Most characteristic of the brain of higher mammals is the great development of the *neopallium*, a mantle of grey matter that first appeared as a small area in the front part of the cerebral hemispheres in some reptiles and in mammals has expanded over the surface of the deeper, "primitive" vertebrate brain. The surface area of the neopallium is vastly increased in many mammals by a complex pattern of folding (Fig. 2-8). A new development in placental mammals is the *corpus callosum*, a large concentration of nerve

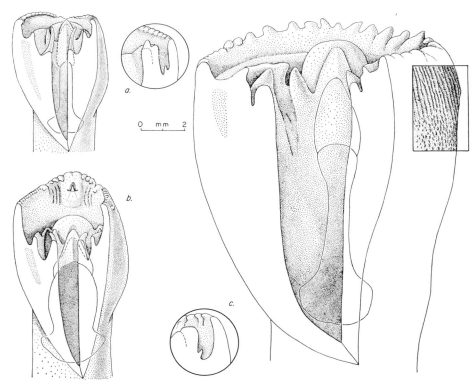

FIGURE 2-7. Ventral views of the penises of several species of New Guinean murid rodents, showing the complex structure of the organ in these mammals. (From Lidicker, 1968.)

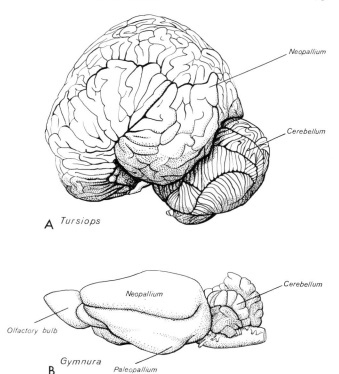

FIGURE 2-8. The brains of a porpoise (A) and a tree shrew (B). The neopallium is greatly enlarged and highly convoluted in the specialized and intelligent porpoise, but is relatively small and smooth-surfaced in the primitive tree shrew. (*Tursiops* after Kruger, L., in Norris, K. S.: *Whales, Dolphins and Porpoises,* originally published by the University of California Press, 1966, redrawn by permission of the Regents of the University of California. *Gymnura* after Romer, A. S.: *The Vertebrate Body,* W. B. Saunders Company, 1970.)

fibers that pass between the two halves of the neopallium and provide communication between them.

The unique behavior of mammals is largely a result of the development of the neopallium, which functions as a control center that has come to dominate the original brain centers. Sensory stimuli are relayed to the neopallium, where much motor activity originates. Present actions are influenced by past experience; learning and "intelligence" are important. The size of the brain relative to total body size seems not always to be a reliable guide to intelligence, for brain size apparently need not increase in proportion to increases in body size to maintain intelligence. The degree of development of convolutions on the surface of the neopallium is perhaps a better indication of intelligence.

Sense Organs. The sense of smell is acute in many mammals, probably in part as a result of the development of turbinal bones in the nasal cavities (Fig. 2–9). The olfactory bulbs and olfactory lobes form a great part of the

brain in some insectivores and are reasonably large in carnivores and rodents. The sense of smell is poorly developed and the olfactory part of the

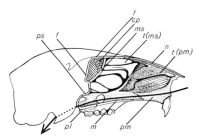

FIGURE 2–9. Cutaway view of the nasal chamber of the Abert's squirrel (*Sciurus aberti*), showing the complicated arrangement of turbinal bones. The entire right half of the nasal part of the skull is removed, exposing the left side of the nasal chamber. The arrow shows the main path air follows from the external to the internal nares, but some air circulates through the upper part of the chamber and over the turbinal bones. Abbreviations: *cp,* cribriform plate of the mesethmoid bone (through which the branches of the olfactory nerve pass out of the braincase); *f,* frontal; *m,* maxillary; *ms,* mesethmoid; *n,* nasal; *pl,* palatine; *pm,* premaxillary; *ps,* presphenoid; *t(ms),* turbinals connected to the mesethmoid; *t(pm),* turbinals connected to the premaxillary.

brain is strongly reduced in whales and the higher primates; the olfactory system is absent in porpoises and dolphins (Kruger, 1966:247).

The sense of hearing is highly developed in mammals, and no other vertebrates seem to depend so heavily on this sense. Mammals alone have an external structure (the *pinna*) that serves to intercept sound waves; the pinnae may be extremely large and elaborate in some mammals, particularly in bats (Fig. 6–26; p. 92). Pinnae are missing (presumably secondarily lost) in some insectivores, phocid seals, and cetaceans. The external auditory meatus, the tube leading from the pinna to the tympanic membrane, is typically long in mammals, and in cetaceans is extremely long. The middle ear is an air-filled chamber that houses the three *ossicles*, and is typically enclosed by a bony bulla (Fig. 2–10). The mammalian cochlea is more or less coiled. (Some variations in the structure of the mammalian ear are discussed on p. 409).

The eye of mammals resembles that of most amniote vertebrates. In most nocturnal mammals the *tapetum lucidum* is well developed. This is a reflective structure within the choroid that improves night vision by reflecting light back to the retina. This reflection accounts for the "eye shine" when a rabbit's eyes are picked up by the beams of headlights at night. Although in most mammals the eyes are well developed, in some insectivores and some cetaceans they are strongly reduced. In such species the eyes are only able to differentiate between light and dark and may serve primarily to aid the animals in maintaining the appropriate nocturnal or diurnal activity cycles (Herald et al., 1969; Lund and Lund, 1965).

Most mammals have *vibrissae*. These are the whiskers on the muzzle and the long, stiff hairs that are present on the lower legs of some mammals. The vibrissae are tactile organs, and those on the face probably enable nocturnal species to detect obstacles near the face. The vibrissae on the muzzle generally arise from a structure termed the *mystacial pad*, and are controlled by a complex of muscles (Fig. 2–11).

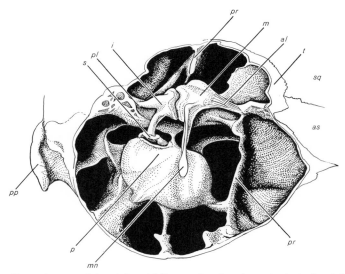

FIGURE 2–10. Lateral view of the right middle ear chamber (anterior is to the right) of the Abert's squirrel (*Sciurus aberti*), with the auditory bulla largely removed. The complex partitioning of the bulla, the positions of the ear bones and the inner ear, and the ligamentous bracing of the malleus and incus are shown. In life the manubrium of the malleus rests against the tympanic membrane. Abbreviations: *al*, anterior ligament (of the malleus); *as*, alisphenoid; *i*, incus; *m*, malleus; *mn*, manubrium; *p*, periotic; *pl*, posterior ligament; *pp*, paroccipital process; *pr*, partitions of bulla; *s*, stapes; *sq*, squamosal; *t*, tympanic.

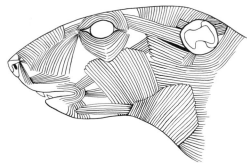

FIGURE 2–11. The superficial facial muscles of the cotton rat (*Sigmodon hispidus*); these muscles partly control facial expression. (After Rinker, 1954.)

Digestive System. Salivary glands are present in mammals, and in some ant-eating species they are specialized for the production of a mucilaginous material that makes the tongue sticky. The stomach is a single sac-like compartment in most species, but is complexly subdivided in ruminant artiodactyls, in cetaceans, and sirenians (Fig. 2–12). In herbivorous species digestion is frequently accomplished partly by micro-organisms that inhabit the stomach or *caecum* (the caecum is a blind sac that opens into the posterior end of the small intestine).

Muscular System. The mammalian limb and trunk musculature has been highly plastic. Different evolutionary lines have developed muscular patterns beautifully adapted to diverse modes of locomotion. Cetaceans are the fastest marine animals, certain carnivores and ungulates are the most rapid runners, and bats as fliers are more maneuverable than birds. Some muscular specializations favoring specific types of locomotion are described in the chapters on the orders of mammals. Especially notable in mammals is the great development of dermal musculature. In many mammals these muscles form a sheath over most of the body and allow the skin to be moved. These dermal muscles have differentiated and have moved over much of the head (Fig. 2–11); these facial muscles control many essential actions. In mammals there are no more vital voluntary muscles than those that encircle the mouth; these function during suckling and are among the first voluntary muscles to be subjected to heavy use. Facial muscles move the ears, close the eyes, and control the subtle changes in expression that are so important in the social lives of many primates.

THE SKELETON

General Features. The mammalian skeleton differs from that of the reptile in several basic ways; all may well be related to the active style of life of mammals. The mammalian skeleton is more completely ossified (Fig. 2–13), a feature perhaps associated with the need for well braced attachments for muscles. Considerable fusion of bones has also occurred, as, for example, in the pelvic girdle. The skeleton has become simplified in mammals; this development seemingly increases the flexibility of the axial skeleton and

FIGURE 2–12. The four-chambered stomach of a ruminant artiodactyl. As the animal feeds, it swallows the vegetation, which is then stored in the rumen (I). While the animal rests, it regurgitates the food from the rumen and "chews its cud" (remasticates the food). The food then goes to the reticulum, omasum, and abomasum, where digestion is aided by a diverse microbiota. (After Storer, T. I., and Usinger, R. L.: *General Zoology*, 4th ed. Copyright 1965 by McGraw-Hill Book Company. Used with permission of McGraw-Hill Book Company.)

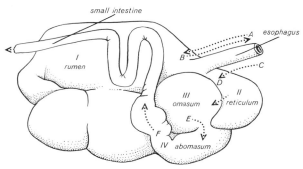

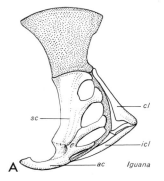
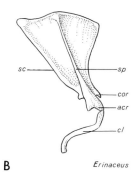

A   *Iguana*   B   *Erinaceus*

FIGURE 2–13. Lateral view of the right side of the pectoral girdles of a lizard (A) and a mammal (B), showing the greater ossification and simplification of the structure in the mammal. Abbreviations: *ac*, anterior coracoid; *acr*, acromion process; *cl*, clavicle; *cor*, coracoid process; *icl*, interclavicle; *sc*, scapula; *sp*, spine of scapula.

allows the limbs greater speed and range of movement. The greater range of movement is of particular advantage to arboreal creatures, which many early mammals may have been. The simplification of the skeleton may have also been advantageous in terms of metabolic economy—the less bone, the less energy invested in its development and maintenance. Further, selection may have favored a light skeleton in the interest of quick movement with relatively little expenditure of energy.

To an animal as active as a mammal, well-formed articular surfaces on limb bones and solid points of attachment for muscles are highly advantageous

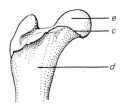

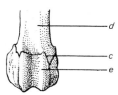

FIGURE 2–14. The proximal (above) and distal (below) ends of the right femur of a young hedgehog (*Erinaceus europaeus*), showing the epiphyses (*e*), the diaphysis (*d*), and the intervening cartilaginous zone (*c*).

during the period of growth of the skeleton as well as during adult life. Mammals have abandoned the pattern of bone growth typical of reptiles. In many reptiles, skeletal growth may continue throughout much of life. Growth in reptiles occurs at the ends of limb bones by ossification of the deep parts of a persistently growing cartilaginous cap; such a pattern clearly limits the establishment of a well formed joint. In mammals, however, skeletal growth is generally restricted to the early part of life. The articular surfaces and some points of attachment of large muscles become well formed and ossified early, while rapid growth is still under way. Growth continues at a cartilaginous zone where the end of the bone and its articular surface, the *epiphysis*, joins the shaft of the bone, the *diaphysis* (Fig. 2–14). When full growth is attained, this cartilaginous zone of growth becomes ossified, fusing the epiphysis and diaphysis. Because within a given species this fusion usually occurs at a certain age, the degree of closure of the "epiphyseal line" is useful in estimating the age of a mammal.

The Skull (Fig. 2–15). The braincase of the mammalian skull is large. In addition to its primary function of protecting the brain, it provides a surface from which the temporal muscles originate. (In many mammals these are the most powerful muscles that close the jaws.) A *sagittal crest* increases the area of origin for the temporal muscles in many mammals; the *lambdoidal crest* gives origin to the temporal mus-

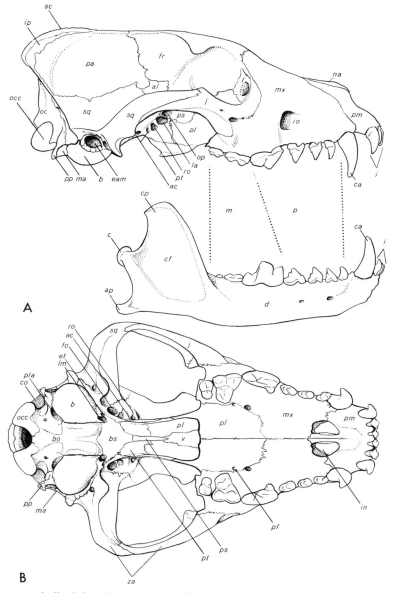

FIGURE 2–15. Skull of the African hunting dog (*Lycaon pictus*), showing bones, foramina, and teeth. Abbreviations: *ac*, alisphenoid canal; *al*, alisphenoid; *ap*, angular process; *b*, auditory bulla; *bo*, basioccipital; *bs*, basisphenoid; *c*, condyle; *ca*, canines; *cf*, coronoid fossa; *co*, anterior condyloid foramen; *cp*, coronoid process; *d*, dentary; *eam*, external auditory meatus; *et*, eustachian tube; *fo*, foramen ovale; *fr*, frontal; *i*, incisors; *in*, incisive foramen; *io*, infraorbital foramen; *ip*, interparietal; *j*, jugal; *l*, lacrimal; *la*, anterior lacerate foramen; *lm*, medial lacerate foramen; *m*, molars; *ma*, mastoid; *mx*, maxillary; *na*, nasal; *oc*, occipital; *occ*, occipital condyle; *op*, optic foramen; *p*, premolars; *pa*, parietal; *pf*, posterior palatine foramen; *pl*, palatine; *pla*, posterior lacerate foramen; *pm*, premaxillary; *pp*, paroccipital process; *ps*, presphenoid; *pt*, pterygoid; *ro*, foramen rotundum; *sc*, sagittal crest; *sq*, squamosal; *v*, vomer; *za*, zygomatic arch.

cles and some cervical muscles. The *zygomatic arch* is usually present as a structure that flares outward from the skull. It serves to protect the eye and to provide origin for the masseter muscles; it forms the surface with which the condyle of the dentary (lower jaw) articulates. The zygomatic arch may be reduced or lost, as in some insectivores and cetaceans, or may be enlarged in those groups in which the masseter muscles largely supplant the temporals as the major jaw muscles (Fig. 10–1; p. 148). The mammalian skull has a secondary palate (see p. 27 and Fig. 3–4), and there are *turbinal bones* within the long nasal cavities (Fig. 2–9).

A number of *foramina* (openings) perforate the braincase and allow passage of the cranial nerves (Fig. 2–16). In some rodents, the infraorbital foramen, through which blood vessels and a branch of the fifth cranial nerve (the trigeminal) pass, is enormously enlarged in association with specializations of the masseter muscles (Fig. 10–21B; p. 167). The incisive foramina, present in the palates of many mammals (Fig. 2–15), house an olfactory organ (Jacobson's organ) that allows the "smelling" of the contents of the mouth. These olfactory organs are widespread among vertebrates; a snake puts the tips of its forked tongue

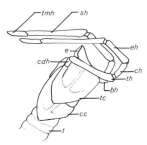

FIGURE 2–16. The hyoid apparatus and associated structures of the domestic cat. Abbreviations: *bh*, basihyal; *cc*, cricoid cartilage; *cdh*, chondrohyal; *ch*, ceratohyal; *e*, epiglottis; *eh*, epihyal; *sh*, stylohyal; *t*, trachea; *tc*, thyroid cartilage; *th*, thyrohyal; *tmh*, tympanohyal. (After Taylor, W. T., and Weber, R. J.: *Functional Mammalian Anatomy.* Copyright © 1951 by Litton Educational Publishing Inc. Reprinted by permission of Van Nostrand Reinhold Company.)

against this part of the palate after "testing" its immediate environment.

Sounds that cause vibration of the tympanic membrane are mechanically transmitted by the three ear ossicles (Fig. 2–10) through the air-filled chamber to the inner ear. The footplate of the stapes fills an opening into the inner ear, and, acting like a piston, transforms the movements of the ossicles to vibrations of the fluid within the cochlea. The inner ear, with the cochlea and semicircular canals, is contained by the periotic bone, which is generally covered by the squamosal bone but is exposed as the mastoid bone in some mammals (Fig. 2–15). The auditory bulla is formed by the expanded tympanic bone or by the tympanic plus the entotympanic, a bone found only in mammals. The bulbous tympanic bullae in many mammals look like structures remembered too late and hurriedly stuck to the skull, but the bullae are highly modified in some species in connection with specialized modes of life (see Fig. 10–11, p. 156; also p. 409).

The lower jaw is formed by the *dentary*. This bone typically has a coronoid process, on which the temporalis muscle inserts, a coronoid fossa, in which the masseter muscles insert, and an angular process, to which a jaw-opening muscle (the digastricus) attaches. In some herbivores, in which the masseter is enlarged at the expense of the temporalis, the coronoid process is reduced or absent and the posterior part of the dentary becomes dorsoventrally broadened (Fig. 9–1A; p. 141).

Several skeletal elements in the throat region are highly modified remnants of the gill arches of fish. These structures, the *hyoid apparatus*, support the trachea, the larynx, and the base of the tongue (Fig. 2–16), and are often braced anteriorly against the auditory bullae.

Teeth. The earliest known vertebrates, which were jawless, had bodies encased in bony plates. When some of the arches that supported the gill apparatus in primitive vertebrates became modified into jaws and jaw supports, teeth developed on the bony plates that bordered the mouth. Teeth are an extremely important vertebrate feature of long standing, and mammals have carried the specialization of teeth further than has any vertebrate group.

In mammals, teeth occur on the premaxillary, maxillary, and dentary bones (Fig. 2–15). The typical mammalian dentition is *heterodont;* that is to say, it consists of teeth that vary in both structure and function. The teeth are designated as *incisors, canines, premolars,* and *molars.* The *cheek teeth* (premolars and molars) are usually the most highly specialized teeth, and serve to grind up food in preparation for digestion. Characteristically, two sets of teeth appear in an individual's lifetime. The *deciduous* dentition develops early and consists of incisors, canines, and premolars—but no molars. These "milk teeth" are lost and replaced by permanent teeth as the animal matures. The permanent dentition consists of a second set of incisors, canines and premolars; it also includes the molars, which have no deciduous counterparts. The deciduous dentition of some species bears little resemblance to the permanent dentition (Fig. 2–17).

The number of teeth of each type in the dentition is designated by the *dental formula.* This is written as the number of teeth of each kind on one side of the upper jaw over the corresponding number in one lower jaw. Such a formula is incisors 3/3, canines 1/1, premolars 4/4, molars 2/3. Because the teeth are always listed in this

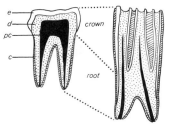

FIGURE 2–18. Generalized sections of mammalian teeth, showing the structure and materials. The black area is the pulp cavity (*pc*), the stippled part is dentine (*d*), the cross-hatched part is cement (*c*), and the unshaded areas are enamel (*e*). The molar on the left is similar to that of primates and is low crowned; the molar on the right is similar to that of a horse and is high crowned.

order, the formula may be shortened to 3/3, 1/1, 4/4, 2/3. (The skull in Fig. 2–15 has this dental formula.) The dental formula lists the teeth of only one side; therefore, the total number of teeth in the formula must be doubled to give the total number of teeth in the dentition. As an example, the arrangement for man is 2/2, 1/1, 3/3, 2/2 × 2 = 32. The basic maximum number of teeth in placental mammals is 44 (3/3, 1/1, 4/4, 3/3), but marsupials commonly have more than this number. The number of teeth is frequently reduced, and a few placentals completely lack teeth; some specialized placentals, most notably odontocete cetaceans, have more than 44 teeth and have *homodont* dentitions (those in which all teeth are alike).

The mammalian tooth typically consists of *dentine,* a bone-like material, covered by a layer of hard, smooth *enamel,* which is largely calcium phosphate. The tooth is bound to the jaw by *cement,* and this relatively soft material may also form part of the crown of the tooth (Fig. 2–18). Most mammals have teeth that are *brachyodont,* or short-crowned, and growth ceases after the tooth is fully grown and the pulp cavity in the root closes. Many herbivores that subject their teeth to rapid wear—resulting from abrasion by silica in grass and by soil particles that adhere to plants—have *hypsodont* or high-crowned teeth (Fig. 2–18). As a further adaptation to abrasive food, in

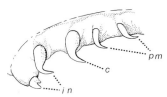

FIGURE 2–17. The deciduous upper dentition (of the left side) of a Neotropical fruit-eating bat (*Artibeus lituratus*). Abbreviations: *c,* canine; *in,* incisors; *pm,* premolars. (From Vaughan, T. A., in Wimsatt, W. A.: *Biology of Bats,* Academic Press, 1970.)

Bunodont   *Lophodont*   *Selenodont*

FIGURE 2–19. Three major types of right upper molariform teeth. Anterior is to the right; the outer edge of each tooth is toward the top. The crosshatched parts are dentine.

some mammals some teeth (and in some rodents all of the teeth) retain open pulp cavities in the roots and grow continuously. These are termed *evergrowing teeth.* The roots of mammalian teeth are often divided; primitively, the upper molars have three roots and the lower molars two. Incisors and canines are single-rooted, and premolars may have one or two roots. The dentitions of herbivores usually serve only two functions: the incisors, or the incisors and the canines, clip vegetation and the cheek teeth grind the food. Between these teeth there is usually a space called a *diastema* (Fig. 10–1; p. 148).

The crowns and the *occlusal surfaces* of the cheek teeth (the surfaces of the teeth that contact their counter-

parts of the opposing jaw—the surfaces the dentist generally attacks when putting in a filling) vary structurally in response to the demands of different diets (Fig. 2–19). In omnivores (some rodents, some carnivores, some primates, and pigs) the molars are *bunodont;* the cusps form separate, rounded hillocks that crush and grind food. In herbivores the molars may be *lophodont,* with cusps forming ridges, or may be *selenodont,* with cusps forming crescents; in these cases the teeth finely section and grind vegetation. In the dental batteries of many insectivores, bats, and carnivores are *sectorial* teeth. These have blade-like cutting edges that section food by shearing against the edges of their counterparts in the opposing jaw (as do

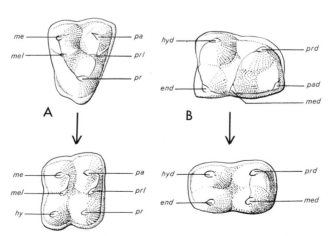

FIGURE 2–20. Basic cusp pattern of mammalian molars. The teeth on the left (A) are right upper molars; those on the right (B) are left lower molars. The upper pair of teeth represents the primitive cusp pattern; this was modified in some evolutionary lines by the addition of a cusp (hypocone) in the upper tooth, and the loss of a cusp (paraconid) in the lower tooth, yielding more or less quadrate teeth (lower pair) adapted to omnivorous or herbivorous diets. Abbreviations: *end,* entoconid; *hy,* hypocone; *hyd,* hypoconid; *me,* metacone; *med,* metaconid; *mel,* metaconule; *pa,* paracone; *pad,* paraconid; *pr,* protocone; *prd,* protoconid; *prl,* protoconule. (After Romer, A. S.: *Vertebrate Paleontology,* 3rd ed., The University of Chicago Press, 1966.)

FIGURE 2–21. The pattern of occlusion between upper and lower molars of a primitive mammal. The lower molars are shaded. Anterior is to the right and the outer edges are above. (After Romer, A. S.: *Vertebrate Paleontology*, 3rd ed., University of Chicago Press, 1966.)

the fourth upper premolar and the first lower molar in the dentition in Fig. 2–15).

The terminology of the cusps of the mammalian molar is based on a *tribosphenic* (three-cusped) pattern common to many primitive mammals and detectable even among many modern mammals. The primitive upper molar is roughly triangular, with the triangle marked by three major cusps, the *protocone*, the *paracone*, and the *metacone*, and often by other less prominent cusps (Fig. 2–20). The apex of the triangle (the protocone) points inward. The lower molar has two sections, an anterior *trigonid* and a posterior *talonid*. The trigonid is triangular; the apex of the triangle, the *protoconid*, points outward, and the *paraconid* and *metaconid* form the inner edge. The talonid has two major cusps, the *hypoconid* and *entoconid* (Fig. 2–20). The way in which the upper and lower molars occlude provides both a shearing and a crushing action. The anterior surface and part of the posterior surface of the trigonid of the lower molar and the adjacent surfaces of the upper molar shear past one

another, and the protocone section of the upper molar fits into the depression in the talonid to provide a crushing effect (Fig. 2–21).

Axial Skeleton. The vertebral column has five well differentiated sections: *cervical, thoracic, lumbar, sacral,* and *caudal.* Anterior to the sacrum, the axial skeleton is most flexible in the cervical and lumbar sections. Only the anteroposteriorly compressed thoracic vertebrae bear ribs. In some groups, such as edentates and bats, the rigidity of the ribcage is greatly enhanced by a broadening of the ribs. The first two cervicals are highly modified, the sacral vertebrae are more or less fused to support the pelvic girdle, and considerable differentiation of the vertebrae of each region is typical (Fig. 2–22). Usually from 25 to 35 presacral vertebrae are present. All mammals, with the exception of several edentates and the manatee (Sirenia), have seven cervical vertebrae.

The sternum is well developed and solidly anchors the ventral ends of the ribs, helping to form a fairly rigid ribcage. The sternum is not highly variable, but in some bats departs strongly from the typical mammalian plan (Fig. 6–36; p. 99).

Limbs and Girdles. In most terrestrial mammals the main propulsive movements of the limbs are fore and aft; the toes point forward and the limbs are roughly perpendicular to the ground. In the most highly cursorial species (*cursorial* mammals are those adapted for running), the joints distal to the hip and shoulder tend to limit

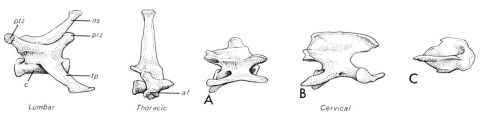

FIGURE 2–22. Vertebrae of the gray fox (*Urocyon cinereoargenteus*), showing the great structural variation in the parts of the vertebral column. The vertebrae are viewed from the right side; anterior is to the right. A, fifth cervical vertebra; B, axis (second cervical); C, atlas (first cervical). Abbreviations: *af*, articular facet for the capitulum of the rib; *c*, centrum; *ns*, neural spine; *prz*, pre-zygapophysis; *ptz*, post-zygapophysis; *tp*, transverse process.

A

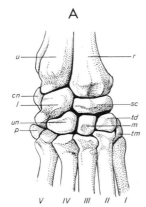

B

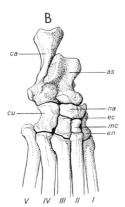

FIGURE 2–23. Primitive patterns of the podials (foot bones) of mammals. A, the carpus of a hedgehog (*Erinaceus europaeus*); B, the tarsus of the wolverine (*Gulo luscus*). The centrale, a carpal element that in some mammals with primitive limbs lies proximal to the trapezoid and magnum, is missing in the hedgehog. Abbreviations: *as*, astragalus; *ca*, calcaneum; *cn*, cuneiform; *cu*, cuboid; *ec*, ectocuneiform; *en*, entocuneiform; *l*, lunar; *m*, magnum; *mc*, mesocuneiform; *na*, navicular; *p*, pisiform; *r*, radius; *sc*, scaphoid; *td*, trapezoid; *tm*, trapezium; *u*, ulna; *un*, unciform. The metacarpals and metatarsals are numbered.

movement to a single plane. This allows reduction of whatever musculature does not control flexion and extension, and results in a lightening of the limbs. The mammalian pelvic girdle has a characteristic shape, with the *ilium* projecting forward and the *ischium* and *pubis* extending backward (Fig. 3–5, p. 26); these bones are solidly fused. In the shoulder girdle of placentals the *coracoid* and *acromion* are usually reduced to small processes on the scapula, and the reptilian interclavicle is gone (Fig. 2–13); the clavicle is reduced or absent in some cursorial species.

In the *manus* (hand) and *pes* (foot) of mammals there is a standard pattern of bones (Fig. 2–23), but many variations

on this basic theme (some of which are described in the chapters on orders) occur among mammals with specialized types of locomotion such as flight (bats), swimming (cetaceans and pinnipeds), or rapid running (ungulates, rabbits, some carnivores). The primitive mammalian number of digits (5), and the basic phalangeal formula of two phalanges in the thumb (pollex) and first digit of the hind limb (hallux) and three phalanges in the remaining digits (2-3-3-3-3), are retained by many mammals. Common specializations involve the loss of digits, reduction in the numbers of phalanges, or occasionally the addition of phalanges (*hyperphalangy*) as in whales and porpoises.

CHAPTER 3

MAMMALIAN ORIGINS

Mammals stemmed from an ancient reptilian lineage that arose at the start of the reptilian radiation some 300 million years ago. "The mammal-like reptiles, constituting the subclass Synapsida, were among the earliest to appear of known reptilian groups and had passed the peak of their career before the first dinosaur appeared on the earth" (Romer, 1966:173). From a late Carboniferous (Table 3–1 lists the geological periods) divergence from other reptile evolutionary lines, mammal-like reptiles became the most important reptile group in the Permian and early Triassic. In the Triassic, however, perhaps because of competition from other reptile groups and changing climatic conditions, synapsids dwindled in importance and were nearly extinct by the close of this period. Primitive late Triassic mammals were the unspectacular descendants of the declining synapsid line. Throughout the remainder of the Mesozoic, the "Golden Age" of reptiles, mammals persisted and continued to evolve along lines established early in synapsid history. A modest Mesozoic mammalian radiation occurred, but only after the disappearance of the ruling reptiles in the late Cretaceous did mammals begin the remarkable adaptive burst that led to their dominant position throughout most of the Cenozoic Era.

MAMMAL-LIKE REPTILES

General Skull Characteristics

The skull of mammal-like reptiles is characterized by a single opening low in the temporal part, with the postorbital and squamosal bones meeting above the opening. From a primitive reptilian skull with no temporal openings, a skull type still retained by turtles (Fig. 3–1), various patterns of perforation of the temporal part of the skull developed among early reptiles (Fig. 3–2). The openings are thought by some to have developed originally to increase the freedom for expansion of the adductor muscles of the jaw; these muscles primitively attached in-

23

Table 3–1. The Geological Time Scale for the Span of Time Since Life Became Abundant. The Mammals or Ancestors of Mammals Typical of Each Period are Shown. (After Romer, A. S.: *The Vertebrate Body,* 4th ed., W. B. Saunders Company, 1970.)

ERA	PERIOD	ESTIMATED TIME SINCE BEGINNING OF EACH PERIOD (IN MILLIONS OF YEARS)	EPOCH	MAMMALS AND MAMMALIAN ANCESTORS
Cenozoic	Quaternary	2+	Recent	Modern species and subspecies; extirpation of some mammals by man.
			Pleistocene	Appearance of modern species or their antecedents; widespread extinction of large mammals.
	Tertiary	65	Pliocene	Appearance of modern genera.
			Miocene	Appearance of modern subfamilies
			Oligocene	Appearance of modern families.
			Eocene	Appearance of modern orders.
			Paleocene	Primitive marsupials and placental mammals dominant.
Mesozoic	Cretaceous	130		Appearance of marsupials and placentals.
	Jurassic	180		Archaic mammals.
	Triassic	230		Therapsid reptiles
Paleozoic	Permian	280		Appearance of therapsid reptiles (from which mammals evolved).
	Carboniferous	350		
	Devonian	400		
	Silurian	450		
	Ordovician	500		
	Cambrian	570		

side the solid temporal part of the skull. Other workers have held that a selective advantage was gained by the reduction in the weight of the skull due to the temporal openings. According to another explanation, the stresses on the skull were greatest around the periphery of the temporal area, whereas the middle of the area was not

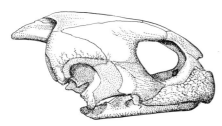

FIGURE 3–1. An anapsid skull (sea turtle, *Chelonia*). Note the unbroken shield of bone in the temporal region.

needed for bracing of the jaw area or for muscle attachment. Therefore, the loss of bone in the middle of the temporal "shield," when solid bone in this area was no longer of adaptive importance, contributed importantly to metabolic economy by reducing the mass of bone in the skull that had to be developed and maintained (Fox, 1964). In any case, the general trend in progressive mammal-like reptiles was clearly toward the enlargement of the temporal opening and the movement of the origins of the jaw muscles from the inner surface of the temporal shield to the braincase and to the zygomatic arch, the remnant of the lower part of the original temporal shield. In some advanced mammal-like reptiles the postorbital bar, a vestige of the anterior part of the temporal shield, was lost.

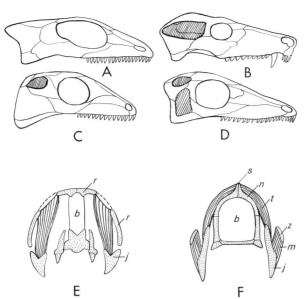

FIGURE 3–2. Diagrammatic views of skulls of reptiles showing the different arrangements of temporal openings: A, "anapsid" skull (no temporal opening); B, "synapsid" skull (postorbital and squamosal meeting above opening); C, "euryapsid" skull (postorbital and squamosal meeting below the opening); D, "diapsid" skull (two temporal openings). Cross sections of skulls showing (diagrammatically) the attachments of the jaw muscles: E, pelycosaur, with the jaw muscles originating within the remaining parts of the temporal shield (r); the sides of the braincase are cartilaginous. F, mammal, with the jaw muscles originating on the new and completely ossified braincase (n), on the saggital crest (s), and on the zygomatic arch (z), a remnant of the original skull roof. Abbreviations: b, brain cavity; j, lower jaw; m, masseter muscle; n, new braincase formed partly by extensions of bones that originally formed the skull roof; r, skull roof; t, temporal muscles; z, zygomatic arch.

Evolution of Mammal-Like Reptiles

The subclass Synapsida is divided into two orders, the primitive Pelycosauria, a largely Permian group, and a more advanced group, the Therapsida, of the late Permian and Triassic. The skull of *Ophiacodon* (Fig. 3–3A) illustrates several pelycosaurian features. The temporal opening is fairly small; the postorbital bar is present; the shield-like structure of the temporal region is still evident; the teeth are not strongly heterodont; and the dentary is not greatly enlarged at the expense of other bones of the lower jaw. *Dime-*

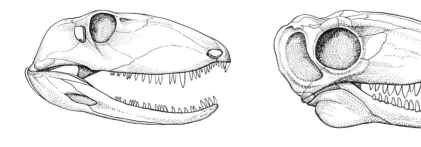

FIGURE 3–3. Skulls of synapsid reptiles. A, *Ophiacodon* (order Pelycosauria); B, *Phthinosuchus* (order Therapsida). (After Romer, A. S.: *Vertebrate Paleontology*, 3rd ed., The University of Chicago Press, 1966.)

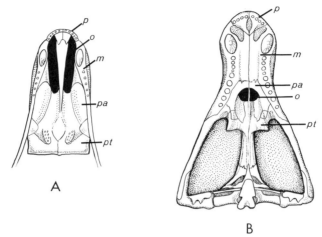

FIGURE 3–4. Palatal views of skulls of synapsid reptiles. A, *Scymnognathus* (order Therapsida); note that the internal nares (*o*) open into the anterior part of the mouth. B, *Cynognathus* (order Therapsida); note that the maxillaries (*m*) and palatines (*pa*) have extended medially, forming a shelf that shunts air from the external nares to near the back of the mouth. Abbreviations: *m*, maxillary; *o*, internal narial opening; *p*, premaxillary; *pa*, palatine; *pt*, pterygoid. (After Romer, A. S.: *Vertebrate Paleontology*, 3rd ed., The University of Chicago Press, 1966.)

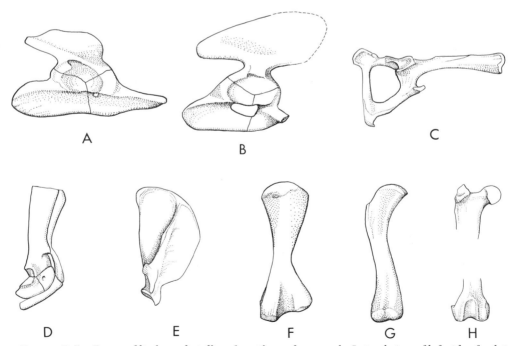

FIGURE 3–5. Bones of limbs and girdles of reptiles and mammals. Lateral view of left side of pelvis in: A, *Dimetrodon* (order Pelycosauria); B, *Cynognathus* (order Therapsida); C, *Erinaceus* (order Insectivora). Lateral views of the right scapulae of: D, *Kannemeyeria* (order Therapsida); E, *Lynx rufus* (order Carnivora). Anterior views of the left femora of: F, *Ophiacodon* (order Pelycosauria); G, a cynodont (order Therapsida); H, *Lynx rufus* (order Carnivora). (All drawings of reptiles after Romer, A. S.: *Vertebrate Paleontology*, 3rd ed., The University of Chicago Press, 1966.)

trodon, an advanced and predaceous pelycosaur, is of special interest because of the greatly elongated neural spines of the thoracic and lumbar vertebrae that formed a "sail" over the body. Presumably skin stretched between the spines. If *Dimetrodon* was able to control the blood supply to the sail, this structure might have served as an effective heat-dissipating or heat-absorbing device because of its large surface area. This structure may therefore represent a remarkably early and bizarre attempt at temperature regulation by pelycosaurian mammal-like reptiles.

Of more direct importance to our consideration of the origin of mammals is the order Therapsida, from which mammals arose. The therapsids are structurally diverse, and span the morphological gap between fairly primitive reptiles and animals of nearly mammalian grade. Despite the complexity introduced by a number of lines of descent and by tangential specializations, several important anatomical trends leading from the typical reptilian organization toward the mammalian pattern are apparent in therapsids. The temporal shield was reduced by an expansion of the temporal opening, and the origins of the jaw musculature were largely on the braincase and zygomatic arch; two occipital condyles replaced the single reptilian condyle. The maxillaries and palatines extended backward and toward the midline, forming a secondary palate (Fig. 3–4); the dentition became more strongly heterodont; and the dentary became progressively larger as the other (typically reptilian) jaw elements became smaller. Ribs were reduced or lost on the cervical and lumbar vertebrae; the spraddle-legged reptilian limb posture was abandoned and the limbs moved beneath the body; the pectoral and pelvic girdles were strongly altered (Figs. 3–5, 3–6); and a basic pattern involving a standard phalangeal formula (2-3-3-3-3) and a simplified series of carpal and tarsal bones was established.

Without departing essentially from the picture presented by vertebrate paleontologists on the basis of present fossil evidence (see Romer, 1968:148–166; 1969), stages in the evolution of mammals can be illustrated by considering the genera *Phthinosuchus, Cynognathus,* and *Probainognathus* (Figs. 3–3B, 3–7). The most primitive of these, *Phthinosuchus,* is known primarily by skull and jaw material from Permian strata in Russia, and is a therapsid reptile within the basal suborder Phthinosuchia. This type had a moderately large temporal opening and weakly heterodont dentition, but lacked such advanced features as a secondary palate. The dentary was not greatly enlarged, and the palate, as in many primitive reptiles, bore teeth. The limbs of this genus are poorly known, but probably the trend toward abandonment of the reptilian limb posture was under way.

The phthinosuchians gave rise to more advanced mammal-like reptiles included in the suborder Theriodontia, many members of which were progressive carnivores. This group, or more specifically the theriodont infraorder Cynodontia, is seemingly the basal stock from which mammals arose. The cynodont *Cynognathus,* known from the lower and middle Triassic beds of Africa, has many advanced and mammal-like features, and illustrates an early stage in the transition between reptiles and mammals. *Cynognathus* was a slender and, for the Triassic, a highly cursorial quadruped. The skull of this reptile retains a small pineal opening, a feature still present in many modern reptiles, and the postorbital bar is broad. In addition, however, a number of advanced features are evident, perhaps the most important being the secondary palate (Fig. 3–4B). This structure is formed by an inward growth of the premaxillary, maxillary, and palatine bones, and is a plate lying beneath the original roof of the mouth and braced along the midline by a bar formed by the fused vomers. This new palate forms a chamber that shunts air from the external nares, at the front of the snout, to

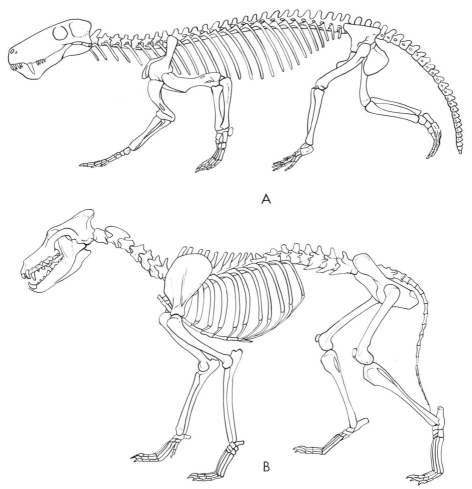

FIGURE 3–6. A, the skeleton of *Lycaenops*, a Permian mammal-like reptile (after Colbert, 1949). B, the skeleton of the dire wolf (*Canis dirus*), a "modern" mammal (after Stock, 1949, courtesy of the Los Angeles County Museum of Natural History).

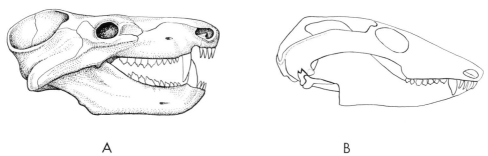

FIGURE 3–7. Skulls of therapsid reptiles. A, *Cynognathus* (after Romer, A. S.: *Vertebrate Paleontology*, 3rd ed., The University of Chicago Press, 1966); B, *Probainognathus* (after Romer, A. S.: Cynodont reptile with incipient mammalian jaw articulation. *Science*, 166:881–882. Copyright 1969 by The American Association for the Advancement of Science).

the internal narial openings, toward the back of the mouth. In homoiotherms such a bypass is of critical importance in allowing the animal to continue to breathe and support a high metabolic rate while masticating a mouthful of food. Another cynodont feature may have been related to homoiothermy. The reduction of lumbar ribs, and the retention of a sturdy thoracic rib cage, may have been correlated with the development of a muscular diaphragm, which in turn was associated with the efficient respiration necessary for homoiothermy.

Additional advanced features include a large temporal opening, a sagittal crest from which the temporal muscles must partly have originated, two occipital condyles rather than the usual single reptilian condyle, and greatly enlarged dentaries in the lower jaw. The dentition is heterodont, and the cheek teeth, as in most mammals, are multi-cusped. Although the vertebral column is still primitive and ribs occur on all but the caudal vertebrae, the limbs extend beneath the body with little of the reptilian splaying outward of the legs; the stride was seemingly directed largely fore and aft. Correlated with the change in limb posture is an alteration of the structure of the pelvis: the ilium is shifted forward and the pubis and ischium are moved backward (Fig. 3–5). In addition, the heads of the humerus and femur are altered in association with the new angle at which the femur meets the pelvis (Fig. 3–5). The soft anatomy of *Cynognathus* is, of course, not known, but a trend toward homoiothermy may have been well advanced in this genus and in cynodonts in general. The slender limbs of *Cynognathus* suggest that this was an active creature, and the well-developed palate reflects a need for continuous respiration during feeding, a feature critical to the maintainance of a fairly high metabolic rate.

A handy and much-used landmark in reptilian-mammalian evolution is the structure of the jaw articulation.

Whereas in reptiles this articulation is between the quadrate bone of the skull and the articular bone of the lower jaw, in mammals the squamosal and dentary bones form the articulation. A generally accepted character used in the recognition of early mammals, then, is the dependence on the squamosal-dentary articulation, with the quadrate and articular forming no part of the joint. Fossil evidence, much of which has been assembled in recent years, illustrates a succession of stages in the abandonment of the quadrate-articular jaw articulation. As might be expected, however, the more complete the fossil record, the more arbitrary any criterion for separating advanced mammal-like reptiles and primitive mammals becomes. The two mammal-like reptiles we have discussed, *Phthinosuchus* and *Cynognathus*, retain the reptilian type of articulation, but a more advanced and structurally transitional genus, *Diarthrognathus*, is known from late Triassic sandstones of South Africa. In this form the squamosal and dentary are in contact, but a quadrate-articular joint is also retained. Although this genus is clearly near the reptile-mammal boundary structurally, it occurs at a time from which mammals are already known, and does not belong to the group (Cynodontia) thought to be ancestral to mammals. A recently described genus, however, is not only structurally intermediate between reptiles and mammals with regard to jaw articulation, but appears to be remarkably close to the main evolutionary stem leading to mammals.

The cynodont genus *Probainognathus*, from middle Triassic beds in Argentina, is almost perfectly intermediate in jaw structure between reptiles and mammals. A functional articulation between the quadrate (which is small and insecurely attached to the squamosal) and the reduced articular is still retained, but the structure of the dentary, and a depression in the squamosal that resembles an incipient glenoid fossa, are clear evidence that a functional squamosal-dentary articula-

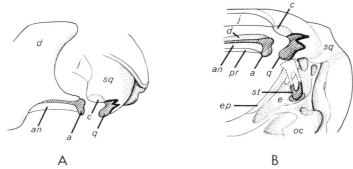

FIGURE 3–8. Bones surrounding the jaw articulation in *Probainognathus* (order Therapsida). A, lateral view; B, ventral view. (Anterior is to the left in both drawings.) Abbreviations: *a*, articular; *an*, angular; *c*, incipient condyle (depression into which the articular surface of the dentary fits); *d*, dentary; *ep*, epipterygoid; *j*, jugal; *oc*, occipital condyle; *pe*, periotic; *pr*, prearticular; *q*, quadrate; *sq*, squamosal; *st*, stapes. (After Romer, A. S.: Cynodont reptile with incipient mammalian jaw articulation. *Science*, 166:881–882. Copyright 1969 by The American Association for the Advancement of Science.)

tion was established (Fig. 3–8). This form seems close to the main line of evolution toward mammals, or may actually be an ancestor of late Triassic mammals (Romer, 1969). Of further interest is a series of contacts between the articular, quadrate, and stapes bones in *Probainognathus* that roughly foreshadow the articulations between the mammalian bones to which they gave rise: the malleus, incus, and stapes, respectively (Fig. 3–8). Of couse, these bones in mammals no longer function in connection with jaw support, but are situated in the middle ear and serve to transmit vibrations mechanically from the tympanic membrane to the oval window of the inner ear.

EARLY MAMMALS

The fossil record is too incomplete to serve as a basis for a clear understanding of the reptile-mammal transition in the late Triassic. A late Triassic intermediate type known only by skull material might have a squamosal-dentary jaw suspension and on this basis be classified as a mammal. However, if this entire creature—bones, flesh, and fur (or scales)—could be taken from a freezer and dissected, many of the characteristic mammalian features might be found to be missing and our

intermediate type might with greater justification be considered a reptile. The reptile-mammal boundary is therefore difficult to draw on the basis of scattered fossil material. As an additional complication, the reptile-mammal boundary may have been crossed by several phyletic lines, and the class Mammalia may have a polyphyletic origin, although current opinion seems to favor the idea that all mammals other than monotremes are monophyletic, or nearly so. In any case, a firm picture of this fascinating evolutionary transition is not presently available, and divergent opinions occur among authorities on details of early mammal-

Table 3–2. Classification of Non-eutherian Mammals.

Class Mammalia
  Subclass Prototheria
      Order Monotremata
  Subclass Uncertain
      Order Triconodonta
      Order Docodonta
  Subclass Allotheria
      Order Multituberculata
  Subclass Theria
    Infraclass Trituberculata
      Order Symmetrodonta
      Order Pantotheria
    Infraclass Metatheria
      Order Marsupialia
    Infraclass Eutheria (including all of the
        orders of "placental" mammals)

ian evolution. Despite these uncertainties, there is fairly general agreement that Mesozoic mammals can be divided into several groups that resulted from a modest Late Triassic and Jurassic radiation. Table 3–2 contains a classification of non-eutherian mammals (from Romer, 1966:379) that includes the orders of Mesozoic mammals to be discussed.

One of the oldest and most primitive groups is the order Triconodonta, consisting of roughly 10 genera spanning much of the Mesozoic from the late Triassic to the Lower Cretaceous. These mammals may have had an origin from mammal-like reptiles separate from that of other mammals, and are probably a peripheral evolutionary line that left no descendents. Triconodonts were predaceous; the largest genus was nearly the size of a house cat. The dentition is heterodont, with as many as 14 teeth in a dentary. The canines are large, and typically the molars have three cusps arranged in a front-to-back row (Fig. 3–9A, B).

The order Docodonta is represented by several primitive genera known from the Upper Triassic and Upper Jurassic. Members of this group have roughly quadrate teeth, with the cusps not aligned anteroposteriorly (Fig. 3–9C, D). The braincase and postcranial skeleton seem to be on a reptilian level of development.

The order Multituberculata contains the earliest mammalian herbivores. This group left no ancestors or derivatives and is a specialized evolutionary side branch apart from the main line of mammalian evolution. It was, however, one of the most successful mammalian orders. Multituberculates are known from the late Jurassic to the Eocene, a span of some 100 million years. In the late Cretaceous and the Paleocene these animals were common, and their disappearance in the Eocene may have been related to the rapid development of rodents, which probably had a similar mode of life. Multituberculates clearly resembled Recent rodents in many ways. In multituberculates, as in rodents, the lower jaw is short, deep, and heavily built;

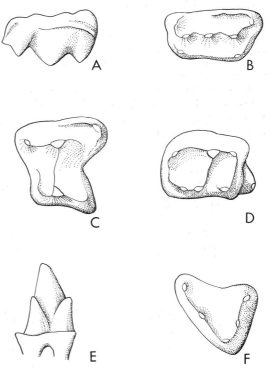

FIGURE 3–9. Teeth of Mesozoic mammals. Right upper molar of a triconodont (order Triconodonta): lateral view (A) and occlusal view (B). Occlusal views of right upper molar (C) and left lower molar (D) of *Docodon* (order Docodonta). Lateral view of lower left molar (E) and occlusal view of upper right molar (F) of a symmetrodont (order Symmetrodonta). (After Romer, A. S.: *Vertebrate Paleontology*, 3rd ed., The University of Chicago Press, 1966.)

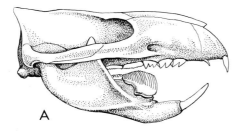

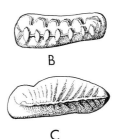

FIGURE 3–10. *Ptilodus* (order Multituberculata). Skull (A) and occlusal views of upper (B) and lower (C) cheek teeth. (After Romer, A. S.: *Vertebrate Paleontology*, 3rd ed., The University of Chicago Press, 1966.)

there are two incisors above and two below; a diastema is present in front of the premolars, and the cheek teeth are highly specialized. Typical of multi-tuberculates are molars with parallel rows of cusps and specialized, blade-like lower premolars (Fig. 3–10). The large number of cusps in the mul-tituberculate molars served to section and grind plant material, and paral-leled the development of complicated patterns of cusps and ridges typical of the molars of many modern rodents. The large olfactory bulbs of the brain of multituberculates (known from fos-sil brain casts) and the large incisive foramina indicate that these animals relied heavily on the sense of smell, as do such "primitive" Recent mammals as marsupials and some insectivores. The postcranial skeleton of multi-tuberculates was primitive.

Another Mesozoic group, the order Symmetrodonta, is known from the Upper Triassic to the Lower Cre-taceous. Symmetrodonts are among the oldest mammals, but are perhaps related to the most advanced subclass of mammals (Theria). Symmetrodonts were predators, and the molar crown pattern is marked by three fairly sym-metrically situated cusps (Fig. 3–9E, F).

Because it is generally accepted that therian mammals evolved from the order Pantotheria, this group is of par-ticular interest. Pantotheres occur mainly in late Jurassic beds, and have several features thought to be traceable to therian mammals. In the pantothere, for example, the profile of the ventral border of the dentary is in-terrupted by an angular process similar to that found in therians. A dental land-mark of importance in pantotheres is a posterior "heel" on the lower molar (Fig. 3–11B). This lobe is absent from the lower molars of other pre-therian mammals, but is represented in many therians by the talonid, the posterior section of the lower molar (Fig. 3–11D). The shape of the anterior tri-

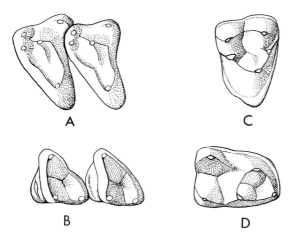

FIGURE 3–11. Right upper molars (A) and left lower molars (B) of a Jurassic pantothere (order Pantotheria). Compar-able teeth (C and D) in a primitive Eocene eutherian mammal (*Omomys*, a tarsier-like mammal, order Primates). (After Romer, A. S.: *Vertebrate Paleontology*, 3rd ed., The University of Chicago Press, 1966.)

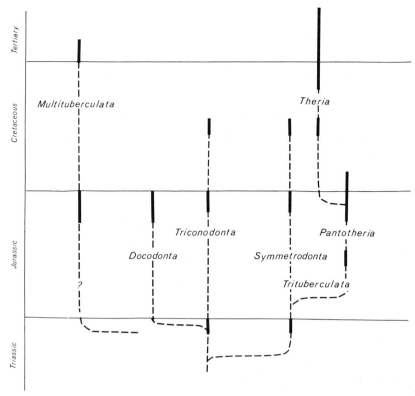

FIGURE 3–12. The relationships of the Mesozoic orders of mammals as envisioned by F. R. Parrington. Solid black lines indicate the extent of the fossil record. (After Parrington, 1971.)

gonid section of the pantothere lower molar also closely resembles the comparable part of this tooth in some therians. The triangular upper molar of pantotheres roughly resembles that of some primitive therians (Fig. 3–11A, C), but no scheme expressing homologies of cusps between these two groups has gained general acceptance. Seemingly, the medial cusp in pantotheres is the protocone of therians; the large central cusp may be homologous to the fused paracone and metacone of early therians. The relationships of the Mesozoic orders of mammals as outlined by Parrington (1971) are shown in Figure 3–12.

The Jurassic, and to some extent the Cretaceous, were times of experimentation among various groups of mammals. Natural selection, partly in the forms of predation by an imposing array of reptilian carnivores and competition with other mammals and with reptiles, "guided" many changes in mammalian structure. Physiological changes not indicated by fossil material were also doubtless of critical importance. Structural plans evolved and proved workable for differing lengths of time, and some Mesozoic groups appear to be sterile side branches from the main mammalian evolutionary line. A striking example of this latter type is the multituberculates, a specialized group that persisted into the Cenozoic. As mentioned previously, this group seemingly could not survive the competition offered by rapidly developing eutherian mammals and disappeared, leaving no modern derivatives. During the Mesozoic period of evolutionary trial and error, the basic mammalian structural plan was perfected, and in the pantotheres we see an evolutionary stage apparently just

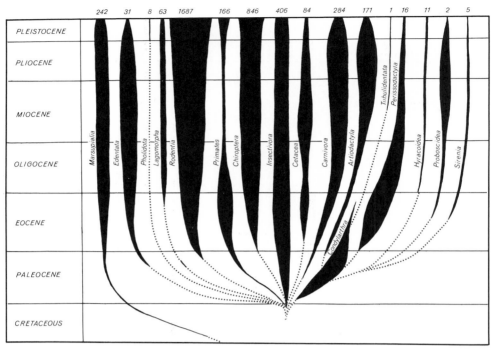

FIGURE 3–13. Times of appearance and relative importance (shown by breadths of black areas representing orders) of the therian orders of mammals. The estimated numbers of species in each order appear at the top of the figure. (Largely after Romer, A. S.: *Vertebrate Paleontology*, 3rd ed., The University of Chicago Press, 1966.)

below that of early therian mammals. With the appearance of the marsupials in the early Cretaceous (Slaughter, 1968) and the placentals in the late Cretaceous, the two most successful "modern" groups of mammals were established. The extinction of the dinosaurs at the end of the Cretaceous, and the prevalence at this time of climatic conditions making homoiothermy highly advantageous, placed therian mammals, the marsupials, and placentals in a position to inherit the earth. Mammals underwent a rapid adaptive radiation at the beginning of the Cenozoic; by the Eocene most of the mammalian orders were established (Fig. 3–13; Table 3–1).

CHAPTER 4

CLASSIFICATION OF MAMMALS

Despite their remarkable success, and perhaps largely because of their greater size and their consequent inability to exploit large numbers of restricted ecological niches, mammals are much less diverse than are most invertebrate groups. Roughly 1000 genera and some 4060 species of mammals are currently recognized. In contrast, there are an estimated 900,000 species of insects, 30,000 of protozoans, and 104,000 of mollusks.

Chapters 5 through 14 consider the orders and families of mammals that are listed in Table 4–1. For each order and family such features as size of the group, present geographic distribution, time of appearance in the fossil record, structural characters, and brief life history notes are given. Whenever appropriate, morphology is related to function. Specializations for certain modes of life or for specific locomotor abilities are frequently discussed in the sections considering the animals that possess these adaptations. Thus, adaptations for aquatic life are discussed in the chapters on cetaceans and on pinniped carnivores, adaptations for flight are considered in the

section on bats, and cursorial adaptations are examined in the chapter on ungulates.

I devote considerable attention to the orders and families of mammals not because I wish to put primary stress on the taxonomic aspect of mammalogy, but rather as an attempt to provide the student with sufficient information (which, if it is not in his head, is at least at his fingertips) on the various kinds of mammals to make the discussions of the biology of mammals meaningful. The edge of a student's interest is often dulled if he must deal with information about animals with which he is completely unfamiliar. It seems pointless to me to discuss population cycles of microtines, for example, if a student has only a vague idea of what a microtine is. In the chapters on orders, then, I have tried to provide enough information, in the forms of both descriptions and illustrations, to enable the student to gain some familiarity with the orders, families, and, in some cases, subfamilies of mammals. This should serve as a background for the chapters on selected aspects of the biology of mammals.

Table 4–1. A Classification of Recent Mammals.
The Numbers of Species are Approximations.

CLASSIFICATION	COMMON NAMES
Subclass Prototheria	
Order Monotremata (3 species)	
Fam. Tachyglossidae	Spiny anteaters
Ornithorhynchidae	Duck-billed platypus
Subclass Theria	
Infraclass Metatheria	
Order Marsupialia (242 species)	
Fam. Didelphidae	Opossums
Dasyuridae	Marsupial "mice," "rats," and "carnivores"
Notoryctidae	Marsupial "moles"
Peramelidae	Bandicoots
Caenolestidae	Rat opossums
Phalangeridae	Cuscuses and phalangers
Petauridae	Gliders
Burramyidae	Pigmy possums
Tarsipedidae	Honey possum
Phascolarctidae	Koala
Vombatidae	Wombats
Macropodidae	Kangaroos and wallabies
Infraclass Eutheria	
Order Insectivora (406 species)	
Fam. Erinaceidae	Hedgehogs
Talpidae	Moles
Tenrecidae	Tenrecs
Chrysochloridae	Golden moles
Solenodontidae	Solenodons
Soricidae	Shrews
Macroscelididae	Elephant shrews
Tupaiidae	Tree shrews
Order Dermoptera (2 species)	
Fam. Cynocephalidae	Flying lemurs
Order Chiroptera (853 species)	
Fam. Pteropodidae	Old World fruit-eating bats
Rhinopomatidae	Mouse-tailed bats
Emballonuridae	Sac-winged bats
Noctilionidae	Bull-dog bats
Nycteridae	Hollow-faced bats
Megadermatidae	False vampire bats
Rhinolophidae	Horseshoe bats
Phyllostomatidae	Leaf-nosed bats
Mormoopidae	Moustached bats
Desmodontidae	Vampire bats
Natalidae	Funnel-eared bats
Furipteridae	Smoky bats
Thyropteridae	Disc-winged bats
Myzapodidae	Sucker-footed bats
Vespertilionidae	Common bats
Mystacinidae	Short-tailed bats
Molossidae	Free-tailed bats
Order Primates (166 species)	
Fam. Lemuridae	Lemurs
Indridae	Indrid lemurs
Daubentoniidae	Aye-ayes
Lorisidae	Lorises, galagos
Tarsiidae	Tarsiers
Cebidae	New World monkeys
Callithricidae	Marmosets
Cercopithecidae	Old World monkeys
Pongidae	Great apes and gibbons
Hominidae	Man

Table 4–1. A Classification of Recent Mammals.
The Numbers of Species are Approximations. (*Continued*)

CLASSIFICATION	COMMON NAMES
Order Edentata (31 species)	
Fam. Myrmecophagidae	Anteaters
Bradypodidae	Tree sloths
Dasypodidae	Armadillos
Order Pholidota (8 species)	
Fam. Manidae	Scaly anteater
Order Lagomorpha (63 species)	
Fam. Ochotonidae	Pikas
Leporidae	Rabbits and hares
Order Rodentia (1687 species)	
Fam. Aplodontidae	Mountain "beaver"
Sciuridae	Squirrels and marmots
Geomyidae	Pocket gophers
Heteromyidae	Kangaroo rats and pocket mice
Castoridae	Beavers
Anomaluridae	Scaly-tailed squirrels
Pedetidae	Springhaas
Cricetidae	New World rats, mice, hamsters, muskrats, gerbils
Spalacidae	Mole rats
Rhizomyidae	Bamboo rats
Muridae	Old World rats and mice
Gliridae	Dormice
Platacanthomyidae	Spiny dormice
Seleviniidae	Dzhalmans
Zapodidae	Jumping mice
Dipodidae	Jerboas
Hystricidae	Old World porcupines
Erethizontidae	New World porcupines
Caviidae	Guinea pigs, Patagonian "hare"
Hydrochoeridae	Capybaras
Dinomyidae	Pacarana
Heptaxodontidae	
Dasyproctidae	Agoutis and pacas
Chinchillidae	Chinchillas and viscachas
Capromyidae	Hutias
Myocastoridae	Nutria
Octodontidae	Octodonts, tuco-tucos
Abrocomidae	Chinchilla rats
Echimyidae	Spiny rats
Thryonomyidae	Cane rats
Petromyidae	Dassie rats
Bathyergidae	Mole rats
Ctenodactylidae	Gundis
Order Mysticeti (10 species)	
Fam. Balaenidae	Right whales
Eschrichtiidae	Gray whales
Balaenopteridae	Rorquals
Order Odontoceti (74 species)	
Fam. Ziphiidae	Beaked whales
Monodontidae	Norwhal and beluga
Physeteridae	Sperm whales
Platanistidae	River dolphins
Stenidae	
Phocoenidae	Porpoises
Delphinidae	Ocean dolphins

Table 4–1. A Classification of Recent Mammals.
The Numbers of Species are Approximations. (*Continued*)

CLASSIFICATION	COMMON NAMES
Order Carnivora (284 species)	
Fam. Canidae	Wolves, foxes and jackals
Ursidae	Bears
Procyonidae	Raccoons, ring-tailed cats, etc.
Mustelidae	Skunks, badgers, weasels, otters, wolverine
Viverridae	Civets, genets and mongooses
Hyaenidae	Hyaenas, aardwolf
Felidae	Cats
Otariidae	Sea lions and fur seals
Odobenidae	Walrus
Phocidae	Earless seals
Order Tubulidentata (15 species)	
Fam. Orycteropodidae	Aardvark
Order Proboscidea (2 species)	
Fam. Elephantidae	Elephants
Order Hyracoidea (11 species)	
Fam. Procaviidae	Hyraxes
Order Sirenia (5 species)	
Fam. Dugongidae	Dugongs and sea cows
Trichechidae	Manatees
Order Perissodactyla (16 species)	
Fam. Equidae	Horses, asses, zebras
Tapiridae	Tapirs
Rhinocerotidae	Rhinos
Order Artiodactyla (171 species)	
Fam. Suidae	Swine
Tayassuidae	Javelinas, peccaries
Hippopotamidae	Hippos
Camelidae	Camels, llamas
Tragulidae	Chevrotains
Cervidae	Deer, elk, moose, caribou, etc.
Giraffidae	Giraffe, okapi
Antilocapridae	Pronghorn
Bovidae	Bison, antelope, gazelles, sheep, goats, cattle, etc.

CHAPTER 5

NON-EUTHERIAN MAMMALS: MONOTREMES AND MARSUPIALS

Monotremes and marsupials can conveniently be considered apart from the rest of the mammals. Monotremes and marsupials are primitive in a variety of ways, and both have a reproductive pattern very different from that of other mammals: monotremes lay eggs; marsupials bear tiny and poorly developed young, and most have a chorio-vitelline placenta that differs from the chorio-allantoic placenta of "placental" mammals (see page 347). The classification of the major groups of mammals reflects the phylogenetic isolation of the monotremes and marsupials. Monotremes belong to the subclass Prototheria. Both marsupials and placentals are put within the subclass Theria, but the marsupials are placed within one infraclass (Metatheria), and the placentals in another (Eutheria).

ORDER MONOTREMATA

Although represented today by but three genera, each with a single species, and therefore comprising an unimportant segment of the Recent mammalian fauna, monotremes are of great interest for several reasons. Morphologically, they closely resemble no other living mammals, and they possess some features more typical of reptiles than of mammals. Monotremes lay eggs and incubate them in bird-like fashion, yet they have hair and suckle their young. These unusual animals are the sole surviving members of the subclass Prototheria, a group that may have diverged from other mammalian lines of descent in the late Triassic. The order Monotremata includes the family Tachyglossidae (echidnas or spiny anteaters), that occurs in Austra-

FIGURE 5–1. Skull of the spiny anteater (*Tachyglossus aculeatus*). Length of skull 111 mm.

lia, Tasmania, and New Guinea, and the family Ornithorhynchidae (duck-billed platypus), restricted to eastern Australia and Tasmania.

Morphology

Many structural features distinguish monotremes from other mammals. The monotreme skull is uniquely bird-like in appearance (Fig. 5–1); it is toothless, except in young platypuses; the sutures disappear early in life; and the elongate and beak-like rostrum (Fig. 5–2) is covered by a leathery sheath (this sheath is horny in birds). The lacrimal bones are absent and the jugals are small or absent, whereas these bones are present in most therian mammals; evidences of prefrontals and postfrontals, typically reptilian elements that are missing in therian mammals, occur on the frontals of monotremes. There is no auditory bulla, but the chamber of the middle ear is partially surrounded by oval tympanic rings.

Monotremes have cervical ribs, and the thoracic ribs lack tubercles, processes that occur on the ribs of most other mammals and are braced against the transverse processes of the vertebrae. The limb girdles of mono-

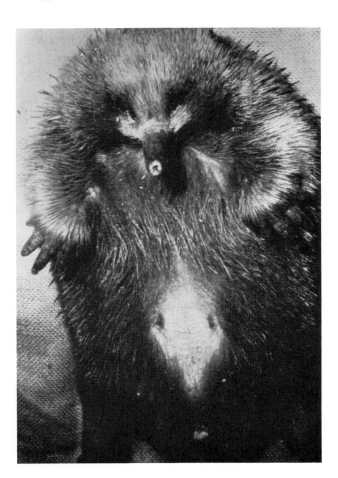

FIGURE 5–2. Ventral view of a live echidna (*Tachyglossus aculeatus*), showing the beak-like rostrum and the poorly developed pouch (typical of the non-breeding season), and the tufts of hairs at the mammary "lobules." (Photograph by M. L. Augee.)

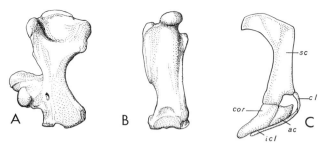

FIGURE 5–3. Bones of monotremes. Left humerus (A) and right femur (B) of the spiny anteater (*Tachyglossus aculeatus*). Pectoral girdle (C) of the duck-billed platypus (*Ornithorhynchus anatinus*). Abbreviations: *ac*, acromion; *cl*, clavicle; *cor*, coracoid; *icl*, interclavicle; *sc*, scapula. (Pectoral girdle of the platypus after Romer, A. S.: *Vertebrate Paleontology*, 3rd ed., The University of Chicago Press, 1966.)

tremes are especially distinctive. The pectoral girdle contains an interclavicle, clavicles, precoracoids, and coracoids (Fig. 5–3), and provides a far more rigid connection between the shoulder joint and the sternum than that characteristic of therian mammals. Large epipubic bones extend forward from the pubes.

As put by Howell (1944:26), no monotremes "by any strength of the imagination might be considered cursorial." Monotremes have retained a limb posture that is similar in many ways to that of a reptile. In monotremes this posture is associated with limited running ability. The proximal segments of the limbs, the humerus and the femur, extend roughly laterally from the body, rather than being more or less vertically oriented to the ground as in terrestrial therians, and the heads of these bones are so formed that the limbs cannot be brought into a typical mammalian posture. In the Australian echidna (*Tachyglossus aculeatus*), rotational movements of the humerus and femur, rather than fore and aft movements as in most mammals, are largely responsible for the propulsion strokes of the limbs (Jenkins, 1970). The humerus moves only slightly fore and aft during propulsion; the femur moves fore and aft through an arc of about 15°, and its distal end moves downward with each stroke. In reptiles the "splayed" limb posture involves contact of the fore and hind feet with the ground well to the side of the shoulder and hip joints,

respectively. The limb postures in the echidna partially depart from this pattern because the forearm angles medially and the manus (forepaw) is roughly ventral to the shoulder joint, and in the hind limb the pes is roughly ventral to the knee. When the echidna is in motion, its body is elevated well above the ground in non-reptilian fashion. Despite the advances in limb posture in the echidna over the reptilian condition, locomotion is slow and appears labored and awkward.

Reproduction

The monotreme reproductive system and reproductive pattern are completely unique among mammals. Eggs are laid, and these are telolecithal (the yolk is concentrated toward the vegetal pole of the ovum) and meroblastic (early cleavages are restricted to a small disc at the animal pole of the ovum), as are birds' eggs. Only the left ovary is functional in the platypus (Asdell, 1964:2), as in most birds, but both ovaries are functional in echidnas. Shell glands are present in the oviducts. A cloaca is present, to the ventral wall of which the penis is attached in males. The testes are abdominal, and seminal vesicles and prostate glands are absent. The female echidna temporarily develops a pouch during the period of incubation and caring for the young, but the platypus never develops a pouch. The mammae lack nipples, and the young suck milk from two areas (lobules) in the pouch in the echidna (Fig. 5–2), or from the abdominal fur in the platypus.

Paleontology

The fossil record of monotremes consists of Recent genera from the Australian Pleistocene and a single species from the middle Tertiary that Stirton et al. (1967) tentatively called a monotreme, and contributes little to our knowledge of the phylogeny of the group. The occurrence of the Tertiary species suggests that monotremes may have been more diverse at one time than they are now. Possible relationships between monotremes and various Mesozoic reptilian and mammalian groups have been discussed (Kermack and Musset, 1958; Kermack, 1963; Simpson, 1959; Romer, 1968: 166, 167), but there is no widely accepted agreement on relationships. Monotremes probably arose from a different reptilian stock than that from which marsupials and placental mammals evolved, and they may therefore have undergone a long, independent development. The extremely specialized habits of monotremes and their isolation from competition with continental placental animals may have contributed importantly to their survival. Van Deusen (1969) pointed out, however, that *Tachyglossus* thrives over much of Australia despite fire, drought, dingos, and other placentals, and that in New Guinea *Zaglossus* is affected severely only by habitat destruction and hunting by man.

Family Tachyglossidae. Members of this group have robust bodies covered with short, sturdy spines (Fig. 5–4) that are controlled by unusually powerful panniculus carnosus muscles. *Zaglossus*, the New Guinea echidna, weighs from 5 to 10 kg., and the Australian spiny anteater *(Tachyglossus)*, from about 2.5 to 6 kg. The braincase of the echidna is moderately large, and the cerebrum is convoluted. The rostrum is slender and beak-like (Figs. 5–1, 5–2), the den-

tary bones are slender and delicate, and the long tongue is protrusile and covered with viscous mucus secreted mostly by the enlarged submaxillary salivary glands. Food is ground between spines at the base of the tongue and adjacent transverse spiny ridges on the palate. The pinnae are moderately large. The limbs are powerfully built and are adapted for digging. The humerus is highly modified by broad extensions of the medial and lateral epicondyles that provide unusually large surfaces for the origins of some of the powerful muscles of the forearm (Fig. 5–3). In *Zaglossus* the number of claws is variable; some individuals have only three claws front and rear, whereas some have a full complement of five claws (Van Deusen, 1969). In *Tachyglossus* all digits have stout claws. The ankles of male echidnas (and of some females) bear medially directed spurs; their function is not known.

These animals have highly specialized modes of life. They are powerful diggers and can rapidly escape predators by burrowing. Food consists largely of termites, ants, and a variety of other invertebrates. Foraging involves turning over stones and digging into termite and ant nests, and the prey is captured by the sticky tongue.

Usually one leathery-shelled egg is laid, and incubation in the temporary pouch lasts from seven to ten days. The young is helpless when hatched and remains in the pouch until the spines develop.

The Australian spiny anteater is known to be a true hibernator and will become torpid in response to cold and lack of food. During periods of torpor in experimental animals the body temperature (5.5°C) was close to the ambient temperature (5.0°C), and the heart rate dropped to seven beats per minute (Augee and Ealey, 1968). Experimental animals were able to arouse spontaneously. One individual could dig slowly when its body temperature was only 10.5°C and its heart rate was seven beats per minute.

Family Ornithorhynchidae. Com-

FIGURE 5–4. Two species of monotremes: Above, the Australian spiny anteater (*Tachyglossus aculeatus*). Below, the New Guinea spiny anteater (*Zaglossus bruijni*). (The photograph of *Tachyglossus* is by M. L. Augee; that of *Zaglossus* is by Hobart M. Van Deusen.)

pared to echidnas, the duck-billed platypus is small, weighing from roughly 0.5 to 2 kg. Some structural features of the platypus are associated with its semiaquatic mode of life. The pelage is dense, and, as in the muskrat (*Ondatra*), the underfur is woolly. The external auditory meatus is tubular, as in the beaver (*Castor*). The eye and ear openings (pinnae are absent) lie in a furrow that is closed by folds of skin when the animal is submerged. The feet are webbed, but the digits retain claws that are used for burrowing. The web of the forefoot extends beyond the tips of the claws, and is folded back against the palm when the animal is digging or when it is on land. The ankles of the male platypus have grooved and medially directed spurs that are connected to poison glands. In a case reported by Walker (1968:6), a man spurred by a platypus suffered immediate intense pain and did not regain full use of the injured hand for months. The braincase of the platypus, in contrast to that of tachyglossids, is small, and the cerebrum lacks convolutions. Although the young have teeth, the gums of adults are toothless and are covered by persistently growing, horny plates. Anteriorly, the occlusal surfaces of the plates form ridges that serve to chop food; posteriorly, the plates are flattened crushing surfaces. Some additional mastication is accomplished by the flattened tongue, which acts against the palate. The elongate rostrum bears a flattened, leathery bill that seems to have remarkable tactile ability.

The platypus inhabits a variety of waters, including mountain streams, slow-moving and turbid rivers, lakes, and ponds, and is primarily a bottom feeder. Aquatic crustaceans, insect larvae, and a wide variety of other animal material, and some plants, are taken during dives that are roughly a minute in duration. The platypus takes refuge in burrows dug into banks adjacent to water. Periods of inactivity, or perhaps hibernation, occur in the winter (Bourliere, 1956:195). The female digs a burrow up to 50 feet in depth, in which the eggs are laid on a nest of moist leaves. Usually two eggs are laid, and these adhere to one another. The burrow is kept plugged during the ten-day incubation period, after which the young, extremely rudimentary at hatching, are suckled for about five months.

Populations of the platypus were seriously declining at one time and the animals appeared to be facing extinction, but due at least in part to protective legislation enacted by the Australian government, the platypus now seems out of danger.

ORDER MARSUPIALIA

Marsupials and placentals are representatives of two evolutionary lines that have been separate since the early Cretaceous or Jurassic (Slaughter, 1968; Clemens, 1968; Lillegraven, 1969). As a result of their long, independent histories, marsupials differ structurally from placentals in many ways. In certain structural features, and in their semi-arboreal habits and omnivorous-insectivorous diet, the more primitive living marsupials probably resemble in some ways the Mesozoic mammals from which Cenozoic mammals were derived. The brain, and seemingly the reproductive pattern, are more primitive in marsupials than in placentals, and perhaps largely due to these differences marsupials have usually been replaced by placentals whenever species of these two groups were in competition. Today only two important strongholds for marsupials remain, the Australian region (including Australia, Tasmania, New Guinea, and nearby islands) and the Neotropics (including southern Mexico, Central America, and most of South America). When isolated from placentals for long periods of time, marsupials have undergone remarkable adaptive radiations, and most marsupials have functional counterparts among placentals.

Morphology

The marsupial skull frequently has a small, narrow braincase, housing small cerebral hemispheres with simple convolutions. The auditory bullae, when present, are usually formed largely by the alisphenoid bone, rather than by the tympanic bone as in most placentals. The marsupial palate characteristically has large vacuities (Fig. 5–5), and the angular process of the dentary is inflected. The dentition is unique in that, except in the family Vombatidae, there are never equal numbers of incisors above and below; the cheek teeth are primitively 3/3 premolars and 4/4 molars, and the primitive eutherian number of 44 teeth is often exceeded.

Marsupials often have highly specialized feet associated with specialized types of locomotion (Fig. 5–6).

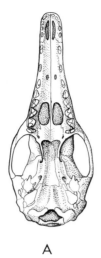
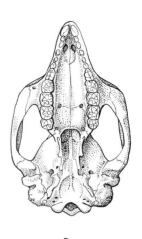

FIGURE 5–5.   Ventral views of skulls of marsupials. A, New Guinea bandicoot (*Peroryctes raffrayanus*, order Peramelidae; length of skull 82 mm.); B, ring-tailed possum (*Pseudocheirus corinnae*, order Phalangeridae; length of skull 97 mm.). (After Tate and Archbold, 1937.)

A

B

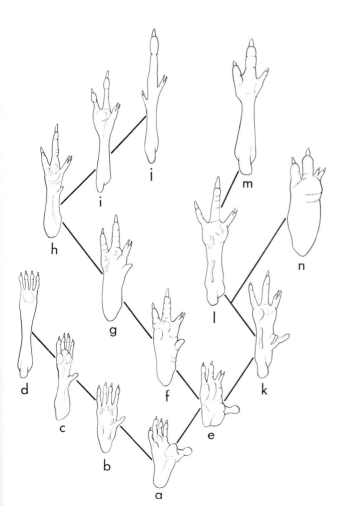

FIGURE 5–6.   Ventral views of right hind feet of marsupials, showing patterns of specialization associated with various styles of locomotion. The presumed basic arboreal type is represented by *a*, the foot of a didelphid. The lines indicate possible evolutionary pathways leading toward greater specialization. Feet of the following kinds are shown. Dasyuridae: *b, Phascogale; c, Sminthopsis; d, Antechinomys.* Petauridae: *e, Pseudocheirus.* Peramelidae: *f, Perameles* sp.; *g, Peroryctes* sp.; *h, Peroryctes* sp.; *i, Macrotis; j, Chaeropus.* Macropodidae: *k, Hypsiprymnodon; l, Potorous; m, Macropus; n, Dendrolagus.* (After Howell, A. B.: *Speed in Animals,* The University of Chicago Press, 1944.)

The unusual patterns of specialization in the feet of marsupials are probably a result of an arboreal heritage and the early development of an opposable first digit and a relatively long fourth digit. As in the monotremes, epipubic bones extend forward from the pubic bones.

Reproduction

The marsupial reproductive pattern is distinctive, and is of special interest because certain features, such as the structure of the fetal membranes (described in Chapter 18), may resemble an evolutionary stage through which eutherian mammals passed. Most female marsupials have a marsupium (an abdominal pouch) or marsupial folds, within which the nipples occur. The number of nipples varies from two, as in the family Notoryctidae and some members of the Dasyuridae, to 19, as in some members of the Didelphidae. Individual variation in the number of nipples often occurs within a species. The female reproductive tract is bifid, i.e., the vagina and uterus are double (Fig. 5–7). In all but the family Notoryctidae, which is adapted for digging, the testes are contained in a scrotum which is anterior to the penis.

The gestation period is characteristically short (Table 18–1, page 348), and the young are tiny and rudimentary at birth. Most marsupials, with the exception of the Peramelidae, have a chorio-vitelline placenta. Although newborn marsupials are blind, naked, and delicate, the forelimbs, and doubtless certain parts of the nervous system, are precociously well developed, and are of critical importance in aiding the young in making their way at birth from the vulva to the marsupium. When the worm-like young reaches the pouch it attaches to a nipple, which soon responds by enlarging within the mouth of the young. Each young is thus firmly attached to a nipple, and remains so for a period of time greatly exceeding the gestation period. The weight of the young marsupial when it leaves the pouch, and that of the placental when newly born, are roughly the same in species of comparable adult size (Sharman, 1970).

Paleontology

Marsupials are first known from the lower Cretaceous of North America (Slaughter, 1968), near the height of the reign of the ruling reptiles. At this time the only surviving groups of primitive Jurassic mammals (discussed in Chapter 3) were the multituberculates and the symmetrodonts. Seemingly, the niches previously filled by such Jurassic types as triconodonts and symmetrodonts were being occupied in the late Cretaceous by the two dominant mammalian groups today, the marsupials and the placentals. These groups probably descended from late Triassic pantotheres. Cretaceous marsupials and placentals were small ani-

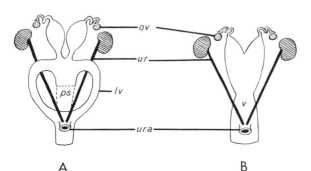

FIGURE 5–7. Diagrams of the female reproductive tracts of marsupials (A) and placentals (B). Abbreviations: *lv*, lateral vagina; *ov*, ovary; *ps*, pseudovaginal canal; *ura*, urethra; *ur*, ureter; *ut*, uterus; *v*, vagina. (After Sharman, G. B.: Reproductive physiology of marsupials. *Science*, 171:443–449. Copyright 1970 by The American Association for the Advancement of Science.)

mals with omnivorous or insectivorous feeding habits, but they provided the ancestral stock for the remarkable mammalian radiation that occurred early in the Cenozoic, after the disappearance of the ruling reptiles.

The most primitive marsupial family, and perhaps the stem group from which all other marsupials evolved, is the Didelphidae. Didelphids occurred in North America in the early Cretaceous, and have a nearly continuous fossil record there through the Cenozoic. A single didelphid genus is recorded from the European Eocene to Miocene. Today, as well as during most of the Cenozoic, Australia and South America have been the two centers of marsupial abundance, and the histories of their marsupial faunas are parallel to some extent. In both continents the marsupial radiation occurred under partial or complete isolation from competition with eutherians.

Biogeography

According to Simpson (1965:157, 158), marsupials probably reached Australia well before placentals, probably in late Cretaceous or early Paleocene. They presumably arrived *via* "sweepstakes" dispersal (see p. 297), and underwent a spectacular adaptive radiation. Six living marsupial families, including 66 genera, and two extinct families, Thylacoleonidae and Diprotodontidae, resulted from this radiation. *Thylacoleo*, the only known member of the Thylacoleonidae, was a Pleistocene form roughly the size of an African lion; the third premolars were modified into shearing blades. There are divergent opinions on the function of these teeth, and it is not known whether the animal was carnivorous or frugivorous. The Diprotodontidae is represented by several Pleistocene genera, of which *Diprotodon* is especially noteworthy. This animal reached a length of 11 feet, roughly resembled an enormous wombat (Vombatidae), and is the largest known

marsupial. The fossil record of marsupials is meager, and except for members of these two extinct families, all Australian fossil marsupials are near relatives of living types. An interesting speculation regarding the history of the land mammals of Australia has been presented by Simpson (1965:160). He suggests that an Australian radiation of Jurassic monotremes may have resulted in a predominantly monotreme fauna that was almost completely replaced by the more progressive marsupials during a late Cretaceous and early Tertiary radiation of these animals.

Didelphoid marsupials probably dispersed from North America to South America in the late Cretaceous and early Paleocene, before a general invasion of placentals, and during much of the Tertiary marsupials were partly free of placental competition (see p. 310). The didelphoid ancestral stock underwent an impressive adaptive radiation, and many adaptive zones exploited in the Tertiary of North America and Eurasia by placentals were filled at this time in South America by marsupials. A group of carnivorous marsupials, the Borhyaenidae, known from the Eocene to the Pliocene, paralleled placental carnivores to a remarkable degree. *Borhyaena*, for example, was a Miocene predatory form with a dog-like skull (Fig. 5–8). Even more remarkable, however, was *Thylacosmilus* (Fig. 5–8), a Pliocene form that was strikingly similar to placental saber-toothed cats. Presumably *Thylacosmilus* killed large prey, perhaps notoungulates, by stabbing with the sabers. This same style of killing was used by the saber-toothed cats when dealing with artiodactylan and perissodactylan prey. Some fossil members of the living South American marsupial family Caenolestidae had highly specialized dentitions (Fig. 5–9), and some presumably had shrew-like, insectivorous feeding habits.

Family Didelphidae. Structurally, didelphids are the most primitive and

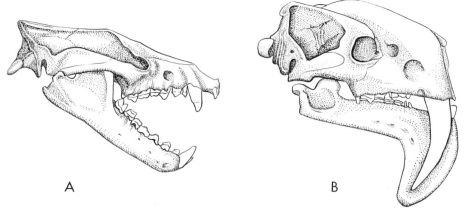

A   B

FIGURE 5–8. Skulls of members of the extinct marsupial family Borhyaenidae. A, *Thylacosmilus* (length of skull 232 mm.); B, *Borhyaena* (length of skull 230 mm.). (After Romer, A. S.: *Vertebrate Paleontology*, 3rd ed., The University of Chicago Press, 1966.)

FIGURE 5–9.   Jaw of a Miocene caenolestid, showing the highly specialized, trenchant cheek tooth. (After Romer, A. S.: *Vertebrate Paleontology*, 3rd ed., The University of Chicago Press, 1966.)

FIGURE 5–10.   A didelphid marsupial (*Didelphis marsupialis*). In the United States this animal is common in the southeast and in many areas along the Pacific coast. (San Diego Zoo photograph.)

generalized marsupials, and are the oldest known family, dating from the early Cretaceous. The Didelphidae includes 12 Recent genera with 66 species, and occurs from southeastern Canada, in the case of the American opossum (*Didelphis marsupialis*; Fig. 5–10) to southern Argentina, in the case of the Patagonian opossum (*Lestodelphis halli*).

In these New World opossums the rostrum is long (Fig. 5–11A), the braincase is usually narrow, and the sagittal crest is prominent. The dental formula is 5/4, 1/1, 3/3, 4/4 = 50. The incisors are small and unspecialized, and the canines are large. The upper molars are basically tritubercular with sharp cusps, and the lower molars have a trigonid and a talonid (Fig. 5–12A, B).

Except for the opposable and clawless hallux in all species, a feature probably inherited from arboreal ancestral stock, and the webbed hind feet in the water opossum (*Chironectes minimus*), the feet are unspecialized, with no loss of digits or syndactyly.

(Syndactyly is the condition in which two digits are attached by skin, as shown by a number of examples in Figure 5–6.) The foot posture is plantigrade. A marsupium is present in some didelphids, but is represented by folds of skin protecting the nipples in others, and is absent from some. The tail is long and is usually prehensile.

Although they occupy a wide range of habitats, didelphids are primarily inhabitants of tropical or subtropical areas, where they are often locally abundant. Most didelphids are partly arboreal and are omnivorous. The water opossum, however, is largely aquatic, is an accomplished diver, and is carnivorous. In this species both sexes have a marsupium. In the male the scrotum lies in this pouch; in the female, a sphincter can tightly close the entrance to the marsupium, allowing the animal to swim while the young remain in the pouch.

The small mouse opossum (*Marmosa*), a widespread neotropical didelphid, is one of the most abundant small

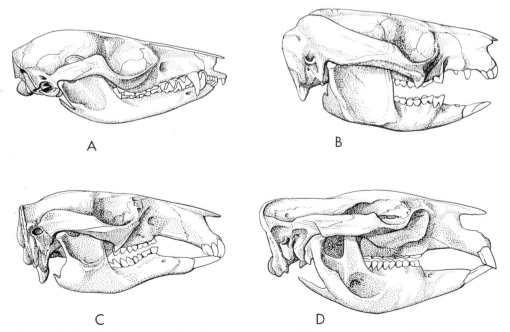

A

B

C

D

FIGURE 5–11.   Skulls of marsupials. A, mouse opossum (*Marmosa canescens*, Didelphidae; length of skull 35 mm.); B, brush-tailed possum (*Trichosurus vulpecula*, Phalangeridae; length of skull 87 mm.); C, wallaby (*Wallabia bicolor*, Macropodidae; length of skull 135 mm.); D, wombat (*Vombatus ursinus*, Vombatidae; length of skull 180 mm.).

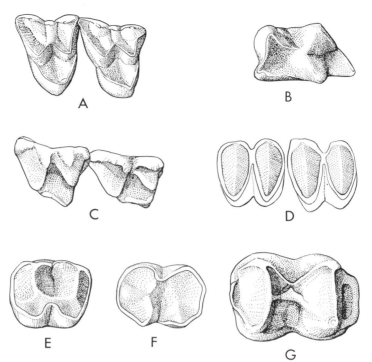

FIGURE 5–12. Occlusal views of molars of marsupials. Second and third right upper molars (A) and third lower left molar (B) of a mouse opossum (*Marmosa canescens,* Didelphidae); C, second and third upper right molars of the New Guinea bandicoot (*Peroryctes raffrayanus,* Peramelidae; after Tate and Archbold, 1937); D, second and third right upper molars of a wombat (*Vombatus ursinus, Vombatidae*); second upper right molar (E) and third lower left molar (F) of a phalanger (*Trichosurus vulpecula,* Phalangeridae); G, second upper right molar of a wallaby (*Wallabia bicolor,* Macropodidae).

FIGURE 5–13. Four members of the family Dasyuridae: A, a marsupial "rat" (*Dasyuroides byrnei*); B, a marsupial "mouse" (*Sminthopsis crassicaudata*); C, a marsupial "mouse" (*Antechinus stuartii*); D, the native "cat" (*Dasyurus viverrinus*). (The photographs of *Dasyuroides* and *Antechinus* are by Jeffrey Hudson; those of *Sminthopsis* and *Dasyurus* are by Anthony Robinson.)

mammals in some parts of Mexico. Although it is mouse-like in general appearance, it seems to be largely insectivorous in some areas, at least during the summer (Smith, 1971). Poorly defined folds of skin protect the nipples of *Marmosa*, and the young simply hang on to the nipples and the mother's venter as best they can. One wonders that any young survive.

Family Dasyuridae (Fig. 5–13). The Dasyuridae and the Didelphidae are seemingly closely related, and each family has characteristics thought to resemble those of primitive Cretaceous marsupials. Dasyurids are more progressive than didelphids, however, both dentally and with regard to limb structure. The fossil record is too incomplete to give evidence of the lines of descent of Australian marsupials, but the dasyurids are thought to be the most direct descendants of the marsupials that originally colonized Australia, probably in the late Cretaceous. Although the earliest known dasyurid is from the Australian middle Tertiary, the family probably arose at a far earlier time. Recent members of this family include 20 genera and 50 species, and the geographic range includes Australia, New Guinea, Tasmania, the Aru Islands, and Normanby Island (Van Deusen and Jones, 1967:60).

Many of the major characters of dasyurids are shared by other marsupials, but several features are diagnostic of the former. The dental formula is 4/3, 1/1, 2-4/2-4, 4/4 = 42-50; the incisors are usually small and either pointed or blade-like, the canines are large and have a sharp edge, and the molars have three sharp cusps adapted to an insectivorous and carnivorous diet. The forefoot has five digits, and the hind foot has four or five digits. The hallux is clawless and usually vestigial, and is absent in some cursorial genera (Figs. 5–6, 5–14). There is no syndactyly. The foot posture is plantigrade in many species, but the long-limbed jumping marsupials, such as *Antechinomys* and the cursorial and carnivorous native "cat" (*Dasyurus viverrinus;* Fig. 5–

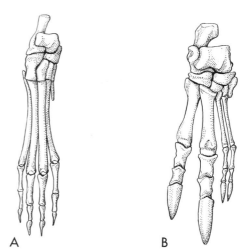

FIGURE 5–14. The feet of two marsupials. A, a terrestrial species, the Australian native "cat" (*Dasyurus viverrinus,* Dasyuridae); B, an arboreal species, a tree kangaroo (*Dendrolagus* sp., Macropodidae). (After Marshall, 1972.)

13) and Tasmanian "wolf" (*Thylacinus cynocephalus;* Fig. 5–15), are digitigrade. The marsupium is often absent; when present, it is often poorly developed and opens posteriorly. The tail is long and well furred, is conspicuously tufted in some species, and is never prehensile. Dasyurids range from the size of a shrew (*Antechinus*) to that of a medium-size dog (*Thylacinus*).

A wide variety of terrestrial habitats are occupied by dasyurids, and a few species are arboreal. A remarkably diverse array of marsupials are grouped within the Dasyuridae. The smaller species fill the feeding niche occupied in Eurasia and North America by shrews (Soricidae), and resemble these animals in the possession of long-snouted heads and unspecialized limbs. A group of rat-sized types seems adapted to preying on insects and small vertebrates, and the desert-dwelling genus *Antechinomys* has long slender limbs and a long tufted tail, and uses a rapid, bounding, quadrupedal gait. Another group, the native "cats," consists of somewhat weasel-like dasyurids that weigh from

FIGURE 5–15. The Tasmanian "wolf" (*Thylacinus cynocephalus*, Dasyuridae), which may now be extinct. (Photograph courtesy of Edwin H. Colbert.)

roughly 0.5 to 3 kg. and prey on a variety of small vertebrates. These "cats" are agile and effective predators, and, although primarily terrestrial, are capable climbers. The largest marsupial carnivores are the Tasmanian devil (*Sarcophilus harrisii*) and the Tasmanian "wolf." The Tasmanian devil is a stocky, short-limbed dasyurid, weighing from roughly 4.5 to 9.5 kg.; it is now restricted to Tasmania. It is a persistent scavenger, but will also kill a wide variety of small vertebrates. The Tasmanian wolf is dog-like in both size and general build (Fig. 5–15); it has long limbs and a digitigrade foot posture. Although it is treated here as a dasyurid, *Thylacinus* has been separated from the dasyurids and put in the family Thylacinidae by Ride (1970:226). Now extinct in New Guinea, on the Australian mainland, and very possibly in Tasmania, *Thylacinus* is able to prey on such large animals as the larger species of wallabies. The most divergent dasyurid is the banded anteater, a small, long-snouted animal considered by some to be the sole representative of the family Myrmecobiidae. The teeth are small and widely spaced in the long tooth row, and the long, protrusile tongue is used in capturing termites. The total number of teeth (52) is greater than that of any other terrestrial mammal.

Family Notoryctidae. This remarkable family is represented by a single species of marsupial "mole" that inhabits arid parts of northwestern and south-central Australia. Many of the diagnostic characters of these mouse-sized animals are adaptations for fossorial life. The eyes are vestigial, are covered by skin and lack lenses, and, as indicated by the specific name *Notoryctes typhlops* (*typhlops* means blind in Greek), are not functional. The ears lack pinnae. The nose bears a broad, cornified shield, and the nostrils are narrow slits. The dental formula is usually 4/3, 1/1, 2/3, 4/4 = 44, but the incisors vary in number. The incisors, canines, and all but the last upper premolar are unicuspid; the paracone and metacone of the upper molars form a prominent single cusp, and the lower molars lack a talonid. As an adaptation serving to brace the neck when the animal forces its way through the soil, the five posterior cervical vertebrae are fused. The forelimbs are robust, and the claws of digits three and four are remarkably enlarged and function together as a spade, whereas the other digits are reduced. The central three digits of the hind feet have enlarged claws; the small first digit has a nail, and the fifth digit is vestigial. The marsupium is partially divided into two compartments, each with a single nipple. The fur is long and fine textured, and varies in color from silvery white to yellowish red. There is no fossil record of notoryctids.

These animals use their powerful forelimbs and the armored rostrum to

FIGURE 5–16. Two peramelid marsupials: A, a New Guinea bandicoot (*Peroryctes raffrayanus;* photograph by Stanley Oliver Grierson); B, a "rabbit" bandicoot (*Macrotis lagotis;* photograph by Anthony Robinson).

force their way through soft, sandy soil. When the "mole" forages near the surface, the soil is pushed behind the animal and no permanent burrow is formed. The food is predominantly invertebrate larvae, but a variety of animal material and some vegetable material have been eaten by notoryctids in captivity. In contrast to moles (Order Insectivora; family Talpidae), which characteristically inhabit moist soils, notoryctids inhabit sandy desert regions supporting scattered vegetation.

Family Peramelidae (Fig. 5–16). Members of this family are called bandicoots, and are characterized in general by an insectivore-like dentition and a trend toward specialization of the hind limb for running. Eight Recent genera, represented by 22 Recent species, are known, mainly from Australia, Tasmania, and New Guinea. Some species of bandicoots have been extirpated or have become uncommon over parts of their former range due, apparently, to the grazing of livestock, to brush fires, and to the introduction of various placental mammals.

The smaller bandicoots are rat-sized; the largest species weighs roughly 7.0 kg. The dental formula is 4-5/3, 1/1, 3/3, 4/4 = 46 or 48. The incisors are small, and the molars are tritubercular (Fig. 5–12C) or quadritubercular. The rostrum is slender (Fig. 5–16), and the ears of some species resemble those of rabbits. The marsupium is present and opens to the rear, and bandicoots, alone among marsupials, have a chorio-allantoic placenta (see page 347). Although often long, the tail is not prehensile. The fourth digit of the foot is always the largest, and the remaining digits are variously reduced (Fig. 5–17). The hind foot posture is usually digitigrade, and the hind limbs are elongate. The opposable hallux, probably inherited by peramelids from an arboreal ancestral stock, is rudimentary or may be lost. The second and third digits are joined as far as the distal phalanges by an interdigital membrane, and the muscles of these digits are partially fused, allowing them to act only in unison (Jones, 1923). An ex-

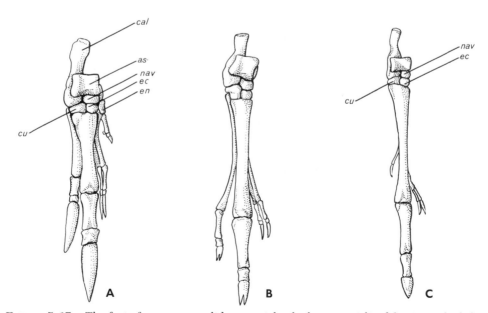

FIGURE 5–17.   The feet of some peramelid marsupials; the least specialized foot is on the left and the most specialized is on the right. A, a long-nosed bandicoot (*Perameles* sp.); B, a rabbit bandicoot (*Macrotis* sp.); C, the pig-footed bandicoot (*Chaeropus ecaudatus*). Abbreviations: *as*, astragalus; *cal*, calcaneum; *cu*, cuboid; *ec*, ectocuneiform; *en*, entocuneiform; *na*, navicular. The digits are numbered. (After Marshall, 1972.)

treme degree of cursorial specialization occurs in the pig-footed bandicoot
(*Chaeropus ecaudatus*), in which the
forelimb is functionally didactyl and
the hind foot is functionally monodac-
tyl during running. The second and
third digits of the forelimb are large
and clawed; the first and fifth are absent, and the fourth is vestigial. The
curious appearance of the forefoot is
the source of the animal's common and
generic names (*Chaeropus* means
pig-footed in Greek).

The structure and function of the
specialized peramelid hind foot is
unique and has been described in detail by Marshall (1972). In mammals,
extreme reduction in the number of
digits is invariably associated with
good running ability. Most highly cursorial ungulates have retained only the
third digit (in the case of the horse) or
digits three and four (in the case of
some antelope). A similar trend occurs
in peramelids. Probably due partly to
an early development of syndactyly involving the second and third digits,
and the use of these digits for grooming, the general trend in the cursorial
peramelids is toward the reduction of
all digits but the fourth, and a great enlargement of this digit (Fig. 5–17).
These specializations are accompanied by an alteration in the structure
and function of the tarsal bones. The
ectocuneiform bone makes broad contact with the proximal end of the fourth
metatarsal and partially supports this
digit, a character unique to peramelids. The mesocuneiform is lost, and
the weight of the body is borne mainly
by the cuboid, ectocuneiform, navicular, and astragulus bones. The calcaneum does not serve a major
weight-bearing function, but of course
serves as a point of insertion for the extensors of the foot, muscles of great importance in locomotion.

Horses, antelope, and peramelids
provide beautiful examples of a similar
functional problem, in this case that of
refining running ability, being solved
by different structural means. In the
horse only the third digit is retained,
and it is supported largely by the ectocuneiform, the navicular, and the astragalus (Fig. 14–4, page 224). In the
pronghorn antelope only two digits are
retained, and the cannon bone (the
fused third and fourth metatarsals) is
supported largely by the fused cuboid
and navicular and the fused mesocuneiform and ectocuneiform (Fig. 14–5);
the calcaneum is no longer a
weight-bearing element. In the most
cursorial peramelid (*Chaeropus*). the
fourth digit is greatly enlarged and is
supported, as outlined above, by the
cuboid, navicular, ectocuneiform, and
astragalus (Fig. 5–17). Marshall (1972)
points out, however, that the structure
of the hind limb of peramelids is not
entirely modified for running, perhaps
because of the burrowing tendencies
of these animals. The fibula is large,
and movement at the ankle joint is not
restricted to a single plane as it is in
most cursorial mammals. (Cursorial adaptations are discussed in more detail
on page 220).

Bandicoots are largely insectivorous,
but small vertebrates, a variety of invertebrates, and some vegetable material are also taken. Some species take
refuge in nests that they build of sticks
and other plant debris; all species of
Macrotis dig burrows in which they
hide during the day. *Chaeropus* is reported by Jones (1923:171) to squat
like a jackrabbit beneath saltbushes
(*Atriplex*) in semi-open areas; in
jackrabbit fashion, *Chaeropus* depends
on its speed to escape enemies, but
will seek shelter in hollow logs when
chased by dogs. The limb structure of
Chaeropus is considerably more specialized in some ways than that of rabbits, but the styles of locomotion and
methods of escape are similar. Unfortunately, *Chaeropus* may be extinct.

Family Caenolestidae. Recent
members of this family (three genera
and seven species) bear the common
name of rat opossum, and are restricted
to parts of the Andes Mountains of
northern and western South America.
The earliest known caenolestids are
from the Eocene of South America,

FIGURE 5-18. A caenolestid marsupial (*Lestoros inca*). This individual was taken in Peru, at an elevation of 3530 m. (Photograph by John A. W. Kirsch.)

and in the Oligocene and Miocene a diverse group of caenolestids, including some highly specialized types (Fig. 5-9), appeared. The Recent caenolestids are representatives of a conservative evolutionary line that dates at least from the early Eocene. There are three separated and seemingly relict populations of caenolestids: the northern population occupies parts of western Venezuela, Colombia, and Ecuador; the intermediate population occurs in a section of the Andes in southern Peru; and the southern population occupies Chiloe Island and a nearby coastal area of south-central Chile.

The rat opossums are rat- or mouse-sized marsupials that, because of their elongate heads and small eyes, resemble shrews (Fig. 5-18). The skull is elongate and the brain is primitive: the olfactory bulbs are large, and the cerebrum lacks fissures. The dental formula is 4/3-4, 1/1, 3/3, 4/4 = 46 or 48; the first lower incisors are large, and the remaining lower incisors, the canine, and the first premolar are unicuspid. The atlas bears a movable cervical rib. The feet are five-toed and unspecialized. The tail is long but not prehensile, and there is no marsupium.

Caenolestids appear to be ecological homologues of the shrews, but, in contrast to shrews, the rat opposums may be a declining group. They are largely terrestrial, but are agile climbers, eat mostly insects, and occur primarily in mesic, forested areas. The limbs and style of locomotion are unspecialized, as in shrews.

Phalangeroid Marsupials: Phalangeridae; Petauridae; Burramyidae; Tarsipedidae; Phascolarctidae

The phalangeroid marsupials are considered here to include the Australian families listed, which are referred to by such names as possum, ringtail, cuscus, glider, noolbender, and koala. Until recently all of the structurally and behaviorally diverse phalangeroid marsupials were included in a single family, Phalangeridae, but Ride (1970) divided this family into the families listed above; I am following his system of classification.

The phalangeroid marsupials occupy a geographic range that includes much of Australia and many of its coastal islands, Tasmania, New Guinea and nearby coastal islands, Celebes, Timor, and Ceram; to the east of New Guinea, they inhabit the Bismarck and Louisiade archipelagos, D'Entrecasteaux Group, Solomon Islands, and, by introduction, New Zealand. These are the most widely distributed Australian marsupials, a fact probably explained by their arboreal habits and the relatively high probability of their "island hopping" on rafts of vegetation. Phalangeroids are first known from the Eocene of Tasmania.

The phalangeroid skull is usually broad and somewhat flattened, and the lower tooth row is usually interrupted by a diastema (Figs. 5-11B, 5-19A). The dental formula is 2-3/1-2, 1/0, 1-3/1-3, 3-4/3-4 = 24-40. At most three upper incisors are present. The first lower incisor is long and robust, and

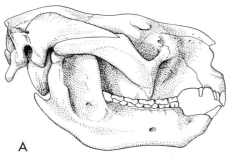
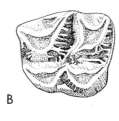

A   B

FIGURE 5–19.   A, the skull of the koala (*Phascolarctos cinereus*, Phascolarctidae; length of skull 132 mm.). B, occlusal view of the second molar, upper right tooth row, showing the crescentic areas of dentine, exposed by wear, and the complex pattern of furrows.

often projects forward; the second and third lower incisors are absent or vestigial. The molars are either moderately flat-crowned and tubercular or have ridges (Figs. 5–12E, F; 5–19B); in the single case of *Tarsipes*, they are small and peg-like.

The limbs are adapted for climbing. The hands have five clawed digits; the first and second digits in some species are opposable to the remaining digits. The first digit of the hind foot is large, opposable, and clawless; digits two and three are syndactylous, and digits four and five are robust (Fig. 5–6E). The tail is long (with the exception of the koala, *Phascolarctos cinereus*, in which it is vestigial) and is often prehensile. The marsupium is present, and in all but the koala it opens anteriorly.

Family Phalangeridae.   In this family are the possums and cuscuses, a group of primarily arboreal animals. Three genera and about 10 species are known. The brush-tailed possum (*Trichosurus vulpecula*) is one of the most familiar of Australian mammals, for it frequently maintains resident populations in suburban areas, where it often seeks shelter in roofs of houses and feeds on cultivated plants.

These marsupials are of moderate size, ranging in weight from approximately 1 to 6 kg. The skull is broad and has deep zygomatic arches (Fig. 5–11B). The molars are bilobed with rounded cusps (Fig. 5–12E, F). As adaptations to arboreal life, the hands and feet are large and have a powerful grasp, and the tail is prehensile. The cuscuses have short ears, woolly fur, and an odd, teddy-bear-like appearance (Figs. 5–20, 5–21).

Members of this family mostly inhabit wooded areas, but the adaptable brush-tailed possum also occupies treeless areas, where it takes refuge in rocks or in the burrows of other mammals. Phalangerids are omnivorous, and are known to take a wide variety of plant material as well as insects, young birds, and bird's eggs. The brush-tailed possum is solitary and has a sternal scent gland, considerably larger in males, which produces a musky smell that is used in the scent marking of objects within the animal's territory. This marsupial is one of Australia's most valuable fur bearers. During 1959 some 107,500 brush-tailed possum skins were marketed in Victoria.

Family Petauridae (Fig. 5–22). This family includes the ringtails, so named because of their prehensile tails, and the greater and lesser gliding possums, some of the handsomest and most remarkable of all marsupials. Eight genera and some 25 species are currently recognized.

Most members of this family are fairly small; weights range from about 100 gm. to 1.5 kg. The skull is broad, and the four-cusped molars have fairly sharp outer cusps forming a roughly W-shaped ectoloph. The tail is prehensile in some petaurids and long and bushy in others. Some species are strikingly marked (Fig. 5–22B, C). The

FIGURE 5–20.   Two views of a New Guinea cuscus (*Phalanger maculatus*, Phalangeridae). Note the prehensile tail with the "traction ridges" on the bare distal part of its ventral surface. (Photographs by Stanley Oliver Grierson.)

FIGURE 5–21. Two New Guinea cuscuses (Phalangeridae): A, *Phalanger vestitus* (photograph by Stanley Oliver Grierson); B, *P. orientalis* (photograph courtesy of Hobart M. Van Deusen).

FIGURE 5–22. Three members of the family Petauridae: A, a ring-tailed possum (*Pseudocheirus forbesi*); B, a gliding possum (*Petaurus breviceps*); C, a striped possum (*Dactylopsila trivirgata*). (Photographs by Stanley Oliver Grierson.)

gliders (*Petaurus* and *Schoinobates*) have furred membranes that extend between the limbs and function as lifting surfaces for gliding. In these gliders the claws are sharp and recurved, like those of a cat, and increase the ability of the animal to cling to the smooth trunks and large branches of trees.

The petaurids are nocturnal and arboreal creatures and occur in wooded areas. (Figure 5–23 shows a habitat of the gliders.) Ringtails are nocturnal and are strictly herbivorous, eating both leaves and fruit. They make conspicuous nests ("dreys") of leaves and twigs in the dense scrub of eastern and southeastern Australia. Curious specializations, similar to those of the pri-

mate *Daubentonia*, occur in two petaurid genera of "striped possums." In *Dactylopsila*, and to a more advanced degree in *Dactylonax*, the fourth digit of the hand is elongate and slender, and its claw is recurved. In addition, the incisors are robust and function roughly as do those of rodents. Striped possums tear away tree bark with their incisors, and extract insects from crevices and holes in the wood with the specialized fourth finger and the tongue. The conspicuous striped color pattern of *Dactylopsila* (Fig. 5–22C) is of interest inasmuch as it is associated, as in skunks, with a powerful, musky scent.

The gliders are strikingly similar in gliding style and ability to flying squir-

FIGURE 5–23. Rain forest in eastern Victoria, Australia. This community is inhabited by two species of lesser gliding possums (*Petaurus*) and by greater gliding possums (*Schoinobates volans*). (Photograph by Diana Harrison.)

rels (*Glaucomys*), and some can glide over 100 m. *Schoinobates*, the greater glider, is remarkable in having perhaps the most specialized marsupial diet; its food is entirely leaves and blossoms, chiefly those of the eucalyptus trees. Sugar gliders (*Petaurus breviceps;* Fig. 5–22B) live in family groups, and scent marking plays an important role in the social organization of the group. Each individual has a particular odor recognized by other individuals. The cohesion of the group is also aided by mutual scent marking, for all members of the group become permeated with the scent of the group's dominant males (Schultz-Westrum, 1964).

Family Burramyidae. The type genus of this family was known for many years only from Pleistocene fossil material; finally, in 1966, in a ski lodge on Mt. Hotham in Victoria, a representative of the genus was found alive, and more recently it has been found at other localities. This family contains five genera and eight species of small, mouse-like marsupials, called pigmy possums.

These diminutive marsupials are from about 60 to 120 mm. in head and body length, are delicately built, and have large eyes and mouse-like ears (Fig. 5–24). The tail is long and prehensile in all species, and has a lateral fringe of hairs in the pigmy glider (*Acrobates pygmaeus*). This species has a narrow gliding membrane that is bordered by a fringe of long hairs. Traction between the digits and the trunks and branches of trees is increased in the pigmy glider by expanded pads at the tips of the fingers and toes; the surfaces of the pads have "traction ridges" that further increase the clinging ability of these animals.

Members of this family are restricted to wooded areas. They are apparently insectivorous-omnivorous, but the feeding habits of some members of the group are not known. As in some small placental mammals, in some bur-

FIGURE 5–24. A pigmy possum (*Cercartetus concinnus*, Burramyidae). (Photograph by Anthony Robinson.)

ramyid marsupials the ability to become torpid during cold weather is well developed. Pigmy gliders become torpid in their nests on cold days, and the tails of the members of two genera of pigmy possums (*Cercartetus* and *Eudromicia*) become greatly enlarged with fat as winter approaches and these animals undergo periods of torpor.

Family Tarsipedidae. This family contains but one species, the highly specialized slender-nosed honey possum or noolbender (*Tarsipes spencerae*). This remarkable animal's many specializations obscure its relationships to other marsupials, and its taxonomic position has long been uncertain.

Tarsipes is small, only about 15 to 20 gm. in weight, and has a long, prehensile tail. The pelage is marked by three longitudinal stripes on the back. The rostrum is long and fairly slim, and the dentary bones are extremely slender and delicate. The cheek teeth are small and degenerate, and only the upper canines and two medial lower incisors are well developed. The snout is long and slender and the long tongue has bristles at its tip (these specializations are similar to those of some nectar-feeding bats of the family Phyllostomatidae). All digits but the syndactylous second and third digits of the hind feet have expanded terminal pads resembling to some extent those of the primate *Tarsius*.

Honey possums occur in forested and shrub-grown areas. Like the hummingbirds and nectar-feeding bats, honey possums feed on nectar, pollen, and to some extent on small insects that live in flowers. The long, protrusile tongue is used to probe into flowers. *Tarsipes* can climb delicately over even the insecure footing of clusters of flowers at the ends of branches, and often clings upside down to flowers while feeding. Although the animal is still common in some areas today, the expansion of agriculture in the southwestern part of West Australia is restricting the honey possum's range.

Family Phascolarctidae. The familiar koala or native "bear" (*Phascolarctos cinereus*) is the sole member of this family. This highly specialized herbivore is restricted to some wooded parts of southeastern Australia.

The tufted ears, odd-looking naked nose, and chunky tail-less form make the koala one of the most distinctive of Australian marsupials (Fig. 5–25). These are fairly large marsupials; the adults range from 8 to 10 kg. in weight. The skull is broad and sturdily built, and the dentary bones are deep and robust (Fig. 5–19A). The roughly quadrate molars have crescentic ridges (Fig. 5–19B), and there is a diastema in both the upper and lower tooth rows between the cheek teeth and the anterior teeth. Branches are grasped between the first two and the last three fingers of the hand, and between the

FIGURE 5–25. The koala (*Phascolarctos cinereus*, Phascolarctidae). (Photograph by Diana Harrison.)

clawless first digit and the remaining digits of the foot; the long, curved claws aid in maintaining purchase on smooth branches.

Koalas are fairly sedentary and feed on only a few species of smooth-barked eucalyptus trees. Maturation of a koala takes considerable time. A single young is born and is carried in the pouch for six months, after which it rides on its mother's back for a few more months. The young koala is dependent on its mother for a year, and sexual maturity is not reached until three or four years of age. According to Ride (1970:88), koalas grunt in pig-like fashion when feeding at night, and when alarmed make continuous wails. Koalas lived in southwestern Western Australia during the late Pleistocene, but no longer occur there even though suitable habitat is present.

Family Vombatidae. This family is represented by two genera and four species. These animals, known as wombats, are completely herbivorous, and show remarkable structural con-

vergence toward rodents. Due to the efforts of man, wombats have become scarce or absent over much of their former range, and now are restricted to parts of eastern and southern Australia, Tasmania, and the islands between Australia and Tasmania.

Wombats are stocky animals with small eyes and rodent-like faces (Fig. 5–26), and reach over 35 kg. in weight (Troughton, 1947:144). The skull and dentition bears a striking resemblance to those of some rodents (Figs. 5–11D, 5–12D). The skull is flattened, the rostrum is relatively short, and the heavily built zygomatic arches flare strongly to the sides. The area of origin of the anterior part of the masseter muscle is marked by a conspicuous depression in the maxillary and jugal that is similar to the comparable depression in the maxillary and premaxillary of the beaver (*Castor*; see Fig. 10–13). The dental formula is 1/1, 0/0, 1/1, 4/4 = 24; all teeth are rootless and ever-growing. Only the anterior surfaces of the incisors bear enamel, and

FIGURE 5–26.   A wombat (*Vombatus ursinus*, Vombatidae). (Photograph by Anthony Robinson.)

the incisors and the first premolars are separated by a wide diastema. The molars are bilophodont (Fig. 5–12D). As in rodents, the coronoid process of the dentary is reduced and the masseter muscle, rather than the temporalis, is the major muscle of mastication.

The limbs are short and powerful, and the foot posture is plantigrade. The forefeet are five-toed; all digits have broad, long claws. The hallux of the hind foot is small and clawless, but the other digits have claws. Digits two and three of the hind feet are partly syndactylous. The tail is vestigial. The marsupium opens posteriorly and contains one pair of mammae.

This family is first known from the late Tertiary (Gill, 1947), and both Recent genera have fossil species. The Pleistocene trend toward large size that is apparent in many other mammalian groups also occurred in the Vombatidae, as evidenced by the huge Pleistocene "wombat" *Phascolonus*.

Wombats are capable burrowers. Their burrows are up to 30 m. in length and are of sufficient diameter to admit a small child. Wombats dig burrows in the open or beneath rock piles as do marmots (*Marmota*), their rodent counterparts. Level or mountainous terrain supporting dry or moist sclerophyll forests or grassland is inhabited. The food of wombats is largely grass, but includes bark, roots, and fungi (Troughton, 1947:139, 140). The widespread destruction of wombats in settled areas is not desirable from a biologist's viewpoint, but seems inevitable because of "conflict of interests" with man. The openings to wombat burrows are hazardous to large livestock, and wombats are locally destructive to crops.

Family Macropodidae (Fig. 5–27). Members of this familiar marsupial group, which includes the kangaroos, euros, and wallabies, are the ecological equivalents of such ungulates as antelope. Both macropodids and ungulates are cursorial, and both have highly specialized limbs. Further, both groups are herbivorous, and have skulls and dentitions specialized for this mode of feeding. The present distribution of the 19 Recent genera and approximately 47 Recent species of macropodids includes New Guinea, Bismarck Archipelago, the D'Entrecasteaux Group, Australia, and by introduction, some islands near New Guinea and New Zealand. The family Macropodidae appears first in the middle Tertiary (late Oligocene or early Miocene) of Australia. Wallabies and a kangaroo are known from the late Ter-

FIGURE 5-27. Three kinds of macropodid marsupials: A, great grey kangaroo (*Macropus giganteus*); B, pademelon (*Thylogale billardierii*); C, red-necked wallaby (*Macropus rufogriseus*). (Photographs by Diana Harrison.)

tiary, and in the Pleistocene unusually large macropodids occurred.

Living macropodids vary tremendously in size and structure. The musky rat kangaroo (*Hypsiprymnodon moschatus*) weighs only 500 g., whereas the great grey kangaroo (*Macropus giganteus*), the largest living marsupial, reaches 2 m. in height and approximately 90 kg. in weight. The marsupium is usually large and opens anteriorly. The macropodid skull is moderately long and slender, and the rostrum is usually fairly long (Fig. 5-11C). The dental formula is 3/1, 1-0/0, 2/2, 4/4 = 32 or 34. The upper incisors have sharp crowns; their long axes are oriented more or less front to back, and they meet the blade-like occlusal surfaces of the lower incisors

(Fig. 5-11C). This specialized arrangement serves a cropping function similar to that of the front teeth of many ungulates. There is a broad diastema between the incisors and the premolars. The molars are quadritubercular and bilophodont (Fig. 5-12G). In many macropodids the last molar does not erupt until well after the animal becomes adult. A unique situation occurs in the little rock wallaby (*Peradorcas concinna*), in which nine molars may erupt in succession. Usually four or five molars are functional at one time, and replacement is from the rear as the molars are successively lost from the front.

Macropodids are highly specialized for jumping. The forelimbs are five-toed and usually small; they are

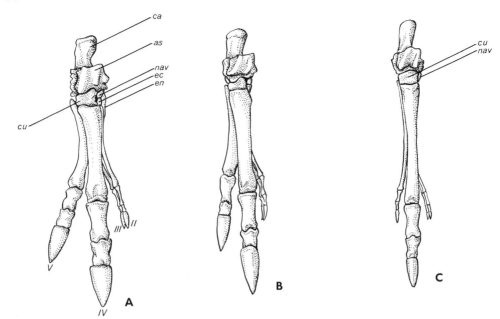

FIGURE 5–28. The feet of some macropodid marsupials; the least specialized foot is on the left, and the most specialized is on the right. A, a scrub wallaby (*Thylogale* sp.); B, a kangaroo (*Macropus* sp.); C, the red kangaroo (*Megaleia rufa*). Abbreviations: *as*, astragalus; *ca*, calcaneum; *cu*, cuboid; *ec*, ectocuneiform; *en*, entocuneiform; *na*, navicular. The digits are numbered in A.

used for slow movement on all fours or for food handling (Frith and Calaby, 1969). The hind limbs are elongate, especially the second segment. The hallux is missing in all but *Hypsiprym-nodon*, and digits two and three are small and syndactylous; the fourth is the largest digit, and the fifth is often also robust (Fig. 5–28). The unusual pattern of digital reduction and the dominance of the fourth digit in the most highly cursorial Australian marsupials are perhaps due to their arboreal ancestry. In these ancestors the foot was five-toed and the hallux was opposable; the fourth was the longest remaining digit, and the foot was adapted to grasping branches. With specialization of the foot for running or hopping, the hallux was lost and the longest toe, the fourth, became the most important digit. In most macropodids the foot is functionally two-toed during rapid locomotion, which is characteristically bipedal, but in *Megaleia* the foot is functionally one-toed (Fig. 5–28). In the macropodid tarsus there is no contact be-

tween the ectocuneiform and the fourth metatarsal (in contrast to the arrangement in the Peramelidae; see Fig. 5–17). Because the hind limb posture of macropodids is basically plantigrade, the calcaneum is an important weight-bearing element of the tarsus (which it is not in the digitigrade Peramelidae). The macropodid tail is usually long and robust, and functions in the more specialized species as a balancing organ and as the posterior "foot" of the tripod formed by the plantigrade hind feet and tail, on which the animal can sit when not in motion.

Several macropodid genera depart from the familiar structural pattern of kangaroos and from the grazing or browsing habit. *Hypsiprymnodon*, a muskrat-sized inhabitant of rain forests and riparian situations, has a tail of modest length and retains all of the digits of the hind foot (Fig. 5–6K). The hind limbs are not greatly elongate, and the animal uses quadrupedal rather than saltatorial locomotion. Animal material forms a large share of the food of this seemingly primitive macropodid. The tree kangaroos (*Dendrolagus*, Macropodinae) are primarily arboreal. This mode of life is reflected by the large and robust forelimbs with strong, recurved claws; by the hind limbs, which are not strongly elongate; and by the short, broad hind foot (Fig. 5–14B). Saltation, typical of terrestrial kangaroos, has not been completely abandoned by tree kangaroos; not only are these animals agile climbers, but they leap between trees and from trees to the ground. Their food is large fruit and leaves.

The running ability of the larger kangaroos (*Macropus*) is impressive. Speeds on level terrain of roughly 50 km./hr. are attained, and leaps covering distances of 8.5 m. and heights of 3 m. have been reported (Troughton, 1947:213). The highly developed jumping ability of macropodids allows these animals to move easily for long distances between scattered sources of water or forage and to escape enemies by erratic leaps. These abilities, rather than the capacity for great speed, are probably of primary adaptive importance. Saltation may have been developed by small forms ancestral to kangaroos as a means of erratic escape in open areas. This style of locomotion, serving this function, is known in a number of desert-dwelling rodents. According to Howell (1944:247), the kangaroo's "method of traveling by saltation was hardly begun for the purpose of ultimate speed. Rather has it built speed into the locomotor pattern that was already established, probably for some other purpose." The locomotion of rock wallabies (*Petrogale*) is adapted to the rocky country they inhabit. According to Jones (1924:231) their movements are spectacular: "There seems to be no leap it will not take, no chink between boulders into which it will not hurl itself."

CHAPTER 6

INSECTIVORES, DERMOPTERANS, AND BATS

This chapter deals with insectivores and some of their descendants. An insectivore ancestry is generally accepted for bats, and on the basis of limited fossil material and anatomical evidence based on Recent species the flying lemurs (Dermoptera) are also thought to have an insectivore derivation. The flying lemurs are customarily put in a separate order, as by Simpson (1945:53-54). But Miller (1906) classified them as bats, whereas Romer (1968:179) saw little reason for separating flying lemurs from insectivores and listed the Dermoptera as a suborder of the Insectivora (1966:380). I have followed recent tradition in considering the Dermoptera to be a separate order, but because the chiropterans and dermopterans probably arose from insectivore stock I feel justified in discussing these groups in the same chapter.

ORDER INSECTIVORA

Insectivores comprise today the third largest order of mammals, con-

taining roughly 77 genera and 406 species. They have an unusually wide geographic distribution: they occur throughout much of both hemispheres, but are absent from most of the Australian region, all but the northernmost part of South America, and the polar regions.

Insectivores are the most direct living descendants of primitive eutherian mammals, and they are an ancient order. Two insectivore-like groups, the Deltatheridiidae and the Leptictidae, are known from the late Cretaceous. These groups are characterized by contrasting types of dentitions and are early members of two separate evolutionary lines. The deltatheridian dentition is distinctive in having the paracone and metacone close together and situated well in from the outer edge of the tooth (Fig. 6–1A); the talonid of the lower molar is narrow (Fig. 6–1B). These mammals were probably ancestral to a primitive carnivorous group called creodonts. In the leptictids, however, the paracone and metacone lie well apart, near the outer edge of

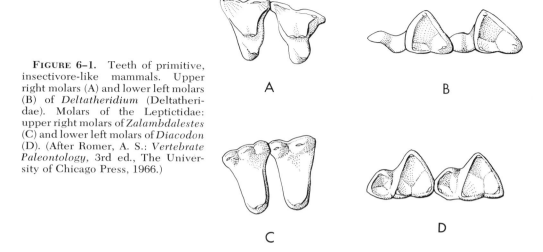

FIGURE 6–1. Teeth of primitive, insectivore-like mammals. Upper right molars (A) and lower left molars (B) of *Deltatheridium* (Deltatheridae). Molars of the Leptictidae: upper right molars of *Zalambdalestes* (C) and lower left molars of *Diacodon* (D). (After Romer, A. S.: *Vertebrate Paleontology*, 3rd ed., The University of Chicago Press, 1966.)

the upper tooth (Fig. 6–1C), and the talonid of the lower molar is broad (Fig. 6–1D); these cusp patterns were perhaps ancestral to those of later "true" insectivores. The skull of *Deltatheridium*, except for the unusual cusp pattern, is of a primitive eutherian type, with a continuous tooth row, large canines, a complete zygomatic arch, large orbits, a rostrum of moderate length, and a small braincase (Fig. 6–2B), whereas the leptictid skull (e.g., *Zalambdalestes*, Fig. 6–2A) has some of the same types of specializations that are typical of later, specialized insectivores.

Morphology

Because the order Insectivora has long been used as a catch-all taxon to

which have been assigned many primitive mammals of doubtful affinities, an anatomical diagnosis of the group is difficult. The order Insectivora has been separated by many authorities into two major divisions (or suborders), the lipotyphlans, including the families Erinaceidae, Talpidae, Tenrecidae, Chrysochloridae, Solenodontidae, and Soricidae, and the menotyphlans, including the Macroscelididae and Tupaiidae. This division is based on a series of morphological differences that indicate a long separation of the lines of descent represented by these two groups.

Because of resemblances between primitive prosimian primates, such as lemurs, and the family Tupaiidae, some authors have placed this family under the order Primates (Simpson,

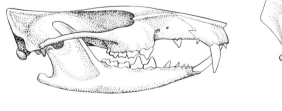

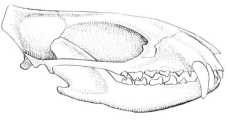

FIGURE 6–2. Skulls of primitive, insectivore-like mammals. A, *Zalambdalestes* (length of skull 50 mm.); B, *Deltatheridium* (length of skull 45 mm.). (After Romer, A. S.: *Vertebrate Paleontology*, 3rd ed., The University of Chicago Press, 1966.)

1954:61; Walker, 1968:395). Evidence based on studies of the nervous system, however, indicates no close tupaiid-primate relationship (Campbell, 1966), and paleontological evidence as interpreted by Szalay (1968) indicates that tupaiids were derived from the Insectivora and not from the Primates. The taxonomic scheme used here—the retention of the menotyphlan families within the Insectivora—is upheld by Findley (1967:103), who states that "acceptance of the menotyphlans as a natural taxon and the grouping of them with insectivores seem justified as a matter of practical expediency."

Questions of taxonomy aside, the structural contrasts between the lipotyphlans and menotyphlans indicate that these groups occupy contrasting adaptive zones. Broadly speaking, the lipotyphlans have primitive brains and depend more on olfaction than on vision; they usually have specialized dentitions; their limbs are unspecialized and they pursue a generalized, quadrupedal locomotion. Menotyphlans, in contrast, have more progressive brains and acute vision; the dentitions are specialized along less insectivorous lines, and locomotion is less generalized (macroscelidids hop and many tupaiids climb).

For all Recent lipotyphlous insectivores the following characters are reasonably diagnostic: the tympanic bone is annular, no auditory bulla is present, and the entotympanic bone is absent; the tympanic cavity is often partially covered by processes from adjacent bones; the olfactory bulbs are longer than the rest of the brain and are largely interorbital; the eyes and the optic foramina are usually small; the jugal is reduced or absent and the zygomatic arch is incomplete in some groups; the orbitosphenoid is mainly anterior to the braincase; the teeth have sharp cusps, and usually the primitive placental crown pattern is recognizable; the anterior dentition is often specialized by the enlargement and specialization of the incisors and the reduction of the canines; the limbs are usually unspecialized and are never adapted to saltation.

Menotyphlans are characterized by the following features: the auditory bulla is complete and the entotympanic is large; the olfactory bulbs are shorter than the rest of the brain and do not extend between the orbits; the eyes and the optic foramina are enlarged; the jugal is large and the zygoma are complete; the orbitosphenoid forms part of the braincase; the hind limbs are greatly elongate and adapted to hopping (Macroscelididae; Fig. 6–3C) or are slightly elongate, with the manus and pes enlarged, and are adapted to climbing (Tupaiidae; Fig. 6–3B).

Family Erinaceidae. Members of the family Erinaceidae, the hedgehogs, are the most primitive lipotyphlans. The family is represented

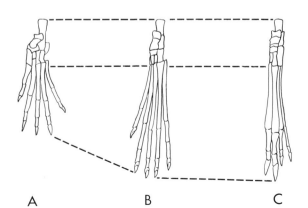

FIGURE 6–3. Left hind feet of several insectivores, showing differing degrees of elongation of the metatarsals and phalanges. A, gymnure (*Echinosorex*, Erinaceidae); B, a tree shrew (*Tupaia*, Tupaiidae); C, an elephant shrew (*Rhynchocyon*, Macroscelididae). (After Evans, 1942.)

A   B   C

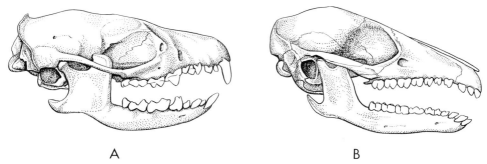

A   B

FIGURE 6–4. Skulls of insectivores. A, a hedgehog (*Erinaceus* sp., Erinaceidae; length of skull 32 mm.); B, an elephant shrew (*Elephantulus* sp., Macroscelididae; length of skull 32 mm.; after Hill and Carter, 1941).

today by 10 genera and 14 species; they occur in Africa, non-wooded parts of Eurasia, southeastern Asia, and the island of Borneo. Erinaceids are first known from the Oligocene, and fossil material is known from the Oligocene to the Pliocene in North America, and from the Oligocene to the Recent in the Old World. The family Adapisoricidae, containing primitive relatives of the hedgehogs, is represented from the Cretaceous to the Oligocene in both hemispheres.

Erinaceids vary from the size of a mouse to that of a small rabbit (1.4 kg.). The eyes and pinnae are moderately large, and the snout is usually long. The zygomatic arches are complete. The dental formula is 2-3/3, 1/1, 3-4/2-4, 3/3 = 36-44. The first upper and, in some species, the first lower incisors are enlarged, but the front teeth (Fig. 6–4A) never reach the degree of specialization typical of shrews. In hedgehogs the upper molars have simple, nonsectorial cusps, with the paracone and metacone near the outer edge; the hypocone completes the quadrate form of the upper tooth (Fig. 6–5A). Both the trigonid and talonid of the lower molars are well developed (Fig. 6–5B). The molars are thus better adapted to an omnivorous than to an insectivorous diet. The feet retain five digits in all but one genus, and the foot posture is plantigrade. An obvious specialization is the possession of spines in members of the subfamily Erinaceinae (Fig. 6–6). In these animals the sheet of muscle beneath the skin (pan-

niculus carnosus) is greatly enlarged and controls the erection of the spines.

In various parts of their wide range, hedgehogs occupy deciduous woodlands, cultivated land, and tropical and desert areas. A wide variety of food is taken, but animal material is preferred. Some members of this family protect themselves by rolling into a tight ball with the spines erected. Members of the subfamily Erinaceinae are probably heterothermic. (Heterothermic animals can regulate their body temperature physiologically, but temperature is not regulated precisely or at the same level at all times.) Hibernation occurs in the widespread genus *Erinaceus*, and estivation is practiced by the desert species *Paraechinus*

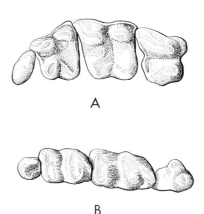

A

B

FIGURE 6–5. Cheek teeth of a hedgehog (*Erinaceus* sp., Erinaceidae). Fourth premolar and three molars of the upper right (A) and lower left (B) tooth rows.

Figure 6–6. An Ethiopian hedgehog (*Paraechinus aethiopicus*). (Photograph by Ron Garrison; San Diego Zoo photo.)

aethiopicus. A related species from India, *P. micropus*, has survived in captivity for periods of from four to six weeks without food or water (Walker, 1968:133), and this species and *Hemiechinus auritus* are known to have winter periods of dormancy in India. The ability to undergo periods of torpor, involving a lowering of body temperature (hypothermia) and a low metabolic rate, may be developed in all members of the subfamily Erinaceinae as a means of achieving metabolic economy during critical periods of stress due to cold, heat, or shortages of food.

Family Talpidae. This family includes a group of small rat- or mouse-sized animals usually referred to as moles. These predominantly burrowing insectivores (15 genera and 22 species) occur in parts of North America, Europe, and Asia. The European fossil record of talpids begins in

the late Eocene; talpids are known first in the New World from the Oligocene. Apparently, the anatomical modifications typical of Recent fossorial genera were attained early, for the Recent European genus *Talpa* is first known from the Miocene.

The head and forelimbs of most talpids are modified for fossorial (burrowing) life. The zygomatic arch is complete, the tympanic cavity is not fully enclosed by bone, and the eyes are small and often lie beneath the skin. The snout is long and slender, the ears usually lack pinnae, and the fur is characteristically lustrous and velvety. The dental formula is 2-3/1-2, 1/0-1, 3-4/3-4, 3/3 = 34-42. The first upper incisors are inclined backward (Fig. 6–7B), and the upper molars have W-shaped ectolophs (Fig. 6–8C). In the fossorial species, the forelimbs are more or less rotated from the usual orientation typical of terrestrial mammals

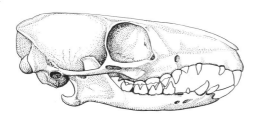
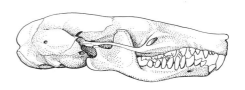

A   B

Figure 6–7. Skulls of insectivores. A, a tree shrew (*Tupaia* sp., Tupaiidae; length of skull 52 mm.); B, the eastern mole (*Scalopus aquaticus*, Talpidae; length of skull 37 mm.).

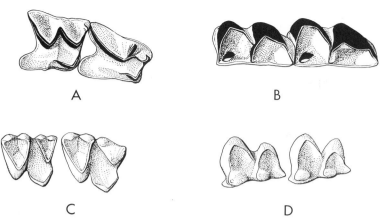

FIGURE 6–8. Cheek teeth of insectivores. The fourth upper right premolar and first molar (A) and the first two lower left molars (B) of the vagrant shrew (*Sorex vagrans*, Soricidae). The pigmented parts of the teeth are shown in black. The first and second upper right molars (C) and the comparable lower left molars (D) of the eastern mole (*Scalopus aquaticus*, Talpidae).

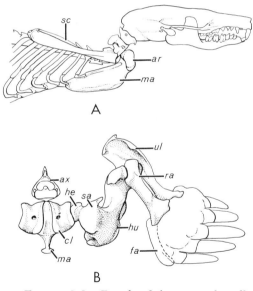

FIGURE 6–9. Details of the pectoral girdle and forelimb of the eastern mole (*Scalopus aquaticus*, Talpidae). A, side view of the pectoral girdle. B, anterior view of part of the pectoral girdle and the forelimb, with the shoulder joint slightly disarticulated to show the head of the humerus and its secondary articular surface. The head of the humerus articulates with the glenoid fossa of the scapula, and a second articulation (involving considerably larger surfaces) occurs between the secondary articular surface of the humerus and an articular surface of the clavicle. Abbreviations: *ar*, articular surface of the clavicle; *ax*, axis; *cl*, clavicle; *fa*, position of the falciform bone; *he*, head of the humerus; *hu*, humerus; *ma*, manubrium of the sternum; *ra*, radius; *sa*, secondary articular surface of the humerus; *sc*, scapula; *ul*, ulna.

in such a way that the digits point to the side, the palms face backward, and the elbows point upward (Fig. 6–9B). In addition, the phalanges are short, the claws are long, and the clavicle and humerus are unusually short and robust. The scapula is long and slender (Fig. 6–9A, B) and serves both to anchor the forelimb solidly against the axial skeleton and to provide advantageous attachments for some of the powerful muscles that pull the forelimb backward. The anteriormost segment of the sternum (the manubrium) is greatly enlarged and extends foreward to beneath the base of the skull. These specializations serve both to increase the area for attachment of the large pectoralis muscles, and to move the shoulder joint forward and allow the forepaws to remove or loosen soil beside the snout. The clavicle is short and broad, and provides a large secondary articular surface for the humerus (Fig. 6–9A, B). The double articulation of the shoulder joint, with articular contacts between the humerus and the scapula and clavicle, provides an unusually strong bracing for this joint during the powerful rotation of the humerus that accompanies the digging stroke of the forelimb. In some genera the falciform bone is large and serves to increase the breadth of the forepaw and to brace the first digit (Fig. 6–9B). One talpid, the Asiatic shrew mole (*Uropsilus*), lacks fossorial or aquatic specializations and resembles a shrew in general form. The two members of the Old World subfamily Desmaninae are modified for semi-aquatic life; they have webbed forefeet, and the greatly enlarged hind feet are webbed and bear a fringe of stiff hairs that increase the effectiveness of the hind feet as paddles. These animals also have strange flexible snouts that are used to probe for food.

Fossorial talpids occur typically in moist and friable soils in forested, meadow, or streamside areas, and feed largely on animal material. A species that occurs in the eastern United States (*Scalopus aquaticus*), however, locally penetrates the moderately dry sandhill prairies of eastern Colorado, where the characteristic ridges of soil appear only during wet weather. In most areas these ridges are a common evidence of the presence of moles, and are made by the animals as they travel just beneath the surface by forcing their way through the soil. Soil from deep burrows is deposited on the surface in more or less conical "mole hills." The burrows of pocket gophers (*Geomys bursarius*) are used to some extent by *Scalopus*, as indicated by the occasional capture in Colorado of moles in traps set for pocket gophers. The semi-aquatic Old World desmanines live along the banks of lakes, ponds, or streams, and feed largely on aquatic invertebrates. Their burrows open beneath the surface of the water and extend upward to a nest chamber above the water level. Moles are not known to hibernate or estivate.

Family Tenrecidae. The tenrecs are a group of insectivores that vary widely in structure and in habits. Various tenrecs bear a general resemblance to such diverse mammals as shrews, hedgehogs, muskrats, mice, and otters. The family includes 11 genera and 23 species, and occupies Madagascar, the Comoro Islands, and west-central Africa. The meager fossil record of tenrecs indicates little about their evolution. The oldest fossil records are from the Miocene of East Africa, and Pleistocene fossils are known from Madagascar.

Tenrecs vary from roughly the size of a shrew to that of a cottontail rabbit. The snout is frequently long and slender. The jugal is absent, the eye is usually small, and the pinnae are conspicuous. The tympanic bone is annular and the squamosal forms part of the roof of the tympanic cavity. The anterior dentition varies between species; the first upper premolars are never present, and the molars are 3/3 in all but *Tenrec* (4/3) and *Echinops* (2/2). The upper molars have crowns that are triangular in occlusal view, and only in one genus (*Potamogale*) is a W-shaped ectoloph present in these teeth.

An unusually broad array of adaptive types occur within the Tenrecidae. *Tenrec* roughly resembles a coarse-pelaged, long-snouted opossum (Fig. 6–10), and has spines interspersed with soft hairs. It is omnivorous. *Echinops* (Fig. 6–11), *Hemicentetes* (Fig. 6–12), and *Setifer* are also spiny, and the latter two genera resemble hedgehogs (Erinaceidae) closely. In these two genera the panniculus carnosus muscle is powerfully developed, and enables the animals to erect the spines. It also contributes to the ability of these animals to roll into a ball. The feet and head are tucked beneath the body during this protective movement, and "the sphincter muscles running around the body at the junction of the spiny dorsum and the hairy venter permit the spiny dorsal skin to be drawn together, thus enclosing the animal in an impregnable shield of spines" (Gould and Eisenberg, 1966). These authors found that newborn *Echinops* and *Setifer* reacted to being disturbed by rolling into a ball. *Hemicentetes* has a group of 14 to 16 specialized quills on the middle of the back that rub together when underlying dermal muscles are twitched and produce sounds in a variety of patterns of repetition. Differences in these sounds depend on differences in associated be-

FIGURE 6–10. A tenrec (*Tenrec ecaudatus*, Tenrecidae). (Photograph by J. F. Eisenberg and Edwin Gould.)

FIGURE 6–11. A Madagascar "hedgehog" (*Echinops telfairi*, Tenrecidae). (Photograph by J. F. Eisenberg and Edwin Gould.)

FIGURE 6–12. A streaked tenrec (*Hemicentetes semispinosus*, Tenrecidae). (Photograph by J. F. Eisenberg and Edwin Gould.)

havior of the animals (Gould; 1965) and may be used in intraspecific communication.

Members of the subfamily Oryzorictinae are probably largely insectivorous, and several genera resemble shrews. *Limnogale* has webbed feet and a somewhat laterally compressed tail, and is rat-like in general appearance. This animal is aquatic and feeds on aquatic invertebrates. The two members of the subfamily Potamogalinae resemble otters, and are semi-aquatic. *Potamogale* is the largest living insectivore (approximately 600 mm. in length). It has small, non-webbed feet; the large, laterally compressed tail is used for swimming. This animal eats aquatic invertebrates, fish, and amphibians. *Micropotamogale*, a considerably smaller animal, resembles a shrew. It swims by the use of its partially webbed feet, and presumably has a diet similar to that of *Potamogale*.

Some tenrecs are known to become torpid during seasons of food shortage (Eisenberg and Gould, 1970).

Family Chrysochloridae. Another variation on the fossorial insectivore theme is typified by chrysochlorids, the golden moles. These animals resemble "true" moles (Talpidae), but even more closely resemble in fossorial adaptations and in function the marsupial "moles" (Notoryctidae). The five genera and roughly 11 species comprising the family Chrysochloridae occur widely in southern Africa, where they occupy forested areas, savannas, and sandy deserts. The earliest fossil chrysochlorids from the Miocene of East Africa resemble Recent species, and these and Pleistocene fossil material give no evidence of the derivation of the group.

Golden moles have modes of life similar to those of the fossorial members of the Talpidae, and possess some parallel adaptations as well as some contrasting structural features. The ears of golden moles lack pinnae, and the small eyes are covered with skin. The pointed snout has a leathery pad at its tip. The zygoma are formed by elongate processes of the maxilla, and the occipital area includes bones, the tabulars, not typically found in mammals. The skull is rather abruptly conical rather than flattened and elongate as in many insectivores. An auditory bulla is present and is formed largely by the tympanic bones. The dental formula is usually 3/3, 1/1, 3/3, 3/3 = 40. The first upper incisor is enlarged, and the molars are basically tritubercular and lack the stylar cusps and the W-shaped ectoloph typical of talpids. The permanent dentition of golden moles emerges fairly late in life. The forelimbs are powerfully built and the forearm rests against a concavity in the rib cage. The fifth digit of the hand is absent, and digits two and three usually have huge pick-like claws. The forelimbs are not rotated as are those of talpids but more or less retain the usual mammalian posture with the palmar surfaces downward.

Golden moles are adept burrowers and probably use the leathery nose pad in forcing their way through the soil. Both deep and shallow burrows are constructed; the depth of the burrows may depend on the amount of soil moisture. The roofs of shallow burrows in sandy soil frequently collapse, leaving a furrow in the sand as a trace of the former burrow. The burrows of rodents are commonly used by golden moles. The diet of golden moles consists mostly of invertebrates; two desert-dwelling genera (*Cryptochloris* and *Eremitalpa*) also eat legless lizards. In sandy deserts *Eremitalpa* occasionally forages for wind-blown insects on the surface in furrows in the sand (Fritz Eloff, 1967).

Family Solenodontidae. Represented today by but two genera and two species, the solenodons seem to be relict types that are unable to survive in competition with other placentals recently introduced into their ranges. Solenodons occurred in sub-Recent and Recent times in Cuba, Haiti, and Puerto Rico, but are now restricted to Haiti (*Solenodon paradoxus*) and to Cuba, where a declining and endangered population of *Atopogale cubanus* occurs. The extinct genus *Nesophontes* occupied the West Indies at least until the arrival of the Spaniards. The introduction by man of the house rat (*Rattus*), the mongoose (*Herpestes*), and dogs and cats into the West Indies, and the extensive clearing of land for agricultural purposes, probably caused the rapid decline of the solenodons. These animals are now rare in most areas, and hopes for their survival under natural conditions seem dim.

Solenodons are roughly the size of a muskrat, and have the form of an unusually large and big-footed shrew. The five-toed feet and the moderately long tail are nearly hairless. The snout is long and slender, the eyes are small, and the pinnae are prominent. The zygomatic arch is incomplete, no auditory bulla is present, and the dorsal profile of the skull is nearly flat. The dentition is 3/3, 1/1, 3/3, 3/3 = 40. The first upper incisor is greatly enlarged and points backward slightly; the second lower incisor has a deep lingual groove that may function to transport the toxic saliva that empties from a duct at the base of this tooth. The upper molars lack a W-shaped ectoloph, and are basically tritubercular. A sharp and blade-like (trenchant) ridge is formed by a high crest at the outer edge of each molar.

Solenodons are generalized, omnivorous feeders that prefer animal material. They often find food by rooting with their snouts or by uncovering animals with their large claws. Solenodons are rather archaic creatures that seem to have little competitive ability. Their distribution on islands is probably the key to their continued, if tenuous, survival.

Family Soricidae. Members of this family, the shrews, are among the smallest and least conspicuous of mammals. In many areas they are the most numerous insectivores, however, and they have the widest distribution of any insectivoran family. The family Soricidae is represented today by some 24 genera and 291 species, and occurs throughout the world except in the Australian area, most of South America, and the polar areas. Soricids appear in the Oligocene in both Europe and North America. Because soricids are rare as fossils, their early evolution is obscure; they may have evolved from an early erinaceid ancestral stock (Dawson, 1967:20).

Shrews are small: the smallest weighs only 2 g. (the smallest living mammal), and the largest weighs roughly 35 g., the size of a large mouse. The snout is long and slim, the eyes are small, and the pinnae are usually visible. The feet are five-toed; except for fringes of stiff hairs on the digits in semi-aquatic species, and enlarged claws in semi-fossorial forms, they are unspecialized. The foot posture is plantigrade or semi-plantigrade. The narrow and elongate skull usually has a flat dorsal profile (Fig. 6–13B); there is no zygomatic arch or tympanic bulla, and the tympanic bone is annular (Fig. 6–13A). The specialized dentition consists of from 26 to 32 teeth; the dental formula of *Sorex* is 3/1, 1/1, 3/1, 3/3 = 32. In the subfamily Soricinae the teeth are pigmented; the first upper incisor is large, hooked, and bears a notch and projection resembling those on the upper mandible of a falcon (Fig. 6–13B). Behind the first upper molar is a series of small unicuspid teeth (presumably incisors, a canine and premolars); P^4 is large and has a trenchant ridge; and the upper molars have W-shaped ectolophs (Fig. 6–8A). Both the trigonid and talonid of the lower molars are well developed (Fig. 6–8B), and the first lower incisor is greatly enlarged and procumbent (leaning forward).

Because they are unusually small, shrews can exploit a unique mode of foraging. Many shrews patrol for insects in spaces beneath logs, beneath fallen leaves and other plant debris, and in the narrow spaces and crevices beneath rocks. Surface runways of rodents and burrows of rodents may also be used as feeding routes. Due to their style of foraging, shrews are seldom observed, even in areas where they are

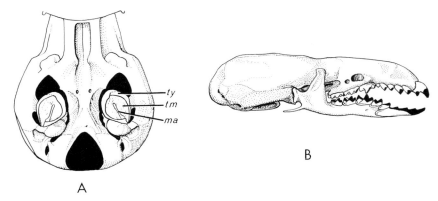

FIGURE 6–13. The skull of the vagrant shrew (*Sorex vagrans*, Soricidae; length of skull 17 mm.). A, ventral view of the basicranial region, showing the annular (ring-like) tympanic bone (*ty*), the tympanic membrane (*tm*), and the malleus (*ma*). B, side view, showing the pincer-like anterior incisors. The pigmented parts of the teeth are shown in black.

common. Although the shrew is typically associated with moist conditions, some species, such as the gray shrew (*Notiosorex crawfordi*) of the southwestern United States and the piebald shrew (*Diplomesodon pulchellum*) of southern Russia, inhabit desert areas. Aquatic adaptations occur in some species which are capable of diving and swimming and feed mainly on aquatic invertebrates. One of the most aquatic species is the Tibetan water shrew (*Nectogale elegans*), which inhabits mountain streams and feeds primarily on fish. In this species the streamlined shape is enhanced by the strong reduction of the pinnae, and the digits and feet have fringes of stiff hairs that greatly increase the effectiveness of the appendages as paddles. The distal part of the tail is laterally compressed and the edges bear lines of stiff hairs. *Blarina,* of the eastern United States, secretes toxic saliva that presumably aids in subduing prey (Pearson, 1942).

Family Macroscelididae. This menotyphlan group, containing the elephant shrews, is alone among insectivores in being specialized for hopping. The common name of elephant shrew refers to the vague resemblance between the trunk of an elephant and the long, flexible snouts of these shrews. The present distribution includes Morocco, Algeria, and the part of Africa south of the southern end of the Red Sea. There are five genera represented by 28 living species. The sparse fossil record of elephant shrews is limited to Africa. One extinct species of the living genus *Rhynchocyon* is known from the Miocene, and an extinct genus is recorded from the Pliocene.

Elephant shrews vary from the size of a mouse to that of a large rat, and have large eyes and ears. The distal segments of the limbs are long and slender, the hind limbs being somewhat longer, and the tail is moderately long. These animals have the cranial features listed for menotyphlans at the beginning of this chapter. In contrast to the Tupaiidae, the large orbits are never bordered by a complete postorbital bar (Fig. 6–4B). The dental formula is 1-3/3, 1/1, 4/4, 2/2-3 = 36-42; the last upper premolar is the largest molariform tooth. The molars are quadrate and have four major cusps. The forelimbs have five digits; the hind limbs have four or five toes.

Elephant shrews inhabit open plains, savannas, brushlands, and forests. The diet is primarily insects, and some species concentrate on ants and termites. Some plant material is also taken. These animals are quadru-

FIGURE 6-14. A tree shrew (*Tupaia longipes*, Tupaiidae). (Photograph by M. W. Sorenson.)

pedal and semi-plantigrade when moving slowly, but become bipedal and digitigrade when running. Elephant shrews can move rapidly by a series of long bounds, and among insectivores are the most highly specialized for rapid running. Runways are used to some extent. In contrast to most other small mammals, elephant shrews are partly diurnal.

Family Tupaiidae. Members of this family, called tree shrews, roughly resemble small, long-snouted squirrels (Fig. 6-14), and occur from India, through Burma, to the islands of Sumatra, Borneo, and the Philippines. This family is represented solely by its Recent members (five genera and 15 species). In addition to the cranial features listed for menotyphlan insectivores at the beginning of this chapter, tree shrews are characterized by well-developed postorbital processes that join the zygoma (Fig. 6-7A). The dental formula is 2/3, 1/1, 3/3, 3/3 = 38; the upper incisors resemble canines, and the upper canine is reduced. The upper molars have trenchant, W-shaped ectolophs (Fig. 6-15A); and

the lower molars retain the basic insectivore pattern (Fig. 6-15B). The limbs are pentadactyl and the digits have strongly recurved claws. The long tail is heavily furred in 12 species, is tufted in one species, and is covered with short hairs in two species.

Tree shrews occupy deciduous forests and forage both in the trees and on the ground. They are opportunistic feeders, and utilize a variety of foods, but animal material and fruit are preferred. These animals are diurnal and are characteristically highly vocal. The mountain tree shrew occurs in social groups in which a rigid dominance hierarchy is apparent, whereas in parts of Borneo other species occupying lowland areas do not form social groups (Sorenson and Conaway, 1968). These authors report an interesting breeding behavior in *Tupaia montana*. The male emits a shrill call when ejaculation occurs. This call is probably an advertisement to other males of the receptivity of a female in estrus, and results in the sharing of a receptive female by males that rank high in the dominance hierarchy.

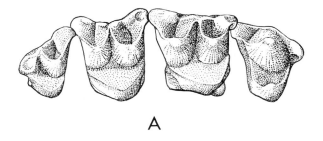

A

FIGURE 6–15. Cheek teeth of a tree shrew (*Tupaia* sp., Tupaiidae). Fourth premolar and molars of upper right (A) and lower left (B) tooth rows.

B

ORDER DERMOPTERA

Members of this order are generally called flying lemurs because of their lemur-like faces and their ability to glide between trees. One family, Cynocephalidae, with but one genus (*Cynocephalus*) and two species, represents the order. The distribution includes tropical forests from southern Burma and southern Indochina, Malaya, Sumatra, Java, Borneo, and nearby islands, to southern Mindanao and some of the other southern islands of the Philippine group. An extinct dermopteran family (Plagiomenidae) is recorded in North America from the late Paleocene and early Eocene, but the family Cynocephalidae is known only from the Recent.

Members of the family Cynocephalidae are of modest size (roughly 1 to 1.75 kg.), and have large eyes and faces that resemble those of Old World fruit bats or some prosimian primates. The brownish, chestnut, or gray pelage is irregularly blotched with white. The molars have retained a basically three-cusped insectivore pattern, and the type of occlusion between upper and lower cheek teeth suggests that shearing rather than grinding is the ac-

tion, an unusual situation considering that dermopterans are largely herbivorous. The anterior dentition is specialized: the lateral upper incisor is caniniform, and the first two lower incisors are broad and pectinate (comblike). The unusual lower incisors are used to groom the fur, but also function to scrape leaves during feeding. The dental formula is 2/3, 1/1, 2/2, 3/3 = 34.

A broad, furred membrane extends from the neck, starting just behind the ear, to near the ends of the digits of the hand, from the hand and forelimb to the body and to near the ends of the digits of the foot, and connects the hind foot and the tail to its tip, forming a tail membrane (uropatagium) that resembles that of a bat (Fig. 6–16B). The hands and feet retain five digits which bear needle-sharp, curved claws that serve to clutch branches. As in bats, the neural spines of the thoracic vertebrae are short, the sternum is keeled, the ribs are broad, the radius is long, and the distal part of the ulna is strongly reduced. The great lengthening of the intestine typical of herbivorous mammals is well illustrated by flying lemurs. *Cynocephalus*, which has a head-and-body length of

FIGURE 6–16. A, a flying lemur (*Cynocephalus volans*) gliding between trees in a tropical forest in Mindanao (Philippine Islands). B, a flying lemur with a young animal clinging with needle-sharp claws to the bare skin where the gliding membrane joins the body. (Photographs by Charles H. Wharton, copyright © 1948 National Geographic Society.)

81

only about 410 mm., has an intestinal tract approaching 4 m. in length, some nine times the head-and-body length (Wharton, 1950:272). The cecum, a blind diverticulum at the proximal end of the colon, is greatly enlarged (to about 480 mm. in length) and is divided into compartments. This chamber harbors microorganisms that help break down cellulose and other relatively indigestible carbohydrates, and its enlargement is usually associated with an herbivorous diet (as in many rodents).

Flying lemurs are nocturnal and are slow but skillful climbers, but they are nearly helpless on the ground. They perform glides of distances well over 100 m. in traveling to and from feeding places (Fig. 6–16A). The diet includes leaves, buds, flowers, and fruit. Winge (1941:145) reports that the enlarged tongue and specialized lower incisors are used in cow-like fashion in picking leaves. According to Wharton (1950), flying lemurs seek refuge during the day in holes in trees, and several individuals may occupy the same den. He reports that while traveling along branches and feeding these animals invariably remain upside down. The distribution of flying lemurs is being restricted in some areas by the clearing of forests for agriculture by man, and in some regions the animals are hunted for their meat and their fur.

ORDER CHIROPTERA

Bats are a remarkably successful group today, and comprise the second largest mammalian order (behind the Rodentia); approximately 168 genera and 853 species of living bats are known. Bats are nearly cosmopolitan in distribution, being absent only from arctic and polar regions and some isolated oceanic islands. Although bats are frequently abundant members of temperate faunas, they reach their highest densities and greatest diversity in tropical and subtropical areas. In certain Neotropical localities, for example, there are more species of

bats than of all other kinds of mammals together. Bats occupy a number of terrestrial environments including temperate, boreal and tropical forests, grasslands, chaparral, and deserts. Because man-made structures often provide excellent roosting sites and agricultural areas provide high insect populations, bats are doubtless more abundant in some areas now than they were before these areas were occupied by man. The poorly controlled use of the insecticide DDT, however, has apparently caused drastic and alarming local reductions in some bat populations.

Two rather sharply differentiated suborders of bats are recognized. The suborder Megachiroptera includes the family Pteropodidae, the Old World fruit bats, and the suborder Microchiroptera includes all the other 16 families of bats. Microchiropterans are nearly cosmopolitan in distribution and are largely insectivorous. Two functional contrasts between the megachiropterans and the microchiropterans are of particular importance. Megachiropterans are not known to hibernate, and maintain their body temperatures within fairly narrow limits by physiological and behavioral means, whereas many microchiropterans are heterothermic, and some hibernate for long periods. In addition, whereas microchiropterans use echolocation as their primary means of orientation and can fly and capture insects in total darkness, most megachiropterans use vision and therefore are helpless in total darkness. One exception is the megachiropteran *Rousettus,* in which the ability to echolocate perhaps evolved independently. *Rousettus* uses clicks made by the tongue as the basis for its acoustical orientation, whereas all microchiropterans use ultrasonic pulses produced by the larynx.

Echolocation, a means of perceiving the environment even in total darkness (see Chapter 21), and flight, allowing great motility, have been two major keys to the success of bats. These abilities enable bats to occupy at night many of the niches filled by birds during the day. In addition, the remarkably maneuverable flight of bats facilitates a mode of foraging for insects that birds have never exploited. Heterothermy, allowing bats to hibernate or to operate at a lowered metabolic output during part of the diel cycle, has enabled these animals to occupy areas only seasonally productive of adequate food and to utilize an activity cycle involving only nocturnal or crepuscular foraging periods. The metabolic economy resulting from hibernation and from lowered metabolism during part of the diel cycle has affected the longevity of some bats. For their size, some microchiropteran bats are remarkably long-lived. *Myotis lucifugus*, a small bat weighing roughly 10 gm., may live as long as 24 years (Griffin and Hitchcock, 1965).

Many of the most important diagnostic features of bats are adaptations for flight. The bones of the arm and hand (with the exception of the thumb) are elongate and slender (Fig. 6–17), and flight membranes extend from the body and the hind limbs to the arm and the fifth digit (plagiopatagium), between the fingers (chiropatagium), from the hind limbs to the tail (uropatagium), and from the arm to the oc-

cipito-pollicalis muscle (propatagium, Fig. 6–18B). In some species the uropatagium is present even when the tail is absent. The muscles bracing the wing membranes are often well developed and serve to anchor a complex network of elastic fibers (Fig. 6–18A). Rigidity of the outstretched wing during flight is partly controlled by the specialized elbow and wrist joints, at which movement is limited to the anteroposterior plane. In most microchiropteran species the enlarged greater tuberosity of the humerus locks against the scapula at the top of the upstroke (Fig. 6–19), allowing the posterior division of the serratus anterior muscle, which tips the lateral border of the scapula downward, to help power the downstroke of the wing (Fig. 6–20). The adductor and abductor muscles of the forelimb raise and lower the wings and are therefore the major muscles of locomotion; a contrasting arrangement occurs in terrestrial mammals, in which the flexors and extensors provide most of the power for locomotion. The distal part of the ulna is reduced in bats, and the proximal section usually forms an important part of the articular surface of the elbow joint (Fig. 6–21). The clavicle is present and articulates proximally with the en-

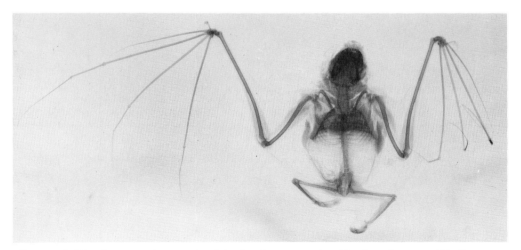

FIGURE 6–17. An x-ray photograph of the big fruit-eating bat (*Artibeus lituratus*, Phyllostomatidae), showing the great elongation of the bones of the arm and hand. (From Vaughan, T. A., in Wimsatt, W. A.: *Biology of Bats*, Academic Press, 1970.)

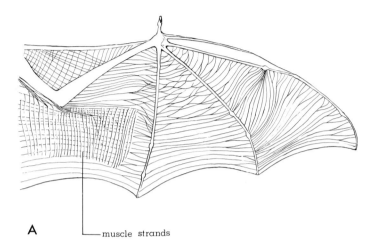

A   └─muscle strands

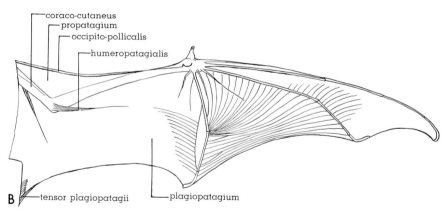

┌─coraco-cutaneus
  ┌─propatagium
    ┌─occipito-pollicalis
      ┌─humeropatagialis

B ├─tensor plagiopatagii   └─plagiopatagium

FIGURE 6–18.   Ventral views of the wings of two bats, showing the parts of the wing and muscles and elastic fibers that brace the membranes. A, the big fruit-eating bat (*Artibeus lituratus*, Phyllostomatidae); note the muscle strands that reinforce the plagiopatagium and the system of elastic fibers. This broad-winged bat does not remain on the wing for long periods. B, the western mastiff bat (*Eumops perotis*, Molossidae). This narrow-winged bat is a fast and enduring flier. (*Artibeus* from Vaughan, T. A., in Slaughter, B. H., and Walton, D. W.: *About Bats*, Southern Methodist University Press, 1970. *Eumops* from Vaughan, T. A., in Wimsatt, W. A.: *Biology of Bats*, Academic Press, 1970.)

FIGURE 6–19. Anterior view of the left shoulder joint of a free-tailed bat (*Molossus ater*) at the top of the upstroke of the wing (A) and during the downstroke (B). The greater tuberosity of the humerus (*b*) locks against the scapula at the top of the upstroke, transferring the responsibility for stopping this stroke to the muscles binding the scapula to the axial skeleton. During the downstroke the greater tuberosity of the humerus moves away from its locked position. This type of action and this type of shoulder joint also occur in the Vespertilionidae and other advanced families of bats. The acromion process (*a*) and the coracoid process (*c*) of the scapula are shown. (From Vaughan, T. A., in Slaughter, B. H., and Walton, D. W.: *About Bats*, Southern Methodist University Press, 1970.)

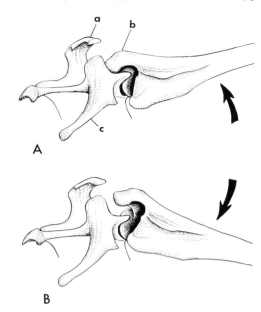

A

B

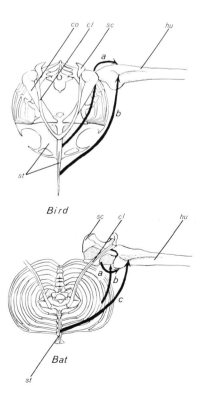

Bird

Bat

FIGURE 6–20. Anterior views of the thorax and part of the left forelimb of a bird and a bat, with some of the major muscles controlling the wing-beat cycle shown diagramatically. In the bird the supracoracoideus muscle (*a*) raises the wing and the pectoralis muscle (*b*) powers the downstroke; both muscles originate on the sternum. In the bat the downstroke is primarily controlled by three muscles, the subscapularis (*a*), serratus anterior (*b*), and pectoralis (*c*). Only the pectoralis originates on the sternum. Many muscles power the upstroke in bats. Abbreviations: *cl*, clavicle; *co*, coracoid; *hu*, humerus; *sc*, scapula; *st*, sternum. (From Vaughan, T. A., in Slaughter, B. H., and Walton, D. W.: *About Bats*, Southern Methodist University Press, 1970.)

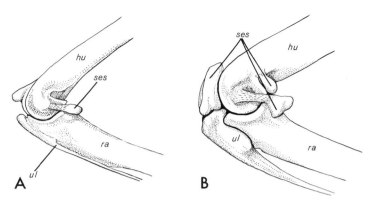

FIGURE 6–21.   Lateral view of the right elbow of a myotis (*Myotis volans*, Vespertilionidae) and of a free-tailed bat (*Molossus ater*, Molossidae). Abbreviations: *hu*, humerus; *ra*, radius; *ses*, sesamoid; *ul*, ulna. (From Vaughan, T. A., in Wimsatt, W. A.: *Biology of Bats*, Academic Press, 1970.)

larged manubrium and distally with the enlarged acromion process and enlarged base of the coracoid process (Fig. 6–20). The hind limbs are either rotated to the side 90° from the typical mammalian position and have a reptilian posture during quadrupedal locomotion, or they are rotated 180°, have a spider-like posture, and are used primarily to suspend the animal upside down from a horizontal support. The fibula is usually reduced, and support for the uropatagium, the calcar (see Fig. 6–22), is usually present.

The evolution of the pattern of muscular control of the wing-beat cycle typical of microchiropteran bats has seemingly been strongly influenced by their means of perceiving their environment. Microchiropterans (but not megachiropterans) detect obstacles and their insect prey by means of echolocation, which enables detailed recognition of objects only at close range, usually within several feet. The ability to fly slowly and to maneuver rapidly when prey or obstacles are detected at close range is therefore of critical importance. In contrast, birds use vision for more long-range perception of their environment and have relatively little need for extremely maneuverable flight. In both groups similar trends toward rigidity of the axial skeleton and lightening of the wings occur, but many of the muscular and skeletal specializations that enable these animals

to control their wings differ between the groups. The pectoral girdle in birds is braced solidly by a tripod formed by the clavicula and coracoids, anchored to the sternum, and by the nearly blade-like scapula, resting almost immovably against the ribcage. The pectoralis and supracoracoideus muscles, both of which originate on the sternum, supply nearly all of the power for the wing beat. In bats nearly the reverse mechanical arrangement occurs: the scapula is braced against the axial skeleton by the clavicle alone, and the job of powering the wing beat is shared by many muscles (Fig. 6–20). This division of labor is made possible partly by the freedom of the scapula to rotate on its long axis. In bats the pectoralis, the subscapularis, the posterior division of the serratus anterior, and the clavodeltoideus muscles control the downstroke of the wings; only the pectoralis originates on the sternum. The muscles of the deltoideus and trapezius groups, and the supraspinatus and infraspinatus muscles, largely power the upstroke.

A morphological trend of critical importance to bats and all other flying animals is toward the reduction of weight of the wings. Propulsion is obtained in all flying animals by movements of the wings, and the kinetic energy produced by such movements depends upon the speed of the wing and its weight. The amplitude of a

FIGURE 6–22. Photographs of bats in flight. A, the Mexican big-eared bat (*Plecotus phyllotis*); B, the pallid bat (*Antrozous pallidus*). Note the tremendously large ears of *Plecotus*, and the calcar (visible in both bats), a bone that extends back from the foot and braces part of the lateral border of the tail membrane. Some of the elastic fibers that reinforce the membranes can also be seen. (Photographs by J. Scott Altenbach.)

stroke and its speed are progressively greater toward the wing tip. Consequently, reduction of the weight of the distal parts of the wing results in a reduction of the kinetic energy developed during a wing stroke. A considerable advantage in metabolic economy is thus gained, for as less kinetic energy is developed during each stroke, less energy is necessary to control the wings. In addition, light wings can be controlled with speed and precision during the extremely rapid maneuvers used when bats chase flying insects. Reduction of the weight of the wings has been furthered in bats by many specializations. Movement at the elbow and wrist joints is limited to one plane, thus eliminating musculature involved in rotation and bracing at these joints. In addition, the work of extending and flexing the wings is transferred from distal muscles (of the forearm and hand) to large proximal muscles (pectoralis, biceps, and triceps), thereby allowing a reduction in the size of the distal musculature. Certain forearm muscles are made nearly inelastic by investing connective tissue, and because of this modification and specializations of their attachments these muscles "automatically" extend the chiropatagium with extension at the elbow joint, or flex the chiropatagium with flexion at this joint (Fig. 6–23).

Because of their small size, delicate structure, nonterrestrial habits, and oc-

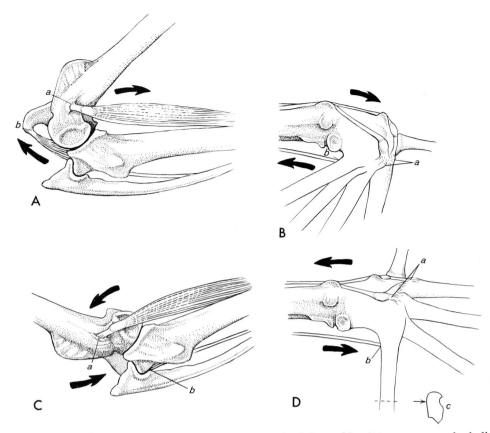

FIGURE 6–23.   Lateral views of the elbow joint of the leaf-chinned bat (*Mormoops megalophylla*), showing the "automatic" flexion and extension of the fingers caused by certain forearm muscles in many advanced bats. Flexion of the elbow joint (A) moves the origins of the extensor muscles (*a*) *toward* the wrist and the origin of one flexor muscle (*b*) *away* from the wrist. Because the flexor muscle is largely inelastic, with flexion at the elbow (B) the distal tendon of the flexor (*b*) pulls on the fifth digit and tends to flex the fingers. With extension at the elbow joint (C), the origin of the extensor muscles (*a*) is moved away from the wrist and the origin of the flexor muscle (*b*) moves toward the wrist. This action results in pulling the extensor tendon (*a*) toward the elbow (D) and releasing tension on the flexor tendon (*b*), thus extending the fingers. In part D the complex cross-sectional shape of the fifth digit (*c*) is shown. (From Vaughan, T. A., in Slaughter, B. H., and Walton, D. W.: *About Bats*, Southern Methodist University Press, 1970.)

currence mostly in tropical areas where fossilization seldom occurs, bats are rare as fossils. Consequently, the evolution of bats is poorly known. The earliest undoubted fossil bat (*Icaronycteris index*, Fig. 6–24) is from early Eocene beds in Wyoming. On the basis of this beautifully preserved specimen, Jepsen (1966) described the extinct family Icaronycteridae. Although this bat has several primitive features, such as claws on the first two digits of the hand and fairly short, broad wings, its basic limb structure is

that of modern bats. The upper molars of *Icaronycteris* have the W-shaped ectoloph typical of most insectivorous bats, and this bat has been put in the suborder Microchiroptera. Late Eocene and early Oligocene deposits in France have yielded the earliest records of the modern microchiropteran families Emballonuridae, Megadermatidae, Rhinolophidae, and Vespertilionidae. Megachiropterans appear first in the Oligocene of Italy.

The Eocene appearance of a bat clearly beautifully adapted for flight,

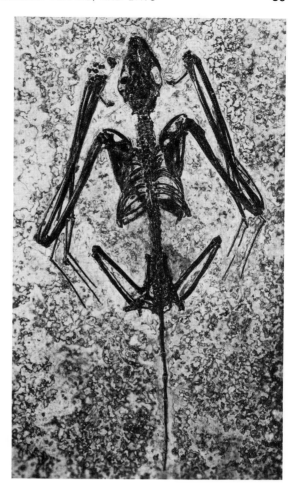

FIGURE 6–24. A beautifully preserved early Eocene bat (*Icaronycteris index*). (From Jepsen, G. L., in Wimsatt, W. A.: *Biology of Bats,* Academic Press, 1970.)

the Oligocene appearance of many modern families, and the assignment by paleontologists of fossils from the late Eocene to the still-living genus *Rhinolophus* indicate an early origin of bats. Although no clear fossil evidence bears on the matter, a Paleocene origin of bats seems probable, and a late Cretaceous divergence of chiropteran ancestors from primitive insectivore stock seems possible. The last word on the origins and evolution of bats has not been written. Jepsen, in his excellent discussion of the evolution of bats (1970:22), summed up the status of our knowledge: "At present bat history has a completely open end, in the distant past, that only more fossils can close."

SUBORDER MEGACHIROPTERA

Inasmuch as only one family represents this suborder today, the descriptions given below for the family Pteropodidae characterize the suborder Megachiroptera.

Family Pteropodidae. Most pteropodids are fruit eaters, and many species are called flying foxes because of their fox-like faces and large size. These bats are abundant and often conspicuous members of many tropical biotas in the Old World. This family is represented by 40 Recent genera and 149 Recent species. Pteropodids occur widely in tropical and subtropical regions from Africa and southern Eura-

FIGURE 6–25. A megachiropteran bat (*Rousettus angolensis*) from central Africa. Note the use of the foot and wings in manipulating the banana. (Photographs by Alvin Novick.)

sia to Australia, and on many south Pacific islands as far east as Samoa and the Caroline Islands.

Members of this family are often large, up to 150 cm. (nearly 5 feet!) in wing spread, and differ from members of the suborder Microchiroptera in many ways. The face is usually fox-like, with large eyes, usually a moderately long snout with a simple, unspecialized nose pad, and simple ears lacking a tragus (Fig. 6–25). (The tragus, a fleshy projection of the anterior border of the ear opening, may be seen on a member of the Vespertilionidae in Fig. 6–30E). The orbits are large, and are bordered posteriorly by well developed postorbital processes which may meet to form a postorbital bar. The rostrum is never highly modified (Fig. 6–26). The dental formula is 1-2/0-2, 1/1, 3/3, 1-2/2-3 = 24-34. The molars are never tuberculosectorial with W-shaped ectolophs, as in most microchiropterans, but are low, moderately flat-crowned,

more or less quadrate, and lack stylar cusps (Fig. 6–27). The teeth are adapted basically to crushing fruit. The wing is primitive in having two clawed digits, and the greater tuberosity of the humerus is not enlarged (Fig. 6–28) to make contact with the scapula at the top of the upstroke. The tail is typically short or rudimentary.

Broadly speaking, pteropodids utilize two types of food. Most members of the subfamily Pteropodinae are fruit eaters, whereas members of the subfamily Macroglossinae eat mostly nectar and pollen. The fruit eaters as a rule are large bats with fairly robust or moderately reduced dentitions. The jaws in these species are usually fairly long, or, in some species that presumably eat hard fruit, the jaws are short and the teeth and dentary bones are unusually robust. The fruit bats often roost in trees in large colonies (Fig. 6–29); in the case of the African genus *Eidolon* as many as 10,000 individuals have been observed roosting together. Fruit bats occasionally travel long distances during their nocturnal foraging, and *Pteropus* regularly flies at least 15 km. from roosting sites to feeding areas (Breadon, 1932; Ratcliffe, 1932). The fruit eaters are usually not particularly maneuverable fliers, but have a steady, direct style of flight. They are adroit at clambering in vegetation, where the clawed first and second digits come into play.

The pteropodids that eat nectar and

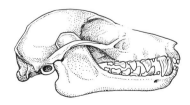

FIGURE 6–26. The skull of a megachiropteran bat (*Pteropus* sp., Pteropodidae). Length of skull 62 mm.

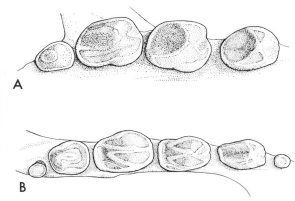

FIGURE 6-27. The cheek teeth of a megachiropteran bat (*Pteropus* sp.). A, right upper tooth row, showing two molars and two premolars. B, lower left tooth row, showing three premolars and three molars. (From Vaughan, T. A., in Wimsatt, W. A.: *Biology of Bats*, Academic Press, 1970.)

pollen are small by comparison with their fruit-eating relatives, and have long slender rostra, strongly reduced cheek teeth, and delicate dentary bones. The tongue is long and protrusile and has hair-like structures at its tip to which pollen and nectar adhere. Some species roost in groups in caves, and some roost solitarily in vegetation. Flight is slow and maneuverable. *Syconycteris*, a nectar feeder that occurs in the Australian region, flies fairly near the ground, can maneuver adroitly through thick vegetation, and is able to hover easily (Van Deusen, 1967).

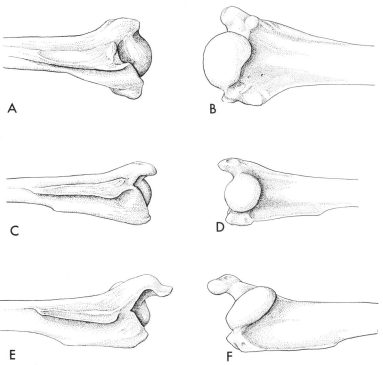

FIGURE 6-28. The proximal end of the right humerus in three bats. Anterior views are on the left and posterior views are on the right. A and B, *Pteropus* sp (Pteropodidae); C and D, *Myotis lucifugus* (Vespertilionidae); E and F, *Molossus ater* (Molossidae). (From Vaughan, T. A., in Wimsatt, W. A.: *Biology of Bats*, Academic Press, 1970.)

FIGURE 6-29. Part of a flying fox "camp" in northeastern Australia. (Photograph by Roger E. Carpenter.)

SUBORDER
MICROCHIROPTERA

Recent members of this suborder are usually small. The eyes are often small, the rostrum is usually specialized, and the nose pad and lower lips may be modified in a variety of ways (Fig. 6-30). The ears have a tragus in all but members of the family Rhinolophidae, are usually complex, and are frequently large (Fig. 6-30E). The postorbital process is usually small. Dentitions vary tremendously, but most microchiropterans (except the Desmodontidae, and some members of the Phyllostomatidae) have tuberculosectorial molars; the upper molars have a W-shaped ectoloph with strongly developed stylar cusps, and in the lower molars the trigonid and talonid are roughly equal in size (Fig. 6-31). In many insectivorous species and in some frugivorous members of the Phyllostomatidae, one or more premolars above and below are caniniform, and in some insectivorous species the premaxillae are separate (Fig. 6-32).

The flight apparatus of the microchiropterans is more progressive than that of the megachiropterans. In microchiropterans the second digit does not bear a claw and lacks a full complement of phalanges, and its tip is connected by a ligament to the joint between the first and second phalanges of the third digit. During flight this connection allows the second digit to brace the third digit, which forms much of the leading edge of the distal part of the wing, against the force of the airstream. The greater tuberosity of the humerus is usually enlarged and locks against a facet on the scapula at the top of the upstroke of the wings (Fig. 6-19). The size of the tail and uropatagium are variable Figs. 6-22, 6-33). The shape of the wing varies according to foraging pattern and style of flight. In general, slow, maneuverable fliers have short, broad wings, whereas rapid, enduring fliers have long, narrow wings (Fig. 6-18).

Since their divergence from primitive insectivore stock, perhaps in the Cretaceous or early Paleocene, microchiropteran bats have undergone a remarkable adaptive radiation. Sixteen Recent families of microchiropterans and approximately 128 genera are now recognized. This large number of families and genera reflects the great structural diversity and widely contrasting modes of life that occur within this suborder.

FIGURE 6–30. Faces of some microchiropteran bats. A, big-eared leaf-nosed bat (*Macrotus water-housii*, Phyllostomatidae), an omnivore; B, long-nosed bat (*Leptonycteris sanborni*, Phyllostomatidae), a nectar and pollen feeder; C, leaf-chinned bat (*Mormoops megalophylla*, Mormoopidae), an insectivore; D, wrinkle-faced bat (*Centurio senex*, Phyllostomatidae), a fruit-eater; E, spotted bat (*Euderma maculata*, Vespertilionidae), an insectivore; F, guano bat (*Tadarida brasiliensis*, Molossidae), an insectivore. (Photograph of *Euderma* by Robert J. Baker.)

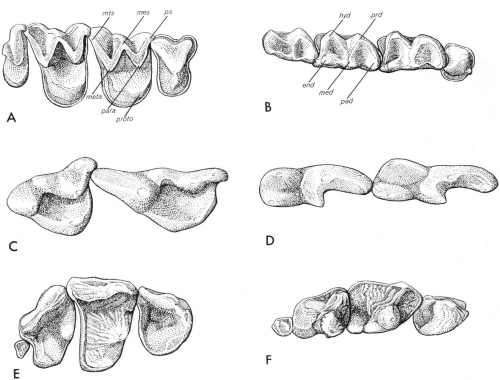

FIGURE 6–31. The upper right and lower left cheek teeth of three bats. A and B, an insect-eating vespertilionid (*Lasiurus cinereus*); A, the upper fourth premolar and three molars and B, the comparable lower teeth. C and D, a nectar-feeding phyllostomatid (*Leptonycteris sanborni*): C, second and third upper molars and D, the comparable lower teeth. E and F, a fruit-eating phyllostomatid (*Artibeus jamaicensis*): E, upper fourth premolar and three molars and F, the comparable lower teeth. Abbreviations: *end*, entoconid; *hyd*, hypoconid; *med*, metaconid; *mes*, mesostyle; *meta*, metacone; *mts*, metastyle; *pad*, paraconid; *para*, paracone; *prd*, protoconid; *proto*, protocone; *ps*, parastyle.

Family Rhinopomatidae. Members of this small family, containing but one genus with three species, occur in northern Africa and southern Asia east to Sumatra. These animals are called mouse-tailed bats because of the long tail that is largely free from the uropatagium. No fossil representatives of the family are known.

These bats are considered to be the

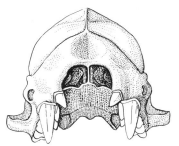 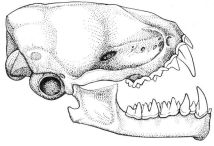

FIGURE 6–32. The skull of the hoary bat (*Lasiurus cinereus*, Vespertilionidae): left, anterior view, showing the emarginate front of the palate; right, side view, showing the shortened rostrum typical of some insect feeding bats. Length of skull 17 mm.

FIGURE 6–33. A nectar-feeding bat (*Leptonycteris sanborni*) in flight, showing some positions of the wings during slow or hovering flight. (Photographs by J. Scott Altenbach.)

most primitive members of the Microchiroptera. The premaxillaries resemble those of megachiropterans in being separate from one another, and their palatal portions are much reduced. The second digit of the hand, in contrast to the arrangement in all other microchiropterans, retains two well developed phalanges. Perhaps the clearest indication of the primitiveness of these bats is the structure of the shoulder joint. In contrast to the situation in most microchiropterans, the greater tuberosity of the humerus is small and does not lock against the scapula at any point in the wing-beat cycle. Other rhinopomatid features include laterally expanded nasal chambers, no fusion of cervical, thoracic or lumbar vertebrae, and a complete fibula.

The dentition is adapted to an insectivorous diet. The molars are tuberculosectorial; the upper molars have W-shaped ectolophs of the usual microchiropteran type. The dental formula is 1/2, 1/1, 1/2, 3/3 = 28. These are fairly small bats (the length of the head and body is up to 80 mm.) with slender tails whose length approaches that of the head and body. The eyes are large, and the anterior bases of the large ears are joined by a fold of skin across the forehead.

Mouse-tailed bats are insectivorous and typically occupy hot, arid areas. They roost in a wide variety of situations, including fissures in rocks, houses, ruins, and caves; one species roosts in large colonies in some Egyptian pyramids. Although locally common, mouse-tailed bats over much of their range are outnumbered by other types of bats, and compared to other microchiropteran families are not an important group today. Rhinopomatids perhaps hibernate in some areas. Large deposits of subcutaneous fat occur in the abdominal area and around the base of the tail in individuals from some localities. These bats tolerate body temperatures as low as 22° C and can spontaneously rewarm themselves (Kulzer, 1965). The structure of the rhinopomatid kidney suggests a remarkable ability to concentrate urine in the interest of water conservation, a specialization of considerable adaptive importance for a group that inhabits arid areas.

Family Emballonuridae. This family contains a variety of small bats that are frequently called sac-winged or sheath-tailed bats. Twelve genera and 44 species are currently recognized, and the wide geographic range of emballonurids includes the Neotropics (much of southern Mexico, Central America, and northern South America), most of Africa, southern Asia, most of Australia, and the Pacific Islands east to Samoa. The earliest fossil emballonurid is from the Eocene or Oligocene of Europe.

These small bats combine a number of primitive features with several noteworthy specializations. In the possession of postorbital processes and reduced premaxillaries that are not in contact with one another, emballonurids resemble pteropodids. In addition, the shoulder joint and elbow joints are primitive. An advancement over the rhinopomatids is the retention of only the metacarpal in the second digit, and the flexion of the proximal phalanges of the third digit onto the dorsal surface of the third metacarpal is a specialization also found in some advanced families of bats. External obvious specializations include a glandular sac in the propatagium in some genera and the emergence of the tail from the dorsal surface of the uropatagium. The nose is simple; that is to say, it lacks leaf-like structures or complex patterns of ridges and depressions. In addition to the more common gray and dark brown species of emballonurids, some species of one genus (*Saccopteryx*) have handsome whitish stripes on the back, and one (*Diclidurus*) is pure white.

These insectivorous bats typically inhabit tropical or subtropical areas, where they use a great variety of roosting sites. Emballonurids occupy houses, caves, culverts, rock fissures, hollow trees, vegetation, or the undersides of rocks and dead trees for daytime retreats, and usually roost in colonies. These bats are often fairly tolerant of well lighted situations. A nursery colony of *Balantiopteryx plicata* that I observed in eastern Nayarit, Mexico, occupied during the day a shaded but only partially enclosed chamber beneath a jumble of huge granite boulders at the base of a cliff. At night these bats rested periodically in a series of shallow grottos at the base of the cliff. In some areas emballonurids probably forage mainly over water. A distinctive feature of some emballonurids is the glandular sac in the propatagium. Recent work on one species of emballonurid (*Saccopteryx bilineata*) has shown that this sac, especially well developed in males, is used during stereotyped displays during the breeding season (see page 340).

Family Noctilionidae. Although this family is not important in terms of numbers of species (it contains but two species of one genus), it is of special interest because one species is structurally and behaviorally highly specialized for eating fish. Noctilionid bats are often referred to as bull-dog bats or fishing bats. They occupy the Neotropics from Sinaloa, Mexico, and the West Indies to northern Argentina in South America. There is no fossil record of noctilionids.

Both structurally and in general external appearance, noctilionids are distinctive. They are fairly large (from roughly 20 to 75 gm. in weight and up to two feet in wingspread), and the heavy lips, somewhat resembling those of a bull-dog, the pointed ears, and the simple nose make the face unmistakable (Fig. 6–34). The dorsal pelage varies in color from orange to dull brown, and a whitish or yellowish

FIGURE 6–34. The face of a fishing bat (*Noctilio labialis*, Noctilionidae). (Photograph by N. Smythe and F. Bonaccorso.)

stripe is usually present from the interscapular area to the base of the tail. The hind limbs and feet are remarkably large, especially in *Noctilio leporinus*, and the feet have sharp, recurved claws. The premaxillae are complete, and in adults the two maxillae are fused together and are fused with the premaxillae, forming a strongly braced support for the enlarged upper medial incisors. The dental formula is 2/1, 1/1, 1/2, 3/3 = 28. The teeth are robust, and the molars are tuberculosectorial.

The seventh cervical vertebra is not fused to the first thoracic, the shoulder joint and elbow joint are primitive, and the second digit of the hand has a long metacarpal and a tiny, vestigial phalanx. The pelvis is powerfully built, with the ischia strongly fused together and fused to the posterior part of the laterally compressed, keel-like sacrum. The tibia and hind foot of *N. leporinus* have a series of unusual specializa-

tions to be considered with the mode of foraging of the animal.

The feeding habits of the two species of *Noctilio* differ (Brown and Hooper, 1968). *Noctilio labialis* eats largely insects, which it seems to catch over water; *N. leporinus*, however, is a markedly atypical microchiropteran in that it eats largely fish. The style of foraging of this species is now known to involve the use of the hind claws as gaffs (Bloedel, 1955). This bat recognizes concentrations of small fish or single fish immediately beneath the surface of the water by detecting (by means of echolocation) the ripples or breaks in the surface that these fish create (Suthers, 1965, 1967). The bat skims low over the water and drags the feet in the water, with the limbs rotated so that the hook-like claws are directed forward. (This involves rotation of the hind limbs 180° from the typical mammalian position.) When a small fish is "gaffed," it is brought quickly from the water and grasped by the teeth. From 30 to 40 small fish were captured in this fashion per night by a *N. leporinus* under laboratory conditions. A series of modifications of the hind limb are clearly advantageous in allowing this animal to pursue efficiently its specialized style of foraging. The long calcar, which is roughly as long as the tibia, the calcaneum, the digits and claws, and the distal part of the tibia are all strongly compressed so that they are streamlined with respect to their direction of movement when they are dragged through the water. During the foraging sweeps the short tail is raised; the blade-like calcar is pulled craniad and is clamped against the flattened side of the tibia. In this way the large uropatagium is brought clear of the water and the streamlined calcar and tibia knife through the water, producing a minimum of drag.

Noctilionids roost during the day in groups in hollow trees and fissures, caves, and occasionally buildings. *Noctilio leporinus* is seemingly most common in tropical lowland areas, frequently occurring along coasts where they forage along rivers or streams, over mangrove-lined marshes and ponds, or over the sea. In western Mexico in the dry season I have taken individuals as they foraged over small, disconnected ponds in a nearly dry stream bed. These ponds supported large numbers of small fish.

Family Nycteridae. Members of this small family (including 13 species of one genus) are called hollow-faced bats. These bats occur in Madagascar, Africa, the Arabia-Israel area, the Malay Peninsula, and parts of Indonesia including Sumatra, Java, and Borneo. No fossil nycterids are known.

Externally, these fairly small bats can be recognized by their large ears, moderate or small size, and distinctive "hollow" face. The skull has a conspicuous interorbital concavity (Fig. 6–35) that is probably associated with the "beaming" of the ultrasonic pulses used in echolocation. This concavity is connected to the outside by a slit in the facial skin. The dental formula is 2/3, 1/1, 1/2, 3/3 = 32, and the molars are tuberculosectorial. Postcranially, these bats combine primitive and specialized features. The shoulder joint and elbow joint are fairly primitive, but the retention of only the metacarpal of the second digit of the hand, and the reduction of the number of phalanges of the third digit to two, are obvious specializations. The pectoral girdle is modified in the direction of enlargement and strengthening of the bracing of the sternum, a pattern parallel to

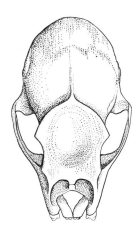 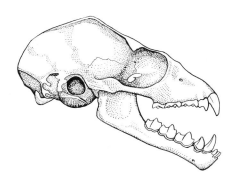

FIGURE 6–35. The skull of a slit-faced bat (*Nycteris thebaica*, Nycteridae): left, dorsal view, showing the depression in the forehead; right, side view, showing the flattened profile. Length of skull 19 mm. (After Hill and Carter, 1941.)

the trend in birds toward the strengthening of the pectoral girdle. The sternum in nycterids is robust and the mesosternum is strongly keeled; the manubrium is broad, the first rib is unusually strongly built, and the seventh cervical and first thoracic vertebrae are fused. This general pattern also occurs in the family Megadermatidae (a family to which the nycterids are closely related), and reaches its most extreme development in the family Rhinolophidae (Fig. 6–36). Because the specializations of the pectoral girdle in these bats parallel to some extent the roughly similar modifications of this girdle in birds, they might be thought to be associated with a progressive structural trend in bats. Actually, it is doubtful that this is the case. Some of the most advanced and successful families of bats have less bird-like pectoral girdles than those in the microchiropteran families listed above, but have modifications of the shoulder and elbow joints and forelimb musculature that provide for a more efficient utilization for flight of the typical chiropteran

structural plan. Perhaps the nycterid-megadermatid-rhinolophid pectoral girdle is associated with a foraging style typified by short intervals of flight. In any case, this style of pectoral girdle seems to be a divergent type and does not represent a progressive morphological trend common to most "advanced" microchiropterans.

Hollow-faced bats inhabit tropical forests and savanna areas, and seem to feed largely on arthropods that are picked from vegetation or from the ground. Flying insects form part of the diet, but flightless arachnids such as spiders and scorpions are also important food items. These bats remain on the wing only for fairly short intervals, for they retire to a resting place to eat their larger prey. Nycterids roost in a variety of situations, and some are even known to occupy burrows made by procupines and aardvarks.

Family Megadermatidae. This is not a large family, consisting of but four genera and five species. These bats are known as false vampires, an inappropriate title as they neither resemble

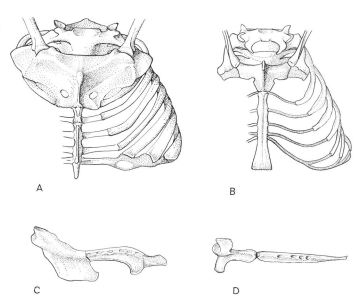

FIGURE 6–36.   Ventral views of the thorax and lateral views of the sternum of a rhinolophid (*Hipposideros commersoni;* A and C) and a vespertilionid (*Myotis yumanensis;* B and D). Note the highly specialized sternum of *Hipposideros,* to which the first two ribs are fused. (From Vaughan, T. A., in Wimsatt, W. A.: *Biology of Bats,* Academic Press, 1970.)

vampires nor feed on blood. These bats occur in tropical areas in central Africa, southeastern Asia including Indonesia, the Philippines, and Australia. The fossil record of megadermatids is scanty; the earliest fossil is from the Eocene or Oligocene of Europe.

These are fairly large, broad-winged bats. The largest species has a wing-spread approaching 1 meter. The ears are large and are connected across the forehead by a ridge of skin. The snout bears a conspicuous "nose-leaf," and the eyes are large and prominent. The premaxillae and upper incisors are absent; the upper canines project forward and have a large secondary cusp. The molars are tuberculosectorial; the dental formula is 0/2, 1/1,1-2/2, 3/3 = 26 or 28. In *Megaderma*, and to a still greater extent in *Macroderma*, the W-shaped ectoloph of the upper molars is modified by the partial loss of the commissures connecting the mesostyle to the paracone and metacone. This trend is toward the development of an anteroposteriorly aligned cutting blade, and may be associated with the carnivorous habits of these genera. The shoulder and elbow joint are primitive; the second digit of the hand has one phalanx, and the third has two phalanges. The pectoral girdle has specializations similar to those of the nycterids, but the strengthening of the pectoral girdle is carried further in megadermatids. The manubrium of the sternum is relatively broader in megadermatids than in the nycterids, and is fused with the first rib and the last cervical and first thoracic vertebrae into a robust ring of bone. The megadermatid sternum is moderately keeled. The tail is very short or absent.

These bats occur in tropical forests and in tropical savannas, often near water, and utilize a variety of foods. Of the five species of megadermatids, three are known to be carnivorous, one is insectivorous, and the feeding habits of one are not known. The Australian ghost bat (*Macroderma gigas*), an unusually large, pale colored megadermatid, feeds on a variety of small vertebrates. In some areas it seems to feed largely on other bats. The ghost bat, and related species in southeast Asia, frequently consume their prey while hanging from the ceilings of spacious covered porches or verandas of large homes, and detract from the gracious atmosphere by littering the floors with feet, tails, and other discarded fragments of frogs, birds, lizards, fish, bats, and rodents. The carnivorous species of megadermatids have unusually large eyes and may hunt partly by sight. Another megadermatid, the insectivorous-carnivorous *Lavia frons*, uses a style of foraging similar to that of a flycatcher. This fully diurnal bat hangs from a branch, apparently scanning the vicinity visually; when an insect is sighted the bat makes a short flight to capture the insect and returns to a perch to eat it. There is no evidence that *Lavia* uses echolocation. *Lavia* is a delicate, maneuverable flier, but when foraging probably is in flight only for short periods. Megadermatids roost in many types of places, from crevices, caves, and buildings, in the case of most species, to sparse, occasionally sunlit vegetation, in the case of *Lavia*.

Family Rhinolophidae. This is a large and successful Old World family, with 10 genera and approximately 127 species. Its members are often called horseshoe bats because of the complex and basically horseshoe-shaped cutaneous ridges and depressions on the nose. The geographic distribution includes much of the Old World from western Europe and Africa to Japan, the Philippines, Indonesia, Melanesia, and Australia. This may be an extremely ancient family, for some late Eocene fossils from Europe have been assigned to the living genus *Rhinolophus*.

Because of the unique and complex face, rhinolophids are one of the most unmistakable groups of bats. The ears are usually large, but lack a tragus, and the eyes are small and inconspicuous. The tail is of moderate length in some species, but is small or rudimentary in others. The pectoral girdle is remarkable because it represents the extreme development of the trend (that occurs also in the Nycteridae and Megadermatidae) toward powerful bracing and enlargement of the sternum. In the most extreme manifestation of this trend the seventh cervical vertebra, the first and second thoracics, the first and most of the second rib, and the enormously enlarged and shield-like manubrium of the sternum are fused into a powerfully braced ring of bone (Fig. 6–36). The shoulder joint has a moderately well developed locking device. In some rhinolophids all but the last two lumbar vertebrae are fused; a similar specialization occurs in the Natalidae (Fig. 6–37). The pelvis is uniquely modified by enlargement of the anterior parts and an accessory connection between the ischium and the pubis. These unusual pelvic specializations may be in response to the need for specialized hind limb musculature to control the hind limbs. When these bats roost they often hang upside down, and the hind limbs are rotated 180° from the usual mammalian posture so that the plantar surfaces face forward. The extreme adaptations for strengthening the pectoral and pelvic girdles that are typical of rhinolophids occur to a comparable degree in no other family of bats.

Horseshoe bats are common in many areas, and in Germany the "Hufeisennase" is a familiar inhabitant of attics and church steeples. These bats have wide environmental tolerances; various species inhabit temperate, subtropical, tropical, and desert regions. Rhinolophids hibernate in some parts of their range, and characteristically rest or hibernate with the body enshrouded by the wing membranes. One species is known to be migratory. The food is largely arthropods, and the style of foraging resembles that of the nycterids and some megadermatids. Horseshoe bats pick spiders and insects from vegetation or capture flying insects in mid-air, and *Rhinolophus ferrumequinum* was observed to alight on the ground and capture flightless arthropods (Southern, 1964). *Rhinolophus fumigatus* hangs from branches of trees and makes short forays after passing insects (Shortridge, 1934). Horseshoe bats also share the nycterid and megadermatid habit of retiring to a temporary roost to eat large prey. Seemingly, then, most rhinolophids make short foraging flights and do not remain continuously on the wing while foraging. Perhaps the wing membranes are important in some species in aiding in the capture of insects. Webster and Griffin (1962) demonstrated photographically that one species of rhinolophid is able to capture insects in the chiropatagium. In contrast to many bats that emit pulses used in echolocation from the open mouth, rhinolophids keep the mouth closed during flight; the ultrasonic pulses used in echolocation are emitted through the nostrils and are "beamed" by the complex nasal apparatus (Mohres, 1953). The enlarged nasal openings in the skull (Fig. 6–38) perhaps facilitate the emission of pulses through the nostrils. Most horseshoe bats are colonial, but some are solitary. Many kinds of roosting sites are used; caves, buildings, and hollow trees are generally preferred, but foliage and burrows of large rodents are used by some species.

Family Phyllostomatidae. This is the most diverse family of bats with respect to structural variation, and contains more genera than does any other chiropteran family. Forty-four genera and 121 species are included in the Phyllostomatidae. These Neotropical "leaf-nosed bats" are so named be-

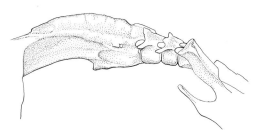

FIGURE 6–37. Lateral view of the left side of the fused lumbar vertebrae of a funnel-eared bat (*Natalus stramineus*, Natalidae). (From Vaughan, T. A., in Wimsatt, W. A.: *Biology of Bats*, Academic Press, 1970.)

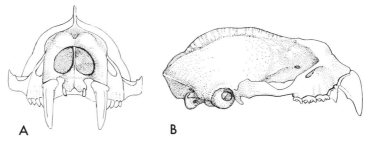

FIGURE 6–38. The skull of *Hipposideros* sp. (Rhinolophidae): A, front view, showing the huge nasal openings; B, side view, showing the "indented" rostrum. Length of skull 32 mm. (From Vaughan, T. A., in Wimsatt, W. A.: *Biology of Bats*, Academic Press, 1970.)

cause of the conspicuous leaf-like structure that is nearly always present on the nose (Fig. 6–30 A). These bats have exploited the widest variety of foods used by any family of bats. Some leaf-nosed bats have retained insectivorous feeding habits, but some are carnivorous and eat small vertebrates, including rodents, birds, and lizards; some eat nectar and pollen, and some are frugivorous. Phyllostomatids are the most important bats in the Neotropics, and occur from the southwestern United States and the West Indies south to northern Argentina. These bats can be traced back to the Miocene of Colombia, South America.

The great structural variation that occurs in the Phyllostomatidae is largely associated with an adaptive radiation into a wide variety of feeding niches. Within the family are some fairly small bats (*Choeroniscus* has a wingspread of roughly 220 mm.) as well as the largest New World bat (*Vampyrum* has a wingspread of over 1,000 mm.). In most species the nose-leaf is conspicuous and is spear-shaped (Fig. 6–30 A), but in a few species the nose-leaf is rudimentary or is highly modified (Fig. 6–30 D). The ears vary from extremely large to small, and a tragus is present. The tail and uropatagium are long in some species, with many stages of reduction and the absence of the tail and uropatagium being represented by various species. Some species have a uropatagium but lack a tail; only *Sturnira* has completely lost the uropatagium.

The wings are typically broad (Fig. 6–18A); the second digit has one phalanx and the third has three phalanges. The shoulder joint has a moderately well developed locking device formed between the greater tuberosity of the humerus and the scapula, but the elbow joint and forearm musculature are primitive, and the forelimb is, for a bat, generalized. Probably all phyllostomatids, whatever their feeding habits, remain on the wing only for short periods during foraging. The forelimbs are not used only for flight but are important in many species in food handling as well as in climbing over and clinging to vegetation (as in the fruit-eating species). The importance of such use of the forelimbs has probably favored the retention in phyllostomatids of limbs more generalized than those of many strictly insectivorous groups of bats. The seventh cervical and first thoracic vertebrae are not fused, and no fusion of elements to form the sturdy "pectoral ring" characteristic of rhinolophoid bats occurs in phyllostomatids. In some leaf-nosed bats, however, the sternum is strongly keeled. The ventral parts of the pelvis are lightly built in most species, but the ilia are robust and are more or less fused to the sacral vertebrae. These vertebrae are fused into a solid mass that becomes laterally compressed posteriorly. The acetabulum is characteristically directed dorsolaterally; the hind limbs are rotated 180° from the usual mammalian orientation and have a spider-like posture. Because of this position of the hind limbs, some phyllostomatids are unable to walk on a horizontal surface and use the hind limbs only for hanging upside down.

All of the Recent leaf-nosed bats probably evolved from an ancestral type that had tuberculosectorial teeth adapted to a diet of insects. Of the six Recent subfamilies, however, only one has retained this type of dentition, and in some species there is no trace of the ancestral pattern. The noteworthy adaptive radiation of phyllostomatids will be traced by considering the dentitions and foraging habits of each subfamily.

The subfamily Phyllostomatinae deviates least from the ancestral structural plan, and some species retain insectivorous feeding habits. This subfamily contains all of the leaf-nosed bats with tuberculosectorial teeth of the ancestral type; however, in some species (*Chrotopterus* and *Vampyrum*, for example) the W-shaped ectoloph of the upper molars is distorted by the reduction of the stylar cusps and the closeness of the protocone, paracone, and metacone. Most members of this subfamily are insectivorous, and some species are known to pick insects either from vegetation or from the ground. On the other hand, a few of the largest phyllostomatines resemble their Old World look-alike ecological counterparts, the megadermatids, in their carnivorous habits. The large phyllostomatine species *Phyllostomus hastatus*, *Trachops cirrhosus*, *Chrotopterus auritus*, and *Vampyrum spectrum* are known to feed on small vertebrates. Beneath the roosts of *V. spectrum*, feathers and the tails of rodents and geckos frequently give indications of feeding preferences. The means by which these carnivorous-omnivorous bats perceive small vertebrates that are clinging quietly to tree trunks or are moving slowly over boulders is not known. Olfaction as well as echolocation may be important, and the large eyes of these bats indicate that hunting may also involve the use of vision. Bats of this type generally have large ears, however,

suggesting highly discriminatory echolocation.

Nectar-feeding is popular among tropical vertebrates (as indicated by the presence of over 300 species of hummingbirds in the American tropics), and has also been adopted by bats of the subfamily Glossophaginae of Mexico and Central and South America, and by bats of the subfamily Phyllonycterinae of the West Indies. These bats feed on the nectar and pollen of a great variety of plants and have many structural features associated with this mode of life. The tongue is long and protrusile, and has a brush-like tip (Fig. 6–39); the rostrum is elongate and the dentaries are slender (Fig. 6–40B). The cheek teeth have largely lost the tuberculosectorial pattern (Fig. 6–31C, D). In nectar-feeding bats the wings are usually broad and the uropatagium is reduced. Although they probably are not able to remain on the wing for long periods of time, as can some insectivorous bats, nectar-feeders can maneuver delicately through dense tropical vegetation and can hover. Flowers are seemingly located by the sense of smell, and these bats feed by hovering and thrusting their long tongues into the flowers. The pollination of many night-blooming Neotropical plants is accomplished by nectar-feeding phyllostomatids, just as many plants in the Old World tropics are pollinated by nectar-feeding pteropodids.

The members of three subfamilies of phyllostomatids — Carolliinae, Sturnirinae and Stenoderminae — are frugivorous. The success of these groups and the richness of this food source in the neotropics is indicated by the fact that within the Phyllostomatidae, the largest and most abundant group of neotropical bats, half of the species (approximately 60 out of 121) are fruit eaters. Several variations on this fruit-eating theme can be recognized. Members of the subfamily Carolliinae have reduced molars with the original tuberculosectorial pattern largely obliterated. These bats apparently prefer ripe, soft fruit, and are known to eat a great variety of fruit. The second frugi-

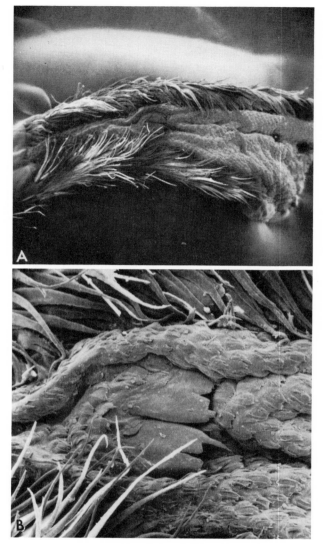

FIGURE 6–39. The tongue of a nectar-feeding bat (*Leptonycteris sanborni*) under 20× (A) and 100× (B) magnification. The tip of the tongue is to the left. (Photographs taken with a scanning electron microscope by Donna J. Howell.)

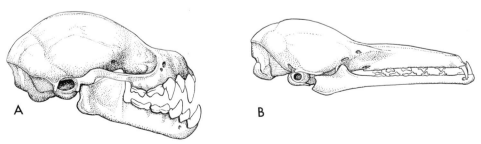

FIGURE 6–40. Skulls of leaf-nosed bats (Phyllostomatidae): A, a fruit-eater (*Artibeus phaeotis;* length of skull 19 mm.); B, a nectar-feeder (*Choeronycteris mexicana;* length of skull 30 mm.).

vorous subfamily, Sturnirinae, is composed of fairly small, often brightly colored bats that have robust molars with no trace of the basic tuberculosectorial pattern. Indeed, their molars strongly resemble those of New World monkeys (Cebidae). Sturnirines eat small and often hard fruit, such as the fruits of low-growing species of nightshade (*Solanum sp.*). The third frugivorous subfamily, Stenoderminae, contains bats with robust teeth that are highly modified for crushing fruit. The upper molars have lost the stylar cusps, and the inner portion is much enlarged and is marked by complex rugosities (Fig. 6–31E, F). The rostrum is short in many species, conferring considerable mechanical advantage for powerful jaw action to the large temporal muscles. Large species of the stenodermine genus *Artibeus* are remarkably abundant in some neotropical areas, and their piercing calls are characteristic sounds of the tropical nights. Often many *Artibeus* of several species concentrate on a single fig tree (*Ficus* sp.) with abundant fruit. In central Sinaloa, Mexico, two students and I camped beneath such a fig tree — but only for one night. The activities of dozens of *A. lituratus, A. hirsutus* and *A. jamaicensis* caused a nearly continuous rain of fruit during the night, and with sunrise came herds of aggressive local pigs to gather from the ground the night's fallout of figs. Stenodermines often eat unripe and extremely hard fruits, and it is perhaps as an adaptation to this type of food that the robust teeth and powerful jaws evolved.

Family Mormoopidae. The three genera and eight species that comprise this family were traditionally considered as members of the family Phyllostomatidae (Miller, 1907:118; Simpson, 1945:57), but recent studies have shown that they differ so markedly from the phyllostomatids as to merit recognition as a separate family (Vaughan and Bateman, 1970; Smith, 1972). All mormoopids can appropriately be called leaf-chinned bats, because in all species a conspicuous,

leaf-like flap of skin occurs on the lower lip (Fig. 6–30C). These bats are largely tropical in distribution, and occur from the southwestern United States and the West Indies south to Brazil.

Leaf-chinned bats are fairly small, and have several distinctive, externally visible specializations. The snout and chin always have cutaneous flaps or ridges (that reach their most extreme form in *Mormoops*, Fig. 6–30C), but a nose-leaf is never present. The ears are moderately large, have a tragus, and vary in shape, but always have large ventral extensions that curve beneath the fairly small eyes. The tail is short and protrudes from the dorsal surface of the fairly large uropatagium. The rostrum is tilted more or less upward (this feature is most extremely developed in *Mormoops*), and the floor of the braincase is elevated. The coronoid process of the dentary is reduced, allowing the jaws to gape widely. The teeth are of the basic insectivorous type; the dental formula is 2/2, 1/1, 2/3, 3/3 = 34.

The number of phalanges in the hand is as in the phyllostomatids, but the shoulder and elbow joints differ markedly from the phyllostomatid pattern. The greater tuberosity of the humerus in mormoopids does not form a well developed locking device with the scapula; the head of the humerus is more or less elliptical, perhaps favoring a specialized wing-beat cycle. The elbow joint is specialized in all species, and in *Mormoops* modifications of the distal end of the humerus and the forearm musculature provide for a highly efficient "automatic" flexion and extension of the hand. The musculature of the hand is reduced and simplified; this serves to lighten the hand and probably favors maneuverability and endurance. The hind limbs do not have the spider-like posture typical of phyllostomatids, but have a reptilian posture that allows leaf-chinned bats to crawl on the walls of caves with considerable agility.

Leaf-chinned bats are among the most abundant bats in many tropical localities, where they are seemingly the major chiropteran insectivores. They are most common in tropical forests, but occur also in some desert areas. Some species appear early in the evening; their insect-catching maneuvers resemble those of their temperate-zone counterparts, the vespertilionids. Leaf-chinned bats usually roost in caves or deserted mine shafts, and may concentrate in large numbers. A colony of *Mormoops* observed by Bernardo Villa-R. (1966:187) in Nuevo Leon, Mexico, contained more than 50,000 bats, and a colony of four species of mormoopids in Sinaloa that Gary C. Bateman and I studied recently contained roughly 750,000 bats. When the bats from the latter colony emerged in the evening they swept down the nearby arroyos and trails in such numbers and at such speeds that one hesitated to move across their path. When they occupy large colonies, these bats seem to disperse several miles from their roosting site to forage at night, and remain continuously on the wing for several hours. Their impact on tropical ecosystems must be great, for the bats in the Sinaloan colony probably consume over 1400 kg. of insects per night. It is not surprising that the bats must disperse over a wide area to forage.

Family Desmodontidae. This family contains the vampire bats, the only mammals that feed solely on blood. Only three genera, each with a single species, comprise this group, but they are widely distributed from northern Mexico southward to northern Argentina, Uruguay, and central Chile. No extinct genera have been found, but *Desmodus* is known from the Pleistocene.

Vampires are fairly small bats. Mexican specimens of the largest and most common species, *Desmodus rotundus*, usually weigh from 30 to 40 gm. The skull and dentition are highly specialized. The rostrum is short and the braincase is high (Fig. 6–41), and in all species the cheek teeth are reduced in both size and number. In *D. rotundus*, the most specialized species, the den-

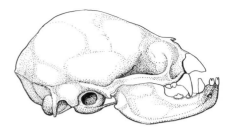

FIGURE 6–41. The skull of a vampire bat (*Desmodus rotundus*, Desmodontidae). Length of skull 24 mm.

tal formula is 1/2, 1/1, 2/3, 0/0 = 20. The upper incisors are unusually large and are compressed and blade-like, as are the upper canines. These teeth have remarkably sharp cutting edges. The cheek teeth are tiny and apparently nonfunctional. Except for the canine, the lower teeth are small (Fig. 6–41). The thumb in *Desmodus* is unusually long and sturdy and contributes an additional segment with three joints to the forelimb during quadrupedal locomotion. The hind limbs are large and robust, and the fibula is not reduced. The proximal part of the femur and the tibia and fibula are curiously flattened and ridged; these irregularities provide large surfaces for the attachment of the powerful hind limb musculature. Vampires can run rapidly and easily and can even jump short distances (Fig. 6–42). Their flight is strong and direct and not highly maneuverable.

The feeding habits of this family are of particular interest. Vampires begin foraging after complete darkness, and have one foraging period per night. *Diaemus* and *Diphylla* prefer the blood of birds, but *Desmodus* feeds on mammals. *Desmodus* alights on the ground near its chosen host, usually a cow, horse, or mule, and climbs up the foreleg to the shoulder or neck. The bat uses its upper incisors and canines to make an incision several millimeters deep from which blood is "lapped" by the tongue. Vampires occasionally feed on the feet of cattle, at which times the bats' ability to jump quickly may enable them to avoid in-

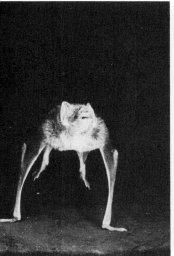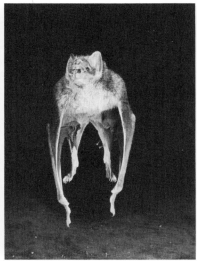

FIGURE 6–42. A vampire bat (*Desmodus rotundus*) leaping. Note the use of the long, robust "thumbs." (Photographs by J. Scott Altenbach.)

jury when the host animal moves its feet. In *Desmodus* the ingestion of blood is facilitated by an anticoagulant in the saliva that retards the clotting of blood. It has been estimated that each bat takes a meal of blood each night that amounts to over 50 per cent of the fasting weight of the bat; a vampire weighing 34 gm., then, takes roughly 18 gm. of blood per night (Wimsatt, 1969). Because of the nightly vampire-caused drain of blood from cattle in certain localities, and because vampires transmit rabies and other diseases, these bats are of great economic importance in many Neotropical areas. Occasionally vampires feed on man.

Renal function in vampires is of great interest. These animals begin urinating soon after they begin to feed, and rapidly lose much of the water taken in with the blood meal. This enables the bats to fly back to their roosts with less expenditure of energy and at less risk from predation than if they were burdened by the full weight of the ingested blood. When back in the roost the bats continue digesting the now partially dehydrated blood and are faced with a problem of excreting large amounts of nitrogenous wastes without losing excessive amounts of

water. At this time, rather than freely excreting water, as was done earlier in the foraging-digestion cycle, the kidney exerts a remarkable ability to concentrate wastes, and highly concentrated urine is excreted. In this notable instance, a tropical mammal that lives in environments that seldom lack accessible water has evolved a kidney surpassing that of many desert mammals in its ability to concentrate urine and to conserve water (Horst, 1971).

Family Natalidae. This small Neotropical family includes a single genus with five species. These bats are commonly referred to as funnel-eared bats, and occur from Baja California, northern Mexico and the West Indies, southward to Colombia, Venezuela, and Brazil. The only fossils are those of the living genus from the Pleistocene and Recent.

These small bats weigh from roughly 5 to 10 gm., and have slender, delicate-looking limbs, broad wing membranes, and a large uropatagium that encloses the long tail. The funnel-shaped ears with a tragus, the simple nose lacking any sort of nose-leaf, and the long, soft pelage that is frequently yellowish or reddish in color are characteristic. The skull has a long, wide rostrum with complete pre-maxillaries, and the braincase is high. The teeth are tuberculosectorial; the dental formula is 2/3, 1/1, 3/3, 3/3 = 38. The humero-scapular locking device is well developed, reduction of the phalanges of the hand is well advanced (the second digit lacks a phalanx and the third has two), and the manubrium of the sternum is unusually broad and has a well developed keel. Some of the most distinctive natalid features, however, are those of the axial skeleton that serve to reduce its flexibility: the thoracic vertebrae are anteroposteriorly compressed and fit tightly together; the ribs are broad and the narrow intercostal spaces are largely spanned by sheets of bone; all except the last two lumbar vertebrae are fused into a solid, laterally compressed, dorsally and ventrally keeled mass (Fig. 6–37); and the sacral vertebrae are mostly fused. As a result of these specializations, the strongly arched thoracolumbar section of the vertebral column is nearly rigid, with movement between this and the sacral section of the column allowed only by the "joint" formed by the last two lumbar vertebrae. These specializations may be associated with the animals' habit of alighting by the hind feet on the ceiling or walls of caves and hanging pendant.

Funnel-eared bats are insectivorous and their foraging flight is slow, delicate and maneuverable. Individuals released in dense vegetation are amazingly adroit at flying slowly through small openings between the interlacing branches of trees and shrubs. These bats inhabit tropical and semi-tropical lowlands and foothills, and typically roost in groups in warm, moist, and deep caves or mines. These are handsome little bats; groups of *Natalus stramineus* scattered over the ceiling of a cave look like bright orange jewels in the beam of a flashlight.

Family Furipteridae. This small family contains but two genera, each with one species. These bats occur in northern South America south to southern Brazil and northern Chile, and in Trinidad. Furipterids, known as smoky bats because of their grey pelage, are seemingly closely related to the Natalidae, Thyropteridae, and Myzopodidae. All these groups share certain structural similarities. No fossil furipterids are known.

Externally, furipterids resemble natalids in the structure of the ears and in their slender build. The shoulder joint and the fused lumbar vertebrae are also similar in these families. Furipterids differ from natalids in minor features of the skull and dentition, such as partially cartilaginous premaxillaries and reduced canines. The furipterid dental formula is 2/3, 1/1, 2/3, 3/3 = 36. The thumb of smoky bats is greatly reduced and is functionless.

These bats apparently are not common, and their habits are poorly known. They are insectivorous, and have been found in caves and buildings. Most of the area inhabited by smoky bats is tropical, but *Amorphochilus* occurs in arid coastal sections of northwestern South America.

Family Thyropteridae. Two small Neotropical species of bats comprise this family. These bats are known as disc-winged bats because of the remarkable sucker discs that occur on the thumbs and feet. These animals, and the one member of the family Myzopodidae, are the only bats and, except for two genera in the order Hyracoidea, the only mammals that have true suction cups. Disc-winged bats occur in southern Mexico, Central America, and South America as far as Peru and southern Brazil. No fossil thyropterids have been recorded.

In general appearance and in many skeletal details these small, delicately formed bats resemble natalids, but the lumbar vertebrae are not fused as in the latter. The skulls of natalids and thyropterids are similar and the dental formulae are the same. The thumb is reduced but retains a small claw, and its first phalanx has a sucker disc. The second digit is short, being represented by only a rudimentary metacarpal, and as a result the membrane between digits two and three is unusually small. The third digit has three bony phalanges. The digits of the feet have only two phalanges each, the third and fourth digits are fused, and the metatarsals bear a suction disc. The discs have a complex structure that allows them to act as suction cups; the bats can cling to smooth surfaces and can even climb a vertical glass surface. A fibrocartilaginous framework braces each disc; the rim of the disc consists of 60 to 80 chambers, each supplied by a sudoriparous gland (sweat gland). These glands improve the tightness of contact with the substrate by insuring that the face of the disc is constantly moistened. The disc itself lacks muscles, but specialized forearm muscles produce suction by cupping the middle of the face of the disc, and release suction by lifting a section of the rim of the disc (Wimsatt and Villa, 1970). These are the most elaborate and efficient suction discs known in mammals.

Disc-winged bats are insectivorous and are restricted to tropical forests. The roosting habits of these bats are highly specialized. They roost only in the young, slightly unfurled leaves of certain tropical plants that are partially or completely shaded by larger trees. Such a roosting site is provided by the "platanillo," which resembles the banana plant. For a few days, while a young leaf of this plant is beginning to unroll, it forms a tube roughly four feet long and an inch or so in diameter with a small opening at its tip. Several disc-winged bats may occupy such a tubular leaf in a head-to-tail row, heads upward, with the sucker discs anchoring the bats to the slippery surface of the smooth leaf. Because the leaf soon unfurls, it is suitable for occupancy for only a few days, and the bats move periodically to new and more suitable leaves.

Family Myzopodidae. The only species representing this family is *Myzopoda aurita*, the sucker-footed bat, a species restricted to Madagascar. No fossils of this family are known.

This bat is probably related to the Natalidae, Furipteridae, and Thyrop-

teridae, as indicated by the structure of the shoulder joint. The lumbar vertebrae are not fused as they are in natalids. The cheek teeth are of the standard tuberculosectorial-insectivorous type, and the dental formula is 2/3, 1/1, 3/3, 3/3 = 38. The ears are very large and the ear opening is partly covered by an unusual mushroom-shaped structure of a sort found in no other bat. The claw of the thumb is rudimentary, and the thumb bears a sucker disc. Only the metacarpal of the second digit is bony; the third digit has three ossified phalanges. The foot bears a sucker disc on its sole, and as in thyropterids each digit has only two phalanges. In *Myzopoda* the metatarsals are fused and all the toes fit tightly against one another.

Myzopoda appears to be rare, and its life history is unknown. Its dentition indicates insectivorous feeding habits.

Family Vespertilionidae. This is the largest family of bats in terms of numbers of species, and is the most widely distributed. Thirty-three genera and approximately 280 species are included in this family, and in temperate parts of the world these are usually by far the most common bats. In the New World, vespertilionids occur from the tree line in Alaska and Canada southward throughout the United States, Mexico, and Central and South America. All of the Old World is inhabited north to the tree line in northern Europe and Asia. Most islands, with the exception of some that are remote from large land masses, support vespertilionids. As can be inferred from their geographic distribution, these bats occupy a wide variety of habitats, from boreal coniferous forests to barren, sandy deserts. In the Neotropics, however, they are greatly outnumbered by bats of other families, particularly by leaf-nosed bats (Phyllostomatidae). Perhaps because of the diversity of habits and structure represented within the Vespertilionidae, no common name for this group is in general use; they are usually simply called vespertilionid bats. This family can be traced back to the middle Eocene in both Europe and North America, but they apparently did not reach Africa and South America until the Pleistocene. The genus *Myotis* is remarkable for its broad geographic distribution, which includes roughly the entire area occuped by the Vespertilionidae, and for its long fossil record, which begins in the middle Oligocene of Europe.

Vespertilionids are rather plain-looking bats that lack the distinctive facial features characteristic of many families. A nose-leaf is rarely present, nor do complex flaps or pads occur on the lower lips. The eyes are usually small. The ears are of moderate or large size, and the tragus is present but differs in shape markedly between species. These bats are usually small, weighing from 4 to 45 gm. The wings are typically broad, and the uropatagium is large and encloses the tail. The shoulder joint is of an advanced type, and provides for a locking of the large greater tuberosity of the humerus (Fig. 6–19) against the scapula at the top of the upstroke of the wing. The elbow joint is also advanced, and the spinous process of the medial epicondyle, which is well developed in many species, enables certain forearm muscles to "automatically" extend and flex the hand (Fig. 6–23). The shaft of the ulna is vestigial, but the proximal portion forms an essential part of the elbow joint. The second digit of the hand has two bony phalanges, and the third digit has three. The fibula is rudimentary. The manubrium of the sternum has a keel, but the body of the sternum has at best a slight ridge. Except in one genus (*Tomopeas*), all presacral vertebrae are unfused. The teeth are tuberculosectorial, and the W-shaped ectoloph of the upper molars is always well developed. The dental formula varies from 1/2, 1/1, 1/2, 3/3 = 28 to 2/3, 1/1, 3/3, 3/3 = 38. The skull lacks postorbital processes; the palatal parts of the premaxillaries are missing and the front of the palate is emarginate (Fig. 6–32). In general, vespertilionids are mostly small, plain bats that are characterized by refinements of the flight apparatus that make them efficient, maneuverable fliers.

Most vespertilionids are insectivorous, and in their ability to capture flying insects they are unexcelled. Most children in Europe and North America gain their first experience with bats by watching vespertilionid bats, silhouetted against the twilight sky, making abrupt turns and sudden dives while pursuing insects. The most commonly used vespertilionid foraging technique is probably also the most demanding: it involves the pursuit and capture of flying insects by bats that remain on the wing throughout most of their foraging periods. The insects are perceived by echolocation (Chapter 21). The bats emit ultrasonic pulses, and locate and follow insects by utilizing the reception and interpretation of echoes of the pulses from the bodies of the insects. Insects are usually followed in their erratic flight by a series of intricate maneuvers by the bat, and are either captured in the mouth or, in the case of some species of bats, are trapped by a wing-tip or by the uropatagium (Webster and Griffin, 1962). This type of foraging demands highly maneuverable flight, and this is the type of flight to which vespertilionids seem best adapted. Styles of foraging vary between different vespertilionids. Whereas some, for example the tree-roosting bats (*Lasiurus*), remain on the wing throughout their foraging, others alight to eat large prey. Some species snatch insects or arachnids from leaves or pounce on them on the ground. The pallid bat (*Antrozous pallidus*), a common species in the southwestern United States, feeds on such large terrestrial arthropods as scorpions, Jerusalem crickets (*Stenopelmatus*), and sphinx moths (Sphingidae). This bat often uses porches, shallow caves, or abandoned buildings as places to rest and eat its prey. Often there are accumulations of discarded legs, fragments of exoskeletons, and wings beneath these roosts. Some vespertilionids capture insects from the surface of the water, and several species of *Myotis* capture fish or crustaceans from the water, probably by gaffing the prey with the claws of the feet. Some vespertilionids are known to have an early evening and a predawn foraging period, but the cycles of foraging activity of most species are unknown.

A wide variety of roosting places are utilized by vespertilionids. They adapt well to urban life, and frequently roost during the day in attics of churches or houses, in spaces between rafters of barns or warehouses, or behind shutters or loose boards. Crevices in rocks, spaces beneath rocks or behind loose bark, caves, mines, holes in trees, and foliage are also utilized. Often these bats are colonial, frequently with nursery colonies of females with young occupying one roost and adult males using another; but many species, such as the foliage-roosting bats, roost singly or in small groups. Some species rest for part of the night beneath bridges or in porches or buildings, often in places never used as daytime retreats.

In temperate regions many vespertilionids hibernate. Although the hibernation sites of some species are not known, some well known species hibernate in caves and mines or buildings, and some species migrate fairly long distances to reach favorite hibernacula. Excellent reviews of hibernation (Davis, 1970) and migration in bats (Griffin, 1970) are available. Two small European species are known to migrate over 1,000 miles from Russia to Bulgaria (Krzanowski, 1964). Some vespertilionids hibernate for short periods and may be at least intermittently active in the winter, but others hibernate throughout the winter, except for occasional, short arousals. The weight of much observational evidence indicates that foliage-roosting bats migrate. However, some red bats (*Lasiurus borealis*) remain in cold regions in the central United States throughout the winter and may be active on warm days (Davis and Lidicker, 1956).

Family Mystacinidae. One rather aberrant species, *Mystacina tuberculata,* the short-tailed bat, is the sole member of this family. This species occurs only in New Zealand, and the family has no fossil record.

Mystacina has some characteristics of vespertilionids and some of free-tailed bats (Molossidae), but in general is sufficiently distinct from either group to merit recognition as a member of a separate family. Vespertilionid characteristics of *Mystacina* include the advanced locking shoulder joint, one phalanx in the second digit and two in the third, and the lack of fusion of presacral vertebrae. The skull is roughly like that of vespertilionids, but there is no anterior palatal emargination. The teeth are tuberculosectorial, and the dental formula is 1/1, 1/1, 2/2, 3/3 = 28. The limbs of *Mystacina* resemble in some ways those of molossids: the wing membranes and uropatagium are tough and leathery, the first phalanx of the third digit folds back on the dorsal surface of the metacarpal, the hind foot is unusually broad; the fibula is complete and the hind limb is robust. Unlike vespertilionids or molossids, the tail of *Mystacina* is short and protrudes from the dorsal surface of the uropatagium, and each of the claws of the thumb and foot has a secondary talon at its ventral base.

This unusual bat is insectivorous and captures insects in mid-air or from vegetation. The wing can be folded compactly, owing to the unique pattern of flexion of the third digit, and during quadrupedal locomotion it is partially protected by the leathery proximal part of the plagiopatagium. The limbs are seemingly well adapted to quadrupedal locomotion. These bats are agile and rapid runners and are known to chase insects over the branches of trees. *Mystacina* lives in forested areas, where it roosts during the day in caves and in hollows in trees.

Family Molossidae. Members of this family, the free-tailed bats, are important components of tropical and subtropical chiropteran faunas throughout much of the world. Eleven genera and 88 species are included in the family. Molossids occupy the warmer parts of the Old World, from southern Europe and southern Asia southward, and they inhabit Australia and the Fiji Islands. In the New World they occasionally occur as far north as Canada, but the main range begins in the southern and southwestern United States and the West Indies and extends southward through all but the southern halves of Chile and Argentina.

Structurally, this is a peripheral group of bats; the most extreme manifestations of many of the typically chiropteran adaptations for flight occur in the Molossidae. The greater tuberosity of the humerus is large (Fig. 6–28E, F), and the locking device between it and the scapula is highly developed. The origin of the extensor carpi radialis longus and brevis, and that of the flexor carpi ulnaris, are well away from the center of rotation of the elbow joint and probably act more effectively than in any other bats as "automatic" extensors and flexors of the hand. The wing is typically long and narrow (Fig. 6–18B), with the fifth digit no longer than the radius, and the membranes are leathery because they are reinforced by numerous bundles of elastic fibers. In many species refinements of structure that favor high-speed flight occur. In many Neotropical molossids, for example, the arrangement of the muscles of the forearm is such that the forearm is flattened and streamlined with respect to the airstream during flight. In the interest of rigidity of the outstretched wing during flight, movement at the wrist and elbow joint is strictly limited to one plane. The muscles that brace the fifth digit and maintain an advantageous angle of attack of the plagiopatagium during the downstroke of the wings are large and unusually highly specialized. Except for fusion of the last cervical and first thoracic vertebrae, the presacral vertebrae are unfused. The body of the sternum is not keeled.

The general appearance of molossids is distinctive. The tail extends well beyond the posterior border of the uropatagium when the bats are not in flight, and the fur is usually short and velvety. (In one genus, *Cheiromeles,* the fur is so short and

sparse that the animal appears naked.) The muzzle is broad and truncate, and the thick lips are wrinkled in some species (Fig. 6–30F). Typically, the ears are broad, project to the side, and are like short wings. As viewed from the side, the pinnae are arched and resemble an airfoil of high camber. The ears are frequently braced by thickened borders and are connected by a fold of skin across the forehead. Because of the unique design of the ears, in most species they do not directly face the force of the airstream during flight, an adaptation probably of considerable importance to these fast-flying bats.

The skull is broad, the teeth are tuberculosectorial, and the dental formula varies from 1/1, 1/1, 1/2, 3/3 = 26 to 1/3, 1/1, 2/2, 3/3 = 32. Several characteristically molossid features are associated with the well developed quadrupedal locomotion typical of these bats. The first phalanges of digits three and four flex against the posterodorsal surfaces of their respective metacarpals, providing for the chiropatagium to be folded into a compact bundle, no longer than the forearm, that is manageable when the animals run. The feet are broad, and have sensory hairs along the outer edges of the first and fifth toes. The fibula is not reduced, and the short hind limbs are stoutly built. Within the structural limits of the basic chiropteran plan, these bats have seemingly made the best of two types of locomotion. The highly specialized wings are clearly adapted to fast, efficient flight, whereas the primitive hind limbs have not lost their ability to serve in rapid quadrupedal locomotion.

These insectivorous bats are remarkable for their enduring flight. Whereas most bats fly fairly close to the ground or to vegetation when foraging, many molossids fly high and may move long distances during their nightly foraging. Some populations of guano bats (*Tadarida brasiliensis*), the bats that occur in great numbers in Carlsbad caverns and other large caverns and caves in the southwestern United States, fly at least 50 miles to their foraging areas each night (Davis et al., 1962). The

FIGURE 6–43.   The roosting place of mastiff bats (*Eumops perotis*, Molossidae) in a granite cliff. The animals occupy the space beneath the tongue-shaped slab of rock at the upper right. (From Vaughan, 1959.)

western mastiff bat (*Eumops perotis*) forages over broad areas, and in southern California may on occasion fly at least 2,000 feet above the ground (Vaughan, 1959:22). Because of the temperature inversions that frequently prevail for many nights in this area, in the winter these high-flying bats may be surrounded by air warmer than that at the ground, and may be catching insects that are flying in the warm "strata." Some molossids remain in flight for much of the night; foraging periods of at least six hours have been recorded for some species. The flight of many molossids is unusually fast, and some species rival swifts and swallows in aerial ability. In some areas of Mexico, early-flying mastiff bats (*Molossus ater*) mingle with late-flying flocks of migrating swallows, and the bats seem at least the equals of the swallows in speed and maneuverability. One gets the impression, in fact, that the swallows are hastened to their roosts by the sudden appearance in abundance of their chiropteran counterparts.

Some molossids make spectacular dives when returning to their roosts. The western mastiff bat, for example, often makes repeated high-speed dives and half-loops past the roosting site. This bat returns to its roost in a cliff by diving toward the base of the cliff, pulling sharply upward at the last instant, and entering the crevice with momentum to spare. Several other molossids are known to return to their roosting places by similar maneuvers. Because the wings of many molossids are narrow and have relatively small surface areas relative to the weights of the bats, these animals must attain considerable speed before they can sustain level flight. As a result, some species roost high above the ground in cliffs (Fig. 6–43), buildings, or palm trees, in situations where they can dive steeply downward for some distance in order to gain appropriate flight speed. These species are unable to take flight from the ground.

Most molossids inhabit warm areas. Migration, therefore, is not characteristic of molossids in general. The guano bat, however, is known to make extensive migrations from the United States to as far south as southern Mexico (Villa and Cockrum, 1962). The tremendous deposits of guano that occur in some large caves inhabited by molossids attest to the effect that large colonies of these bats must have on insect populations in some areas.

CHAPTER 7

ORDER PRIMATES

Primates have been most successful in tropical and subtropical areas, where today they pursue mostly arboreal modes of life. Some anthropoids, such as baboons and chimpanzees, have become partly or mostly terrestrial, but only man has become fully bipedal. Approximately 47 genera and 166 species of primates are living today, of which 16 genera and 63 species are in the New World. The primitive primates, such as lemurs, lorises, and tarsiers, are included in the suborder Prosimii; the more progressive monkeys, apes, and man are in the suborder Anthropoidea.

The evolution of the principal anatomical features of primates was influenced strongly by arboreal life. Stereoscopic vision, the dextrous, grasping hand, and the remarkable agility and muscular coordination of primates are seemingly the result of arboreal life. These features were in large part responsible for the primate trend toward enlargement of the brain. A good general description of primates is given by Anderson (1967:151): primates are "eutherian mammals having generalized limb structure, primitively arboreal habits, omnivorous diet and comparatively unspecialized teeth; grasping with mobile digits and possessing freely movable limbs; phylogenetically replacing claws with nails and developing enlarged and sensitive pads on digits; reducing nose and sense of smell, enlarging eyes and improving vision, enlarging brain, and progressively improving placentation."

Most primates are omnivorous, and many species seem to be largely opportunistic feeders. Because soft foods are usually taken, the molars of primates are largely bunodont and brachyodont, and have the quadrate form typical of molars of generalized feeders. Early in the evolution of primates a hypocone was added to the upper molar, and the paraconid of the lower molar disappeared, leaving a basically four-cusped pattern (Fig.

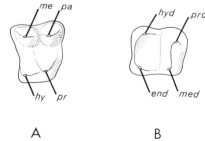

A   B

FIGURE 7-1. A diagrammatic representation of the basic four-cusped crown pattern of primates. A, a right upper molar; B, a left lower molar. Abbreviations: *end*, entoconid; *hy*, hypocone; *hyd*, hypoconid; *me*, metacone; *med*, metaconid; *pa*, paracone; *pad*, paraconid; *pr*, protocone; *prd*, protoconid.

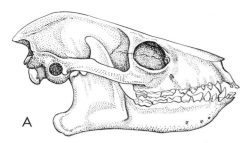 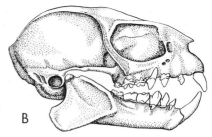

FIGURE 7–2. Skulls of fossil primates. A, an Eocene lemur (*Notharctus*, Adapidae; length of skull 75 mm.); B, an Eocene tarsier-like primate (*Tetonius*, Anaptomorphidae; length of skull 46 mm.). (After Romer, A. S.: *Vertebrate Paleontology*, 3rd ed., The University of Chicago Press, 1966.)

7–1). The primate trend toward short-ening of the rostrum is probably re-lated to the importance of stereoscopic vision and the lack of importance of the sense of smell rather than to a need for greater mechanical advantage for jaw musculature.

Primates comprise one of the oldest eutherian orders, dating from the late Cretaceous (Van Valen, 1965:743). The fossil record of primates has consider-able gaps, some of which result from the primate preference for tropical or subtropical environments. In these areas conditions are generally not suit-able for the fossilization of animals. Primates arose from primitive insec-tivores, and some early primate fami-lies departed so little from the structur-al plan of insectivores as to make ordinal assignment to either Insec-tivora or Primates difficult. The pri-mate divergence from insectivores in-volved a trend toward a fruit-eating or leaf-eating habit and was accompanied by changes in dentition that favored thorough mastication and partial oral digestion of food (Szalay, 1968). The Eocene, Old World family Adapidae (Fig. 7–2A) may have been the basal group that gave rise to most prosi-mians, but the lack of an adequate fos-sil record for most prosimian families makes any conclusions on prosimian evolution tentative. An origin of the anthropoids from some prosimian group is assumed by many paleonto-logists, but is not documented by fossil evidence. Cebids and callithricids ap-pear to be strictly New World types,

whereas the other anthropoids arose in the Old World and all but hominids have remained there.

SUBORDER PROSIMII

The five families included in this suborder contain an assemblage of mostly arboreal mammals that in some cases bear only a marginal resem-blance to the more "standard" pri-mates (monkeys, great apes, and man) that comprise the suborder Anthropoi-dea. The prosimians are primitive pri-mates, and in the case of several fami-lies occupy restricted geographic areas and pursue specialized modes of life. Even among prosimians, however, the importance of vision and manual dex-terity is apparent.

Family Lemuridae. The lemurs in-habit Madagascar and the nearby Co-moro Islands. Among the 15 Recent species, belonging to five genera, some are arboreal, some are semi-ar-boreal, and some are largely terrestrial. These are the most primitive living primates. The fossil record of lemurids is from Pleistocene and sub-Recent de-posits in Madagascar, where Recent genera as well as enormous extinct forms are known. One extinct giant of presumably arboreal habits had an elongate skull 30 cm. in length (roughly one foot!). The survival of lemurs is perhaps related to their insu-lar distribution; they are not in associa-tion with more progressive primates (except man).

FIGURE 7–3. A lemur (*Lemur fulvus*, Lemuridae). (Photograph by D. Schmidt; San Diego Zoo photo.)

In contrast to the condition in most primates, the cranium of lemurs is elongate and the rostrum is usually of moderate length, giving the faces of some lemurs a fox-like appearance (Fig. 7–3). In more typical primate fashion, the lemurid braincase is large, crests for the origin of the temporal muscles are inconspicuous, and the foramen magnum is directed somewhat downward. The largest lemurs are roughly the size of a house cat; the smallest are the size of a mouse. The dental formula is 0-2/2, 1/1, 3/3, 3/3 = 32-36. The upper incisors are usually reduced or absent, and between those of the two sides is a broad diastema; the lower canine is incisiform and the first lower premolar is caniniform. The molars are basically tritubercular. The pollex and hallux are more or less enlarged and are opposable in all genera (Fig. 7–4A). The pelage is woolly, the tail is long and heavily furred, the limbs are usually slim and the tarsal

bones are not greatly elongated (Fig. 7–5A). Conspicuous color patterns occur in some species.

Lemurs are variously omnivorous, insectivorous, or herbivorous-frugivorous, and, depending on the species, are diurnal or nocturnal. They are agile climbers, and the hands are used both for climbing and for food handling. Some species make great leaps from branch to branch. Some lemurs store fat in preparation for estivation during the dry season. Some lemurs live in social groups of up to 20 individuals, and, as in higher primates, vocalization seems important in maintaining contact between members of a group. Lemurs emit a variety of grunts and calls (Petter, 1965).

Family Indridae. The indrids, often called woolly lemurs, are not a diverse group, including but three genera and four species, and are restricted to Madagascar. There are Pleistocene and sub-Recent records of indrids from

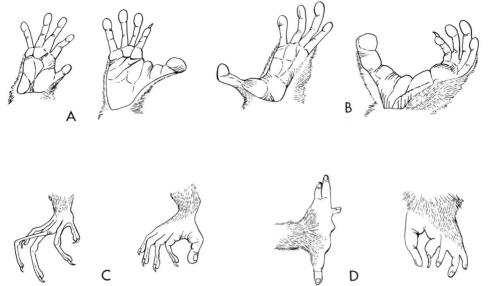

FIGURE 7–4. Hands and feet of some prosimian primates (the hand is on the left in each pair). A, a lemur (*Lemur mongoz*, Lemuridae); B, an indrid (*Propithecus diadema*, Indridae); C, an aye-aye (*Daubentonia madagascariensis*, Daubentoniidae); D, a potto (*Arctocebus calabarensis*, Lorisidae).

Madagascar, and matching the extinct, huge lemurids are extinct indrids that are also of great size.

These animals are fairly large (up to 900 mm. in head-and-body length); two genera have shortened rostra and monkey-like faces; the snout is fairly long in the other genus. The dental formula is 2/2, 1/0, 2/2, 3/3 = 30. The upper incisors are enlarged and the first lower premolar is caniniform. The hands and feet are highly modified for grasping branches during climbing (Fig. 7–4B). The pelage is conspicuously marked in some species, and the tail is long in three species.

These primates are largely herbivorous, and their leaf-eating habits

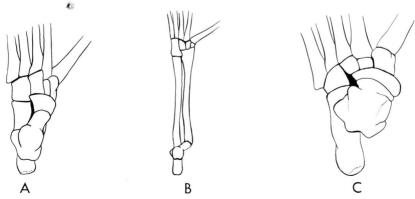

FIGURE 7–5. Dorsal views of the left feet of several primates, showing the tarsal bones. A, a lemur (*Lemur* sp., Lemuridae); B, a tarsier (*Tarsius*, Tarsiidae); C, a gorilla (*Gorilla gorilla*, Pongidae). Note the remarkable elongation of the calcaneum and the navicular in the tarsier. (Redrawn by permission of Quadrangle Books from *The Antecedents of Man* by W. E. LeGros Clark, copyright © 1959, 1962, 1971 by W. E. LeGros Clark.)

resemble those of the Neotropical howler monkey (*Alouatta;* Cebidae) and the African colobus monkey (*Colobus;* Cercopithecidae). Indrids are typically fairly slow, deliberate climbers. The hind limbs are long relative to the front limbs, and when traveling on the ground these primarily arboreal and diurnal animals proceed by a series of hops. The hands are used for climbing and for handling food, but manual dexterity seems limited and food is often picked up in the mouth. One genus is solitary or occurs in pairs; the other two genera typically live in small bands. A specialized laryngeal apparatus enables *Indri* to produce loud, resonant calls. These howls are given with greatest frequency in the morning and evening, as are the calls of the howler monkey, and perhaps function in maintaining territorial boundaries between neighboring bands. All indrid species are vocal to some extent.

Family Daubentoniidae. This family is represented by only one highly specialized Recent species (*Daubentonia madagascariensis*) with the common name of aye-aye. This nocturnal animal occurs locally in northern Madagascar, where it is restricted to dense forests and stands of bamboo. The fossil record of the Daubentoniidae consists of sub-Recent fossils from Madagascar of an extinct species that was larger than the surviving aye-aye.

Aye-ayes weigh approximately 2 kg.; they have prominent ears and a long, bushy tail. The skull and dentition are remarkably specialized, and depart strongly from the usual primate plan. The skull is short and moderately high. The orbit is prominent and faces largely forward; the postorbital bar and zygomatic arch are robust, and the rostrum is short and deep (Fig. 7-6A). The dentition differs from the basic primate type both in the extensive loss of teeth and in the strong specialization of the teeth that are retained. The dental formula is 1/1, 0-1/0, 1/0, 3/3 = 18 or 20. The canine is often absent, and the cheek teeth have flattened crowns with no clear cusp pattern. The laterally compressed incisors are greatly enlarged, wear to a sharply beveled edge because only the anterior surfaces are covered with enamel (as in rodents), and are ever-growing. Because of the shape of the teeth and the presence of a diastema between the incisors and the cheek teeth, *D. madagascariensis* was first described as a rodent. The hand is unique among primates. The digits are clawed, and all but the non-opposable pollex are long and slender; the third digit is remarkably slender (Fig. 7-4C). In the hind paw, the hallux is opposable and bears a nail, but the other digits are clawed.

Aye-ayes are nocturnal and are mainly insectivorous. They are arboreal, and are capable of making

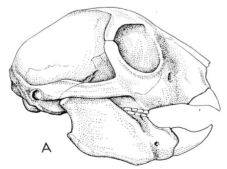

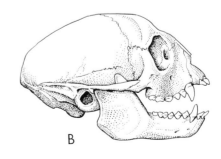

FIGURE 7-6.   Skulls of primates. A, an aye-aye (*Daubentonia madagascariensis*, Daubentoniidae; length of skull 90 mm.); B, a marmoset (*Saguinus geoffroyi*, Callithricidae; length of skull 51 mm.). (*Saguinus* is after Hall, E. R., and Kelson, K. R.: *The Mammals of North America,* copyright © 1959 by the Ronald Press Company, New York.)

graceful leaps between branches. Their foraging technique is noteworthy. The elongate third finger is used to tap on wood harboring wood-boring insects; the aye-aye then listens carefully for insects within the wood, and the remarkable third digit is used for removing adult and larval insects from holes or fissures in the wood. When necessary the powerful incisors tear away wood to enable the third digit to reach insects in deep burrows. Surprisingly, this strange mode of foraging is shared by two Australian genera of marsupials of the family Petauridae. In *Dactylonax*, the most specialized of these marsupials, the front incisors are modified and the manus is specialized along lines parallel to those in the hand of *Daubentonia*, except that the fourth rather than the third digit is the probing finger in *Dactylonax*. As is the case with many mammals that occupy limited areas, the future of the aye-aye seems dim. Because *Daubentonia* is restricted to continuous, heavy forests, its continued survival depends largely on the extent to which land is not cleared for agriculture in northern Madagascar.

Family Lorisidae. The lorises are more widely distributed than are the primitive primates of Madagascar, and are locally common. Lorises occur in Africa south of the Sahara, in India, Ceylon, and Southeast Asia, and in the East Indies. The fossil record of lorisids is scanty, but it suggests that these animals evolved in the Old World and have never occurred elsewhere. A Pliocene form is known from Asia, and lorises can be traced back to the Miocene of Africa on the basis of fragmentary material.

The eyes face forward in the lorisids (Fig. 7–7), rather than more or less to the side as in the lemurids, and the rostrum is short. Lorisids are arboreal, and their locomotor styles range from the usually methodical hand-over-hand climbing of the tailless or short-tailed lorises and pottos (subfamily Lorisinae) to the rapid climbing and jumping typical of galagos (subfamily Galaginae). Lorisids vary from the size

FIGURE 7–7. A loris (*Loris tardigradus,* Lorisidae). (Photograph by Ron Garrison; San Diego Zoo photo.)

of a rat to that of a large squirrel. The braincase is globular, the facial part of the skull is often short and ventrally placed, and the anteriorly directed orbits are separated by a thin interorbital septum. The dental formula is 1-2/2, 1/1, 3/3, 3/3 = 34 or 36. The upper incisors are small, the lower canine is incisiform, and the molars are basically quadritubercular (Fig. 7–9A, B).

The manus and pes are specialized in a variety of ways for clutching branches. In the genus *Arctocebus* an odd, pincer-like hand has been developed by the reduction of digits two and three and a change in the postures of the remaining digits; the first digit of the pes is opposable and is frequently greatly enlarged (Fig. 7–4D). Circulatory adaptations in the appendages provide for an increased blood supply to the digital flexor muscles that are used in gripping branches during ex-

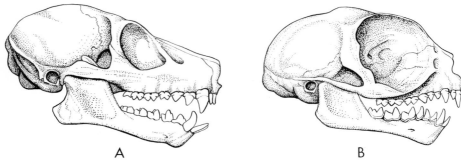

FIGURE 7-8. Skulls of prosimian primates. A, a galago (*Galago* sp., Lorisidae; length of skull 65 mm.); B, a tarsier (*Tarsius spectrum*, Tarsiidae; length of skull 36 mm.). (Redrawn by permission of Quadrangle Books from *The Antecedents of Man* by W. E. LeGros Clark, copyright © 1959, 1962, 1971 by W. E. LeGros Clark.)

tended periods of contraction. These same circulatory modifications, involving the formation of a *rete mirabile,* are also important in this and many other mammals in conserving body heat. (A *rete mirabile* is a complex meshwork of small arteries and veins that are intertwined so that the warm blood pass-

ing to an appendage is cooled by the cooler blood coming from the appendage, and the cooler blood from the appendage is warmed by the arterial blood. One result of this system is the avoidance of much of the metabolic drain that would accompany the warming of drastically cooled blood from a

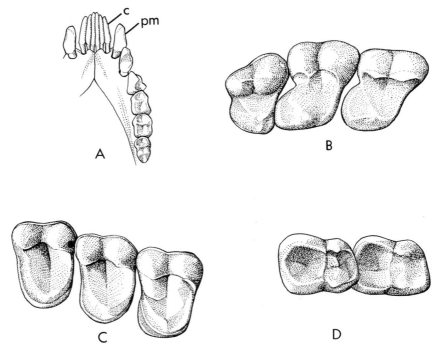

FIGURE 7-9. Teeth of prosimian primates. Teeth of a galago (*Galago* sp., Lorisidae): A, lower right tooth row, showing the incisiform canine (*c*) and the caniniform premolar (*pm*); B, upper right molars. Teeth of a tarsier (*Tarsius spectrum*, Tarsiidae): C, upper right molars; D, first and second lower left molars. (Redrawn by permission of Quadrangle Books from *The Antecedents of Man* by W. E. LeGros Clark, copyright © 1959, 1962, 1971 by W. E. LeGros Clark.)

poorly insulated limb as the blood entered the general blood stream.) These
nocturnal primates are insectivorous
and carnivorous, and prey is usually
captured by the hands after a stealthy
approach. The specialized lorisine
genus *Arctocebus* spends considerable
time upside down, and is reported to
sleep in this position.

Members of the subfamily Galaginae (the skull of which is shown in
Fig. 7–8A), which occur in much of
Africa south of the Sahara Desert, are
accomplished arboreal jumpers. The
hind limbs of galagos are long, and the
elongation of the foot segment has involved mostly the tarsus. As a result of
these modifications, and due to the
enlarged and disc-like tips of their
digits, the galagos resemble *Tarsius*
(Tarsiidae), an even more highly specialized jumper. Lorisids live singly or
in groups, and in social situations utilize a variety of vocalizations.

Family Tarsiidae. This family is
represented today by three species of
the genus *Tarsius,* and occurs in jungles and secondary growth in southern
Sumatra, in some East Indian islands,
and in some of the Philippine Islands.
Tarsiids are known from the Eocene of
Europe, but there are no fossils representing the remainder of the Cenozoic.

The tarsier is roughly the size of a
small rat, and with its large head and
huge eyes, its long limbs and long tail,
has a distinctive appearance (Fig.
7–10). The most conspicuous cranial
features are the enormous orbits,
which face forward and have expanded
rims and a thin interorbital septum
(Fig. 7–8B). The eye of the tarsier is
apparently adapted entirely to night
vision, for it lacks cones in the retina.
The dental formula is 2/1, 1/1, 3/3, 3/3
= 34. The medial upper incisors are
enlarged, the premolars are simple,
the crowns of the upper molars are
roughly triangular, and the lower
molars have large talonids (Fig. 7–9C,
D). The neck is short, a characteristic
of many saltatorial vertebrates. All but
the clawed second and third pedal
digits have flat nails, and all digits
have disc-like pads (Fig. 7–11A). The
limbs, especially the hind ones, are
elongate; the tibia and fibula are fused.
The trend toward jumping ability
that is apparent in galagos is developed to an extreme degree in the
family Tarsiidae. As in all highly specialized jumpers, the hind foot is
elongate, but in the tarsier the elongation has been unique in involving two
tarsal bones (hence the name *Tarsius*)
rather than metatarsals, as in such

FIGURE 7–10. Tarsiers (*Tarsius*
sp., Tarsiidae). (Photograph by Ron
Garrison; San Diego Zoo photo.)

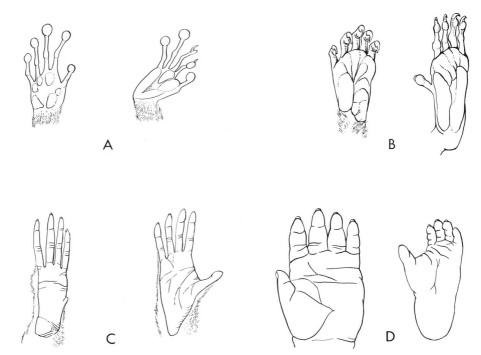

FIGURE 7–11. Hands and feet of four primates (the hand is on the left in each pair). A, a tarsier (*Tarsius spectrum,* Tarsiidae); B, a marmoset (*Callithrix* sp., Callithricidae); C, a woolly spider monkey (*Brachyteles arachnoides,* Cebidae); D, a gorilla (*Gorilla gorilla,* Pongidae). (A, B, and D redrawn by permission of Quadrangle Books from *The Antecedents of Man* by W. E. LeGros Clark, copyright © 1959, 1962, 1971 by W. E. LeGros Clark.)

jumpers as elephant shrews (Fig. 6–3C; p. 78) and kangaroo rats (Fig. 10–12E; p. 157). In *Tarsius* the calcaneum and navicular are greatly elongate (Fig. 7–5B), and the metatarsals are not unusually long in relation to the phalanges (Fig. 7–11A). An important functional end is achieved by this unusual system of foot elongation: because the elongation has occurred in the tarsus, the dexterity and grasping ability of the digits themselves (the metatarsals and phalanges) has not been sacrificed. A reduction of dexterity would have accompanied an elongation of the metatarsals and the resulting functional reduction of the number of segments of the digits. In elephant shrews and kangaroo rats, dexterity and gripping ability of the hind foot is not important, and a more "direct" means of elongation—lengthening of the already somewhat elongate metatarsals—occurred.

Tarsiers are arboreal and nocturnal and feed primarily on insects which they pounce upon and grasp with the hands. Although more highly adapted to leaping than any other primate, tarsiers can walk quadrupedally, can hop or run on their hind legs on the ground, can climb quadrupedally, and can slide down branches (Sprankel, 1965). The tarsier is alone among primates, however, in its ability to leap long distances (over 3 m.) with great precision. After a leap from one branch to another, the landing is frequently mostly bipedal. In association with jumping ability, much of the weight of the tarsier is concentrated in the hind limbs, which together comprise 21 per cent of the total weight of the animal; the musculature of the thighs alone equals 12 per cent of the body weight, largely owing to great enlargement of the quadriceps femoris (Grand and Lorenz, 1968), a powerful extensor of the shank. Tarsiers usually live in pairs, and although they are more nearly silent than most prosimians, they can make a variety of high-pitched sounds.

SUBORDER ANTHROPOIDEA

The five families within this suborder are "higher" primates, and have many progressive features not typical of members of the suborder Prosimii. Anthropoids—monkeys, marmosets, apes, and man—are the most familiar primates and are vastly more important in terms of taxonomic diversity and adaptability than are the relatively primitive prosimians.

Family Cebidae. The New World monkeys all belong to this family, which includes 29 Recent species of 11 genera. Cebids range from southern Mexico, through Central America, to southern Brazil. Cebids first appear in the early Oligocene of South America. Primitive primates ancestral to the cebids may have entered South America from Central America on logs or debris that floated across the stretch of water that separated these land masses during much of the Tertiary.

These are usually small, slim monkeys; the largest cebid is the howler monkey (*Alouatta*), which weighs up to 9 kg. New World monkeys are arboreal and have elongate limbs and curved nails on the digits. The pollex is not opposable and is reduced in some species (Fig. 7–11C), but the hand usually has considerable dexterity. The hallux is strongly opposable. The tail is long in all but one genus and is prehensile in four of the 11 genera. The skull is more or less globular, with a high braincase, and the rostrum is typically short. The orbits face forward, and the nostrils are separated by a broad internarial pad and face to the side (Fig. 7–12), a condition termed platyrrhine. The dental formula is 2/2, 1/1, 3/3, 3/3 = 36, and the lateral pair and medial pair of cusps of the molars are separated by a central anteroposteriorly aligned depression (Fig. 7–13B). Brightly colored and bare patches of skin on the rump (ischial callosities) do not occur in cebids as they often do in Old World monkeys (family Cercopithecidae).

Cebids typically occur in tropical forests, and most are diurnal. They are basically vegetarians; fruit is often preferred, but a wide variety of plant and

FIGURE 7–12. A howler monkey (*Alouatta palliata*, Cebidae). (San Diego Zoo photo.)

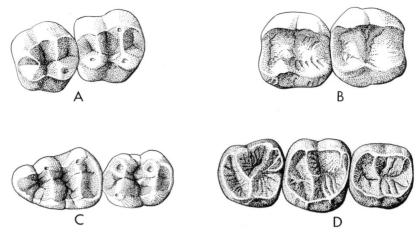

FIGURE 7–13.   Cheek teeth of anthropoid primates. A, second and third upper right molars of the orange crowned mangabey (*Cercocebus torquatus*, Cercopithecidae); B, first and second upper right molars of a saki monkey (*Pithecia monachus*, Cebidae); C, second and third lower left molars of a baboon (*Papio* sp., Cercopithecidae); D, upper right molars of the orangutan (*Pongo pygmaeus*, Pongidae). Note the cross lophs on the teeth of the mangabey and the baboon, and the extra posterior cusp (hypoconulid) on the third lower molar of the baboon. (*Pongo* redrawn by permission of Quadrangle Books from *The Antecedents of Man* by W. E. LeGros Clark, copyright © 1959, 1962, 1971 by W. E. LeGros Clark.)

animal material is eaten. The night monkey (*Aotes trivirgatus*), an inhabitant of Central and South America, is apparently insectivorous and carnivorous and may feed to some extent on bats. Cebids are active, intelligent animals and are adroit climbers; some species move with amazing speed through the trees. For dazzling arboreal ability, the Neotropical spider monkey (*Ateles*) is probably only surpassed by the Old World gibbons (*Hylobates*, Pongidae).

Most cebids are vocal to some extent, and several species have loud, penetrating calls. Outstanding among these is the howler monkey (Fig. 7–12), in which the hyoid apparatus is enlarged into a resonating chamber. The males emit loud "roaring" sounds that carry for long distances through the tropical rain forest. These sounds are seemingly important in keeping the members of a troop together. These troops may include up to 40 individuals. The territories of troops are probably announced and partly maintained by the loud vocalizations. Most cebids are gregarious, the most common social aggregation consisting of a

family group. Some species form unusually large troops; the squirrel monkey (*Saimiri*) occurs in bands of up to 100 animals.

Family Callithricidae.   Members of this family are called marmosets, and are small, often conspicuously marked primates that inhabit tropical forests from Panama and the northern part of South America south to southern Brazil. Four genera and 33 species are known. These are somewhat squirrel-like primates that vary in size from that of a mouse to that of a squirrel. Marmosets are seemingly more closely related to cebids than to any other primate group. The earliest record of marmosets is in the Upper Oligocene.

Marmosets are primarily arboreal, but do not grasp branches as do many other primates and never use brachiation (a mode of arboreal locomotion involving using the hands in swinging from branch to branch). Instead, they use their clawed hands and feet for climbing quadrupedally, and they bound through the trees in squirrel-like fashion. The marmoset skull has a short rostrum with prominent and forward-directed orbits (Fig.

7–6B). The dental formula is 2/2, 1/1, 3/3, 2/2 or 3/3 = 32 or 36. The medial incisors are chisel-like and frequently project forward; the upper molars are approximately triangular and usually have sharp cusps. The body and limbs are slender and the tail is always long. The hallux has a flat nail; all other digits bear laterally compressed claws. The hand resembles that of a squirrel and is slender, and the pollex is not opposable (Fig. 7–11B). Conspicuous ruffs or tufts of fur occur on the heads of some species, and can be erected.

Marmosets are omnivorous. The diet consists mostly of fruit and insects, and lizards and small birds and their eggs may be important foods for some species. Marmosets are social and typically live in small family groups; in some Neotropical areas they are the most common primates. These animals are often extremely vocal and emit a variety of high pitched sounds and piercing "alarm" calls, some of which resemble those of birds that live in the same areas.

Family Cercopithecidae. These are the Old World monkeys, and are the most successful primates in terms of numbers of species (60 Recent species of 11 genera). They occupy a wide range, including Gibraltar, northwest Africa, Africa south of the Sahara, southern Arabia, much of southeastern Asia east to Japan, Indonesia east to Timor, and the Philippine Islands. Among non-hominid primates, cercopithecids have the greatest tolerance for cold climates; some of these primates occupy high forests in Tibet, and others live in northern Honshu, Japan, where winter snows occur. Cercopithecids first appear in the Oligocene of Egypt, and like the living species, fossil forms are known only from the Old World. Just as in many other groups, some Pleistocene cercopithecids reached large sizes; an extinct South African baboon of this epoch reached the size of a gorilla.

In weight, cercopithecids range from 1.5 kg. to over 50 kg., and some species are stocky in build, quite unlike most cebids. In cercopithecids the nostrils are close together and face downward (Fig. 7–14), a condition termed catarrhine. The skull is often robust and heavily ridged, and, compared to cebids, the rostrum is long (particularly in the baboons). The dental formula is 2/2, 1/1, 2/2, 3/3 = 32, as in the apes (Pongidae) and man (Hominidae). The medial upper incisors are often broad and roughly spoon-shaped; the upper canines are usually large and in some species are tusklike, and when the jaws are closed the lower canine rests in a diastema between the upper canine and the last incisor. The first lower premolar is enlarged and forms a shearing blade that rides against the sharp posterior edge of the upper canine (Fig. 7–15A). Most of the molars have four cusps, the outer pair connected to the inner pair by two transverse ridges producing a bilophodont tooth (Fig. 7–13A, C); the last lower molar has an additional posterior cusp, the hypoconulid (Fig. 7–13C).

All of the digits have nails, and the pollex and hallux are opposable except in the strongly arboreal, leaf-eating genus *Colobus*, in which the pollex is vestigial or absent. The tail is vestigial in some species but long in others. Ischial callosities are well developed in many species, and the bare skin is frequently bright red. Bare facial skin may also be red, but is bright blue in the mandrill (*Papio sphinx*). These patches of skin are more brightly colored in the male than in the female in some species. The olfactory epithelium is greatly reduced in cercopithecids, and apparently their sense of smell is rudimentary. The facial muscles are well developed and produce a wide variety of facial expressions. Some cercopithecids are brightly or conspicuously marked. For example, the variegated langur of Indochina (*Pygathrix nemaeus*) has a bright yellow face, a chestnut strip beneath the ears, black and chestnut limbs, a gray body, and a white rump and tail.

Although most cercopithecids are probably largely omnivorous, some are

FIGURE 7–14. Two cercopithecid monkeys: A, Japanese macaques (*Macaca fuscata*); B, mona guenon (*Cercopithecus mona*). Note the dense fur of the macaques; these animals live in northern Honshu, Japan, where snow falls in the winter. (Photograph of *Macaca* by Carl B. Koford; photograph of *Cercopithecus* is a San Diego Zoo photo.)

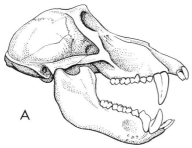
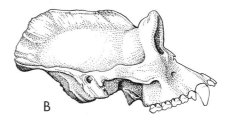

FIGURE 7-15. Skulls of anthropoid primates. A, a baboon (*Papio* sp., Cercopithecidae); length of skull 200 mm. B, a gorilla (*Gorilla gorilla,* Pongidae); length of skull 320 mm.

adapted to an herbivorous diet. Members of the subfamily Colobinae (the arboreal langurs and colobus monkeys) are herbivorous and frugivorous, and some species seem to feed primarily on leaves. The baboons (*Papio, Theropithecus*) have probably been the most successful terrestrial cercopithecids, and some species seldom climb trees even to seek shelter at night, preferring to assemble in large groups (up to 750 individuals) on cliffs (Kummer, 1968:310). Baboons live in social "troops" that may include up to 200 individuals, and each troop has a well developed dominance hierarchy. The remarkably complicated social behavior of baboons and of certain other Old World monkeys is reasonably well known (see Jay, 1968; De Vore, 1965). Interesting contrasts between the behavior of the baboons and the equally terrestrial patas monkey (*Cercopithecus patas*) are discussed by Hall (1968:114). The baboon is highly vocal, lives in fairly large troops controlled by several dominant males, and is prone to noisy, rough, and aggressive interactions. In contrast, the patas monkey usually maintains "adaptive silence," but has a repertoire of soft calls, lives in small troops, each with a single adult male that serves as a watchdog, is rarely aggressive, and never fights. The patas monkey has a slim greyhound-like build and is the fastest runner of all primates, having been timed at a speed of 55 km./hr. Adaptations for speed in this animal include elongation of the limbs and the carpals and tarsals, shortening of the

digits, reduction of the pollex and hallux, and the development of palmar and plantar pads. This remarkably cursorial primate has a quiet mode of life, usually attempts to escape detection, and depends on its speed to escape danger. In these respects the patas monkey is the primate counterpart of the small antelope. The noisy and considerably less cursorial baboon troop, however, frequently depends on its aggressive dominant males to confront and discourage a predator, and terrestrial locomotion is relatively unimportant as a means of escaping enemies.

Sexual dimorphism is pronounced in both the baboons and the patas monkey, as it is in many primates. The male baboon weighs roughly 33 kg., the female 16.5 kg.; the male patas monkey averages 13 kg., and the female, 6.5 kg. (Hall, 1968:114). Probably all cercopithecids are basically social, and vocalizations and facial expressions play central roles in social interactions. The life span of these monkeys is long: a Chacma baboon (*Papio ursinus*) lived in captivity for 45 years, and life spans of 20 or 25 years in the wild may be common (Walker, 1968:447).

Family Pongidae. The gibbons and great apes are included in this family, members of which occur in equatorial Africa, Southeast Asia, Java, Borneo, Sumatra, and the Mentawi Islands. Eight Recent species of four genera are known; all species are restricted to tropical forests. The pongid record begins in the Oligocene of Egypt, when the gibbon-like genus *Pro-*

pliopithecus appears. Judging from the fossil record, pongids evolved in Africa, reached Europe in the Miocene and Asia in the Pliocene, and never occurred in the New World.

The family Pongidae includes two subfamilies that differ markedly in structure and in modes of life. The gibbons, subfamily Hylobatinae, are arboreal and are the most rapid and spectacular climbers and brachiators of all mammals. Gibbons are relatively small, weighing from 5 to 13 kg., have remarkably long arms, and use the hands like hooks rather than as grasping structures during brachiation. The great apes, subfamily Ponginae, vary from 48 to 270 kg. (nearly 600 lb.!) in weight, and have robust bodies and powerful arms. The hands and feet are similar to those of man, but the hallux is opposable (Fig. 7–11D). The pongid skull is typically robust, and in older animals is marked by bony crests and ridges; it is long relative to its width (Fig. 7–15B). The teeth are large. The dental formula is 2/2, 1/1, 2/2, 3/3 = 32, as in cercopithecids and man. The incisors are broad and the premaxillae and anterior parts of the dentaries are broadened to accommodate them. The canines are large and stoutly built, but are never tusklike. The upper molars are quadrangular and basically four-cusped, and the lower molars have an additional posterior cusp (hypoconulid). In contrast to cercopithecids, a trend toward elongation of the molars does not occur in pongids, and the molars lack well defined cross ridges (Fig. 7–13D). The tooth rows are parallel and the mandibular symphysis is braced by a bony shelf (the "simian shelf"). The forelimbs are longer than the hind limbs and the hands are longer than the feet; all digits bear nails. Pongids have no tails. The thorax is wide and the scapula has an elongate vertebral border. Adaptations allowing advantageous muscle attachments during erect or semierect stances include lengthening of the pelvis and enlargement and lateral flaring of the ilium.

Regarding structural details, locomotor ability, brain size, and level of intelligence, the great apes are closer to man than are any other mammals.

Pongids are largely vegetarians, but some are occasionally carnivorous. The chimpanzee (*Chimpansee*), for example, occasionally catches and eats the colobus monkey. Arboreal locomotion in pongids involves brachiation in some species. The gorilla (*Gorilla*) and chimpanzee are mostly terrestrial, and although capable of bipedal stance and limited bipedal locomotion, are mostly quadrupedal. The behavior of pongids and cercopithecids has been studied intensively in recent years. Good general references on primate behavior include De Vore (1965) and Jay (1968). Owing primarily to the efforts of man, who acts almost as if his survival and prosperity were threatened by his next of kin, some of the great apes are dangerously close to extinction. Destruction of habitat and killing of the animals themselves, fostered by man's anachronistic feeling that his position as the dominant form of life justifies any form of exploitation of his environment, has led to serious reductions of some primate populations. Perhaps no more than 2,500 wild orangutans (*Pongo pygmaeus*) survive today; these are restricted to parts of the islands of Sumatra and Borneo. As a sad stroke of irony, an important drain on the declining populations of orangs resulted from their capture and exportation to European and American zoos, institutions dedicated in part to the preservation of vanishing species.

Family Hominidae. Man is the only living member of the family Hominidae. In man the skull has a greatly inflated cranium, housing a large cerebrum, and the rostral part of the skull is virtually absent. The foramen magnum is beneath the skull, a feature associated with an upright stance. The dentition is not as robust as in the pongids: the incisors are less broad, the canines typically rise but slightly above adjacent teeth, and the cheek teeth are less heavily built. The premolars are usually bicuspid. The upper molars have four cusps; the first lower molar has five cusps, the second has four, and the third has five. The dental formula, 2/2, 1/1, 2/2, 3/3 = 32, occurs in most individuals, but one or more of the posterior molars (the "wisdom teeth") may not appear. The tooth rows are not parallel, as they are in pongids, nor is the simian shelf present in the mandible. The pollex, but not the hallux, is opposable. With a change in the posture and use of the forelimbs, the thorax has become broad and the scapulae have come to lie dorsal to the ribcage, as in bats, rather than lateral to the ribcage, as in most mammals. As in the case of many primates, in humans the males are considerably larger than the females. If the same standards used in the classification of other mammalian orders were applied to the primates, man would be included with the gibbons and great apes in a single family. "Differences of no greater magnitude than those separating the hominids and the pongids characterize subfamilies in some other orders of mammals" (Anderson, 1967:177). The fossil record of hominids can be traced either to the Miocene or to the Pleistocene, depending upon the authority followed.

The astounding growth of the populations of man throughout much of the world is leading to progressively more acute problems that threaten, if not the very survival of man, at least his present style of life. It is obviously essential that men of different races learn to accept and appreciate living and working together as equals, and that effective birth control measures be followed in most parts of the world in order to halt population growth. Man is clearly creating an environment in which he is ill suited, both psychologically and physically, to live. In his attempt to improve his lot, man is actually changing his environment in ways that make it less able to support life over a long period. Our present exploitation and modification of the environment cannot long continue. The adaptive ability of man is being put to more critical tests today than ever before; the future of man may well be determined by choices he makes within the next few decades.

CHAPTER 8

EDENTATES; SCALY ANTEATERS; THE AARDVARK

The orders of mammals considered in this chapter (Edentata, Pholidota, Tubulidentata) are not brought together because they share a common ancestry; indeed, they are probably rather distantly related to one another. They also seem to be isolated phylogenetically from other orders of mammals. But as a result of convergence (not closeness of relationship!) there are some striking structural similarities between these three orders, one of the most obvious being the loss or reduction of teeth. Primarily as a matter of convenience, these aberrant and unrelated orders are discussed in the same chapter.

ORDER EDENTATA

Although the edentates are not of great importance today (the order Edentata contains but 14 living genera and 31 species), they are remarkably interesting animals because of their unique structure, their large and bizarre fossil types, and because of the probable influence of the Tertiary sep-

aration of South America and North America on their evolution.

The living members of the order Edentata share a series of distinctive morphological features. Extra zygapophysis-like (xenarthrous) articulations (Fig. 8–1) brace the lumbar vertebrae; the incisors and canines are absent; the cheek teeth, when present, lack enamel, and each has a single root. The tympanic bone is annular; the brain is small and the braincase is usually long and cylindrical; the cora-

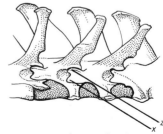

FIGURE 8–1. Three lumbar vertebrae (viewed from the left; anterior is to the left) of the nine-banded armadillo (*Dasypus novemcinctus*), showing the "xenarthrous" articulations (x) supplementing the normal articulations between zygapophyses (z).

A   B

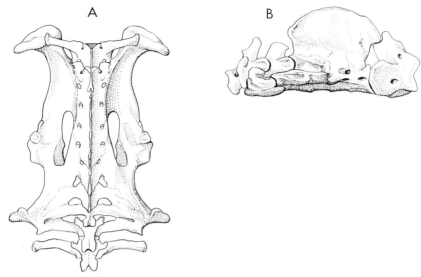

FIGURE 8–2.   Parts of the skeleton of the nine-banded armadillo (*Dasypus novemcinctus*): A, dorsal view of the pelvic girdle, showing the great degree of fusion of vertebrae with the ilium and ischium; B, lateral view of the cervical vertebrae (anterior is to the right). The axis and cervicals three, four, and five are fused.

coid process is unusually well developed and the clavicle is present; the ischium is variously expanded and specialized (Fig. 8–2) and usually forms an ischio-caudal, as well as an ischio-sacral, symphysis. The hind foot is typically five-toed, and the forefoot has two or three predominant toes with large claws. Major edentate structural trends are toward reduction and simplification of the dentition, specialization of the limbs for such functions as digging and climbing, and rigidity of the axial skeleton. Perhaps the use of the powerful forelimbs when the animals are in the bipedal posture, which is adopted frequently during feeding or defense, has influenced the evolution of the rigid axial skeleton and the bracing of the pelvis.

Edentates, strictly a New World group, probably originated in North America, where members of the primitive suborder Palaeanodonta are known from the Paleocene. Palaeanodonts were probably descendants of the stock that gave rise to the more specialized suborder of edentates, the Xenarthra, and disappeared from the fossil record of North America in the

Oligocene, apparently without reaching South America. Edentates probably migrated to South America in the late Cretaceous. The earliest fossils from this area are from the Paleocene, and are members of the family Dasypodidae (armadillos), the most successful surviving family. A diverse array of armadillos developed in South America; one Pleistocene type was as large as a rhinoceros. The early dasypodids were armored with ossified dermal scutes, as are all modern species, and perhaps are the basal group from which other xenarthrans radiated.

North and South America were seemingly separated in the Cretaceous, and were not joined again until late Pliocene. During the early Tertiary, the same interval when palaeanodonts disappeared from North America, xenarthrans underwent a radiation in South America free from competition with the other mammalian orders that were becoming established in North America. Several evolutionary lines of xenarthrans arose. The Glyptodontidae represent one line that probably evolved from dasypodid stock, and appeared in late Eocene.

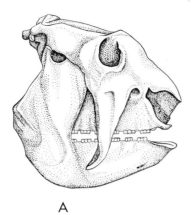

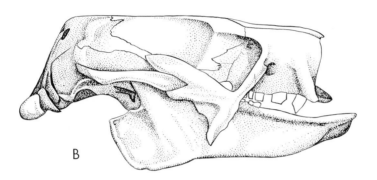

A

FIGURE 8–3. Skulls of extinct edentates: A, *Glyptodon*, length of skull 560 mm.; B, *Paramylodon*, length of skull 510 mm. (After Romer, A. S.: *Vertebrate Paleontology*, 3rd ed., The University of Chicago Press, 1966.)

B

These large creatures, some of which were nine feet long, had unusual deep skulls (Fig. 8–3A). Many of the unique structural features of the glyptodonts are associated with their development of a turtle-like carapace composed of many fused polygonal scales. The limbs are distinctive and highly specialized, and the last two thoracic vertebrae and the lumbar and sacral vertebrae are fused into a massive arch that, together with the ilium, support the carapace.

Additional evolutionary lines are represented by various ground sloths of several extinct families. Ground sloths first appear in South American Oligocene deposits. These animals were herbivores; their teeth lacked enamel and were ever-growing. The family Megatheriidae appears in the Oligocene of South America, and includes *Megatherium*, a massive ground sloth

larger than an elephant, as well as smaller types four to eight feet in length. These animals were covered with hair, lacked the upper canine-like tooth, and walked on the outer edges of the unusually specialized hind feet. A second family of ground sloths, the Megalonychidae, is closely related to the megatheriids, but differs from that group in having the anterior-most cheek teeth modified into "canines." *Megalonyx*, a Pleistocene genus that reached the size of a cow, was widely distributed in North America. The remains of smaller species of megalonychids have been found in the West Indies in association with human artifacts. These sloths seemingly survived into the Recent. A third family of ground sloths, the Myodontidae, appeared in the Oligocene, and is characterized in part by the development of upper "canines" (Fig. 8–3B) and re-

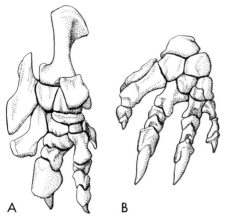

FIGURE 8-4. The right pes of *Nothrotherium* (A) and the right manus of *Paramylodon* (B). (After Romer, A. S.: *Vertebrate Paleontology*, 3rd ed. The University of Chicago Press, 1966.)

markably robust limbs. Protection was afforded some members of this group by round dermal ossicles embedded in the presumably thick skin. A trend toward large size is apparent in the mylodonts, as in other ground sloths.

The glyptodonts, megatheriids, megalonychids, and mylodonts underwent much of their evolution in the Tertiary in isolation from the progressive North American mammalian fauna. Nevertheless, when the land bridge between the Americas was re-established in the Pliocene, the glyptodonts and ground sloths were remarkably successful in invading North America. These edentates did not suffer mass extinctions, as did the South American marsupials, as a result of competition with invading placentals. The plains-dwelling mylodont *Paramylodon* was widespread in North America and is the most common edentate in the Pleistocene deposits of Rancho La Brea in Los Angeles. This edentate had large claws on digits two and three (Fig. 8-4B), an arrangement similar to that in the living armadillo (Fig. 8-5B). The common megatheriid *Nothrotherium* occurred in North America and South America in the Pleistocene, and its remains from Gypsum Cave in Nevada include bones, skin, and hair. *Nothrotherium* probably walked on the sides of its highly modified hind feet (Fig. 8-4A), as did other large ground sloths. Dung of this animal, preserved in caves in Nevada and Arizona, contains yucca. This ground sloth persisted into the

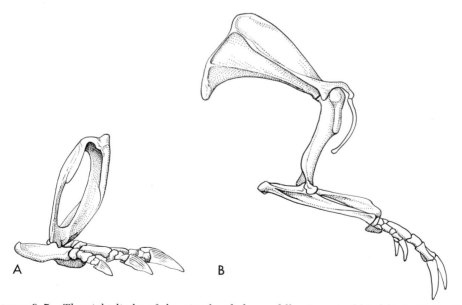

FIGURE 8-5. The right limbs of the nine-banded armadillo: A, part of hind limb; B, forelimb. The flattening of the bones of the forearm and shank increases the surface area for attachment of muscles, and the elongation of the olecranon and the calcaneum give added mechanical advantage for power to the muscles that insert on them.

Recent in arid parts of the southwestern United States, and probably became extinct several thousand years ago.

Highly developed protective devices (as in Dasypodidae) and narrow, specialized feeding habits (Bradypodidae and Myrmecophagidae) have perhaps been important in allowing edentates to survive under competitive pressure from more "advanced" eutherians. It is probably significant that the edentate stronghold is in tropical areas; these regions often provide last refuges for declining groups that were formerly more widespread and abundant.

Family Myrmecophagidae. Members of this group, the anteaters, are highly specialized for feeding on ants and termites. Anteaters occur in tropical forests of Central and South America south to Argentina, and are represented by four Recent species of three genera. Although not important in terms of numbers of species, anteaters are often common in suitable habitats and appear to "own" their narrow feeding niche in the Neotropics.

The most obvious structural features of anteaters are associated with their ability to capture insects, to dig into or tear apart insect nests, and to climb. The skull is long and roughly cylindrical (Fig. 8–6A), the zygoma are incomplete, and the long rostrum contains complex, double-rolled turbinals. Teeth are absent, and the dentary is long and delicate. The long, vermiform tongue is protrusile, and is covered with sticky saliva secreted by the fused submaxillary and parotid salivary glands. These glands are situated in the neck and are enormously enlarged. The forelimbs are powerfully built; the third digit is enlarged and bears a stout, recurved claw, and the remaining digits are reduced. The giant anteater (*Myrmecophaga*) walks on its knuckles with its toes partly flexed, whereas the other anteaters (*Cyclopes* and *Tamandua*), which are fully or partly arboreal, walk on the side of the manus with the claws toed inward. It is

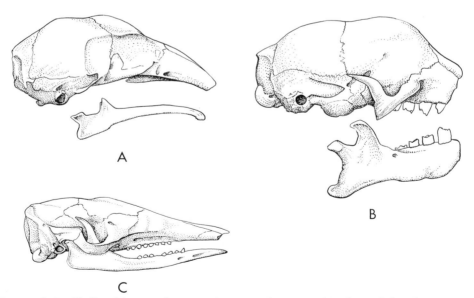

FIGURE 8–6. Skulls of living edentates: A, two-toed anteater (*Cyclopes didactylus*, Myrmecophagidae; length of skull 46 mm.); B, three-toed sloth (*Bradypus griseus*, Bradypodidae; length of skull 76 mm.); C, nine banded armadillo (*Dasypus novemcinctus*, Dasypodidae; length of skull 95 mm.). (*Cyclopes* and *Bradypus* after Hall, E. R., and Kelson, K. R.: *The Mammals of North America*, copyright © 1959 by The Ronald Press Company, New York.)

of interest to observe that the unusual posture of the manus of *Myrmecophaga* during terrestrial locomotion is essentially the same as that used by *Nothrotherium,* an extinct megatheriid ground sloth. In anteaters the pes has four or five clawed digits and has a plantigrade posture. The size of anteaters ranges from that of a squirrel (in *Cyclopes;* 350 g.) to that of a large dog (in *Myrmecophaga;* 25 kg.). *Myrmecophaga* is covered with long, coarse fur, and the non-prehensile tail has long hairs that hang downward, but in the other anteaters, the fur on the body and tail is shorter and the tail is prehensile.

Anteaters use the powerful forelimbs to expose ants and termites by tearing apart their nests; the insects are captured by the long tongue, are swallowed whole, and are ground up by the thickened pyloric portion of the stomach. All of the anteaters pursue a slow and fairly awkward type of terrestrial locomotion. In the two genera that climb trees, the claws of the manus are used as grappling hooks or are used for grasping as the animals travel along branches in a hand-over-hand fashion. *Cyclopes* is nocturnal and entirely arboreal, and forages for insects high in the trees. The two species of *Tamandua* are largely arboreal but are terrestrial to some extent, and are mostly nocturnal. *Myrmecophaga* is entirely terrestrial and seems largely diurnal. The defensive behavior of these animals is unusual. When threatened, anteaters stand on the hind limbs and use the tail or the back braced against a support to form a solid tripod; they slash or grasp at an enemy with the claws of the manus. The power of the forelimbs and the unusual size of the claws provide these animals with a formidable defense. Of particular interest is the fact that, as in the Recent anteaters, the large members of the extinct families Mylodontidae, Megatheriidae, and Megalonychidae had large claws on the strong forelimbs; the hind limbs and pelvic girdles were specialized in directions indicating considerable bipedal ability; and the tail was

sturdy and may have functioned as a brace during bipedal stance. These large edentates may well have defended themselves against saber-tooth cats (*Smilodon*) and other large Pleistocene predators much as Recent anteaters protect themselves from today's less imposing predators.

Family Bradypodidae. The tree sloths are strange animals (see Fig. 8–7) that are so highly modified for a specialized form of arboreal locomotion that they have nearly lost the ability to move on the ground. The six Recent species of tree sloths belong to two genera, and range from Central America (Honduras) through the northern half of South America to northern Argentina. These animals primarily occupy tropical rain forest.

The adaptive zone of tree sloths is quite different from that of the anteaters and involves strictly arboreal habits and an entirely herbivorous diet. The bradypodids differ strongly from the myrmecophagids, especially in skull characteristics. The tree sloth skull is short and fairly high, with a strongly reduced rostrum. The zygomatic arch is robust but incomplete,

FIGURE 8–7.   A two-toed tree sloth (*Choloepus hoffmanni*). (San Diego Zoo photo.)

and its jugal portion bears a ventrally projecting jugal process similar to that present in many extinct edentates (Fig. 8–6B). The premaxillaries are greatly reduced and the turbinals are complexly rolled, as in myrmecophagids. Five maxillary and four or five mandibular teeth are present; the anteriormost teeth in *Choloepus* are caniniform and are kept sharp by abrasion between the posterior surface of the upper tooth and the anterior surface of the lower tooth. The persistently growing teeth are roughly cylindrical, and have central cores of soft dentine surrounded successively by hard dentine and cement. A departure from the usual mammalian pattern of seven cervical vertebrae occurs in the bradypodids, in which from six to nine occur, the number differing between species and in some cases even between individuals of the same species. Xenarthrism (cf. Fig. 8–1) is strongly developed in the thoracic and lumbar vertebrae, and as in some extinct ground sloths the coracoid and acromion processes of the scapula are united. The externally visible digits do not exceed three in number and, except for the long and laterally compressed claws, are syndactylous (bound together). Tree sloths are not large, weighing from 4 to 7 kg., and are covered with long, coarse hair. This fur provides a habitat for algae, which grow in the flutings on the surfaces of the hairs during the rainy season and tint the fur green. In addition, the adults of two genera of moths (*Bradypodicola* and *Cryptoses; Pyralididae, Microlepidoptera*) hide in large numbers in the dense pelage. In tree sloths the tail is rudimentary or is short.

These remarkably specialized animals are strictly herbivorous, eating primarily leaves and seldom descending to the ground. Climbing is done in an upright position by embracing a branch, or by hanging upside down and moving along hand-over-hand. Tree sloths spend considerable time hanging upside down, and at times sleep in this position. In both genera

the forelimbs are greatly elongate and are considerably longer than the hind limbs, and when the animals are on the ground the limbs splay out to the side. The animals are unable to support their weight on their limbs and progress on the ground by slowly dragging their bodies forward with the forelimbs. In some tropical basins, where the ground is inundated during part of the rainy season, tree sloths disperse by swimming.

Although tree sloths occupy tropical environments, in which temperature extremes virtually never occur, they possess some adaptations typical of animals that are subjected to cold stress in boreal areas. In tree sloths, and in anteaters and armadillos, the limbs have retia mirabilia (see page 120). By performing counter-current heat exchange, these specialized vascular bundles allow the limbs, which are unusually long in the case of tree sloths, to become considerably cooler than the body without causing a serious cooling of the body or an unusual metabolic drain. In addition, tree sloths are insulated by long, fairly dense pelage, consisting of long guard hairs and short, fine-textured underfur, a type of pelage characteristic of inhabitants of cold climates. What is the adaptive importance to tropical animals of such specializations for the retention of body heat? Apparently edentates, and perhaps tree sloths especially, are extremely sensitive to cold. Tree sloths are probably heterothermic, with body temperatures varying from 28° to 35° C; when they are inactive, they have difficulty maintaining a constant body temperature. Tree sloths have been observed to shiver at ambient temperatures as high as 27° C (80° F)! The specializations in tree sloths for the retention of body heat, therefore, enable these extremely cold-sensitive animals to maintain non-lethal temperatures in the body during cool Neotropical nights. It is not surprising that tree sloths range northward no further than the tropical forests of Honduras.

Family Dasypodidae. Members of

FIGURE 8–8. A naked-tailed armadillo (*Cabassous centralis*). (Photograph by Lloyd G. Ingles.)

this family, which include all of the armadillos, are remarkable for the protective armor that occurs in all species. In terms of numbers of species and breadth of distribution, this is the most successful edentate family. Twenty-one Recent species of armadillos of nine genera are known, and the geographic range includes Florida (by introduction), much of the south-central United States as far north as Kansas, Mexico, Central America, and nearly all of South America to near the southern end of Argentina. Armadillos are common in many areas and occupy tropical forests and savannas, semi-deserts, and temperate plains and forests.

The most obvious and unique structural feature of armadillos is the jointed armor. This consists of bony scutes covered by horny epidermis (Fig. 8–8). The scutes form plates that occur in a variety of patterns, but always present is a head shield and a series of plates protecting the neck and much of the body. Sparse hair usually occurs on the flexible skin between the plates and on the limbs and the ventral surface of the body. Individuals of some species can curl into a ball so that their limbs and vulnerable ventral surfaces are largely protected by the armor. The largest species, the giant armadillo (*Priodontes giganteus*),

weighs up to 60 kg.; the smallest, the pigmy armadillo (*Chlamyphorus truncatus*), is roughly the size of a small rat (120 g.).

The skull is often elongate and is dorsoventrally flattened; the zygomatic arch is complete and the mandible is slim and elongate (Fig. 8–6C). The teeth are borne only on the maxillary (except in one species), are nearly cylindrical, and vary from 7-9/7-9 to 25/25. The teeth are frequently partially lost with advancing age. The axial skeleton is fairly rigid and is partially braced against the carapace: the second and third cervical vertebrae, and in four species other cervicals as well, are fused (Fig. 8–2B); 8 to 13 sacral and caudal vertebrae form an extremely powerfully braced anchor for the pelvis (Fig. 8–2A); xenarthral articulations between thoracic and lumbar vertebrae (Fig. 8–1) produce a rigid vertebral column; and elongate metapophyses from the lumbar vertebrae brace but do not contact the carapace. Bracing of the carapace by the skeleton is remarkably extensive in some species, but the skeleton does not actually contact the carapace. For example, in the familiar nine-banded armadillo (*Dasypus novemcinctus*), the species that enters the United States, the carapace is partially supported by prominent dorsolateral processes from the ilium and ischium (Fig. 8–2A), and by the modified tips of the neural spines of all the thoracic and lumbar vertebrae. In this species the powerfully developed panniculus carnosus muscles and a broad ligament from the expanded dorsolateral flange of the ilium attach to the inner surface of the carapace and bind it to the body, but no bony contact is made between the carapace and the skeleton. In this species the ribcage is made rigid by a broadening of the ribs, by heavy intercostal muscles, and by ossification of the greatly enlarged parts of the ribs that in most mammals are the costal cartilages.

In all armadillos the limbs are powerfully built, and the fore and hind feet bear large, heavy claws (Fig. 8–5). The feet are five-toed in all but one genus, and the foot posture is usually plantigrade. The tibia and fibula are fused proximally and distally and are highly modified for giving origin to powerful muscles of the shank (Fig. 8–5A). The femur has a prominent third trochanter. Retia mirabilia occur in the limbs.

Compared to other edentates, armadillos are more generalized in their feeding and their locomotion. Most species feed primarily on insects, but a variety of invertebrates, small vertebrates, and vegetable material is also taken. All armadillos are at least partly adapted for digging, and some species are highly fossorial. In some species the forelimbs are unusually powerful and the manus is somewhat like that of a mole. Such a fossorial creature is the pigmy armadillo (Chlamyphorus truncatus), which utilizes a style of digging seemingly unique among mammals. The soil is dug away and pushed beneath the animal by the long claws of the forepaws, and the hind feet rake the soil behind the animal. The pelvic scute is then used to pack the soil behind the body. During the packing the front limbs push the animal backward and the hind quarters vibrate rapidly from side to side (Rood; 1970). No permanent burrow is formed. Running is limited in armadillos by a

structural pattern that achieves power but no speed; some species can be run down by a man. The "pichi" (Zaedyus pichiy) was observed in southern Argentina by Simpson (1965:201), who found this armadillo to depend on its sense of smell. A captive individual "seemed to pay practically no attention to anything she saw or heard but lived by and for her nose." The pichi cannot curl into a ball, and escapes from enemies either by burrowing beneath a thorn bush or by clutching a solid surface with the claws and pulling the carapace down against the ground.

As indicated by their wide range, armadillos are more resistant to cold than are tree sloths, but the possession of retia mirabilia in the limbs suggests that armadillos share the general edentate sensitivity to cold. As a further indication, armadillos in the northern parts of their range have been found to suffer as high as 80 per cent winter mortality when prolonged cold spells occurred (Fitch, Goodrum, and Newman, 1952). Despite such mortality, armadillos have extended their range into the south-central United States in the last 90 to 100 years (see Fig. 16–14, p. 319). They were formerly restricted to extreme southern Texas.

ORDER PHOLIDOTA

The pangolins, or scaly anteaters, comprise a single family, Manidae, and are represented today by a single genus (Manis) with eight species. Pangolins occur in tropical and subtropical parts of the southern half of Africa and in much of southeast Asia. Their fossil record is poor, and while it documents the probable Oligocene occurrence of these animals in Europe, it contributes little to our knowledge of the evolution or relationships of the group. The paucity of fossils, and the limited number of Recent forms, suggest that pangolins have never been abundant or diverse.

Pangolins are strange-looking creatures that at a glance seem more repti-

FIGURE 8–9. A pangolin (*Manis gigantea*, Pholidota). (Photograph courtesy of the American Museum of Natural History.)

lian than mammalian (Fig. 8–9). They are of moderate size, weighing from about 5 to 25 kg. Although pangolins and South American anteaters (Myrmecophagidae) are not closely related, they share some anatomical features associated with their insect-eating habits. The skull of the pangolin is conical and lacks teeth; the dentaries are small and lack angular and coronoid processes; the tongue is extremely long and vermiform; some muscles of the tongue are unusually long, probably in response to the specialized use of the tongue in feeding, and originate on the xiphisternum rather than on the hyoid. The scales are the most distinctive feature; these cover the dorsal surface of the body and the tail and are composed of agglutinated hair. The manus and pes have long, recurved claws; the pes has five toes and the manus is functionally tridactyl. The walls of the pyloric part of the stomach are thickened. This part of the stomach usually contains small pebbles and seems to function to grind food as does the gizzard of a bird.

The food of pangolins is mostly termites, but ants and other insects are also taken. The pangolin rolls into a ball when disturbed, erects the scales, flails the tail, and moves the large, sharp-edged scales in a cutting motion. Foul-smelling fluid is sprayed from the anal glands of some species as further protection. Pangolins are fairly awkward on the ground, but some are capable climbers.

ORDER TUBULIDENTATA

This order, the aardvarks, includes but one family (Orycteropodidae). The one Recent species (*Orycteropus afer*) occupies Africa south of the Sahara Desert. The earliest record of tubulidentates is from the Miocene of Africa, and in the Pliocene an extinct member of the Recent genus occupied parts of Europe and Asia. Skeletal similarities between the two groups indicate that tubulidentates may have evolved from condylarths (primitive ungulates; order Condylarthra).

Many structural parallels occur between tubulidentates and certain edentates, not because the groups are closely related, but because of similarities in habits. Both tubulidentates and some edentates are powerful diggers and feed on ants and termites. The aardvark weighs up to roughly 82 kg., and the thick, sparsely haired skin provides protection from insect bites.

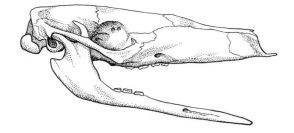

FIGURE 8–10. The skull of the aardvark (*Orycteropus afer*). (After Hatt, 1934.)

The skull is elongate, and the dentary is long and slender (Fig. 8–10). In the adult dentition, incisors and canines are lacking; the cheek teeth are 2/2 premolars and 3/2 molars. Each tooth is rootless and consists of many (up to nearly 1,500) hexagonal prisms of dentine, each surrounding a slender, tubular pulp cavity. The columnar teeth lack enamel but are surrounded by cement. The anteriormost teeth erupt first, and are often lost before the posterior molars are fully erupted. The slender tongue is protrusile. Olfaction is used in finding insects; the olfactory centers of the brain are unusually well developed, and the turbinal bones are remarkably large and complex. The pollex is absent, and the hind foot is five-toed; the robust claws are flattened and blunt.

The powerful forelimbs are used in burrowing and in dismantling termite and ant nests, and the hind limbs thrust accumulated soil from the burrow. Although the foot posture is digitigrade, aardvarks are slow runners and can be run down by a man. Burrows dug by aardvarks are numerous in some areas: Walker (1968:1318) reports some 60 burrow entrances in an area 100 m. by 300 m. These burrows are used as retreats by a variety of mammals, including the warthog (*Phacochoerus africanus*).

CHAPTER 9

ORDER LAGOMORPHA

Although lagomorphs, the rabbits (Leporidae) and pikas (Ochotonidae), are not a diverse group, including but ten genera with 63 Recent species, they are important members of many terrestrial communities and are nearly world-wide in distribution. Considering large land masses, lagomorphs were absent only from the Australian region and from southern South America before recent introductions by man. Lagomorphs occupy diverse terrestrial habitats from the arctic to the tropics, and in many temperate and boreal regions rabbits are subject to striking population cycles marked by periods of great abundance alternating with times of extreme scarcity. In such regions, population cycles of many carnivores are influenced strongly by changes in population densities of rabbits.

Many important diagnostic features of Recent lagomorphs are related to their herbivorous habits and, in the case of leporids, to their cursorial locomotion. Lagomorphs have fenestrated skulls, a feature highly developed in some leporids (Fig. 9–1A). The anterior dentition resembles that of a rodent, but whereas rodents have 1/1 incisors, rabbits have 2/1 incisors; the second incisor is small and peglike, and lies immediately posterior to the

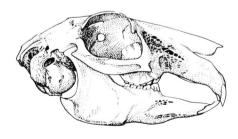

A   B

FIGURE 9–1. A, skull of the antelope jackrabbit (*Lepus alleni*); note the highly fenestrated maxillary and occipital bones. B, anterior part of the skull of the arctic hare (*L. arcticus*); note the procumbent incisors and the receding nasals. These specializations are associated with this animal's habit of using the incisors to scrape away ice and snow to reach food. (*L. arcticus* after Hall, E. R., and Kelson, K. R.: *The Mammals of North America,* copyright © 1959 by The Ronald Press Company, New York.)

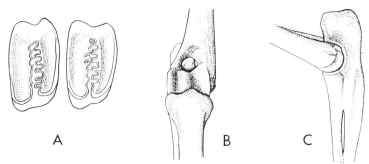

FIGURE 9–2. Leporid structural features. (The drawings are of the antelope jackrabbit, *Lepus alleni.*) A, occlusal view of upper right premolars three and four. B, anterior view of the right elbow joint; movement is limited to a single (anteroposterior) plane by this "tongue and groove" articulation. When full extension of the forearm is reached, a process on the olecranon of the ulna locks into the conspicuous hole in the humerus and braces the joint. C, medial view of the right elbow, showing the tight fit between the articular surface of the humerus and the articular surfaces of the radius and ulna. The radius and ulna are partially fused.

first (Fig. 9–1A). As in rodents, the lagomorph incisors are ever-growing. A long post-incisor diastema is present in lagomorphs, and the canines are absent. The cheek teeth are hypsodont and rootless, and the crown pattern features transverse ridges and basins (Figs. 9–2A and 9–3B). The distance between the upper tooth rows is greater than that between the lower tooth rows, allowing occlusion of upper and lower cheek teeth only on one side at a time and requiring a lateral or oblique jaw action. The masseter muscle is the primary muscle of mastication; the temporalis is small and the coronoid process, its point of insertion, is rudimentary. The clavicle is either well developed (Ochotonidae) or rudimentary (Leporidae), and the elbow joint limits movement to a single (anteroposterior) plane (Fig.

9–2B, C). The tibia and fibula are fused distally; the front foot has five digits and the hind foot has four or five digits. The soles of the feet, except for the distalmost toe pads in *Ochotona*, are covered with hair. The foot posture is digitigrade during running but is plantigrade during slow movement. The tail is short, and in *Ochotona* is not externally evident.

The lagomorphs are a group of mammals with no clear relationships to other Recent eutherian orders. The first fossil record of mammals with lagomorph-like characters is from the late Paleocene of Asia, when the primitive family Eurymylidae appears. (But eurymylids may not be lagomorphs, for they have such non-lagomorph features as 1/1 incisors.) Although rodents and rabbits were once thought to be closely related, the fos-

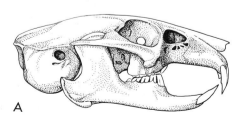 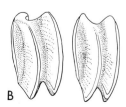

FIGURE 9–3. Skull (left) and occlusal view of the upper right fourth premolar and first molar (right) of the pika (*Ochotona princeps*).

sil evidence indicates no such relationship. It appears, rather, on the basis of cusp patterns, that the earliest rodents (Paramyidae) and the earliest lagomorphs were derived from separate eutherian ancestry. The family Leporidae probably originated in Asia, but underwent most of its early (Oligocene and Miocene) evolution in North America. Leporids became well established in the Old World in the Pliocene, and the advanced subfamily Leporinae arose there. The pikas appeared first in the Oligocene of Eurasia, and spread in the Pliocene to Europe and to North America. The Recent genus *Ochotona* is known from the early Pliocene. In contrast to the leporids, which have remained widespread since the Pliocene, the ochotonids reached their greatest diversity and widest distribution in the Miocene, when they occupied Europe, Asia, Africa, and North America (Dawson, 1967:305), and have declined since. In North America pikas are now of local occurrence in high mountains north of Mexico. They occur more widely in the Old World, where they inhabit eastern Europe and much of northern and central Asia. The environmental tolerances, or perhaps the competitive success, of ochotonids has seemingly changed in the New World since the Miocene. For example, in the Miocene an ochotonid occupied riparian communities in the Great Plains, Rocky Mountain, and Great Basin regions of North America (Wilson, 1960:7); riparian situations in these areas no longer support pikas. Factors influencing the striking post-Miocene decline of the ochotonids are poorly understood.

Why, although they are an old and thriving group, have lagomorphs not undergone a greater adaptive radiation? Perhaps their conservatism is related to the limitations of their functional position as "miniature ungulates." Competition with members of the larger and more diverse order Artiodactyla, a group highly adapted to an herbivorous diet and to cursorial locomotion, may have limited lagomorphs to the exploitation of but a single, limited adaptive zone, although this zone was occupied with great success over broad areas.

Family Ochotonidae. The pikas are represented today by one genus with roughly 14 species. Pikas are less progressive with regard to cursorial adaptations than are the rabbits, and they usually venture only short distances from shelter. Pikas occur in the mountains of the western United States and south-central Alaska, and occur over a wide area in the Old World including eastern Europe and much of Asia south to northern Iran, Pakistan, India, and Burma.

In contrast to rabbits, pikas are small, about 100 to 150 gm.; they have short, rounded ears, short limbs, and no externally visible tail (Fig. 9–4). The ear opening is guarded by large valvular flaps of skin that may protect the ear opening during severe weather. The skull is strongly constricted between the orbits and lacks a supraorbital process; the rostrum is short and narrow. The skull is less strongly arched in ochotonids than in leporids (Fig. 9–3A), and the angle between the basicranial and palatal axes is lower. The maxilla has a large fenestra. The dental formula is 2/1, 0/0, 3/2, 2/3 = 26. The third lower premolar has more than one re-entrant angle, and the re-entrant enamel ridges of the upper cheek teeth are straight (Fig. 9–3B). The anal and genital openings are enclosed by a common sphincter, and males have no scrotum.

In North America, pikas typically occupy talus slopes in the high mountains. These herbivorous animals are characteristic of boreal or alpine situations, and they occur from near sea level in Alaska to the treeless tops of some of the highest peaks in the Rocky Mountains and Sierra Nevada-Cascade chain. In North America, habitat requirements usually include fairly extensive areas of large rocks or irregular boulders adjacent to growths of forbs and grasses. When frightened, pikas seek shelter in the labyrinth of spaces and crevices between rocks, and sel-

FIGURE 9-4. A pika (*Ochotona princeps*) at a lookout point on a rock, with a mouthful of plant material for its hay pile. (Photograph by O. D. Markham.)

dom forage far from such shelter. Large "hay piles" are built each summer in the shelter of large, usually flat-bottomed boulders, and the dried but green material is eaten during the winter when snow covers the ground and little growth of vegetation occurs. In Eurasia, pikas occupy an extensive geographic range and a wide range of habitats, including talus, forests, rock-strewn terrain, and open plains and desert-steppe areas. Unusually large hay piles, weighing up to 20 kg., are made by pikas inhabiting dry areas in southern Russia (Formozov, 1966).

Family Leporidae. The rabbits are a remarkably successful group in terms of ability to occupy a variety of environments over broad areas. Rabbits are now nearly cosmopolitan. Their distribution before introductions by man included most of the New and Old Worlds, and rabbits have been introduced into New Zealand, Australia, parts of southern South America, and various oceanic islands in both the Atlantic and Pacific. Nine Recent genera represented by 49 Recent species are known.

Several major leporid evolutionary trends in structure are recognized by Dawson (1958:6). The cheek teeth have become hypsodont, some of the premolars have become molariform,

and the primitive crown pattern has been modified into a simple arrangement in which most traces of the primitive cusp pattern have been lost. These changes resemble those in some groups of strictly herbivorous rodents. The skull has become arched, and the angle between the basicranial and palatal axes has increased. These changes are associated with a posture involving a greater angle between the long axis of the skull and the cervical vertebrae than that typical of primitive leporids. Trends in limb structure leading to increased cursorial ability include elongation of the limbs and specializations of articulations so that movement is limited to one plane.

The leporid skull (Fig. 9-1A) is more or less arched in profile, and the rostral portion is fairly broad. The maxillae, and often the squamosals, occipitals, and parietals, are highly fenestrated, and a prominent supraorbital process is always present. The auditory bullae are globular and the external auditory meatus is tubular. The dental formula is usually 2/1, 0/0, 3/2, 3/3 = 28; the re-entrant enamel ridges of the upper cheek teeth are usually crenulated (Fig. 9-2A). The clavicle is rudimentary and does not serve as a brace between the scapula and the sternum. The limbs, especially the hind limbs,

FIGURE 9–5. A white-tailed jackrabbit (*Lepus townsendii*) in its partially white winter pelage. This animal is from Colorado, but in more northerly parts of its range this jackrabbit is almost entirely white in winter. (Photograph by George D. Bear.)

are more or less elongate; movement at the elbow joint is limited to the anteroposterior plane (Fig. 9–2B, C). The tail is short. The ears have a characteristic shape: the proximal part of the ear is tubular, and the lower part of the ear opening is well above the skull (Fig. 9–5). The testes become scrotal during the mating season. In some species that inhabit regions with snowy winters, the animals molt into a white winter pelage in the fall (Fig. 9–5) and into a brown summer pelage in the spring. Wild leporids weigh from 0.3 to roughly 5 kg.

Leporids inhabit a tremendous array of habitats, from arctic tundra and treeless and barren situations on high mountain peaks to coniferous, deciduous, and tropical forests, open grassland and deserts. Some species, such as *Sylvilagus palustris* and *S. aquaticus* of the southeastern United States, are excellent swimmers and lead semi-aquatic lives. Leporids are entirely herbivorous and utilize a wide variety of grasses, forbs, and shrubs.

Several species are known to reingest fecal pellets and are thought to obtain essential nutrients (proteins and some vitamins) from material as it passes through the alimentary canal for a second time.

Habitat preference and cursorial ability differ markedly among different species of leporids and are strongly interrelated. Broadly speaking, species with relatively poor cursorial ability, such as *Brachylagus idahoensis* and *S. bachmani* of the western United States, scamper short distances to the safety of burrows or dense vegetation when disturbed, and typically occur in stands of big sagebrush (*Artemesia tridentata*) or dense chaparral, respectively. Other cottontails, such as *S. floridanus* of the eastern and *S. audubonii* of the western United States, are intermediate in cursorial ability and typically inhabit areas with scattered brush, rocks, or other cover, and do not run long distances to reach a hiding place. Representing the extreme in cursorial specialization

FIGURE 9–6. A white-tailed jackrabbit (*Lepus townsendii*) in its "form," a hollowed-out hiding place beneath a bush or other (often scanty) shelter. This animal is in its brown summer pelage. (Photograph by George D. Bear.)

among lagomorphs are the North American jackrabbits (*Lepus californicus, L. townsendii,* and *L. alleni* and its relatives). These animals have greatly elongate hind limbs, have adopted a bounding gait, and occupy areas with limited shelter, such as deserts, grasslands or meadows, where they take shelter in "forms" (Fig. 9–6). Instead of taking cover at the approach of danger, they depend for escape on their running ability. (Adaptations that contribute to running ability are discussed on page 220). Jackrabbits, and other similarly adapted members of the genus *Lepus,* are extremely rapid runners for their size; some attain speeds of up to 70 km. per hour. This speed allows some species to occupy open areas with little cover, where safety depends upon outrunning predators. So strongly entrenched is the habit of seeking safety by running that some speedy leporids only take to shelter as a last resort when injured or exhausted.

Although rabbits are seemingly peaceful and nonaggressive types, they are strong competitors and are remarkably adaptable. In some parts of Australia the extinction or near extinction of certain marsupials is due primarily to competition with introduced European rabbits (*Oryctolagus*). In addition, these prolific rabbits have caused great damage to crops and rangeland, and at various times have been a primary agricultural pest in many parts of Australia as well as in New Zealand, where they were also introduced. The range of environmental conditions to which leporids have adapted is tremendous. Populations of rabbits (*Lepus arcticus*) along the arctic coasts of Greenland use their protruding incisors (Fig. 9–1B) to scrape through snow and ice to reach plants during the long arctic winters, whereas far to the south, in the deserts of northern Mexico, jackrabbits (*Lepus alleni*) maintain their water balance through hot, dry periods by eating cactus and yucca.

CHAPTER 10

ORDER RODENTIA

Rodents comprise the largest mammalian order, including some 34 families, 354 genera, and roughly 1685 species. Rodents are a spectacularly successful group; they are virtually cosmopolitan in distribution and are important members of nearly all terrestrial faunas. They are also a remarkably complicated group with respect to morphological diversity, lines of descent, and parallel evolution of similar features in different groups. Because of these complexities, zoologists have often not agreed on the relationships between families and other taxa. As a result, superfamily and subordinal groupings remain uncertain, and numbers of species and genera listed here are approximations. The terms *sciuromorph, myomorph,* and *histricomorph* have been used repeatedly to designate major taxonomic divisions of the rodents. These terms, which are based on important differences in the structure of the skull and masseter muscles, are entrenched in the literature on mammals. Although they have frequently been given subordinal rank (Sciuromorpha, Myomorpha, Histricomorpha), they are given no taxonomic status in this discussion, but are used simply to designate types of specialization of the zygomassteric structure (Fig. 10–1 and 10–2). The taxonomic system proposed by Wood (1955), involving the recognition of

seven suborders, probably better expresses phylogenetic relationships between families, but it seems cumbersome and overly complex for use in the present discussion. Later in this chapter the terms sciuromorph, myomorph, and histricomorph will be discussed; the classification used here will be almost entirely that of Anderson (1967:208). The systems proposed by Anderson and Wood are compared in Table 10–1.

Morphology

Recent members of the order Rodentia share a series of distinctive cranial features. The upper jaw and lower jaw each bear a single pair of persistently growing incisors, a feature developed early in the evolution of rodents and one that committed them to a basically herbivorous mode of feeding. Because only the anterior surfaces are covered with enamel, the incisors assume a characteristic beveled tip as a result of wear. The occlusal surfaces of the cheek teeth are often complex and allow for effective sectioning and grinding of plant material. The dental formula seldom exceeds 1/1, 0/0, 2/1, 3/3 = 22, and a diastema is always present between the incisors and the premolars. The incisors and canines are always 1/1, 0/0.

147

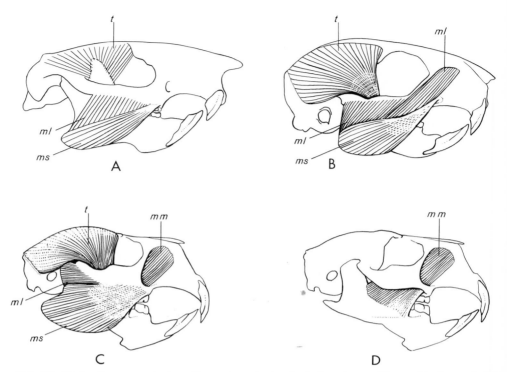

FIGURE 10–1. Zygomasseteric patterns in rodents. A, *Ischryotomus* (Paramyidae), a primitive Eocene rodent; the jaw muscles are restored. (After Wood, 1965.) Note that the masseter muscles originate entirely on the zygomatic arch. B, Abert's squirrel (*Sciurus aberti*, Sciuridae); the anterior part of the masseter lateralis originates on the rostrum and the zygomatic plate. C, porcupine (*Erethizon dorsatum*, Erethizontidae); the anterior part of the masseter medialis originates largely on the rostrum and passes through the enlarged infraorbital foramen. (The temporalis muscle, typically reduced in size in histricomorphs, is unusually large in porcupines.) Abbreviations: *ml*, masseter lateralis; *mm*, masseter medialis; *ms*, masseter superficialis; *t*, temporalis.

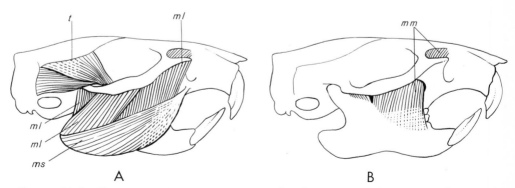

FIGURE 10–2. Zygomasseteric pattern in myomorph rodents. A, a cotton rat (*Sigmodon hispidus*, Cricetidae); the masseter superficialis originates on the rostrum, and the anterior part of the masseter lateralis originates on the anterior extension of the zygomatic arch. B, the superficial muscles have been removed; the masseter medialis originates partly on the rostrum and passes through the narrow infraorbital foramen. (After Rinker, 1954.) Abbreviations are the same as in Figure 10–1.

Table 10–1. Two Systems of Classification of the Rodents.

BASED ON ANDERSON (1967)	WOOD (1955)
"SCIUROMORPHA"	Suborder SCIUROMORPHA
SUPERFAMILY APLODONTOIDEA	Superfam. Aplodontoidea
Aplodontidae	Fam. Aplodontidae
SUPERFAMILY SCIUROIDEA	Superfam. Sciuroidea
Sciuridae	Fam. Sciuridae
SUPERFAMILY GEOMYOIDEA	?SCIUROMORPHA
Geomyidae	Superfam. Ctenodactyloidea
Heteromyidae	Fam. Ctenodactylidae
SUPERFAMILY CASTOROIDEA	Suborder THERIDOMYOMORPHA
Castoridae	Superfam. Anomaluroidea
SUPERFAMILY ANOMALUROIDEA	Fam. Anomaluridae
Anomaluridae	?SCIUROMORPHA or
Pedetidae	THERIDOMYOMORPHA
"MYOMORPHA"	Fam. Pedetidae
SUPERFAMILY MUROIDEA	Suborder CASTORIMORPHA
Cricetidae	Superfam. Castoroidea
Spalacidae	Fam. Castoridae
Rhizomyidae	Suborder MYOMORPHA
Muridae	Superfam. Muroidea
SUPERFAMILY GLIROIDEA	Fam. Cricetidae
Gliridae	Fam. Muridae
Platacanthomyidae	?Muroidea
Seleviniidae	Fam. Spalacidae
SUPERFAMILY DIPODOIDEA	Fam. Rhizomyidae
Zapodidae	Superfam. Geomyoidea
Dipodidae	Fam. Heteromyidae
"HYSTRICOMORPHA"	Fam. Geomyidae
SUPERFAMILY HYSTRICOIDEA	Superfam. Dipodoidea
Hystricidae	Fam. Zapodidae
SUPERFAMILY ERETHIZONTOIDEA	Fam. Dipodidae
Erethizontidae	?MYOMORPHA
SUPERFAMILY CAVIOIDEA	Superfam. Gliroidea
Caviidae	Fam. Gliridae
Hydrochoeridae	Fam. Seleveniidae
Dinomyidae	Suborder CAVIOMORPHA
Heptaxodontidae	Superfam. Octodontoidea
Dasyproctidae	Fam. Octodontidae
SUPERFAMILY CHINCHILLOIDEA	Fam. Echimyidae
Chinchillidae	Fam. Ctenomyidae
SUPERFAMILY OCTODONTOIDEA	Fam. Abrocomidae
Capromyidae	Superfam. Chinchilloidea
Myocastoridae	Fam. Chinchillidae
Octodontidae*	Fam. Capromyidae
Abrocomidae	Superfam. Cavioidea
Echimyidae	Fam. Caviidae
Thryonomyidae	Fam. Hydrochoeridae
Petromyidae	Fam. Dinomyidae
SUPERFAMILY BATHYERGOIDEA	Fam. Heptaxodontidae
Bathyergidae	Fam. Dasyproctidae
SUPERFAMILY CTENODACTYLOIDEA	Fam. Cuniculidae
Ctenodactylidae	Superfam. Erethizontoidea
	Fam. Erethizontidae
	Suborder HYSTRICOMORPHA
	Superfam. Hystricoidea
	Fam. Hystricidae
	Superfam. Thryonomyoidea
	Fam. Thryonomyidae
	Fam. Petromyidae
	Suborder BATHYERGOMORPHA
	Superfam. Bathyergoidea
	Fam. Bathyergidae

*Including the family Ctenomyidae.
Anderson considered the Ctenomyidae as a separate family.

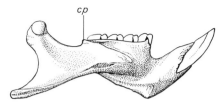

FIGURE 10-3. Dentary bone of the nutria (*Myocastor coypu*, Myocastoridae); note the strongly reduced coronoid process (*cp*). Compare this dentary to that of *Aplodontia*, shown in Figure 10-4.

The glenoid fossa of the squamosal is elongate, allowing anteroposterior and rotary jaw action. The mandibular symphysis has sufficient "give" in many species to enable the transverse mandibular muscles to pull the ventral borders of the rami together and spread the tips of the incisors. The masseter muscles are large and complexly subdivided, and provide most of the power for operating the lower jaw. By comparison with the masseters, the temporal muscles are usually small and their point of insertion, the coronoid process, is reduced, particularly in histricomorphs (Fig. 10-3). In general, rodents have undergone little specialization other than in the head area. Notable exceptions occur among some rodents that inhabit barren, arid regions, and are adapted for jumping.

Paleontology

The earliest records of rodents are from the late Paleocene of North America. These most primitive known rodents, the Paramyidae (Fig. 10-4A), may be the basal group from which all rodents evolved. The structure of the paramyid skull and dentary indicates that the temporalis muscle was large, and that the masseter muscles were not highly specialized and originated entirely from the zygomatic arch (Fig. 10-1A). The dental formula is 1/1, 0/0, 2/1, 3/3 = 22, and the cheek teeth are brachyodont. Rodents underwent an impressive Tertiary adaptive radiation that led to the large number of Recent species. In contrast to the marsupials, edentates, carnivores, and ungulates, rodents seemingly suffered no major Pleistocene or post-Pleistocene extinctions, and today we see a rich rodent fauna.

Rodents have occupied a diverse array of habitats and, although primarily vegetarians, have exploited a wide variety of food, including insects, fish, reptiles, birds, mammals, and above-ground and below-ground parts of trees, grasses, and forbs. Morphological diversity in rodents is associated largely with different feeding habits and with contrasting styles of locomotion. Rodents are variously terrestrial, arboreal, fossorial, saltatorial, and semi-aquatic.

Zygomasseteric Specializations

The evolution of the complex jaw musculature in rodents has doubtless been influenced strongly by the impor-

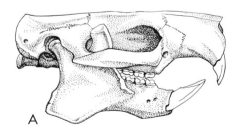
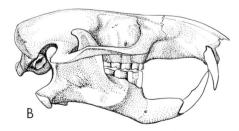

A   B

FIGURE 10-4. Skulls of primitive rodents. A, *Paramys* (Paramyidae), a primitive late Paleocene and early Eocene sciuromorph. Length of skull 89 mm. (After Romer, A. S.: *Vertebrate Paleontology*, 3rd ed., The University of Chicago Press.) B, mountain beaver (*Aplodontia rufa*, Aplodontidae), the most primitive living rodent. Length of skull 68 mm.

tance of rotary or oblique grinding movements as well as by the need for crushing power. The terms *sciuromorph* (squirrel-shaped), *myomorph* (mouse-shaped), and *histricomorph* (porcupine-shaped) refer to contrasting patterns or types of specialization of the masseter muscles, the temporal muscles, the skull, and the mandible. Not every rodent can readily be fitted into one of these groups, nor do experts agree as to how the types evolved. The myomorph pattern, for example, could have evolved from either sciuromorph or histricomorph types, but the weight of evidence favors an histricomorph origin (Klingener, 1964:76,77). For the present discussion, the questions concerning the evolution of the zygomasseteric types and their validity as a basis for classifying rodents can be laid aside in favor of a consideration of these types as simply contrasting solutions to the problem of the control of mastication.

The sciuromorph zygomasseteric structure is seemingly the least specialized type; it involves the least departure from the primitive, entirely zygomatic origins of the masseter muscles. The masseter muscles originate entirely from the zygomatic arch, as in *Aplodontia* (Fig. 10–1A), or part of the masseter lateralis originates from the rostrum in front of the zygomatic "plate" (Fig. 10–1B). In histricomorph rodents the infraorbital foramen is typically enormously expanded and allows passage of the greatly enlarged anterior part of the masseter medialis (Fig. 10–1C). The other divisions of the masseter originate entirely on the zygomatic arch. In myomorph rodents some advantageous aspects of both sciuromorph and histricomorph modifications are utilized (Fig. 10–2). In myomorphs the origins of both the medial and lateral divisions of the masseter have moved forward. The anterior part of the zygomatic arch is not platelike, but the masseter lateralis has a partly rostral origin. The infraorbital foramen is enlarged and part of the masseter medialis, which originates partly from the side of the rostrum on

the maxillary and premaxillary bones, passes through it. In addition, the origin of the superficial division of the masseter is well forward on the rostrum.

The adaptive importance of these different zygomasseteric arrangements is not entirely clear, but advantages over the primitive entirely zygomatic origins for the masseter muscles are apparent. In histricomorph rodents the attachments of the large anterior part of the medial masseter are such as to produce unusually powerful jaw action. Because this muscle originates far forward on the rostrum and inserts well forward on the mandible, it has great mechanical advantage for power and an efficient right-angle attachment on the dentary. The trend toward enlargement and specialization of the anterior division of the medial masseter in histricomorphs is seemingly correlated with a shift in responsibility for powerful jaw adduction from posterior, "less efficient" muscles (such as the temporalis and masseter lateralis), to an anterior muscle with attachments that favor power. The specializations of the zygomatic plate and the anterior placement of the origin of the masseter lateralis in advanced sciuromorphs appear to be a parallel functional trend, but one involving different structural modifications. The myomorph pattern has combined some features of both of the other zygomasseteric types, and has again achieved powerful jaw action, but by a third structural plan, which produces a strong anteriorly directed component to the pull exerted by the masseters.

Sciuromorph Rodents

Included in this group are seven Recent families that can be broadly considered as squirrel-like rodents. Within these families are species variously adapted to fossorial, terrestrial; arboreal, and semi-aquatic modes of life.

Family Aplodontidae. This family is of interest primarily because of the

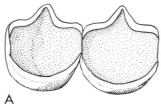
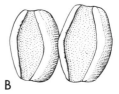

A   B

FIGURE 10–5. Crowns of first two right upper molars of two rodents. A, mountain beaver (*Aplodontia rufa*, Aplodontidae); note the simplified and unique crown pattern. B, Merriam's kangaroo rat (*Dipodomys merriami*, Heteromyidae); note the highly simplified crown pattern. The outer border of the tooth is above; anterior is to the right. The unshaded part is enamel and the stippled part is dentine.

unique, primitive morphological features that characterize its one living member. *Aplodontia rufa*, the mountain "beaver," is restricted to parts of the Pacific Northwest. This animal is roughly the size of a small rabbit, and has a robust, short-legged form. *Aplodontia* is generally regarded as the most primitive living rodent. Its sciuromorph zygomasseteric arrangement is close to that of the ancestral family Paramyidae, with the masseters having an entirely zygomatic origin. The skull is flat and the coronoid process of the dentary is large (Fig. 10–4B). The cheek teeth are ever-growing (a specialized feature) and have a unique crown pattern (Fig. 10–5A). The dental formula is 1/1, 0/0, 2/1, 3/3 = 22.

The earliest records of aplodontids are from the early Miocene of western North America; they spread later to Europe and Asia, but since the middle Pliocene have lived only in the moist, forested parts of the Pacific slope of North America, where they occur today from central California to southern British Columbia. Widespread late Tertiary aridity in North America may have restricted the aplodontids to their present range. *Aplodontia* occurs in small colonies, favors moist areas supporting lush growths of forbs, and often builds its burrows next to streams. The diet includes a variety of forbs and the buds, twigs, and bark of such riparian plants as willow (*Salix*) and dogwood (*Cornus*). On occasion *Aplodontia* builds "hay piles" of cut sections of forbs (Grinnell and Storer, 1924:157). Although usually terrestri-

al, this animal is known to climb to some extent in search of food.

Family Sciuridae. This successful and widespread family includes some 261 Recent species representing 51 genera. Squirrels, chipmunks, marmots (Fig. 10–6), and prairie dogs (Fig. 10–7) belong to this family. Sciurids appeared first in the middle Oligocene of North America, and spread to the Old World by late Oligocene. Ground squirrels, tree squirrels, and flying squirrels appear in the fossil record in the Miocene. Sciurids remain widespread today, being absent only from the Australian region, Madagascar, the polar regions, southern South America, and certain Old World desert areas.

Sciurids are fairly distinctive structurally. The skull is usually arched in profile, and the front of the zygomatic arch is flattened (forming the so-called zygomatic plate) where the anterior part of the masseter lateralis rests against it (Fig. 10–8). The dental formula is 1/1, 0/0, 1-2/1, 3/3 = 20-22. The cheek teeth are rooted and usually have a crown pattern that features transverse ridges. Sciurids have relatively unspecialized bodies: a long tail is retained, and the limbs have no loss of digits or reduction of freedom of movement at the elbow, wrist, and ankle joints. The digits usually have sharp claws. Several semi-fossorial types, including ground squirrels (*Spermophilus*), prairie dogs (*Cynomys*), and marmots (*Marmota*), have variously departed from this plan in the direction of greater power in the forelimbs and, in some cases, reduction of the tail.

FIGURE 10–6. Yellow-bellied marmot (*Marmota flaviventris*, Sciuridae). (Photograph by O. D. Markham.)

FIGURE 10–7. Black-tailed prairie dogs (*Cynomys ludovicianus*, Sciuridae). (Photograph by O. J. Reichman.)

FIGURE 10–8. Ventral (left) and lateral (right) views of the skull of a pocket gopher (Geomyidae) showing the zygomatic plate (*zy*).

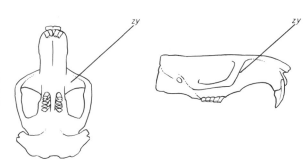

Although sciurids are basically herbivores, a great variety of food is utilized. Tree squirrels occasionally eat young birds and eggs; chipmunks (*Eutamias*) and antelope ground squirrels (*Ammospermophilus*) are seasonally partly insectivorous in some areas. Sciurids are tolerant of a great range of environmental conditions. Some sciurids, such as marmots, prairie dogs, chipmunks, and some ground squirrels, hibernate during cold parts of the year, but the tree squirrels are active throughout the year. Red squirrels (*Tamiasciurus*) occur in mountainous areas, and in Alaska in coniferous forests, where temperatures frequently stay well below 0°F for days at a time, remain active through the winter. The antelope ground squirrel lives under the tremendously different conditions typical of Sonoran deserts. These ground squirrels live in an environment having almost no free water. For the maintenance of water balance, they rely largely on water in such succulent foods as green leaves and insects (see p. 384). *Ammospermophilus* is diurnal, but avoids heat stress in an intensely hot environment by periodically seeking shelter (see p. 375).

Styles of locomotion vary among sciurids; the most specialized style occurs in the flying squirrels. This group of 13 genera, comprising the subfamily Petauristinae, is characterized by gliding surfaces formed by broad folds of skin between the forefoot and hindfoot. These animals are able to glide fairly long distances between trees. The giant flying squirrel (*Petaurista*) of southeast Asia, for instance, can glide up to 450 m., and can turn in mid-air (Walker, 1968:716).

Family Geomyidae. Members of this family, the pocket gophers, are the most highly fossorial North American rodents. Pocket gophers are distributed from Saskatchewan in Canada to northern Colombia in South America. The family includes roughly 40 Recent species in eight genera. Pocket gophers appeared first in the early Miocene of North America. Although they are not restricted to semi-arid habitats today, many of their most characteristic specializations probably evolved in response to the soil conditions and floral assemblages of the semi-arid and plains environments that developed in the Miocene.

The most obvious structural characteristics of pocket gophers were developed in response to fossorial life. These animals are moderately small, weighing from roughly 100 to 900 g. They have small pinnae, small eyes, and short tails. The head is large and broad, and the body is stout. External, fur-lined cheek pouches are used for carrying food. The dorsal profile of the geomyid skull is usually nearly straight, the zygoma flare widely, and in the larger species the skull is angular and features prominent ridges for muscle attachment. The rostrum is broad and robust, and is marked laterally by depressions from which the masseter lateralis muscles take origin. The large incisors often protrude forward, in some species beyond the anteriormost parts of the nasals and premaxillae; the lips close behind the incisors, which are therefore outside the mouth. The dental formula is 1/1, 0/0, 1/1, 3/3 = 20. The cheek teeth are ever-growing, and have a highly simplified crown pattern. There is no loss of digits. The forelimbs are powerfully built and bear large, curved claws; the toes of the forepaw have fringes of hairs that presumably increase the effectiveness of this foot during digging (Fig. 10–9A).

Pocket gophers occupy friable soils in environments ranging from tropical to boreal. These rodents are entirely herbivorous, and eat a variety of above-ground and below-ground parts of forbs, grasses, shrubs, and trees. The extensive burrow systems of pocket gophers provide retreats for many vertebrates. In Colorado, burrows occupied by pocket gophers were frequently found also to harbor tiger salamanders, *Ambystoma tigrinum* (Vaughan, 1961), and burrows abandoned by pocket gophers are used by a variety of reptiles and mammals, and occasionally by burrowing owls (*Speo-*

A   B

FIGURE 10–9. Ventral views of the forefeet of two rodents. A, left manus of the pocket gopher (*Thomomys bottae*, Geomyidae); note the fringes of hairs on the toes. B, right manus of the porcupine (*Erethizon dorsatum*, Erethizontidae); note the pattern of tubercles on the pads that increase traction.

tyto cunicularia). Because pocket gophers keep the entrances to their burrow systems tightly plugged, predators and animals seeking refuge are usually excluded from occupied burrow systems. Pocket gopher burrows provide channels allowing fairly deep penetration of water during periods of snowmelt in mountainous areas, and in some areas they apparently reduce erosion of topsoil. The mounds of soil thrown up by pocket gophers, and the disturbance of the soil, strongly influence vegetation, and often sites disturbed by intense pocket gopher activity can be recognized at a distance by the striking dominance of pioneer plants. In some areas where pocket gophers are abundant, roughly 20 per cent of the surface of the ground is covered with mounds. Because their digging affects plant composition in rangeland and their local preference for alfalfa and other cultivated plants results in great crop damage, pocket gophers are of considerable economic importance. Large amounts of money have been spent by farmers and by Federal agencies on the control of pocket gophers on cultivated land in the western United States.

Family Heteromyidae. Most members of the family Heteromyidae are adapted to arid or semi-arid conditions; kangaroo rats (*Dipodomys;* Fig. 10–10) and pocket mice (*Perognathus*) are, in fact, the most characteristic desert rodents in North America. This family contains 75 Recent species of five genera. Heteromyids are restricted to the New World, where they range from southern Canada, through the western United States, to Equador, Colombia, and Venezuela; they occupy tropical, subtropical, arid, and semi-arid regions. Heteromyids first appeared in the fossil record in the Oligocene of North America. The kangaroo rats (subfamily Dipodomyinae) are known from the Pliocene, when the deserts and semi-arid brushlands that heteromyids now frequently inhabit were widespread in western North America. Certain diagnostic characteristics of kangaroo rats, such as the greatly enlarged auditory bullae (Fig. 10–11) and features of the hind limbs that favor saltation, probably

FIGURE 10–10. A kangaroo rat (*Dipodomys ordii*, Heteromyidae). Note the large head and compact body.

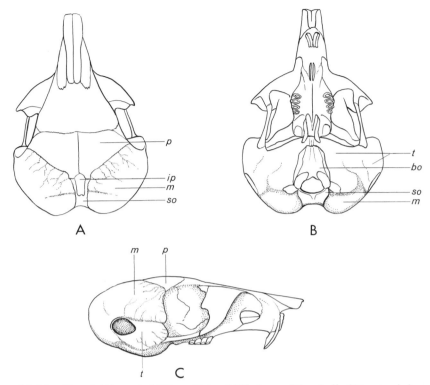

FIGURE 10–11.   Dorsal (A), ventral (B), and lateral (C) views of the skull of Merriam's kangaroo rat (*Dipodomys merriami*, Heteromyidae); note the great enlargement of the auditory bulla, the chamber surrounding the middle ear. Length of skull 45 mm. Abbreviations: *bo*, basioccipital; *ip*, interparietal; *m*, mastoid part of the bulla; *p*, parietal; *sp*, supraoccipital; *t*, tympanic part of the bulla (petrosal and tympanic bones). (After Grinnell, 1922.)

evolved in association with desert or semi-desert conditions.

In general form, heteromyids reflect a trend toward saltation as the primary form of locomotion. These animals usually have large heads, compact bodies, long hind limbs, and long, frequently tufted tails. As in the pocket gophers, fur-lined cheek pouches are present. The skull is delicately built, with thin, semi-transparent bones; the zygomatic arch is slender. The auditory bullae are usually large, and in some genera are enormous, being formed largely by the mastoid and tympanic bones (Fig. 10–11). The enlargement of the bullae in heteromyids (and in jerboas, Dipodidae) greatly increases auditory sensitivity (see p. 409). The nasals are slender and usually extend well forward of the slender upper incisors. The dental for-

mula is 1/1, 0/0, 1/1, 3/3 = 20. The cheek teeth have a strongly simplified crown pattern (Fig. 10–5B) resembling that of pocket gophers, to which heteromyids are seemingly closely related.

The bodies of most heteromyids are specialized for jumping. Such adaptations are most strongly developed in the kangaroo rats and kangaroo mice (*Microdipodops*). In these genera the forelimbs are small, the neck is short, and the tail is long and serves as a balancing organ. The hind limbs are elongate and the thigh musculature is powerful. The hind foot is elongate, but except for the almost complete loss of the first digit in some kangaroo rats (Fig. 10–12E), there is no loss of digits. The cervical vertebrae are largely fused in *Microdipodops*, and in *Dipodomys* they are strongly com-

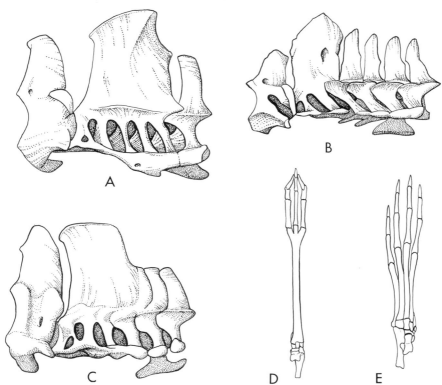

FIGURE 10–12. The cervical vertebrae and hind feet of several saltatorial rodents. A, cervical vertebrae of a jerboa (*Jaculus* sp. Dipodidae); B, of the springhaas (*Pedetes surdaster*, Pedetidae); C, of Heermann's kangaroo rat (*Dipodomys heermanni*, Heteromyidae). (After Hatt, 1932.) D, dorsal view of right hind foot of a jerboa (*Scirtopoda* sp., Dipodidae); note the reduction of digits and the cannon bone formed by metatarsals two, three, and four (after Howell, 1944). E, left hind foot of the desert kangaroo rat (*Dipodomys deserti*, Heteromyidae); note the near loss of the first digit and the elongation of the foot (after Grinnell, 1922).

pressed and partly fused (Fig. 10–12C), producing a short, rigid neck. These species are mostly bipedal when moving rapidly, and when frightened they move by a series of erratic hops.

Most heteromyids live in areas with strongly seasonal patterns of precipitation, and many must cope with annual cycles of weather and plant growth of the sort that occur in deserts. In such areas brief periods of precipitation and long periods of drought are typical. Many small annual plants in deserts are able to make the most of irregular moisture by germinating, growing, and flowering rapidly, and by producing abundant seeds that remain dormant until the next rains. This enormously abundant seed crop is the major food source of heteromyids. Many heteromyid features, such as saltatorial ability and seed gathering habits, evolved in response to the demands of desert or semi-desert climates. Perhaps the most remarkable heteromyid adaptation is their ability to survive for long periods on a diet of dry seeds with no free water. This capability probably does not occur in all heteromyids, nor is it developed to the same degree in all species adapted to dry climates. In the species of kangaroo rats and of pocket mice (*Perognathus*) that occupy the desert, however, this ability is well developed (see p. 390).

Family Castoridae. To this family belong the beavers. The family is represented today by but two species, *Castor canadensis* of the United States, Canada, and Alaska, and *Castor fiber,* of northern Europe and northern Asia. Beavers are remarkable in their ability to modify their environment strongly by building dams and by cutting trees. In addition, because of their economic importance, beavers had a pronounced effect on the history of the United States. Much of the early exploration of some of the major river systems in the western United States was done by men in quest of the valuable pelts of beavers.

The fossil record of beavers begins in the Oligocene, and several lines of descent developed in the Tertiary. One line developed fossorial adaptations, and another led to the bear-sized giant beaver (*Castoroides*) of the North American Pleistocene. Throughout their history castorids have been restricted to the Northern Hemisphere.

Beavers are semi-aquatic, and some of their distinctive structural features are adaptations to this mode of life. The animals are large, reaching over 30 kg. in weight. Their large size is associated with a mass to surface ratio that is more advantageous in terms of heat conservation than those of smaller rodents. In addition, the body is insu-

lated by fine underfur protected by long guard hairs. These are important adaptations in animals that frequently swim and dive for long periods in icy water. The large hind feet are webbed, the small eyes have nictating membranes, and the nostrils and ear openings are valvular and can be closed during submersion.

Because of two structural specializations, beavers can open their mouths when gnawing under water; while they are swimming, they can carry branches in the submerged open mouth without danger of taking water into the lungs (Cole, 1970). The epiglottis is internarial (it lies above the soft palate); it allows efficient transfer of air from the nasal passages to the trachea, but does not allow mouth breathing or panting. Also, the mid-dorsal surface of the back part of the tongue is elevated and fits tightly against the palate and, except when the animal is swallowing, blocks the passage to the pharynx (Cole, 1970).

The tail is broad, flat, and largely hairless. The skull is robust. The zygomasseteric structure is of an advanced sciuromorph type, in which the rostrum is marked by a conspicuous lateral depression (Fig. 10–13B) from which a large part of the masseter la-

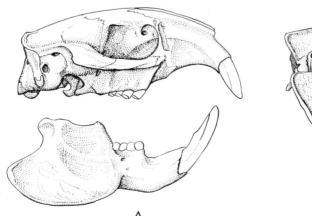

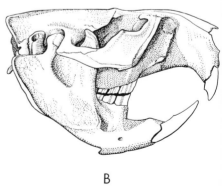

FIGURE 10–13. Two rodent skulls. A, a mole rat (*Cryptomys mechowii*, Bathyergidae); note the large, procumbent incisors that are used for digging (after Hill and Carter, 1941). Length of skull 57 mm. B, a beaver (*Castor canadensis*, Castoridae); note the depression in the side of the rostrum from which the anterior part of the masseter lateralis originates. Length of skull 139 mm.

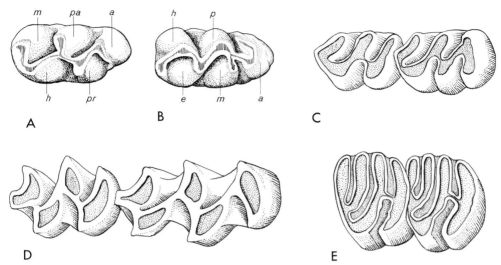

FIGURE 10–14. Crown patterns of rodent molars. Crowns of the first right upper molar (A) and first left lower molar (B) of a harvest mouse (*Reithrodontomys megalotis*, Cricetidae), showing the cusps (after Hooper, 1952). Abbreviations for A: *a*, anterocone; *h*, hypocone; *m*, metacone; *pa*, paracone; *pr*, protocone. Abbreviations for B: *a*, anteroconid; *e*, entoconid; *h*, hypoconid; *m*, metaconid; *p*, protoconid. First two right upper molars of (C) a cotton rat (*Sigmodon alleni*, Cricetidae); (D) a Mexican vole (*Microtus mexicanus*, Cricetidae); and (E) a beaver (*Castor canadensis*, Castoridae). Unshaded areas on the occlusal surfaces are enamel; stippled areas within enamel folds are dentine.

teralis muscle originates. The jugal is broad dorsoventrally, and the external auditory meatus is long and is surrounded by a tubular extension of the auditory bulla. The dental formula is 1/1, 0/0, 1/1, 3/3 = 20. The premolars are molariform, and the complex occlusal pattern features transverse enamel folds (Fig. 10–14E).

Beavers always occur along waterways, and although they are most typical of regions supporting coniferous or deciduous forests, beavers live in some hot desert regions, as, for example, along the lower Colorado River of Arizona and California. There is considerable geographic and local variation in the types of shelter constructed by beavers. Along many rivers in the southwestern and middlewestern United States, beavers dig burrows in the river banks, but in many northern areas and in mountainous regions they built lodges of sticks and mud. Along large rivers the animals are often fairly regularly distributed, but along small streams they are distributed irregularly and may live in colonies consisting of several family groups.

Few mammals equal the ability of the beaver to modify its environment. Many high mountain valleys in the Rocky Mountains have been transformed by beavers from a series of meadows through which a narrow, willow-lined stream meandered, to a terraced series of broad ponds bordered by willow thickets and soil saturated with water. A valley suitable for grazing by cattle before occupancy by beavers is often more suitable afterward for trout and waterfowl. Naturally, these striking environmental changes are welcomed by some and condemned by others. Because the work of beavers tends to reduce erosion caused by rapid runoff during snowmelt and to raise water tables locally, these animals are regarded as valuable in many "natural" areas.

Family Anomaluridae. This family, composed of some 12 Recent species of four genera, includes the scaly-tailed squirrels. These animals occupy forested, tropical, or subtropical parts of western and central Africa, where they are locally common. They resemble the flying squirrels (Sciuridae) in some structural features and in gliding ability.

Increased surface area for gliding is provided in anomalurids by a fold of skin that extends between the wrist and the hind foot, and is supported and extended during gliding by a cartilaginous rod, roughly the length of the forearm, that originates on the posterior part of the elbow. (In "flying"members of the Sciuridae, in contrast, a short cartilaginous brace arises from the wrist.) In anomalurids folds of skin also occur between the ankles and the tail a short distance distal to its base. One genus (*Zenkerella*) does not have a gliding membrane. The anomalurid tail is usually tufted and has a bare ventral area near its base that has two rows of keeled scales. The feet are strong and bear sharp claws.

The anomalurid skull differs from that of a sciurid. The infraorbital canal is enlarged and transmits part of the medial masseter muscle. The dental formula is 1/1, 0/0, 1/1, 3/3 = 20, and the cheek teeth are rooted. The general body form resembles that of tree squirrels, but differences between anomalurids and sciurids involving zygomasseteric structure, dental features, and types of bracing for the anterior part of the lateral fold of skin indicate that scaly-tailed squirrels and flying squirrels of the family Sciuridae are not closely related. No fossil anomalurids are known.

The scaly-tailed squirrels are accomplished gliders, and they use this ability to travel rapidly from tree to tree. The scales beneath the tail seemingly serve as an abrasive surface to help keep the animals from losing traction when climbing or landing, when the tail is braced against a tree. All anomalurids but *Zenkerella* are nocturnal. Anomalurids are apparently largely vegetarians, but little is known of their feeding habits or of other aspects of their biology.

Family Pedetidae. This family is represented by but two Recent species of one genus. The common name for both species is springhaas or spring hare. The distribution of these animals includes central and southern Africa, where they occupy sandy soils in arid or semi-arid regions. Sparsely vegetated areas or sites where the vegetation has been disturbed are often inhabited; the habitat preferences of these rodents thus resemble those of the kangaroo rats of North America.

The spring hares are large rodents, roughly the size of a large rabbit, and are saltatorial. The clawed forefeet are small, but the hind limbs are long and powerful, the fibula is reduced and fused to the tibia, and the hind feet have only four toes. The furred tail has a tufted distal section. The robust skull has an enlarged infraorbital canal through which passes part of the anterior division of the medial masseter. The cervical vertebrae are partly fused (Fig. 10–12B). The dental formula is 1/1, 0/0, 1/1, 3/3 = 20, and the cheek teeth are ever-growing. The ear opening is guarded by a tragus that fits against the ear opening when the animal is digging; it excludes sand and debris.

These animals dig fairly elaborate burrows, from which they emerge at night by a great leap that presumably confounds a waiting predator. They make long bipedal hops when frightened, but are quadrupedal when foraging and when moving slowly. These rodents eat a variety of plant materials including bulbs, seeds, and leaves, and water balance may be maintained in some arid areas by eating succulent vegetation or perhaps insects.

MYOMORPH RODENTS

This division of rodents includes approximately two-thirds of all the Recent species of rodents, and is nearly world-wide in distribution. The myomorph zygomasseteric pattern is seemingly highly adaptive, judging by the great success of such families as Cricetidae and Muridae. Considerable diversity occurs within myomorphs.

Family Cricetidae. This family contains a great variety of mouse-like creatures such as deer mice, woodrats, voles, lemmings, muskrats, and gerbils. A total of roughly 567 Recent species of some 97 genera, approximately one-third of the kinds of living rodents, belong to this family. Cricetids are nearly ubiquitous, being absent only from some islands, Antarctica, and the Australian and Malayan areas. Habitats ranging from arctic tundras to tropical rain forests are occupied; in many terrestrial communities cricetid rodents are the most important small mammals in terms of both their effect on the environment and their importance as a staple food item for many predators.

Most cricetids retain a "standard" mouse-like form, with a long tail and a generalized limb structure. Cricetids vary in size from roughly 10 g. in weight and 100 mm. in total length, as in the pygmy mouse (*Baiomys*), to approximately 1500 g. and 600 mm., as in the muskrat (*Ondatra*). The skull is quite variable in shape, but always has a somewhat enlarged infraorbital foramen that transmits part of the masseter medialis and a branch of the trigeminal nerve (Fig. 10–15A). The masseter lateralis originates partly from the enlarged zygomatic plate. The dental formula is 1/1, 0/0, 0/0, 3/3 = 16, and the molars vary from low-crowned and rooted to high-crowned and ever-growing. The occlusal surface of the cheek teeth is variable, but is based on a pattern of five crests formed by re-entrant enamel folds (Fig. 10–14A, B).

As might be expected from the large number of species within the Cricetidae, a variety of modes of life and structural patterns are represented. By considering three of the subfamilies of cricetids, some of this diversity will be illustrated.

Most species of the subfamily Cricetinae are adapted to a generalized terrestrial or scansorial (climbing) mode of life, but arboreal, semi-aquatic, and fossorial habits also occur. Foods include a wide variety of plant and animal material. Three semi-aquatic Neotropical genera (*Ichthyomys*, *Daptomys*, and *Anotomys*) are fish eaters, and the Asian mole rats (*Myospalax*) are fossorial and feed partly on below-ground parts of plants. The skulls of cricetines are, in general, not highly specialized, and except for such features as fringes of hair on the hind feet or webs between the toes in the case of aquatic forms, and strong and clawed forefeet in the case of the mole rat, the limbs of cricetines are not strongly modified.

The subfamily Microtinae includes the voles, lemmings (Fig. 10–16), and muskrats, a group of rodents that are distributed throughout the Northern Hemisphere. These rodents frequently have short tails, ear openings that are partially guarded by fur, and a chunky, short-legged appearance. The cheek teeth often feature complex crown patterns (Fig. 10–14D) adapted to masticating forbs and

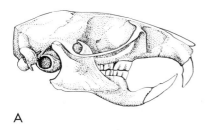

A

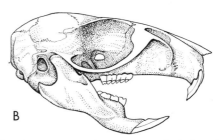

B

FIGURE 10–15. Rodent skulls. A, Stephen's woodrat (*Neotoma stephensi*, Cricetidae); length of skull 42 mm. B, gerbil (*Tatera humpatensis*, Cricetidae); length of skull 38 mm. (B after Hill and Carter, 1941.)

FIGURE 10–16. Norwegian lemmings (*Lemmus lemmus*, Cricetidae). The animal on the right is in a threatening posture; the one on the left is in a submissive posture. (Photograph by Garrett C. Clough.)

grasses. The voles and lemmings undergo remarkable population fluctuations in some areas, and high population densities have been recorded in many areas. In Sweden, for example, densities of 200 to 300 per acre were recorded during an irruption of lemmings (*Lemmus lemmus*). Population cycles of rodents are discussed in Chapter 15 (page 290). Because of the reliance on microtines for food by many northern carnivores, the densities and even distributions of these predators are partially controlled by cycles of microtine abundance (see Table 15–9, p. 289).

The subfamily Gerbillinae includes the gerbils, a group of rodents that resemble jerboas (Dipodidae) and kangaroo rats (Heteromyidae) in being semi-fossorial and more or less saltatorial, and in inhabiting mainly desert regions. Gerbils occur in arid parts of Asia, in the Near East, and in northern Africa. The hind limbs are large, the central three digits are larger than the lateral ones, and the tail is often long and functions as a balancing organ. The skull does not depart strongly

from the general cricetid plan (Fig. 10–15B). In their ability to hop and in their choice of habitats, gerbils resemble heteromyid rodents, but they differ from heteromyids in being diurnal and in feeding more on leaves and insects than on dry seeds. Gerbils maintain water balance in the face of hot and arid conditions partly by eating food with a high water content and by concentrating urine as do heteromyids.

Family Spalacidae. This family contains the mole rats, a group of three Recent species of one genus (*Spalax*). These rodents occur in the eastern Mediterranean region and southeastern Europe, and are strongly fossorial. The eyes are small, and the eye muscles and optic nerve are reduced or absent. The ears are small and the tail is vestigial. Unlike most fossorial rodents, *Spalax* digs primarily with its large incisors and moves soil with its blunt head. As adaptations to this style of digging, the neck and jaw musculature is powerful, the incisors are robust, and the nose is protected by a broad, horny pad. The feet, surprisingly, are not unusually large; the

FIGURE 10–17. Extremes in skull shape in rodents of the family Muridae. A, rock mouse (*Delanymys brooksi*), a rock dwelling omnivore; B, shrew-like rat (*Rhynchomys soricoides*), a rare species that apparently feeds on invertebrates. (After Walker, E. P.: *Mammals of the World*, 2nd ed., The Johns Hopkins Press, 1968.)

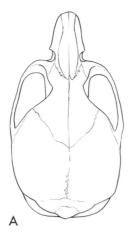
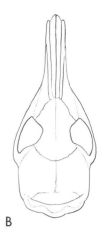

A   B

claws have been described as blunt, round nubbins. These nocturnal rodents burrow in both alluvial and stony soils, and eat both below-ground and above-ground parts of plants.

Family Rhizomyidae. Members of this group, the bamboo rats, occur in Southeastern Asia and in tropical parts of eastern Africa. The family includes three genera represented by 18 species. These specialized rodents have short limbs with nearly nail-like claws, heavily-built bodies, and broad skulls. The sturdy incisors are used in digging. Habitats ranging from dense bamboo thickets to sandy flats are occupied, and some species dig extensive burrows.

Family Muridae. This is the second largest family of rodents. In this group, including the Old World rats and mice, are roughly 98 genera and 457 species. Some murids live in close association with man in situations ranging from isolated farms to the world's largest cities. As a result of introductions by man, these animals have become nearly cosmopolitan in distribution, and are probably the rodents most familiar to man. Murids that are not commensal with man occur in much of Southeast Asia, Eurasia, Australia, Tasmania and Micronesia, and Africa. Tropical and subtropical areas are centers of murid abundance, but these animals have occupied a wide variety of habitats and some genera are highly adapted to specialized modes of life.

Murids range in size from that of a small mouse to that of a large rat. Some Philippine climbing rats (*Phloeomys*) are roughly 800 mm. in length and weigh over 1 kg. Although the tail is usually more or less naked and scaly, it is occasionally heavily furred and bushy. The shape of the skull is variable (Fig. 10–17). The molars are rooted or ever-growing, and usually have crowns with cusps or laminae (Fig. 10–18); great simplification of the crown pattern occasionally occurs. The dental formula is usually 1/1, 0/0, 0/0, 3/3 = 16. In some murids the reduction

FIGURE 10–18. Diagram of the occlusal surfaces of the right upper molar of a murid rodent (*Rattus*). With wear, the cross lophs become lakes of dentine (crosshatched areas) rimmed with enamel. Abbreviations are the same as those in Figure 10–14A.

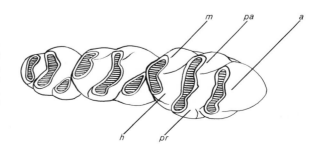

of the cheek teeth has become extreme. The greatest reduction occurs in *Mayermys,* a rare mouse from New Guinea, in which only one molar is retained on each side of each jaw. The zygomasseteric structure is of a myomorph type: the infraorbital foramen is not greatly enlarged, but through it passes some of the anterior part of the medial masseter; the origin of the lateral masseter is partly from the zygoma, and the superficial masseter takes origin from the side of the rostrum. The feet retain all of the digits, but the pollex is rudimentary.

Murids appear fairly late in the fossil record (Pliocene), but the family has been remarkably plastic from an evolutionary point of view. Among living murids amphibious, terrestrial, semi-fossorial, arboreal, and saltatorial types are known. The water rats (*Crossomys*) have greatly reduced ears, large, webbed hind feet, and nearly waterproof fur. These animals live along waterways in New Guinea. At the other extreme is the hopping mouse (*Notomys alexis*), a saltatorial inhabitant of extremely arid Australian deserts. This rodent needs no drinking water, and has the greatest ability of any animal whose water metabolism has been studied to concentrate urine as a means of conserving water (MacMillen and Lee, 1969).

Murids feed on a variety of plant material and on invertebrate and vertebrate animals. In association with the great diversity of feeding habits within the murids, the skull form varies widely within the family as noted earlier (Fig. 10–17), with a shrew-like elongation of the rostrum occurring in some genera.

Extremely high population densities have been recorded for feral populations of some murids often commensal with man. A 35-acre area near Berkeley, California, which had only an occasional house mouse for a number of years after population studies began in 1948, supported 7,000 *Mus* in June of 1961 (Pearson, 1964). Among other factors, the high reproductive rate of *Mus* contributes to its ability to reach high densities quickly. Periodic high populations of murids seemingly do not occur in populations of wild murids in tropical or subtropical areas. In such regions population densities of most wild mammals remain fairly stable. *Mus* and *Rattus* (Fig. 10–19) are not restricted to living with or near man, but now live away from man over broad areas in such regions as Australia and South America.

Commensal murids are of great economic importance. Not only are they instrumental in spreading such serious diseases as bubonic plague and typhus, but the damage they do to stored grains and other foods is tre-

FIGURE 10–19. The Australian bush rat (*Rattus fuscipes*, Muridae). (Photograph by Jeffrey Hudson.)

mendous. In some countries *Rattus* and *Mus* compete effectively and devastatingly with man for food.

Family Gliridae. This family includes the dormice, a group (seven genera with 23 Recent species) of rather squirrel-like, Old World rodents known first from the Oligocene of Europe. These animals are entirely Old World in distribution, occurring in much of Africa south of the Sahara, in England, Europe from southern Scandinavia southward, Asia Minor, southwestern Russia, and Japan.

Dormice are small (up to 325 mm. in length) and most genera have bushy or well-furred tails. The skull has a smooth, rounded braincase, a short rostrum, and large orbits. The dental formula is 1/1, 0/0, 1/1, 3/3 = 20. The crowns of the brachyodont molars have parallel cross ridges of enamel, or in some cases the ridges are reduced and the crowns have basins. The infraorbital foramen is somewhat enlarged and transmits part of the masseter muscle. The limbs and digits are fairly short, and the short claws are curved and are used in climbing. The manus has four toes and the pes has five.

Glirids are typically climbers that occupy trees and shrubs, rock piles, or rock outcrops. Food is largely fruit, nuts, seeds, birds' eggs and nestlings, and insects. Typically the animals are active and breed in the spring and summer, but hibernate in the winter. A unique feature of glirids is their ability to lose and regenerate their tails (Mohr, 1941:63).

Family Platacanthomyidae. To this small family (containing but two genera and two Recent species) belong the spiny dormice, inhabitants of southern India and southern China. The taxonomic position of these poorly known mice is uncertain. They are small, arboreal rodents, and are apparently herbivorous. The zygomasseteric structure is of a myomorph type, and the cheek teeth have parallel, oblique cross ridges. The dental formula is 1/1, 0/0, 0/0, 3/3 = 16.

Family Seleviniidae. This family is represented by one species, which

bears the common name of dzhalman, and is restricted to the Betpak-dala Desert of Russia. This small rodent is insectivorous, and, as is typical of many desert mammals, has greatly enlarged auditory bullae. The dental formula is 1/1, 0/0, 2/0, 3/3 = 20; the cheek teeth are small and short-crowned, and the much-simplified crown pattern features smooth, concave surfaces. This unusual desert rodent may have evolved from dormouse-like (gliroid) ancestors.

Family Zapodidae. Members of this small family, containing but four Recent genera with four species, are called jumping mice (Zapodinae) or birch mice (Sicistinae), and are typically boreal rodents. These mice occur widely in North America (in the western United States they occur almost exclusively in the mountains) south to roughly the southern part of the Rocky Mountains, and in the Old World they inhabit Germany, Norway, Russia, Mongolia, and China. Zapodids appeared in Europe in the Oligocene; the earliest records from North America are from the Miocene.

Zapodids are small, graceful mice, from roughly 10 to 25 g. in weight, with long tails and, in all genera but *Sicista*, elongate hind limbs. The coloration in most species is striking: the belly is white and the dorsum is bright yellowish or reddish brown. Much of the anterior part of the medial masseter muscle originates on the side of the rostrum and passes through the enlarged infraorbital foramen (Klingener, 1964:11). The dental formula is 1/1, 0/0, 0-1/0, 3/3 = 16 or 18; the cheek teeth are brachyodont or "semihypsodont," and have quadritubercular crown patterns with re-entrant enamel folds. The hind limbs in members of the subfamily Zapodinae are elongate and are somewhat adapted for hopping, but unlike more specialized saltatorial rodents, all digits are retained. As an additional contrast with specialized saltators, the cervical vertebrae of zapodids are unfused.

Jumping mice usually inhabit boreal

forests. Some species occur typically in coniferous forests, while others appear in birch stands or in mixed deciduous forests. Usually jumping mice favor moist situations, and *Zapus princeps* of the western United States is most abundant in many areas in dense cover adjacent to streams or in wet meadows. These mice hibernate in the winter and emerge during or after snowmelt. Food consists of a variety of seeds and other vegetable material, but insect larvae and other animal material made up approximately half of the food of *Z. hudsonius* in New York (Whitaker, 1963), and roughly one-third of the diet of *Z. princeps* in Colorado (Weil, 1968).

The development of saltation in a boreal mammal that inhabits dense vegetation is rather unusual, for this style of locomotion has almost without exception evolved in mammals frequenting open, sparsely vegetated situations offering little concealment, where erratic evasion of predators is advantageous. *Zapus* seems to use saltation for brief intervals; a few jumps are usually sufficient to enable it to reach concealment.

Family Dipodidae. This family includes the jerboas, a group remarkable for their extreme adaptations for saltation and for life in arid environments. These specialized rodents occur in arid and semi-arid areas in northern Africa, Arabia, and Asia Minor, and in southern Russia eastward to Mongolia and northeastern China. Dipodids are represented today by 10 genera and 27 species. The family first appeared in the Pliocene of Asia.

Jerboas have compact bodies, large heads, reduced forelimbs, and elongate hind limbs, features associated with saltatorial locomotion. The tail is long and usually tufted, and, as in the New World kangaroo rats (Heteromyidae), the tuft is frequently conspicuously black and white. The posterior part of the skull is broad (Fig. 10–20), owing mostly to the enlargement of the auditory bullae, which are huge in some species. The rostrum is usually short, the orbits are large, and through the enlarged infraorbital canal passes most of the anterior part of the medial masseter, which originates largely on the side of the rostrum. The zygomatic plate is narrow and is below the infraorbital canal. The dental formula is 1/1, 0/0, 0-1/0, 3/3 = 16 or 18; the cheek teeth are hypsodont, and the crown pattern usually involves re-entrant enamel folds. The hind limbs are elongate in all genera, but varying stages of specialization for saltation are represented. In members of the subfamily Cardiocraninae the toes vary in number from three to five, and the metatarsals are not fused. At the other extreme are such genera as *Dipus* and *Jaculus* (subfamily Dipodinae), which represent the greatest degree of specialization of the hind limbs for saltation that occurs in rodents. In these genera only three toes (digits two, three, and four) remain, and the

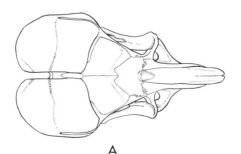
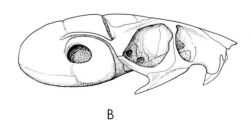

A   B

FIGURE 10–20.   Dorsal (A) and lateral (B) views of the skull of a jerboa (*Salpingotus kozlovi*, Dipodidae). Length of skull 27 mm. Note the greatly enlarged auditory bullae and the general resemblance between this skull and that of the kangaroo rat (Fig. 10–11). (After Allen, G. M.: *Mammals of China and Mongolia*, American Museum of Natural History, 1940.)

elongate metatarsals are fused into a cannon bone (Fig. 10–12D). An additional specialization that occurs in some species is a brush of stiff hairs on the ventral surfaces of the phalanges. The ears of jerboas vary from short and rounded to long and rabbitlike.

Jerboas lead lives that resemble in some ways those of members of the New World family Heteromyidae. Jerboas live in burrows that are frequently kept plugged during the day, a habit that favors water conservation by keeping the humidity in the burrow as high as possible. They are nocturnal, and many species sift seeds from sand or loose soil with the forefeet, although some species depend largely on insects for food. Unlike kangaroo rats, jerboas hibernate during the winter in fairly deep burrows. Locomotion in jerboas is chiefly bipedal, but when moving slowly the forefeet may be used to some extent. When frightened, jerboas move rapidly in a series of long leaps, each of which may cover three meters. Such a rapid, and, more important, erratic mode of escape from predation is especially effective in the barren terrain jerboas occupy.

Comparing the two well known saltatorial rodent families, the Heteromyidae and Dipodidae, the latter is far more highly specialized for saltation. This apparently is not due to the dipodids being the older group and therefore having had more time available for evolutionary change, for he-

teromyids are known from the Oligocene and the dipodids do not appear until the Pliocene. Instead, perhaps, the different degrees of adaptation are related to the habitats occupied by the two groups. Jerboas have probably been restricted throughout their history, as they are now, to deserts or dry plains, where adaptations leading to greater perfection of saltation have been advantageous. Such common heteromyid genera as *Dipodomys* and *Perognathus,* by contrast, occupy habitats ranging from arid deserts to dense chaparral, and in such broadly adapted animals extreme saltatorial modifications would seemingly be disadvantageous.

HISTRICOMORPH RODENTS

Within this division of the rodents 19 families may be included. Among these families three are monotypic (are represented by but one species), six contain a single genus with two or more species, and only four have five or more genera. Although a few histricomorphs live elsewhere, the Neotropics and Africa are the present centers of histricomorph abundance. Great structural diversity occurs within this group (some of which is illustrated in Figures 10–21 and 10–22). Among histricomorphs are the largest rodents and those with the most highly developed cursorial specializations.

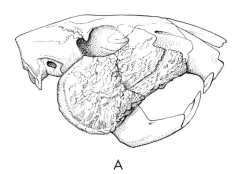
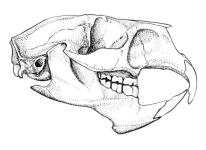

A   B

FIGURE 10–21. Skulls of two histricomorph rodents. A, paca (*Cuniculus paca,* Dasyproctidae); length of skull 150 mm. (after Hall, E. R., and Kelson, K. R.: *Mammals of North America,* copyright © 1959 by The Ronald Press Company, New York). Note the great enlargement of the zygomatic arch. B, porcupine (*Erethizon dorsatum,* Erethizontidae); length of skull 115 mm.

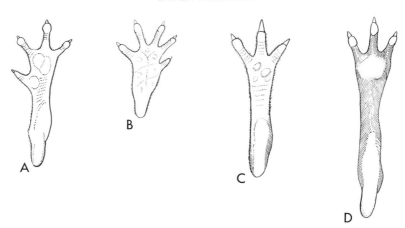

FIGURE 10–22. Ventral views of the left hind feet of some histricomorph rodents. A, chinchilla (*Chinchilla* sp., Chinchillidae); B, degu (*Octodon* sp., Octodontidae); C, agouti (*Dasyprocta* sp., Dasyproctidae); Patagonian cavy (*Dolichotis* sp., Caviidae). (After Howell, A. B.: *Speed in Animals*, The University of Chicago Press, 1944.)

In histricomorphs, the large anterior part of the masseter medialis typically originates in a more or less well-marked fossa on the side of the rostrum and passes through the greatly enlarged infraorbital foramen; insertion is on the dentary, and is largely anterior to the insertions of most other jaw muscles (Fig. 10–1C).

Family Histricidae. To this family belong the Old World porcupines, a widely distributed group of rodents (four genera with 15 Recent species) that resemble the New World porcupines (Erethizontidae) in having quills for protection. Histricids occur throughout most of Africa, in southern and central Italy, in southern Asia and South China, and in Borneo, southern Celebes, Flores, and the Philippines. This is a fairly ancient family, appearing first in the Oligocene of Europe.

These large rodents weigh up to 27 kg. and have a stocky build. The occipital region of the skull is unusually strongly built and provides attachment for powerful neck muscles, and the nasoturbinal, lacrimal, and frontal bones are pneumatic. The dental formula is 1/1, 0/0, 1/1, 3/3 = 20. The hypsodont cheek teeth have re-entrant enamel folds that, with wear, become islands on the occlusal surfaces. Some of the hairs are stiff, sharp spines; in some species open-ended, hollow spines when rattled make a noise that appears to have a warning function. One genus (*Trichys*, of Borneo, Malay Peninsula and Sumatra) lacks stiff spines. The large, plantigrade feet are five-toed, and the soles are smooth.

Histricids eat a wide variety of plant material and locally may damage crops. In contrast to New World porcupines, histricids are terrestrial rather than partly arboreal, and dig fairly extensive burrows that are used as dens. The quills are often conspicuously marked with black and white bands, and this visual signal together with rattling of the quills deters some predators. When threatened, histricids erect the quills and often rush an attacker. Aside from man, the larger cats are the main predators of histricids. Large felids often learn to kill and eat porcupines without being impaled by quills.

Family Erethizontidae. To this small family belong the New World porcupines, a group including four Recent genera with eight living species. These animals are widely distributed in forested areas, and occur from the Arctic Ocean south through much of the forested part of the United States into Sonora, Mexico, in the case of *Erethizon*, and from southern Mexico

 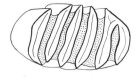

A   B

FIGURE 10–23. Crowns of histricomorph molars. A, upper right molars one and two of the porcupine (*Erethizon dorsatum*, Erethizontidae); B, third lower left molar of the capybara (*Hydrochoerus hydrochoeris*, Hydrochoeridae). The cross hatched areas on the porcupine teeth are dentine; the stippled areas on the capybara tooth are cement. (B after Ellerman, J. R.: *The families and genera of living rodents*, British Museum (Natural History), 1940.)

through much of the northern half of South America, in the case of the other genera. Porcupines are of interest to most people because of their remarkable coat of quills and because the animals often have little fear of man and can be observed easily.

New World porcupines are large, heavily built rodents, weighing up to 16 kg.; all species have quills on at least part of the body. The stiff quills are usually conspicuously marked by dark-colored and light-colored bands, and the sharp tips have small, proximally directed barbs. These barbs make the quills difficult to remove from flesh and aid in their penetration, which may be at the rate of 1 mm. or more per hour. The skull is robust, the rostrum is deep, and the greatly enlarged infraorbital foramen is nearly circular in some species (Fig. 10–21B) and accommodates the highly developed masseter medialis. The dental formula is 1/1, 0/0, 1/1, 3/3 = 20; the rooted cheek teeth have occlusal patterns dominated by re-entrant enamel folds (Fig. 10–23A). New World porcupines have some arboreal adaptations that are lacking in their more terrestrial Old World counterparts. The feet of erethizontids have broad soles that are marked by a pattern of tubercles that increases traction (Fig. 10–9B); in some species the hallux is replaced by a large, movable pad. The toes bear long, curved claws, and the limbs are functionally four-toed. In *Coendou*, a Neotropical genus, the

long tail is prehensile, and curls dorsally to grasp a branch.

Erethizontids appeared in the Oligocene of South America, but are not known from deposits earlier than the Pliocene in North America. As suggested by this evidence, this family underwent its early evolution in South America and became established in North America only after the emergence of the previously inundated Isthmus of Panama in the Pliocene.

New World porcupines utilize a great variety of plant material. *Erethizon* feeds extensively on the cambium layer of many conifers, and in many timberline areas trees missing large sections of bark and cambium give evidence of long-term occupancy by porcupines. Cambium is a staple winter food in the Rocky Mountains, but in summer a variety of plants are taken. Porcupines have been observed "grazing" at this time on sedges (*Carex*) along the borders of meadows. Most species are able climbers, and *Coendou* spends most of its life in the trees. *Erethizon* is inoffensive and at times almost oblivious of humans, but when in danger the animal directs its long dorsal hairs forward exposing the quills, erects the quills, and humps its back. The tail is flailed against an attacker as a last extremity. Surprisingly, *Erethizon* is killed by a variety of carnivores. Some mountain lions learn to flip porcupines on their backs and kill them by attacking the unprotected belly. Occasionally, however, dead or dying carnivores are found with masses of quills penetrating the mouth and face, indicating that learning to prey on porcupines may be a dangerous undertaking. Erethizontids frequently take shelter in rock piles, beneath overhanging rocks, or in hollow logs, but do not dig burrows as do Old World porcupines.

Family Caviidae. This fairly small family, including but five genera and 12 species, contains the familiar guinea pig (*Cavia*), as well as several similar types, and the Patagonian "hare" (*Dolichotis*), an animal remarkable in having many cursorial adaptations similar to those of rabbits. Caviids occur nearly throughout South America, except in Chile and parts of eastern Brazil.

The guinea pig-like caviids (Caviinae) are chunky and moderately short limbed, and weigh from roughly 400 g. to 700 g. *Dolichotis* (Dolichotinae), in contrast, has a rabbit-like form, with long slender legs and feet, and weighs up to approximately 16 kg. All caviids have ever-growing cheek teeth with occlusal patterns consisting basically of two prisms. The dentary has a conspicuous lateral groove into which insert the temporal muscle and the anterior part of the masseter medialis. Although only *Dolichotis* is strongly cursorial, all caviids have certain features typical of cursorial mammals. The clavicle is vestigial, the tibia and fibula are partly fused, and the digits are reduced to four on the manus and three on the pes. Members of the subfamily Caviinae, despite these cursorial adaptations, have a plantigrade foot posture and scuttle about in mouse-like fashion. Locomotion, in *Dolichotis*, however, is of the bounding style typical of jackrabbits. The foot posture of *Dolichotis* during running is digitigrade, and specialized pads beneath the digits (Fig. 10–22D) cushion the impact when the feet strike the ground. In *Dolichotis* the resemblance to rabbits is furthered by a deep, somewhat laterally compressed skull and large ears.

Caviids are herbivorous and social, and occupy habitats ranging from grassland and open pampas to brushy and rocky areas and forest edges. Most caviines are nocturnal or crepuscular, and often live in large colonies, the locations of which are marked by conspicuous series of burrows. *Dolichotis*, an inhabitant of open, arid regions, is diurnal, and large groups of individuals have been observed together on occasion. In some ways *Dolichotis* is a remarkably close ecological counterpart of the cursorial jackrabbits (*Lepus*). *Dolichotis* occurs in an area where *Lepus* is absent.

Family Hydrochoeridae. This family contains the largest living rodent, the capybara (*Hydrochoerus*). The two species of capybara occupy roughly the northern half of South America east of the Andes, and Panama.

Capybaras are large (up to 50 kg. in weight), robust, rather short-limbed rodents with coarse pelage. The head is large and is made especially ungraceful by a deep rostrum and truncate snout. The skull and dentary are similar to those of members of the Caviidae, but the paroccipital processes are unusually long and the teeth are ever-growing. Both M^3 and M_3 are much larger than any other cheek tooth in their respective tooth rows, and these highly specialized teeth are formed by transverse lamellae united by cement (Fig. 10–23B). The tail is vestigial, and the same cursorial features listed for the caviids occur in the capybaras. In the latter, however, the digits are partly webbed and are unusually strongly built, an adaptation that probably allows the support of considerable weight.

Capybaras occur along the borders of marshes or the banks of streams, and forage on succulent herbage. They are largely crepuscular, and although they can run fairly rapidly, they usually seek shelter in the water. Capybaras swim and dive well, and in an extremity remain submerged beneath water plants with only the nostrils above water.

Hydrochoerids first appear in Pliocene deposits of South America. Because caviids and hydrochoerids share a series of morphological characters, such as similar cursorial modifications of the limbs, some authors have regarded these groups as having a common ancestry and have placed the capybaras as a subfamily of the Caviidae (see Ellerman, 1940:237–249; and Landry, 1957:57).

Family Dinomyidae. This family includes a single South American species, *Dinomys branickii*, the pacarana. This seemingly rare animal occupies the foothills of the Andes and adjacent remote valleys in Peru, Colombia, Ecuador, and Bolivia.

The pacarana is a large rodent, up to 15 kg. in weight, with a fairly long tail; the dark brown pelage is marked by longitudinal white stripes and spots. Pacaranas lack the cursorial adaptations of the Caviidae and Hydrochoeridae. Instead, the broad, tetradactyl feet of pacaranas have long, stout claws seemingly adapted to digging, and the foot posture is plantigrade. The clavicle is complete, another departure from the conventional cursorial plan. The unusually hypsodont cheek teeth consist of a series of transverse plates.

These unusual rodents feed on a variety of plant material, and are slow-moving and docile in captivity. Although they appear to be near extinction today, they may have been more successful and diverse in the past. Fields (1957:359, 360) assigned eight fossil genera to the Dinomyidae, the oldest of which appears in Miocene beds in South America.

Family Heptaxodontidae. Several kinds of West Indian mammals became extinct in Recent or sub-Recent times and are known from fragmentary skeletal material from caves or human kitchen middens. Two genera of heptaxodontids, each with a single species, are among this group of recently extinct forms. Heptaxodontids are recorded from Puerto Rico and Hispaniola. These were large rodents, the length of the skull of the largest genus (*Elasmodontomys*) indicating an animal roughly the size of a beaver. The skull is robust, with strongly developed ridges for muscle attachments, and the cheek teeth have four to seven laminae oblique to the long axis of the anteriorly converging tooth row. These rodents were probably terrestrial and herbivorous, and were doubtless eaten by man. Little else is known of their biology.

Family Dasyproctidae. Members of this family, consisting of the agoutis and pacas, occur in the Neotropics from southern Mexico through most of the northern half of South America. Four genera and some 11 species are included in this family. This is an ancient group; the earliest records are from the Oligocene of South America.

These histricomorphs are fairly

FIGURE 10–24. An agouti (*Dasyprocta* sp., Dasyproctidae). (Photograph by Lloyd G. Ingles.)

Figure 10–25. A paca (*Cuniculus paca*, Dasyproctidae). (Photograph by Lloyd G. Ingles.)

large, from approximately 1 to 10 kg. in weight, and are described by some authors as rodents with rabbit-like heads and pig-like bodies (Figs. 10–24, 10–25). The tail is absent or rudimentary. The skull is robust, the incisors are fairly thin, and the occlusal surfaces of the hypsodont cheek teeth often bear isolated, transversely oriented islands surrounded by enamel. Although these animals are compactly built and appear to have short limbs, they have many cursorial specializations. The pacas (*Cuniculus* and *Stictomys;* Cuniculinae) have a digitigrade foot posture, but the forefeet retain four toes and the hind feet have five toes. Members of the subfamily Dasyproctinae, the agoutis (Fig. 10–24), are more highly cursorially modified, however, with tetradactyl forefeet, in which the plane of symmetry passes between digits three and four (as in the Artiodactyla), and elongate, tridactyl hind feet, in which the plane of symmetry passes through digit three (Fig. 10–22C), as in the Perissodactyla. In agoutis, the clavicle is either reduced or absent, and the claws are thick, blunt, and in some species nearly hoof-like. In the paca (*Cuniculus paca*), unique modifications of the skull and mouth allow the animal to produce a resonant, rumbling sound. Resonating chambers are formed by concavities in the maxillaries and by greatly broadened zygomatic arches (Fig. 10–21A); air is forced through associated pouches, producing the unusual sound.

Dasyproctids are herbivorous and typically inhabit tropical forests, where they take refuge in burrows that they dig in the banks of arroyos, beneath roots, or among boulders. Agoutis are rapid and agile runners, and usually travel along well-worn trails that lead from their burrows. The less cursorial pacas frequently live near water, and are prone to seek refuge in the water when threatened. Both agoutis and pacas are hunted and eaten by man.

Family Chinchillidae. One member of this family, the chinchilla (*Chinchilla*), is somewhat familiar to many people because of the publicity given to chinchilla fur farming. The family also includes the viscachas (*Lagidium* and *Lagostomus*). Three genera with eight Recent species represent the family, which occurs in roughly the southern half of South America in the high country of Peru and Bolivia, and throughout much of Argentina to near its southern tip. The fossil record of this group is entirely South American, and extends from the Oligocene to the Recent.

Chinchillids are densely furred histricomorphs of moderately large size (1 to 9 kg.), with long, well furred tails. Mountain viscachas (*Lagidium*) and chinchillas have fairly large ears and a somewhat rabbit-like appearance, whereas the plains viscacha (*Lagostomus*) has short ears. In all species the cheek teeth are ever-growing and the occlusal surfaces are formed by transverse enamel laminae with intervening cement. The incisors are narrow. There are some cursorial adaptations, but the clavicle is retained. The fore-limbs are fairly short and are tetradac-tyl; the hind limbs, however, are long, and the elongate feet have four (in *Chinchilla* and *Lagidium*) or three (in *Lagostomus*) toes.

Chinchillids are herbivorous, and occupy a variety of situations includ-ing open plains (pampas), brushlands, and barren, rocky slopes at elevations ranging from approximately 800 to 6000 m. The mountain viscachas and chinchillas are diurnal and seek shelter in burrows or crevices among rocks. Although adept at moving rap-idly over rocks and broken terrain, they seem not to depend on speed in the open to escape from enemies. The plains viscacha, in contrast, occurs in open pampas areas with little cover, where colonies live in extensive bur-row systems marked by low mounds of earth and accumulations of such debris as bones, droppings of livestock, and plant fragments. In such a habitat cur-sorial ability is highly advantageous; these animals are able to run rapidly with long leaps and to evade a pursuer by abrupt turns. They have consider-able endurance, and can run at speeds up to at least 40 km. per hour. Plains viscachas occupy a largely rabbitless area, and in locomotor ability and foraging habits strikingly resemble some of the larger rabbits (leporids).

Family Capromyidae. Members of this family are called hutias, and are restricted to the West Indies, where living species occupy the Bahama Is-lands, Cuba, Isle of Pines, Hispaniola, Puerto Rico, and Jamaica. These her-bivorous histricomorphs weigh up to 5 kg. and look like unusually large rats (Fig. 10–26). These animals are of little importance today except as interest-ing, and to some people alarming, ex-amples of a group seemingly on its way toward extinction due to the influence of man. The hutias, adapted to the con-ditions on the islands of the West Indies before the coming of European man, were unable to cope with preda-tion by the introduced mongoose (*Her-pestes*) or by man and his dogs. Of the 11 known species of capromyids, four are extinct, one having only recently become extinct, and some of the sur-viving species are apparently in danger of extinction. One living spe-cies, *Capromys nana*, was first de-scribed from bones found in cave de-posits in Cuba, but was subsequently found alive.

Family Myocastoridae. The nutria (*Myocastor coypu*), the only living member of this family, is familiar to

FIGURE 10–26. Bahamian hutias (*Geocapromys ingrahami*, Capromyidae). (Photograph by Garrett C. Clough.)

many people in North America, Europe, and Asia, because this South American rodent has been introduced widely and has thrived in certain areas. In some places it has become a serious pest because of its destruction of aquatic vegetation and crops and its disruption of irrigation systems. In some areas nutrias have caused a deterioration of waterfowl habitat. Costly Federal study and local control of the nutria has become necessary in some parts of the United States. This animal is native to southern South America, from Paraguay and southern Brazil southward. The family is also represented by nine extinct genera ranging from the Miocene to the Recent in South America.

The nutria is large, up to roughly 8 kg., and looks like a rat-tailed beaver (*Castor*). The skull is heavily ridged and has a deep rostrum. The zygomasseteric structure is of an advanced histricomorph type; in association with the reduction of the temporal muscles the coronoid process of the dentary has nearly disappeared, and is represented by a small knob (Fig. 10–3). The hypsodont cheek teeth well illustrate changes in crown pattern that occur with increasing age and wear: the enamel folds become islands under advanced wear (Fig. 10–27). The feet have heavy claws, and a web joins all but the fifth toe of the pes.

Nutrias resemble beavers in some of their habits. They dig burrows in banks, use cleared trails through vegetation, are extremely destructive to plants near their dens, and are skillful swimmers and divers. Nutrias have dense, fine underfur, and have been raised in some fur farms in the United States. Regrettably, animals that have escaped from fur farms have given rise to wild nutria populations in some areas. Most biologists strongly oppose the indiscriminate introductions of such animals as the nutria. The activities of non-native species occasionally result in the alteration of the vegetation, with the resultant disappearance of native species and the destruction, perhaps irretrievably, of the original biotic community.

Family Octodontidae. Octodontid rodents are roughly rat-sized histricomorphs that occupy sparsely vegetated terrain, rocky hills and, locally, cultivated land, in roughly the southern half of South America. Among New World histricomorphs, in this family alone are fossorial adaptations strongly developed. Octodontids occur from southern Peru and Mato Grosso, Brazil, southward, and inhabit mountains to elevations of at least 5000 m. The family includes six Recent genera with a total of approximately 33 species. The family is recorded from the Oligocene to the Recent in South America.

These rodents are unique among histricomorphs in having simplified cheek teeth in which re-entrant folds form occlusal surfaces that are roughly the shape of a figure eight (hence the name octodont). In *Ctenomys* the re-entrant folds are shallow and the figure eight is obliterated. The auditory bullae are enlarged in octodontids, as in many mammals that occupy arid or semi-arid regions. The digits bear sharp, curved claws. In some gen-

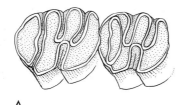   

A   B

FIGURE 10–27. First and second upper right molars of the nutria (*Myocastor coypu*, Myocastoridae); note the tremendous changes in the crown pattern due to wear. A, lightly worn molars; B, heavily worn molars. Stippled areas on the occlusal surfaces surrounded by enamel (unshaded) are dentine.

era the claws and feet are greatly enlarged and the limbs are powerfully built.

The species of the fossorial octodontid *Ctenomys* (commonly called the tuco tuco) are remarkable for their resemblance to the distantly related pocket gophers (Geomyidae). Tuco tucos are ecological, and in many ways structural, counterparts of the pocket gophers. Both groups of rodents are highly fossorial. In both tuco tucos and pocket gophers the head is large and broad and the stout incisors protrude permanently from the lips, the eyes and ears are small, the neck is short and powerfully built, the body is compact, the forelimbs are powerful, the manus has long claws, and the tail is short and stout. In contrast to pocket gophers, tuco tucos have greatly enlarged hind feet with powerful claws, and they lack external cheek pouches. Fringes of hair on the toes of fore and hind feet in tuco tucos are presumably an aid to the animals when they are moving soil. *Spalacopus*, another fossorial octodontid, can also be compared to pocket gophers. Its eyes and ears are not so strongly reduced, however, and its forelimbs are not so powerfully built.

Octodontids are herbivorous, and some species eat such below-ground parts of plants as roots, tubers, and rhizomes. Three of the six genera of octodontids are not fossorially adapted, take shelter among rocks, in brush or in burrows, and forage largely on the surface of the ground; but the other three genera are more or less fossorial. All members of the genus *Ctenomys* dig extensive burrow systems in open, often barren areas, and live in colonies composed of many solitary individuals, each with its burrow systems spaced widely apart from those of its neighbors. An individual typically occupies a given burrow system permanently, but periodically seeks adjacent foraging areas by digging new burrows. Leaves, stems, and roots are eaten, and short forays are made from open burrows to gather food (Pearson, 1959). A different mode of life is typical of *Spalacopus*. In Chile *Spalacopus cyanus* occupies sandy coastal areas where it occurs in colonies, all members of which occupy a common burrow system. The animals feed entirely below ground, and in this area the tubers and underground stems of huilli, a species of lily (*Leucoryne ixiodes*), form the bulk of the diet. *Spalacopus cyanus* is nomadic, an exceptional mode of life for a rodent. When a colony exhausts the supply of roots of huilli at one place, the animals abandon this foraging site and move to a nearby undisturbed area (Reig; 1970). Both *Ctenomys* and *Spalacopus* are unusually vocal (for rodents), and give distinctive calls at their burrow openings. Both use their forelimbs and teeth to loosen soil, and use their large hind feet to throw dirt from the mouth of the burrow. Pocket gophers, on the other hand, are never spontaneously vocal, and use the head, chest and forelimbs to push soil from the burrow.

Family Abrocomidae. Members of this family, the chinchilla rats, occur in bleak, frequently cold and mountainous parts of west-central South America. Their range includes southern Peru, Bolivia, and northwestern Argentina and Chile. The family is represented today by one genus with two species. This family appears first in the South American Miocene. There is no general agreement as to the systematic position of these rodents, and some authors regard them as a subfamily of the Octodontidae.

Abrocoma looks roughly like a large woodrat (*Neotoma*), and reaches over 400 mm. in total length. The pelage is long and dense, somewhat resembling that of the chinchilla. The skull has a long, narrow rostrum, and the bullae are enlarged. The cheek teeth are ever-growing; the upper teeth have an internal and an external enamel fold, while the lowers differ in having two internal folds. The limbs appear short, with short, weak nails. The pollex is absent.

These herbivorous rodents are poorly known. They are seemingly colonial, are scansorial (climbing) to some extent, and usually seek shelter beneath or among rocks.

Family Echimyidae. Members of this important Neotropical family, which includes a variety of roughly rat-sized rodents, are called spiny rats. Most of the 14 Recent genera have flattened, spine-like hairs with sharp points and slender basal portions. Approximately 43 Recent species are recognized. Spiny rats are widely distributed in the Neotropics, occurring from Nicaragua in Central America southward through the northern half of South America to Paraguay and southeastern Brazil.

Echimyids are normally-proportioned rodents with prominent eyes and ears. The tail, which in some genera is longer than the head and body, is lost readily, a feature perhaps of value in aiding escape from a predator. The point of weakness is at the centrum of the fifth caudal vertebra. Among 637 *Proechimys* taken in Panama, 18 per cent were tailless (Fleming, 1970:486). The cheek teeth are rooted, and the occlusal surfaces in most species are marked by transverse re-entrant folds. Aside from the reduction of the pollex, the feet are not highly specialized in most genera. In several arboreal genera (members of the subfamily Dactylomyinae), however, the digits are elongate and partially syndactylous, and when an animal is climbing the first two digits grasp one side of a branch in opposition to the remaining digits, which grasp the other side.

Echimyids are an old group, appearing first in the early Oligocene of South America. Two extinct genera of echimyids are known from skeletal material found in Indian kitchen middens in Cuba and Haiti. These genera seemingly became extinct fairly recently. In the case of the genus from Haiti (*Brotomys*), extinction may have resulted from the introduction of predators by European man.

As far as is known, spiny rats are completely herbivorous. In Panama fruit was the primary food found in the stomachs of many *Proechimys semispinosus* (Fleming, 1970:486). Many species apparently seek food by arboreal foraging. Spiny rats are common in many tropical habitats. They frequently occur in heavily vegetated areas near water, and show no tolerance of dry conditions.

Family Thryonomyidae. One genus with six species comprises this small family, the members of which are known as cane rats. These animals are broadly distributed in Africa south of the Sahara Desert. This family may have arisen as early as the Miocene, but doubt remains as to the taxonomic positions of the Miocene forms. Fossil cane rats from the Pliocene of Asia and Europe, and the occurrence of an extinct species of the Recent genus *Thryonomys* from the central Sahara Desert, indicate that the range of cane rats was once far greater than it is now.

Cane rats are large rodents, from 4 to 6 kg. in weight, and have coarse, grizzled pelage. The snout is blunt and the ears and tail are short. The robust skull has prominent ridges and a heavily built occipital region. The cheek teeth are hypsodont, and the large upper incisors are marked by three longitudinal grooves. The dental formula is 1/1, 0/0, 1/1, 3/3 = 20. The fifth digit of the forepaw is small, and the claws are strong and are adapted to digging.

Cane rats are capable swimmers and divers, and are largely restricted to the vicinity of water, where they take shelter in matted vegetation or in burrows. They are herbivorous and do considerable local damage to crops, particularly sugar cane. Cane rats are prized for food in many parts of Africa. The animals are often taken during organized drives in which dogs are used, or are driven from their hiding places and captured when natives set fire to reeds.

Family Petromyidae. This family includes but a single species, *Petromus typicus*, the dassie rat. This animal is restricted to parts of southwestern Africa and is not represented by fossil material. The taxonomic position of these animals is uncertain.

The dassie rat is a small rodent with a squirrel-like appearance. The rooted, hypsodont cheek teeth have a simplified crown pattern; the dental formula is 1/1, 0/0, 1/1, 3/3 = 20. Structurally, these animals are most remarkable for specializations enabling them to seek shelter in narrow crevices. Such specializations include a strongly flattened skull, flexible ribs that allow the body to be dorsoventrally flattened without injury, and mammae situated laterally, at the level of the scapulae, where the young can suckle while the female is wedged in a rock crevice.

Dassie rats are diurnal and feed largely on seeds and berries. They are restricted to rocky sections of foothills and mountains, where shelter in the form of crevices in rocks is available. These rodents are agile in running over rock, but choose sites for resting or sunning that are close to familiar retreats.

Family Bathyergidae. This family contains the African mole rats, a group of unusual, highly fossorial rodents. The family includes five genera with approximately 20 Recent species. These mole rats occupy much of Africa from Ghana, Sudan, Ethiopia, and Somaliland southward. The earliest fossil records of bathyergids are from the Oligocene of Mongolia.

Bathyergids are small rodents (up to 330 mm. in head and body length) that possess a number of unique structural features, many of which are associated with the animals' burrowing habits. The eyes are small in all species and vision is apparently poorly developed, and the ears lack or nearly lack pinnae. The skull is robust and of an unusual shape; the powerful incisors are procumbent in all species (Fig. 10–13A), and the roots of the upper incisors are above or behind the molars. The lips close tightly behind the incisors so that dirt does not enter the mouth when the animal is burrowing. The cheek teeth are hypsodont but rooted, and typically have a simplified crown pattern. The dental formula is variable (1/1, 0/0, 2/2 or 3/3, 0/0 to 3/3 = 12-28). In *Heliophobius* there are six cheek teeth but not all are functional simultaneously. The zygomasseteric structure is distinctive. The infraorbital foramen transmits little or no muscle, and the anterior part of the zygomatic arch is not modified into a zygomatic plate. The masseter muscles, however, are highly specialized: the large anterior part of the masseter medialis originates from the upper part of the medial wall of the orbit and the superficial part of the masseter lateralis originates partly on the anterior face of the zygoma. The mandibular fossa and angular part of the dentary are greatly enlarged (Fig. 10–13A) and provide a large surface for the insertion of the masseter muscles.

The limbs are robust in all species. The claws are long and curved or moderately long in those species that use the feet as the primary digging tools (*Bathyergus* and *Heliophobius*); they are short and more or less conical in the genera that dig mostly with the incisors (*Georychus*, *Cryptomys*, and *Heterocephalus*). The tail is usually short. The pelage is normal in most species, but in *Heterocephalus glaber* the skin is nearly naked, with only a sparse sprinkling of long, pale-colored hairs.

The African mole rats are basically herbivorous, and utilize largely bulbs, roots, and rhizomes that are reached by burrowing. These animals do considerable local damage to crops. They seldom appear above ground, and typically occupy soft loamy or sandy soils in desert and savanna areas. Mole rats seem to be the African ecological homologues of the North American pocket gophers (Geomyidae). Just as do the pocket gophers, mole rats dig extensive burrow systems that perforate the soil over fairly wide areas; but in contrast to pocket gophers, the forelimbs of all but one species of mole rat (*Bathyergus suillus*) are not highly specialized for digging. The incisors seem more important as digging tools in the mole rats, which inhabit soft soils, than in pocket gophers, which occupy a variety of soil types. It is interesting that in the two North American species of pocket gophers restricted to soft, loamy soils (*Thomomys townsendii* and *T. bulbivorus*), the incisors are unusually procumbent, approaching the condition in the mole rats. Mole rats cache quantities of food material in storage chambers connected to the main burrow systems, a habit also characteristic of most pocket gophers. Largely because of their prodigious burrowing abilities, both mole rats and pocket gophers strongly modify the environments they occupy. The burrows, the soil deposited on the surface of the ground, and the organic material brought beneath the surface of the ground have marked effects on erosion and on the density and fertility of the soil. In addition, many animals that occupy common ground with these burrowing rodents are quick to appropriate abandoned burrows, and such retreats are of critical importance to certain reptiles and amphibians.

Family Ctenodactylidae. Members of this family, commonly called gundis, occupy arid parts of northern Africa from Senegal, French West Africa, on the west, to Somaliland on the east. There are four Recent genera and eight Recent species. The fossil record of this family begins in the Oligocene of Europe and Asia.

These are small, compact, short-tailed rodents with long, soft pelage. The ears are round and short, and are protected from wind-blown debris in some species by a fringe of hair around the inner margin of the pinnae. The eyes are large. The infraorbital canal is enlarged, and through it passes part of the medial masseter muscle. The skull is flattened and the auditory bulla and external auditory meatus are enlarged. The cheek teeth are ever-growing and the crown pattern is simple. The dental formula is 1/1, 0/0, 1/1 or 2/2, 3/3 = 20 or 24. The limbs are short, and the body is carried close to the ground. The manus and pes each have four digits.

These herbivorous rodents occur in arid and semi-arid areas, where they are restricted to rocky situations. They are diurnal and crepuscular, and scurry into jumbles of rock or fissures in rock when threatened. Some members of the genus *Ctenodactylus* are known to "play possum" for long periods of time when in danger.

CHAPTER 11

CETACEANS: WHALES, PORPOISES, DOLPHINS

Cetaceans are notable for being the mammals most perfectly adapted to aquatic life. Of further interest, however, is their frequent position at the top of the food chain in marine environments generally thought to be "dominated" by fish. The large baleen whales are the largest living or fossil animals known, and cetaceans are the fastest creatures in the sea. Remarkable swimming ability, the capability to echolocate, considerable intelligence, and well-developed social behavior have all contributed to the success of cetaceans.

Morphology

All cetaceans are completely aquatic, and their structure reflects this mode of life. The body is fusiform (cigar-shaped), lacks sebaceous glands, is nearly hairless, and is insulated by thick blubber. Regional differentiation of vertebrae is not pronounced and most vertebrae have high neural spines (Fig. 11–1). The clavicle is absent, the forelimbs (flippers) are paddle-shaped, and no external digits or claws are present. The joints distal to the shoulder allow no movement. The proximal segments of the forelimb are short, whereas the digits are frequently unusually long because of the development of more phalanges per digit than the basic eutherian number (Fig. 11–2). The hind limbs are vestigial, do not attach to the axial skeleton, and are not visible externally. The flukes (tail fins) are horizontally oriented. The skull is typically highly modified as a result of the migration of the external nares to the top of the skull. The premaxillaries and

FIGURE 11–1. The skeleton of the Tasmanian beaked whale (*Tasmacetus shepherdi*, Ziphiidae).

179

FIGURE 11–2. Dorsal view of the right forelimb of the bottle-nosed dolphin (*Tursiops truncatus*, Delphinidae).

maxillaries form most of the roof, and the occipitals form the back of the skull. The nasals, frontals, and parietals are telescoped between these bones and form only a minor part of the skull roof (Figs. 11–3 and 11–4). The tympanoperiotic bone (the bone that houses the middle and inner ear) is not braced against adjacent bones of the skull, and is partly insulated from the rest of the cranium by surrounding air sinuses (Fig. 11–5).

Paleontology

The cetacean fossil record is not consistently good, and the progenitors of cetaceans are unknown. Presumably they diverged early from primitive eutherian stock, and the evidence suggests independent lines of descent

for each of the three orders. The most primitive cetaceans comprise the order Archaeoceti, which is known from the middle Eocene to the Miocene. Although these animals were well adapted to aquatic life (the hind limbs are vestigial and the body is elongate), the skull is primitive, and the migration of the external nares to the top of the skull so typical of the advanced orders is not evident (Fig. 11–6). The dentition is heterodont, and the primitive eutherian number of teeth (44) is not exceeded. Some archaeocetes attained large size. The skull of *Basilosaurus*, an Eocene type, is five feet long, and the slim body is 55 feet long.

The earliest records of the order Mysticeti, the baleen whales, are from the middle Oligocene of Europe. The origin, assumed to be from primitive toothed forms, is not known. The earliest known family, Cetotheriidae, had a partially telescoped skull and, as in modern Mysticeti, lacked teeth. The gray whale (*Eschrichtius gibbosus*) is the only surviving member of the family Eschrichtiidae, a group closely related to the extinct Cetotheriidae. The order Odontoceti, the toothed whales, appears in the late Eocene of North

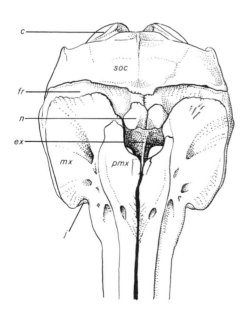

FIGURE 11–3. Dorsal view of part of the skull of a bottle-nosed dolphin (*Tursiops truncatus*, Delphinidae). Note the asymmetry of the bones surrounding the external nares (*ex*). The function of this remarkable asymmetry is not known. Abbreviations: *c*, occipital condyle; *fr*, frontal; *j*, jugal; *mx*, maxillary; *n*, nasal; *pmx*, premaxillary; *soc*, supraoccipital.

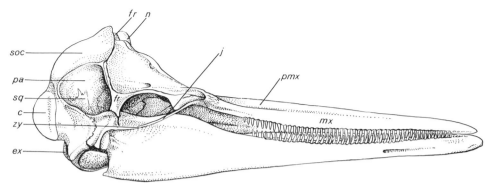

FIGURE 11–4. The skull of a dolphin (*Delphinus bairdii*, Delphinidae); length of skull 475 mm. Note the highly telescoped skull with the maxillary and frontal bones roofing the small temporal fossa. The frontal is barely exposed on the skull roof. Abbreviations: *c*, occipital condyle; *ex*, exoccipital; *fr*, frontal; *j*, jugal; *mx*, maxillary; *n*, nasal; *pa*, parietal; *pmx*, premaxillary; *sq*, squamosal; *soc*, supraoccipital; *zy*, zygomatic arch.

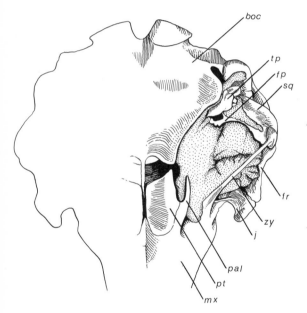

FIGURE 11–5. Ventral view of part of the skull of the grampus (*Grampus griseus*, Delphinidae), showing the large air sinuses (stippled) anterior to the tympanoperiotic bone (*tp*) and partly surrounding it. Abbreviations: *boc*, basioccipital; *fp*, falciform process of the squamosal; *fr*, frontal; *j*, jugal; *mx*, maxillary; *pal*, palatine; *pt*, pterygoid; *sq*, squamosal; *zy*, zygomatic arch. (After Purves, P. E., in Norris, K. S.: *Whales, Dolphins, and Porpoises*, originally published by the University of California Press; redrawn by permission of The Regents of the University of California.)

FIGURE 11–6. Skull of *Prozeuglodon*, a fossil archaeocete from the Eocene. Length of skull approximately 600 mm. Note the lack of telescoping of the skull and the heterodont dentition. (After Romer, A. S.: *Vertebrate Paleontology*, 3rd ed., The University of Chicago Press, 1966.)

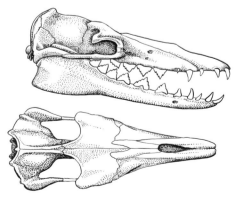

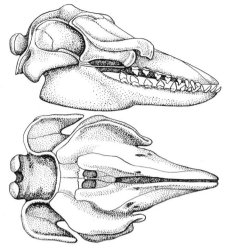

FIGURE 11–7. Skull of *Prosqualodon*, a fossil porpoise from the Miocene. Length of skull approximately 450 mm. A, lateral view; B, dorsal view. The skull is highly telescoped, and the maxillary and frontal bones form a roof over the temporal fossa (as in the modern delphinid shown in Figure 11–4). (After Romer, A. S.: *Vertebrate Paleontology*, 3rd ed., The University of Chicago Press, 1966.)

America. By the Miocene, cetaceans had undergone considerable radiation; 99 of the 173 known cetacean genera, fossil and Recent, are known from this epoch. Advanced odontocetes with highly telescoped skulls, homodont dentitions, and with many more teeth then the primitive eutherian compliment, are known from the Miocene (Fig. 11–7).

The two Recent orders of cetaceans can be distinguished on the basis of a number of morphological characters. Members of the order Mysticeti lack teeth, but have baleen-plates of horny material that grow from the upper jaw and function as a sieve (Fig. 11–8); the ascending processes of the maxillae are narrow, interlock with the frontals, and do not spread laterally over the supraorbital processes; and the respiratory and alimentary canals are not permanently separated. Mysticetes are not known to use echolocation. Odontocetes, in contrast, have simple, peglike teeth, frequently far exceeding the typical eutherian maximum number, or secondarily reduced in extreme cases to 0/1; the ascending processes of the maxillae are large, do not interlock with the frontals, and spread laterally over the supraorbital processes (Figs. 11–3, 11–4); and the respiratory and alimentary tracts are permanently separated by specializations of the glottis and the laryngeal apparatus. The bones surrounding the blowhole in odontocetes, but not in mysticetes, depart from the usual mammalian pattern of bilateral symmetry (Fig. 11–3). Probably all odontocetes use echolocation.

Swimming Adaptations

The Achilles' heel of cetaceans is their need to breathe air, but unlike most other mammals, many cetaceans are able to alternate between periods of eupnea (normal breathing) and long periods of apnea (cessation of breathing). Some whales remain submerged for roughly two hours, but many small delphinids surface to breathe several times a minute. The ability of ce-

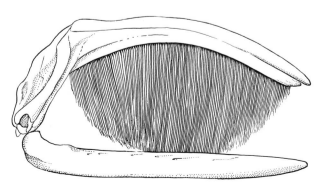

FIGURE 11–8. Skull of the Atlantic right whale (*Eubalaena*, Balaenidae); length of skull roughly 4 m. Note the baleen plates attached to the maxilla.

Table 11–1. Depths at Which Cetaceans Have Been Recorded and Methods by Which Observations Have Been Made. Data from Kooyman (1969:72), Who Cites the Sources of the Observations.

SPECIES	DEPTH (m.)	METHOD OF OBSERVATION
Balaenopteridae		
Fin Whale, *Balaenoptera physalus*	500	Harpooned and collided with bottom
	355	Depth manometer on harpoon
Physeteridae		
Sperm Whale, *Physeter catodon*	900	Entangled in deep-sea cable
	1134	Entangled in deep-sea cable
	520	Echo-sounder
Stenidae		
Rough-toothed dolphin, *Steno bredanensis*	30	Attached to depth recorder
Delphinidae		
North Pacific pilot whale,		
Globicephala scammoni	366	Inferred from feeding behavior
Bottle-nosed dolphin, *Tursiops truncatus*	92	Visual observations from underwater craft
	185	Vocalizations near underwater craft
	170	Trained to activate buzzer

taceans to remain active during periods of apnea probably depends on many adaptations. Rapid gaseous exchange is enhanced by two layers of capillaries in the interalveolar septa. During expiration most of the air can be exhausted from the lungs, and up to 12 per cent of the oxygen from inhaled air is utilized (the corresponding figure for terrestrial mammals is only 4 per cent). Compared to terrestrial mammals, cetaceans have up to twice as many erythrocytes per given volume of blood, and about two to nine times the myoglobin (a molecule able to store oxygen and to release it to tissue) in the muscles. During deep dives, the heart rate drops to roughly half the "surface rate," and vascular specializations allow blood to bypass certain muscle masses. According to Rice (1967:293), the most important physiological adaptations to prolonged submersion include "(1) anaerobic glycolysis, (2) tolerance to high levels of lactic acid, and (3) a relative insensitivity to carbon dioxide." Good discussions of cetacean adaptations to deep diving are given by Elsner (1969), Irving (1966), Kooyman and Andersen (1969), and Lenfant (1969).

Most small odontocetes are seemingly shallow divers, but some large odontocetes and some mysticetes can perform deep dives (Table 11–1). During deep dives cetaceans are subjected to tremendous pressures, for with every 10 m. increase in depth an additional atmosphere of pressure is exerted on their bodies. A cetacean swimming at a depth of only 200 m., probably a commonplace depth for many species, is subjected to 20 atmospheres of pressure, or 294 lbs. per square inch. Because of their lungs and air sinuses, cetaceans have semi-hollow bodies, which tend to collapse during deep dives. Any gases that remain within body cavities are therefore subjected to great pressures, resulting in a decrease in their volumes and an increase in the amounts of gases that go into solution in the body solvents, such as blood. One serious result in man is "bends," or decompression sickness. When man uses equipment that allows him to breathe under water, and undergoes prolonged exposure to high pressures during diving, greater than normal amounts of gases in the lungs are dissolved in the tissues and the blood. If decompression is too rapid, these gases cannot be carried to the lungs rapidly enough to be removed from the body; instead, they quickly leave solution and appear as bubbles in the tissues. Intravascular bubbles may occlude capillaries and result in injury to tissues or even in death.

Cetaceans are not known to have these problems, although they stay for considerable periods at great depths and ascend to the surface rapidly. This is partly due to their ceasing breathing while beneath the water, but is also a result of structural modifications. The following anatomical specializations are seemingly adaptations to deep diving: (1) a large proportion of the ribs lack sternal attachment or lack attachment either to the sternum or to other ribs; (2) the lungs are dorsally situated above the oblique diaphragm; (3) in deep divers the lungs are small and the volume of the air passages is relatively large; (4) the trachea is short and often of large diameter, and the cartilaginous rings bracing the trachea are nearly complete, have small intermittent breaks, or are fused (Slijper, 1962); (5) the bronchioles are reduced in length and the entire system of bronchioles, to the very origin of the alveolar ducts, is braced by cartilaginous rings; (6) the lungs, especially the walls of the alveolar ducts and the septa, contain unusually large concentrations of elastic fibers; (7) in some of the small odontocetes series of myoelastic sphincters occur in the terminal sections of the bronchioles. The adaptive importance of all these features is not yet completely understood. Clearly the specializations of the ribs and the placement of the lungs relative to the diaphragm permit the collapse of the lungs, and bracing of the respiratory passages may aid in the regulation of air in the lungs during diving. Most probably the alveoli fully collapse during deep dives, forcing air into the respiratory passages and the air sinuses that surround the tympanoperiotic bone (Fig. 11–5). Fraser and Purves (1955) hypothesize that this air, plus mucous and fat droplets, forms a foam that absorbs nitrogen (nitrogen dissolves roughly six times as rapidly in some oils as it does in water) and other gases; this prevents the gases from going into solution in the blood stream and leading to decompression problems. This foam is expelled in the "blow" of a cetacean. (The possible function of this foam in connection with echolocation is discussed on p. 406).

Cetaceans are remarkably fast swimmers. Powerful dorsoventral movements of the tail provide propulsion, and the flippers are used for steering. Dolphins have been observed to maintain speeds of about 32 km. per hour for up to 25 minutes, and for short periods to reach 38 km. per hour (Johannessen and Harder, 1960). These authors report that a killer whale (*Orcinus orca*) approached a ship at some 56 km. per hour and cruised around the ship for 20 minutes at speeds in excess of the ship's 38 km. per hour. A blue whale (which weighs some 200,000 pounds) has been observed traveling for 10 minutes at 37 km. per hour (Gawn, 1948), and a pilot whale swam in a tight circle around a tank at the same speed (Norris and Prescott, 1961:343).

This remarkable swimming performance of cetaceans has proven difficult to explain. Recent studies have demonstrated that their speed is not due to muscles that are vastly more powerful than those of other mammals, but to specializations that greatly reduce resistance (drag) as the animals swim. The amount of resistance depends on the type of flow of water over the surface of the body. If the flow is smooth—parallel to the surface—it is said to be *laminar*. When such smooth flow is interrupted by movements of the water that are not consistently parallel to the surface of the body, however, *turbulent* flow occurs. All other conditions being equal, laminar flow creates much less resistance than does turbulent flow. If the bodies of small dolphins were subjected to turbulent flow, swimming at 24 mph would require their muscles to be five times as powerful as those of man (Lang, 1966:426). But assuming flow to be nearly laminar, this speed is approximately what would be expected if their power output were that of a well-trained human athlete. Scientists for many years have attempted without success to design bodies shaped so that flow of air or water over their surfaces is laminar or nearly so. What is the cetacean solution to this problem?

Several factors seemingly contribute importantly to furthering the reduction of resistance as a porpoise swims rapidly (these are discussed by Hertell, 1969:33-49). The body is hairless, and no obstructions except the streamlined appendages break the extremely smooth surface. In addition, the body form of the dolphin is approximately parabolic (lens-shaped); this form creates even less resistance than the rounded (elliptical) head end and tapered body of such a rapid swimmer as a trout. Perhaps most remarkable is the structure of the skin (described by Kramer, 1960). The skin consists basically of two layers. The soft outer layer is 80 per cent water and has narrow canals filled with a spongy material; the stiff inner layer has stringy connective tissue. Pressure differences caused by turbulent flow of water are transmitted to the soft outer skin, which yields in areas of increased pressure and expands where pressure is reduced. The liquid in the skin tends to move parallel to the surface in response to distortion of the skin, but this movement is inhibited by the channels and their spongy filling. These effects result in a strong damping of vibrations caused by turbulent flow and a more complete laminar flow over the body. According to Hertel (1969:40), further reduction of resistance may be associated with movements of the flukes and body during swimming. The story, however, is still incomplete; more must be learned in order for us to fully account for the remarkable high-speed swimming ability of cetaceans.

ORDER MYSTICETI

This order contains the huge baleen whales, which inhabit all oceans. There are but ten Recent species grouped in five genera and three families. In terms of abundance and numbers of species, this order is less important than the Odontoceti.

All mysticetes are filter-feeders. Plankton-rich water is taken into the large mouth and forced out through the plates of baleen by the enormous tongue; the plankton strained from the water by the baleen is swallowed. Under some conditions these whales "skim" plankton from the surface with the head partly exposed, but they may also feed down to several hundred meters below the surface. One species is known to feed on the floor of shallow seas.

Family Balaenidae. These, the right whales, are creatures that were killed in such great numbers during the height of the whaling activities that they are rare today and are protected by international treaty. The family includes three Recent species of two genera, and inhabits most marine waters except tropical and south polar seas. Balaenids are known as fossils from the Miocene to the Recent.

These are large, robust whales that reach about 18 m. in length and over 67,000 kg. in weight. The head and tongue are huge — the head amounts to nearly one third of the total length. The flippers are short and rounded, and the dorsal fin is usually absent. There are more than 350 long baleen plates on each side of the upper jaw; these plates fold on the floor of the mouth when the jaws are closed. No furrows are present in the skin of the throat or chest. The cervical vertebrae are fused, and the skull is telescoped to the extent that the nasals are small and the frontals are barely exposed on the top of the skull.

Right whales feed largely on planktonic crustaceans and molluscs, and are most common near coastlines or near pack-ice. These whales are not highly migratory, and some species remain in far-northern waters throughout the year. According to Rice (1967: 299), Eskimos claim that the bowhead whale (*Balaena mysticetus*) can break through ice nearly one meter thick to reach air.

Family Eschrichtiidae. This family is represented today only by the gray whale (*Eschrichtius gibbosus*), which occupies parts of the North Pacific. There is no fossil record of this family, but there are subfossil records from the North Atlantic (Rice and Wolman, 1971:20).

The gray whale is fairly large, weighing up to about 31,500 kg. and measuring some 15 m. in length, and has a slender body with no dorsal fin. The baleen plates are short, and the telescoping of the skull is not extreme; the nasals are large and the frontals are broadly visible on the roof of the skull. The throat usually has two longitudinal furrows in the skin.

Gray whales perform the longest known mammalian migration. They occupy parts of the North Pacific (the Bering, Chukchi, and Okhotsk seas) in the summer. Here they feed largely on bottom-dwelling crustaceans (amphipods), which they take by stirring up the sediments with their snouts. In late autumn they migrate southward along the coastlines. The western Pacific population winters along the coast of Korea, and the eastern Pacific gray whales winter along the coast of Baja California. Young are born in shallow coastal lagoons in the wintering areas. The round trip distance of the migration is from about 10,000 to 22,000 km. (some 6,500 to 14,500 miles!). Many people each year watch migrating gray whales from a vantage point at Cabrillo National Monument, near San Diego, California.

The future of the gray whale today seems reasonably bright. Driven near extinction by whaling activities between 1850 and 1925, they are now protected by the International Convention for the Regulation of Whaling, and have increased greatly in recent years. Unfortunately, however, the gray whale is being molested to some extent by sightseers in its wintering and breeding areas in bays of Baja California.

Family Balaenopteridae. This group of whales, frequently known as rorquals, includes six Recent species of two genera. The distribution includes all oceans.

These whales vary in size from fairly small (for whales) to extremely large. The huge blue whale (*Balaenoptera musculus*), a giant even among whales, reaches a length of about 31 m. and a weight estimated at some 160,000 kg. In some species of rorquals the body is slender and streamlined, but it is chunky in others. The baleen plates are short and broad, and the skin of the throat and chest is marked by numerous longitudinal furrows. The nasals are small and the frontals are either not exposed or only barely exposed on the skull roof.

These whales feed in cold waters, often near the edges of the ice where upwelling water results in great growths of plankton in summer. Planktonic crustaceans and small schooling fish are eaten. During the northern winter, the Northern Hemisphere populations move southward toward equatorial areas, and during the southern winter, southern populations move northward. Wintering adults do not feed, but live off stored blubber. Breeding occurs in the wintering areas, but because the southern and northern winters are six months out of phase, no interbreeding between populations occurs. The humpback whale (*Megaptera novaeangliae*), an animal given to spectacular leaps, makes remarkably melodious and varied underwater sounds that have been beautifully recorded by Payne (1970) and discussed by Payne and McVay (1971).

Excessive commercial exploitation has resulted in a tremendous decline in the populations of fin whales (*Balaenoptera physalus*), humpbacks, and blue whales. Over broad areas, blue and humpback whales are so scarce as to be "commercially extinct" (Rice, 1967:304), but intense hunting of whales continues over broad areas and populations continue to decline at alarming rates. A bleak prospect seems imminent: within this century man could destroy the blue whale, the largest and one of the most remarkable species of animals that has ever lived.

ORDER ODONTOCETI

The toothed whales, porpoises, and dolphins comprise the most important group of cetaceans. Not only are odontocetes vastly more numerous than mysticetes, both with respect to abundance and diversity, but they are more widely distributed. The order Odontoceti includes about 74 Recent species within 33 genera and seven families, and occurs in all oceans and seas connected to oceans. Members of five families also inhabit some rivers and lakes in North America, South America, Asia, and Africa. Odontocetes are readily observed; they frequently forage close to shore, often make spectacular leaps, and roll repeatedly out of the water; some are prone to ride the bow waves of ships much as man rides shore waves.

The classification of the odontocetes is still uncertain; the arrangement used here is that of Fraser and Purves (1960).

Family Ziphiidae. The beaked whales are widely distributed, occupying all oceans, but are rather poorly known; some species have never been seen alive. Eighteen Recent species of five genera are recognized. The earliest fossil record of ziphiids is from the early Miocene.

These are medium-sized cetaceans with fairly slender bodies. The length varies from 4 to over 12 m., and the weight reaches some 11,500 kg. The snout is usually long and narrow, and in some species the forehead bulges prominently. Only one species (*Tasmacetus shepherdi*) has a large number of teeth; in the others the dentition is strongly reduced. Only two lower teeth on each side occur in the two species of *Berardius;* in all of the remaining ziphiids there is only a single functional lower tooth on each side (Fig. 11–9). In some species the lower jaw is "undershot" and the teeth are outside the mouth. Two to seven cervical vertebrae are fused. The stomach is divided into from four to 14 chambers.

Beaked whales are deep divers that are able to remain submerged for long periods. The North Atlantic bottle-nosed whale (*Hyperoodon ampullatus*) may dive for periods up to two hours. Some species forage in the open ocean well away from coasts. Whereas some species are solitary, others are highly social and travel in schools in which all members surface and dive in synchrony. The primary food is squid, but deep sea fishes are also taken. The North Atlantic bottle-nosed whale is known to make annual migrations, and other species are probably also migratory.

Family Monodontidae. This family contains but two species, the narwhal (*Monodon monoceros*), remarkable for its long, straight, forward-directed tusk, and the beluga (*Delphinapterus leucas*), also called the white whale. They occur in the Arctic Ocean and the Bering and Okhotsk seas, in Hudson

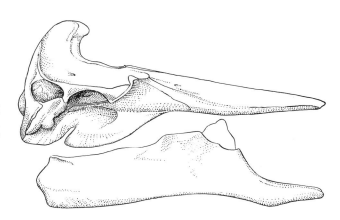

FIGURE 11–9. Skull of a beaked whale (*Mesoplodon* sp., Ziphiidae); length of skull about 590 mm. Note the single large tooth in the dentary.

Bay, in the St. Lawrence River in Canada, and in some large rivers in Siberia and Alaska. Fossil monodontids are known from the Pleistocene and Recent of arctic areas of North America and Eurasia.

These are medium-sized cetaceans; belugas reach about 6 m. in length and 2000 kg. in weight, and narwhals, without the tusk, are a similar length. The facial depression in the skull is large, and the maxillary and frontal bones roof over the reduced temporal fossa (refer to Figs. 11–3 and 11–4); the zygomatic process of the squamosal is strongly reduced; and the cervical vertebrae are not fused. The beluga has 9/9 teeth, and the narwhal has 1/0. One tooth of the male narwhal (usually the left) forms a straight, spirally-grooved tusk up to 2.7 m. long; the corresponding tooth in the other jaw is normally rudimentary.

These gregarious cetaceans are characteristic of northern seas, where in winter they assemble in areas of open water. In summer belugas move far up large rivers. Belugas feed largely on fish, both benthic (bottom dwelling) kinds and those that live at intermediate depths, and squid. Narwhals are seemingly largely pelagic (open-sea dwellers). The function of the narwhal's tusk is not known, but it is this tusk, discovered separate from the whale, that probably gave rise to the myth of the unicorn. Both species are quite vocal, and the trilling sounds made under water by belugas account for their common name of "sea canary."

Family Physeteridae. The sperm whales occur in all oceans, and the giant sperm whale (*Physeter catodon*), of Moby Dick fame, has long been an important species to the whaling industry. Fossil sperm whales are known from the early Miocene.

Physeter is large, attaining a length of over 18 m. and weight in excess of 53,000 kg.; the pigmy sperm whales (*Kogia*) are small, reaching about 4 m. in length and some 320 kg. in weight. The head is huge in *Physeter,* accounting for over one third of the total

length. In both genera the rostrum is truncate, broad, and flat. The facial depression of *Physeter* contains a spermaceti organ, which contains great quantities of oil. The blowhole is toward the end of the left side of the snout. The left nasal passage serves in respiration, but the right one is specialized as a sound producing organ. The upper jaw lacks functional teeth; the lower jaw has some 25 functional teeth on each side in *Physeter,* and from 8 to 16 in *Kogia.* All of the cervicals are fused in *Kogia,* and all but the atlas are fused in *Physeter.*

The habits of *Kogia* are not well known, but those of *Physeter* are better understood, probably because man has persistently hunted this animal for many years. *Physeter* is social and assembles in groups with occasionally as many as 1000 individuals. Schools of females with their calves together with male and female subadults are overseen by one or more large adult males, whereas younger males congregate in "bachelor schools." Some adult males are solitary. Sperm whales generally forage in the open sea at depths where little or no light penetrates (the use of echolocation by *Physeter* is discussed on page 407). Dives to depths of roughly 1000 m. are probably usual, and dives of 1130 m. have been recorded (Heezen, 1957). *Physeter* feeds largely on deep water squids, including giant squids, and a variety of bony fishes, sharks, and skates. Males commonly migrate far north to the edge of the pack-ice in summer, but females remain in temperate and tropical waters. *Kogia* is either solitary or travels in small schools, and feeds largely on cephalopods such as squid and cuttlefish.

Family Platanistidae. This group, the long-snouted river dolphins, is remarkable because its members live largely in rivers. The distribution includes some large river systems in India, the Amazon and Orinoco river systems of South America, coastal waters along the east coast of South America, and Tungt'ing Lake in China. There are four Recent genera, each with a single species, and fossil members are recorded back to early Miocene.

These are small cetaceans, from 1.5 to 2.9 m. in length and from about 40 to 125 kg. in weight. The jaws are unusually long and narrow and bear numerous teeth (from about 26/26 to 55/55); the forehead rises abruptly and is rounded, giving the head an almost bird-like aspect. In *Platanista* there are large maxillary crests (discussed on page 407). The large temporal fossa is not roofed by the maxillary and frontal bones. None of the cervical vertebrae are fused. The eyes of all members of the family are reduced, and presumably food and obstacles are detected largely by echolocation. The Ganges dolphin (*Platanista*), which usually swims on its side (Fig. 11–10), lacks eye lenses and can perhaps only detect light and dark; the eyes of the white-flag dolphin (*Lipotes*) are greatly reduced and vision is presumably poor, and the eyes of the other river dolphins are small but are presumably functional.

These strange cetaceans often in-habit rivers that are made nearly opaque by suspended sediment, and under these conditions echolocation may completely supplant vision. A variety of fishes and crustaceans are eaten, some of which are captured by probing muddy river bottoms. The Amazon dolphins (*Inia*) feed entirely on fish, and during the rainy season may move deep into flooded tropical forests (von Humboldt and Bonpland, 1852). River dolphins are seemingly not as social as are many other cetaceans. Only 14 per cent of Layne's (1958:18) observations of Amazon dolphins were of groups with more than four individuals, and these animals typically did not form closely knit groups. Layne (1958:16) made observations that suggest fairly acute vision above water. Individuals that were approaching a narrow channel used their eyes above water, presumably to scan the banks for danger.

Family Stenidae. This family includes eight Recent species representing three genera. They inhabit much of the east coast of South America and the Amazon and Orinoco river systems

FIGURE 11–10. The Ganges dolphin (*Platanista gangetica*, Platanistidae), which normally swims on its side. (From Herald, E. S., et al.: Blind river dolphin: first side-swimming cetacean, *Science*, 166:1408–1410, 1969. Copyright 1969 by the American Association for the Advancement of Science.)

(*Sotalia*), tropical and warm temperate waters of all oceans (*Steno*), and coastal waters, river mouths, and estuaries of southern Asia and some of the coasts of Africa.

The members of this family resemble the Delphinidae except for certain details of the air-sinus system. The snout is slender and has from approximately 24/24 to 32/32 teeth. These are fairly small cetaceans, weighing from about 50 to 70 kg.

The habits of these animals are very poorly known. Scattered information indicates primarily a fish and cephalopod diet. *Sotalia*, which occurs together with the plantanistid *Inia* in some large rivers in South America, may be more prone than the latter to leave the river channels and penetrate inundated jungles (Layne, 1958:4).

Family Phocoenidae. The members of this family are generally called porpoises. Seven Recent species of three genera are recognized. They occur widely in coastal waters of all oceans and connected seas of the Northern Hemisphere, as well as in some coastal waters of South America and some rivers in southeastern Asia. The earliest fossil record of phocoenids is from the late Miocene.

Phocoenids are small, from about 1.5 to 2.1 m. in length and from roughly 90 to 118 kg. in weight, and have fairly short jaws and no beak. The dorsal fin is either low or absent. The skull resembles that of the Delphinidae, but

FIGURE 11-11. Dorsal view of the head of the Pacific pilot whale (*Globicephala scammoni*, Delphinidae), showing the "blowhole" (the opening of the external nares). (Photograph courtesy of the Marineland of the Pacific.)

FIGURE 11–12.   Bottle-nosed dolphin (*Tursiops truncatus*, Delphinidae) giving birth. (Photograph courtesy of the Marineland of the Pacific.)

has conspicuous prominences anterior to the nares. The teeth of phocoenids are distinctive in being laterally compressed and spadelike; the crowns have two or three weakly developed cusps. The number of teeth varies from 15/15 to 30/30. From three to seven cervical vertebrae are fused.

Some phocoenids (*Phocoena* and *Neophocaena*) inhabit inshore waters, such as bays and estuaries, whereas the swift white-flanked porpoises (*Phocoenoides*) generally inhabit deeper water. Small schools of at least 100 phocoenids may assemble, and crescentic formations associated with feeding have been noted (Fink, 1959). A variety of food is taken, including such cephalopods as cuttlefish and squid, crustaceans, and fish.

Family Delphinidae (Figs. 11–11, 11–12).   This is by far the largest and most diverse group of cetaceans. Because some species come close to shore and roll and jump conspicu-

ously, they are the most frequently observed cetaceans. About 32 Recent species representing 14 genera are known. Delphinids inhabit all oceans and some large rivers in southeastern Asia. Fossil delphinids appear first in the early Miocene.

Small delphinids are roughly 1.5 m. in length and some 100 kg. in weight, but the killer whale (*Orcinus orca*) reaches 9.5 m. in length and at least 7000 kg. in weight. The facial depression of the skull is large, and the frontal and maxillary bones roof over the reduced temporal fossa (Fig. 11–4). The "melon," a lens-shaped fatty deposit that lies in the facial depression, is well developed and gives many delphinids a forehead that bulges prominently behind a beaklike snout. Some delphinids, such as the killer whale, lack a beak and have a rounded profile. The number of teeth varies from 65/58 to 0/2. From two to six cervical vertebrae are fused (Fig. 11–13).

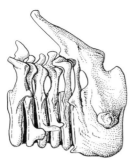

FIGURE 11–13. Cervical vertebrae of a dolphin (*Delphinus bairdii*, Delphinidae). Only the axis and atlas are fused, whereas most of the series are fused in some cetaceans.

Males are typically larger than females, and in some species there is considerable sexual dimorphism in the shapes of the flippers and dorsal fin. Coloration is varied: some species are uniformly black or gray, some have beautiful contrasting patterns of black and white, and still others have colored stripes or spots.

Delphinids characteristically feed by making shallow dives and surfacing several times a minute. They are rapid swimmers, and some species regularly leap from the water during feeding and traveling. In the Gulf of California I observed a bottle-nosed dolphin (*Tursiops truncatus*) leaping completely out of the water and catching mullet in midair, much as a trout catches a fly. Pacific striped dolphins (*Lagenorhynchus obliquidens*) have been trained at Marineland of the Pacific to leap over a wire 4.8 m. (over 15 feet!) above the water. Most small delphinids eat fish and squid, but the killer whale is known to take a great variety of items including large bony fish, sharks, sea birds, sea otters, seals and sea lions, porpoises, dolphins, and whales.

Most delphinids are highly gregarious, and assemblages of approximately 100,000 individuals have been observed. Some groups of delphinids kept in large tanks establish a dominance heirarchy, with an adult male having the highest position (Bateson and Gilbert, 1966). From an underwater vehicle Evans (Evans and Bastian, 1969:448) observed that narrow-snouted dolphins (*Stenella attenuata*) had three major types of groups. The first had five to nine adult females and juveniles; the second consisted of a lone male, occasionally accompanied by a female; and the third included four to eight subadult males. Individuals of the adult female and juvenile group were spaced one above the other, with the topmost individual near the surface and the others occupying successively deeper levels. Members of the subadult male group, in contrast, were spread out horizontally. Similar spatial arrangement of groups has been observed in other species by Evans. Many recent studies have indicated that cetaceans are remarkably intelligent and inventive (see, for example, Tavolga, 1966). Their behavioral adaptability is demonstrated by the observations of Hoese (1971), who watched two bottle-nosed dolphins cooperatively pushing waves onto a muddy shore and stranding small fish. The dolphins rushed up the bank and snatched the fish from the mud before sliding back into the water.

CHAPTER 12

CARNIVORES

Predation in mammals is an ancient and profitable, if not entirely honorable, occupation. Primitive carnivorous mammals (creodonts) appear in the early Paleocene, before the appearance of most of the Recent mammalian orders. Mammalian carnivores probably evolved in response to the food source offered by an expanding array of terrestrial herbivores, and underwent adaptive radiation as herbivores diversified.

The classification of carnivorous mammals used in this discussion is that of Romer (1966; see also Romer, 1968:189, 190), and recognizes the order Creodonta (primitive carnivorous mammals) and the order Carnivora. Included in the Carnivora are the suborders Fissipedia (terrestrial carnivores) and Pinnipedia (aquatic carnivores: seals, sea lions, and walruses). Other schemes of classification have been proposed. Simpson (1945), for example, regarded the Creodonta, Fissipedia, and Pinnipedia as suborders of the order Carnivora, and Van Valen (1966) described a new order, Deltatheridia, into which he put some archaic carnivorous types. Present classifications will probably be modified as new fossil material clarifies our picture of the evolution of the carnivorous mammals.

Most Recent fissiped carnivores are predaceous and have a remarkable sense of smell. Cursorial ability may be limited, as in the Ursidae and Procyonidae, or may be strongly developed, as in the cheetah and some canids. The braincase is large; the orbit is usually confluent with the temporal fossa; the turbinal bones are usually large, and their complex form provides a large surface area for olfactory epithelium. There are usually 3/3 incisors (3/2 in the sea otter, *Enhydra lutris*), and the canines are large and usually conical; the cheek teeth vary from 4/4 premolars and 2/3 molars in long-faced carnivores, such as the Canidae and Ursidae, to 2/2 premolars and 1/1 molars, as in some cats. The fourth upper premolar and the first lower molar are carnassials (specialized shearing blades). The teeth are rooted. The condyle of the dentary and the glenoid fossa of the squamosal are transversely elongate and allow no rotary jaw action and only limited transverse movement. Cursorial adaptations evident in the carpus include the fusion of the scaphoid and lunar bones and the loss of the centrale (Fig. 12–1). The foot posture is plantigrade, as in ursids and procyonids, or digitigrade, as in canids, hyaenids, and felids. Little reduction of digits has occurred; the greatest reduction occurs in the hyaenas and in the African hunting

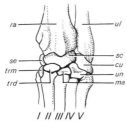

Figure 12–1. Anterior view of the left carpus of the grey fox (*Urocyon cinereoargenteus*). Abbreviations: *cu*, cuneiform; *ma*, magnum; *ra*, radius; *se*, sesamoid; *sc*, scapholunar; *trd*, trapezoid; *trm*, trapezium; *ul*, ulna; *un*, unciform.

dog (*Lycaon pictus*), in which the manus and pes have four toes.

ORDER CREODONTA

The oldest carnivorous mammals, order Creodonta, appeared in the early Paleocene, were the typical carnivores of the Paleocene and Eocene, and persisted in Old World tropical refugia into the early Pliocene. Creodonts probably evolved from primitive Cretaceous types similar to *Deltatheridium* (Fig. 6–2B, p. 68). Some creodonts appear to have retained the insectivorous food habits of their ancestors, whereas some were carnivorous and some were omnivorous. The creodont skull (Fig. 12–2) differs from that of the Carnivora in lacking an ossified auditory bulla, and in either having no carnassial pair or in having M^1 and M_2, or M^2 and M_3, forming the carnassials. The creodont braincase was small, and intelligence was presumably low. The limbs were primitive.

Figure 12–2. The skull of *Sinopa*, an Eocene creodont in which the carnassials are M^2 and M_3; length of skull approximately 150 mm. (After Romer, A. S.: *Vertebrate Paleontology*, 3rd ed., The University of Chicago Press, 1966.)

The feet were usually five-toed and plantigrade, and the limbs were often short; the scaphoid and lunar of the carpus were not fused and the centrale was present; the distal phalanges were fissured, and in some species bore flattened, rather than claw-like, nails.

ORDER CARNIVORA

SUBORDER FISSIPEDIA

Modern terrestrial carnivores, suborder Fissipedia, are advanced, basically predatory types. Fissipeds largely replaced the smaller brained and less cursorial creodonts early in the Tertiary. The basal fissiped family is the Miacidae, which first appears in middle Paleocene beds of North America and is not known after the Eocene. Miacids were small and perhaps mostly arboreal carnivores. Morphologically transitional between the Creodonta and Carnivora in some features, the miacids had the modern carnassial arrangement, but lacked ossified bullae and had separate scaphoid and lunar bones. In contrast to the creodonts, the brain was fairly large, and the distal phalanges were not fissured. Except for the Miacidae, all of the families of the suborder Fissipedia survive today.

Family Canidae (Figs. 12–3, 12–4). Fifteen genera with about 41 Recent species comprise this familiar family. The canids—the foxes, wolves, dogs, and jackals—occupy a great array of environments from the arctic to the tropics. Previous to their dispersal with man, canids occurred nearly world-wide except on most oceanic islands. The dingo (*Canis dingo*) was probably brought to Australia by early man. The canids appeared in the late Eocene in Europe and North America and have occupied these areas continuously to the Recent.

Canids are broadly adapted carnivores, a feature reflected in their morphology. The canid skull typically has a long rostrum (Fig. 12–5A) that

FIGURE 12-3. The black-backed jackal (*Canis mesomelas*), a common canid in East Africa and South Africa. (Photograph by W. Leslie Robinette.)

houses a large nasal chamber with complex turbinal bones, a feature associated with a remarkable sense of smell. Most canids have a nearly complete placental compliment of teeth (3/3, 1/1, 4/4, 2/3 = 42); the canines are generally long and strongly built, and the carnassials retain the shearing blades (Fig. 12-6A, B). The post-carnassial teeth have crushing surfaces, indicating a more flexible diet than that of the more strictly carnivorous cat family (Felidae). The limbs in most species are long, and rotation at the joints distal to the shoulder and hip joints is reduced in the interest of cursorial ability. The clavicle is absent. The feet are digitigrade, and the well developed but blunt claws are non-retractile. The forepaw usually has five toes, and the hind paw has four. The weight of canids ranges from roughly 1 to 75 kg.

As a family, the Canidae are the most cursorial carnivores. Many species forage tirelessly over large areas, and lengthy pursuit is frequently part of the hunting technique. (The canid hunting style is discussed in Chapter 17, p. 324.) The coyote (*Canis latrans*), probably one of the swiftest canids, can run at speeds up to about 65 km.

FIGURE 12-4. A young kit fox (*Vulpes macrotis*) at the entrance to its burrow. This fox occurs in the deserts of the southwestern United States. (Photograph by O. J. Reichman.)

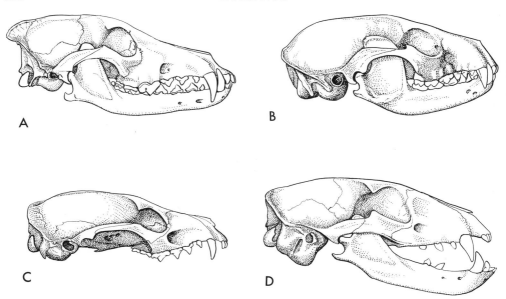

FIGURE 12–5. The skulls of several carnivores. A, the coyote (*Canis latrans*); length of skull 202 mm. B, a raccoon (*Procyon lotor*); length of skull 115 mm. C, a genet (*Genetta victoriae*); length of skull 108 mm. D, the aardwolf (*Proteles cristatus*); length of skull 134 mm. (*Procyon* after Allen, 1924. *Proteles* after Hill and Carter, 1941.)

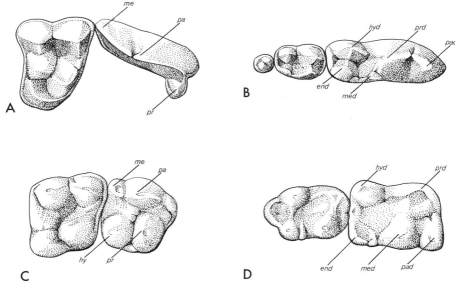

FIGURE 12–6. Occlusal view of some cheek teeth of two carnivores; the right upper teeth are on the left, and the left lower teeth are on the right. The cusps of the carnassials (P^4 and M_1) are labeled. A and B, the coyote (*Canis latrans*); the cheek teeth serve both shearing and crushing functions. C and D, a raccoon (*Procyon lotor*); note that the upper carnassial has a hypocone and that all of the teeth are adapted to crushing. Abbreviations: *end*, entoconid; *hy*, hypocone; *hyd*, hypoconid; *me*, metacone; *med*, metaconid; *pa*, paracone; *pad*, paraconid; *pas*, parastyle; *pr*, protocone; *prd*, protoconid.

per hour. The fact that coyotes in many areas depend partly on jackrabbits for food is an impressive testimonial to this carnivore's speed. Canids typically hunt in open country, and wolves (*Canis lupus*) and the African hunting dog (*Lycaon pictus*) seem to rely more on endurance than on speed when hunting. These canids, and the east Asian dholes (*Cuon alpinus*), habitually hunt in packs and kill larger prey than could be overcome by a solitary hunter. The grey fox (*Urocyon cinereoargenteus*) does not generally forage in open areas, but is amazingly agile and can run rapidly through the maze of stems beneath a canopy of chaparral (Fig. 15–18, p. 272). The food of canids includes vertebrates, arthropods, mollusks, carrion, and many types of plant material. Black-backed jackals have recently become a problem in parts of South Africa because of their extensive feeding on pineapples (Ewer, 1968:30), and coyotes in parts of the western United States feed heavily on such cultivated crops as melons and such non-cultivated plant material as juniper berries and prickly-pear cactus fruit. The average canid is clearly an opportunist; this may in large part account for the great success of this family.

Family Ursidae. The bears are notable for their large size and their departure from a strictly carnivorous mode of life. This family contains six genera and eight species (if the giant panda, *Ailuropoda*, is considered to be

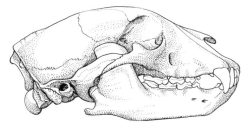

FIGURE 12–7. Skull of a black bear (*Ursus americanus*); length of skull 289 mm. Note the diastema behind the anteriormost cheek tooth and the small upper carnassial (the third tooth in back of the canine).

an ursid). The distribution includes most of North America and Eurasia, the Malay Peninsula, the South American Andes, and the Atlas Mountains of extreme northwestern Africa. Bears inhabit diverse habitats, from drifting ice in the arctic to the tropics, but are most important in boreal and temperate areas.

The bears are an offshoot from the canid evolutionary line, and first appeared in Europe in the middle Miocene. Ursids apparently did not enter North America until late Pliocene. They probably reached South America and northwest Africa in the Pleistocene and Recent, respectively.

The bear skull retains the long rostrum typical of the canids, but the orbits are generally smaller and the dentition is very different (Fig. 12–7). The post-carnassial teeth are greatly enlarged, and the occlusal surfaces are "wrinkled" and adapted to crushing (Fig. 12–8B). On the other hand,

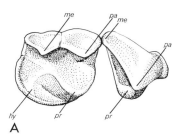 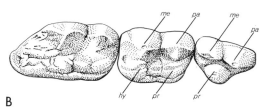

FIGURE 12–8. Occlusal views of some upper cheek teeth of two omnivorous carnivores. A, right upper carnassial (P⁴) and first molar of the hog-nosed skunk (*Conepatus mesoleucus*); note the blade on the carnassial and the broadened crushing surface on M¹. B, right upper carnassial and two molars of the black bear (*Ursus americanus*); note the small, non-trenchant carnassial and the greatly lengthened molars. Abbreviations are as in Figure 12–6.

the first three premolars are usually rudimentary or may be lost, and a diastema usually occurs between premolars. The upper carnassial is roughly triangular because of the posterior migration of the protocone, and is much smaller than the neighboring molars (Fig. 12–8B); both upper and lower carnassials no longer have a shearing function. The dental formula is usually 3/3, 1/1, 4/4, 2/3 = 42, but premolars may be lost with advancing age. The limbs, especially the forelimbs, are strongly built; the plantigrade feet have long, non-retractile claws. There are five toes on each foot. The ears are small and the tail is extremely short. In size bears range from that of a large dog to over 760 kg. (1650 lbs.).

The abandonment of cursorial ability in favor of power of the limbs, and the loss of the shearing function of the cheek teeth in favor of a crushing battery, has accompanied the adoption of omnivorous feeding habits. The strong forelimbs can aid in the search for food by rolling stones or tearing apart logs, and the crushing surfaces of the molars can cope with many kinds of food, from insects and small vertebrates to berries, grass, and pine nuts. Carrion is also avidly sought. The polar bear (*Ursus maritimus*) has a more restricted diet, consisting largely of seals, and the giant panda eats mostly bamboo shoots. In areas with cold winters, bears sleep for much of the winter in caves or other retreats protected from drastic temperature fluctuations. This is not a true hibernation, however, because the temperature and metabolic rate do not drop precipitously.

Family Procyonidae. This family, which includes the familiar raccoon (*Procyon*) and its relatives, probably had a common ancestry with the canids. As with the bears, in procyonids omnivorous feeding habits have become predominant. Some seven genera and 18 species are known. The taxonomic position of the lesser panda (*Ailurus*) is controversial; it is here regarded as a procyonid. Procyonids occupy much of the temperate and tropical parts of the New

World, from southern Canada through much of South America. The lesser panda occurs in south-central China, northern Burma, Sikkim, and Nepal. Procyonids chiefly inhabit forested areas, but the range of one species of ringtail (*Bassariscus astutus*) includes arid desert mountains and foothills. Procyonids are known from the late Oligocene to the Recent in North America, from the late Miocene to the late Pliocene in Europe, and from the Pliocene to the Recent in Asia. They reached South America from North America in the Pliocene.

The structural and functional departure of procyonids from the carnivorous norm has included adaptations favoring both omnivorous feeding habits and climbing ability. Associated with the omnivorous trend has been a specialization of the cheek teeth. The premolars are not reduced, as in the bears, but the shearing action of the carnassials is nearly lost. Instead, the carnassials are high-cusped crushing teeth; a hypocone was added to the upper, and in the lower the talonid was enlarged and broadened (Fig. 12–6C, D). In contrast to the elongate upper molars of bears, those of procyonids are broader than they are long. The dental formula is usually 3/3, 1/1, 4/4, 2/2 = 40. (A procyonid skull is shown in Fig. 12–5B). There are five toes on each foot; the foot posture is usually plantigrade, and the claws are non-retractile or semi-retractile. The limbs are fairly long. The toes are separate, and the forepaw has considerable dexterity in some species and is used in food handling. Tracks left by the man-like hand of the raccoon are familiar to many. The tail is long, is generally marked by dark rings, and is prehensile in the arboreal kinkajou (*Potos flavus*). Procyonids are of modest size, weighing from less than a kilogram to about 20 kg.

The familiar raccoon often takes advantage of man's crops. Corn is a staple food item for midwestern raccoons, and they eat grapes, figs, and melons in parts of California (Grinnell et al., 1937:159,160). In addition, they prey

FIGURE 12–9. A long-tailed weasel (*Mustela frenata*) at timberline in northern Colorado. This weasel has the typical long-bodied, short-legged mustelid form. (Photograph by O. D. Markham.)

on a variety of small vertebrates and some invertebrates. Some tropical procyonids are largely vegetarians. Hall and Dalquest (1963) report that in Veracruz the coati (*Nasua narica*) eats corn, bananas, and the fruit of the coyol palm, and that kinkajous eat mostly fruit. The ringtail, on the other hand, is known to feed mostly on small rodents in some areas. Procyonids reach their greatest diversity and greatest densities in the Neotropics, where the animals are largely arboreal; in some tropical forests several species may occur together. In such areas the nocturnal, quavering cries of kinkajous can be heard regularly.

Family Mustelidae. This large family, with 25 genera and some 70 Recent species, includes the weasels, badgers, skunks, and otters. Mustelids occupy virtually every type of terrestrial habitat, from arctic tundra to tropical rain forests, and they occur in rivers, lakes, and the sea. The distribution is nearly cosmopolitan, but they do not inhabit Madagascar, Australia, or oceanic islands. Mustelids appear in the fossil records of North America and Eurasia in the early Oligocene,

but did not reach South America and Africa until the Pliocene.

These are typically fairly small, long bodied carnivores with short limbs and "pushed in" faces (Fig. 12–9). The skull generally has a long braincase and a short rostrum (Figs. 12–10, 12-11), and the postglenoid process partially encloses the glenoid fossa so that in some species the condyle of the dentary is difficult to disengage from the fossa. Obviously, little lateral and no rotary jaw action is possible. The dentition is quite variable, but is generally 3/3, 1/1, 3/3, 1/2 = 34. The car-

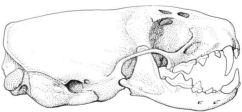

FIGURE 12–10. The skull of the least weasel (*Mustela nivalis*); length of skull 31 mm. Note the unusually long braincase and the short rostrum. The carnassials of this entirely carnivorous species are shearing blades (see also Fig. 12–12).

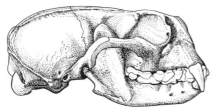

FIGURE 12–11. The skull of the sea otter (*Enhydra lutris*); length of skull 152 mm. The heavy cheek teeth are adapted to crushing marine invertebrates (see also Fig. 12–12).

nassials are trenchant in many species (Fig. 12–12C, D), but have been modified into crushing teeth in others; in the sea otter (*Enhydra lutris*), for example, none of the cheek teeth are trenchant, the carnassials have rounded cusps adapted to crushing, and the post-carnassial teeth (M^1 and M$_2$) are broader than they are long (Fig. 12–12E, F). The first upper molar

is frequently hourglass-shaped in occlusal view (Fig. 12–12C), or may be expanded into a large crushing tooth, as in skunks (Fig. 12–8A). The limbs are usually short, the five-toed feet are either plantigrade or digitigrade, and the claws are never completely retractile. Anal scent glands are usually well developed; they are extraordinarily large in skunks and are used for defense. The tail is generally long, and the pelage may be conspicuously marked, as in skunks and badgers. Some mustelids have beautiful, glossy fur that has considerable value in the fur trade. In size mustelids range from the smallest member of the order Carnivora, a circumboreal weasel (*Mustela nivalis*) which weighs some 35 to 50 gm., to the fairly large sea otter (about 35 kg.).

Mustelids, although basically carnivorous, pursue many styles of feed-

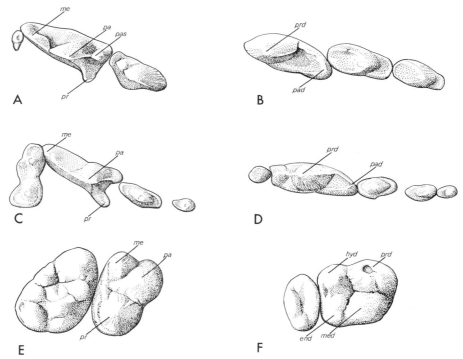

FIGURE 12–12. Occlusal views of the cheek teeth of three carnivores; the right upper teeth are on the left, and the left lower teeth are on the right. The cusps of the carnassials (P^4 and M$_1$) are labeled, and abbreviations are as in Figure 12–6. In A the parastyle (*pas*) is also shown. A and B, the entire sets of cheek teeth of the bobcat (*Lynx rufus*). Note the lack of crushing teeth. Only sectorial teeth are present; the parastyle (*pas*) of P^4 increases the length of its shearing blade, and the loss of the talonid of M$_1$ makes this tooth entirely bladelike. C and D, the entire sets of cheek teeth of the least weasel (*Mustela nivalis*); the crushing teeth, M^1, and M$_2$ are reduced, and the shearing function of the cheek teeth is of major importance. E, P^4 and M^1, and F, M$_1$ and M$_2$, of the sea otter (*Enhydra lutra*). Note that all of the teeth have rounded cusps and are adapted to crushing; the "carnassials" retain no shearing function.

ing. Most mustelids aggressively search for prey in burrows, crevices, or dense cover. Many mustelids are able killers. In Colorado, I observed a male long-tailed weasel (*M. frenata*) killing a young cottontail rabbit (*Sylvilagus audubonii*) roughly twice its own weight. The weasel killed by biting the back of the rabbit's skull repeatedly. According to Errington (1967:24), a mink (*Mustela vison*) "hugs its victim with the forelegs while it scratches violently with the hind legs and bites vital parts, especially about head and neck." Some mustelids, such as the beautiful and graceful marten (*Martes americana*), are swift and agile climbers and feed partly on arboreal squirrels. Otters (subfamily Lutrinae) are semi-aquatic, or almost completely aquatic in the case of the sea otter, and feed on a wide array of vertebrates and invertebrates. The skunks, with no claim to remarkable agility or killing ability, seem to feed on whatever animal material is most readily available, which during the summer is generally insects.

Family Viverridae. This is the largest family in the order Carnivora with regard to numbers of species, and includes the civets, genets, and mongooses. Thirty-six genera and roughly 75 Recent species are recognized, but the taxonomy of this family is still uncertain. Viverrids inhabit much of the Old World; however, the center of their distribution is in tropical and south temperate areas, and they are absent from northern Europe and all but southern Asia, as well as from New Guinea and Australia. This is an old group: it appeared in the late Eocene of Europe, but did not reach Africa until the Pleistocene, and only reached Madagascar in the Recent.

Viverrids are usually small, fairly short-legged and long-tailed carnivores (Fig. 12–13). Like mustelids, some viverrids have well-developed scent glands. The viverrid skull frequently has a moderately long rostrum (Fig. 12–5C). The premolars are large and the carnassials are usually trenchant. The upper molars are tritubercular and are wider than they are long; the lower molars have well-developed talonids. The dental formula is generally 3/3, 1/1, 3-4/3-4, 2/2 = 36-40. The five toes on each foot include a much reduced pollex or hallux. The foot posture is plantigrade or digitigrade, and the claws are partly retractile. The ears are generally small and rounded. Some species are banded, others are spotted, and still others are striped. The smallest viverrid weighs less than a kilogram, and the largest weighs 14 kg. Because of the great morphological variation within this family, it is difficult to frame a defini-

FIGURE 12–13. The banded mongoose (*Mungos mungo*), an African viverrid. (Photograph by W. Leslie Robinette.)

tive structural diagnosis applicable to all species.

Viverrids make their livings in a variety of ways. Most are carnivorous, and eat small vertebrates or insects. Some feed on the gound, and some in trees; some species forage in groups. The palm civets (subfamily Paradoxurinae) are omnivorous and often feed primarily on fruit or other plant material. Some viverrids are semiaquatic and feed largely on aquatic animals, and some, such as the mongoose (*Herpestes*), kill and eat snakes, including certain highly venomous species. The habits of several viverrids are almost unknown; the Congo water civet (*Osbornictis piscivora*), for example, is represented in museums by only a few specimens and has never been seen in the wild by a biologist. Probably no family of carnivores is so poorly known as the viverrids; clearly, much field study could profitably be concentrated on this group.

Family Hyaenidae. Many carnivores will eat carrion if the opportunity arises, but most members of the family Hyaenidae have become specialized for carrion feeding. This is a small family, with but three genera and four Recent species. The distribution includes Africa, southwestern Asia, and parts of India. The Hyaenidae, probably derived from viverrid stock, appeared in Eurasia in the late Miocene and, except for *Chasmaporthetes* (which probably crossed the Bering Strait land bridge and is known from the Pleistocene of North America), has been an entirely Old World family.

Leaving the unusual aardwolf (*Proteles cristatus*) aside temporarily, hyaenids are characterized by rather heavy builds, forelimbs longer than the hind limbs, strongly built skulls, and powerful dentitions. The carnassials are well developed, and all of the cheek teeth have heavily built crowns adapted to bone crushing. The dental formula is 3/3, 1/1, 4/3, 1/1 = 34. The feet are digitigrade, and both the forepaws and the hind paws have four toes that bear blunt, non-retractile claws. The pelage is either spotted (*Crocuta*; Fig. 12–14) or variously striped (*Hyaena*). Hyaenas weigh up to 80 kg.

The aardwolf (subfamily Protelinae), in comparison with the hyaenas (Hyaeninae), is lightly built and has a delicate skull and smaller teeth (Fig. 12–5D). All teeth except the canines are small, and the cheek teeth are simple and conical. The dental formula is generally 3/3, 1/1, 3/2-1, 1/1-2 = 28-32, but frequently some of these teeth are lost (as in the skull shown in Fig. 12–5D). The forefeet have five toes and the hind feet have four. The animal is striped and has a mane of long

FIGURE 12–14. Spotted hyaenas (*Crocuta crocuta*) bringing down a wildebeest (*Connochaetes taurinus*) in East Africa. (Photograph by George B. Schaller.)

FIGURE 12–15. An African lion (*Panthera leo*) standing over the remnants of a zebra, one of the lion's favorite prey species in many areas. (Photograph by W. Leslie Robinette.)

hair from neck to rump, and the tail is quite bushy. The hair of the mane and tail is erected but the mouth remains closed when the animal is threatened and adopts a defensive posture. *Proteles* has abandoned the open-mouthed threat used by most carnivores, which in its case would merely advertise the weakness of the dentition, in favor of extensive erection of the long hair. The aardwolf also releases fluid from the well developed anal glands when attacked.

Hyaenas in some areas specialize in scavenging on the kills of lions and other large carnivores, and are able to drive cheetahs (*Acinonyx*) from their kills. They may also forage around or in villages at night for edible refuse. In their ability to crush large bones they are unsurpassed. But recent studies by Kruuk (1966) and others have shown that the spotted hyaenas are surprisingly able predators. Often hunting in packs of up to 30 animals, these nocturnal hunters can bring down even zebras. Indeed, in the Ngorongoro Crater in Africa a reversal of the usual pattern of interactions between lions and hyaenas has occurred: spotted

hyaenas are better able than the lions to make regular kills, and lions live by driving hyaenas from their kills and eating the carrion (Ewer, 1968:103). The strange aardwolf eats mostly ants, termites, and insect larvae. According to Walker (1968:1264), aardwolves visit carrion not to eat meat but to eat carrion-feeding insects. One is tempted to speculate that the insect-feeding habit may have evolved from a progressive abandonment, on the part of the ancestors of the aardwolf, of feeding on carrion in favor of eating the insects drawn to the carrion.

Family Felidae (Figs. 12–15, 12–16, 12–17). Within this family are four or five Recent genera (depending on whether or not the genus *Lynx* is regarded as separate from *Felis*) with some 37 species. (No general agreement has been reached on felid taxonomy.) The cats are quite a uniform group; all cats, from the pampered house cat (*Felis catus*) to the tiger (*Panthera tigris*), bear a strong family resemblance. This family occurs world-wide, with the exceptions of Antarctica, Australia, Madagascar, and some isolated islands. Of all of the car-

FIGURE 12–16. A leopard (*Panthera pardus*) well concealed against a background of mottled light and shadow. (Photograph by W. Leslie Robinette.)

nivores the cats are the most proficient killers; some species regularly kill prey as large or considerably larger than themselves.

The evolution of the cats, the most specialized carnivores, was rapid, and the evolutionary pattern was unusual. There are differences of opinion concerning the interpretation of the fossil record of felids, but the following broad features of their evolution are agreed upon by many. (The taxonomic scheme used here for the felids is that of Simpson, 1945.) Two lines of descent, represented by the subfamilies Nimravinae and Machairodontinae, were established in the early Oligocene, and members of these groups were probably widely sympatric (occupying the same area at the same time) in North America and Europe from the Oligocene through the Pleistocene. The subfamily Nimravinae includes species with somewhat

FIGURE 12–17. The cheetah (*Acinonyx jubatus*) spends considerable time watching for prey. (Photograph by W. Leslie Robinette.)

enlarged upper canines and small lower canines. A progressive reduction of the upper canines and development of "true cat" features is evident in Miocene and Pliocene nimravines, and the Recent cats (Felinae) perhaps descended from nimravine ancestry in the late Miocene or early Pliocene. Although the structure of members of the felid ancestral stock is not documented by the fossil record, the progenitors of felids probably lacked enlarged upper canines. In the Nimravinae, then, an early trend toward the saber-tooth specialization was seemingly reversed; the upper canines became smaller, and certain specializations of the skull and lower jaw were lost. Forms representing the second line of descent, the Machairodontinae (saber-toothed cats), are first known from the Oligocene, when already they had larger canines than did nimravines (Fig. 12–18). In contrast to the nimravines, however, the saber-tooth specializations became progressively more pronounced in machairodonts until the Pleistocene, when this felid line, together with many of the large ungulates, became extinct.

Many machairodont specializations are adaptations to the unique mode of attack these animals employed. It is generally belived that the sabers were plunged into large prey by powerful downward and forward thrusts of the head, while the cat clung to the prey with the heavily muscled forelimbs. The greatly enlarged mastoid process in *Smilodon* (Fig. 12–19A) is one of many adaptations to the stabbing action. This enlarged process provided a large surface area and unusual mechanical advantage for the origin of the sternomastoideus, the muscle that pulled the snout downward. The receding nasal bones probably allowed the nose pad to be withdrawn so that the animal could breathe while its sabers were deeply imbedded in its prey. The upper carnassial had a prostyle (in addition to the usual parastyle that occurs in Recent felids; see Fig. 12–12A) that increased the length of the shearing blade formed by this tooth.

The felid rostrum is short, an adaptation furthering a powerful bite, and the orbits in most species are large (Fig. 12–19B). The number of teeth is reduced. The typical dental formula is 3/3, 1/1, 3/2, 1/1 = 30, and the anteriormost upper premolar is strongly reduced or lost (as in *Lynx*). The carnas-

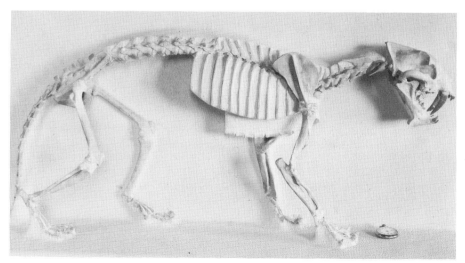

FIGURE 12–18. The skeleton of *Hoplophoneus primaevus* (Machairodontinae), an Oligocene saber-tooth cat. (Courtesy of Robert W. Wilson; South Dakota School of Mines, No. 2528.)

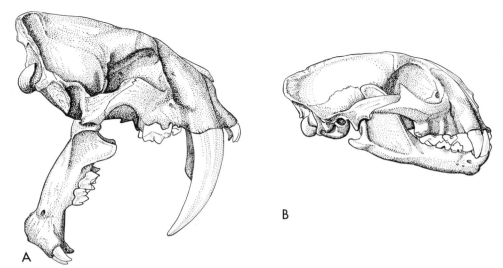

FIGURE 12–19. Skulls of two cats: A, an extinct saber-tooth cat (*Smilodon californicus*); length of skull approximately 300 mm. B, the mountain lion (*Felis concolor*); length of skull 197 mm. (*Smilodon* after Stock, 1949.)

sials are well developed and have specializations that enhance their shearing ability (Fig. 12–12A, B). The foot posture is digitigrade. The forelimbs are strongly built, and the manus can be supinated; the claws are sharp and recurved and are completely retractile, except in the cheetah (*Acinonyx*), in which they are partly retractile. These features of the forelimbs allow cats to clutch and grapple with prey with the forelimbs. Some species are spotted or striped; these color patterns enable the animals to conceal themselves effectively (Fig. 12–16). The weight of cats varies from about that of a domestic cat (3 kg.) to 275 kg., in the case of the tiger.

As described in Chapter 17 (p. 325), cats usually catch prey by a stealthy stalk followed by a brief burst of speed. They are typically sight hunters, and some species spend considerable time watching for prey and waiting for it to move into striking distance (Fig. 12–17). Many kinds of animals are eaten, from fish, mollusks, and small rodents to ungulates as large as buffalo. *Felis planiceps* of southeastern Asia seems to be the only cat with non-carnivorous tendencies; this species prefers fruit (Goodwin, 1954:570).

SUBORDER PINNIPEDIA

Many distinctive morphological features of pinnipeds are adaptations to marine life. Compared to land carnivores, pinnipeds are large (approximately 91 to 3,600 kg.). Large size improves metabolic economy in cold environments because of the favorable mass to surface ratio of large animals (as discussed on page 367). According to Scheffer (1958:8), large size in pinnipeds is primarily an adaptation to a cold environment. The body is insulated by thick layers of blubber. The pinnae (external ears) are either small or absent, the external genitalia and mammary nipples are withdrawn beneath the body surface, the tail is rudimentary, and only the parts of the limbs distal to the elbow and knee protrude from the body surface. As a result, the torpedo-shaped body has smooth contours and creates little drag during swimming. The nostrils are slit-like (Fig. 12–20), are normally closed, and are opened by voluntary effort. The skull is partially telescoped, with the supraoccipital partially overlapping the parietals; the rostrum is usually shortened and the orbits are usually large and encroach on the nar-

FIGURE 12-20. The face of a California sea lion (*Zalophus californianus*, Otariidae). Note the valvular nostrils and the external ear. (Photograph courtesy of the United States Navy.)

row interorbital area. Either one or two pairs of lower incisors are present. The canines are conical; the cheek teeth are homodont (none are modified as carnassials), two-rooted, and usually simple and conical (Fig. 12–21); they vary in total number from 12 to 24. In some pinnipeds cheek teeth are characteristically lost with advancing age. The limbs and girdles are highly specialized. The clavicle is absent, and the humerus, radius, and ulna are short and heavily built; the pollex is the longest and most robust of the five

digits, and forms the leading edge of the wing-like fore flipper. The pelvic girdle is small and is nearly parallel to the vertebral column. The femur is broad and flattened. The first and fifth are the longest digits of the pes, and both the manus and pes are fully webbed.

The reduction of the vertebral zygapophyses and the absence of the clavicle allow the vertebral column and the forelimbs considerable flexibility and freedom of movement; these features may favor rapid maneuvers during the

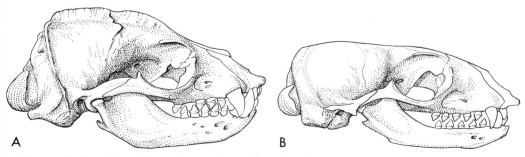

A   B

FIGURE 12–21.   Sexual dimorphism in the skulls of the California sea lion (*Zalophus californianus*). A, an adult male, length of skull 290 mm.; B, an adult female, length of skull 214 mm.

pursuit of prey. Although terrestrial locomotion is characteristically slow and laborious in most species, the importance of terrestrial locomotion when the animals are hauled out on rocks or ice, or are on breeding grounds, has probably limited the extent to which the limbs of pinnipeds have become specialized for swimming.

Although not so completely adapted to aquatic life as are cetaceans, pin-

nipeds make impressively deep dives. Scattered records of seals with depth recording instruments and of individuals caught on the hooks of deep-set fishing lines indicate that some seals reach depths of at least 600 m. (Kooyman, 1969:72). Dives lasting 20 minutes have been recorded for several species of seals (Degerbøl and Freuchen, 1935:210; Scheffer and Slipp, 1944; Backhouse, 1954). The following responses to diving occur in pin-

FIGURE 12–22. Steller sea lions (*Eumetopias jubata*, Otariidae) on Año Nuevo Island, off the coast of California. Above, a bull with cows and young. Below, cows with young. (Photographs by Robert T. Orr.)

nipeds: the heart rate slows to as low as one tenth normal; the metabolic rate and body temperature drop; peripheral vasoconstriction occurs, but normal blood flow to vital organs continues. In addition, their tolerance of CO_2 in the tissues is unusually high, and large amounts of myoglobin in the muscles aid in the storage of oxygen.

Family Otariidae (Figs. 12–20, 12–22). This family, containing six genera and 12 Recent species, includes the eared seals and the sea lions. These animals inhabit many of the coastlines of the Pacific Ocean and parts of the South Atlantic and Indian Oceans. They are common along the Pacific coast of North America. The earliest otariids are known from the middle Miocene, and may have originated in the "food-rich kelp reefs of the North Pacific" (Scheffer, 1958:33).

Otariids differ from other pinnipeds in being less highly modified for aquatic life and better able to move on land. The hind flippers can be brought beneath the body and used in terrestrial locomotion, and well developed nails occur on the three middle digits. A small external ear is present (Fig. 12–20). Males are much larger than females; in the northern fur seal (*Callorhinus ursinus*), males weigh four and a half times as much as females. Considerable sexual dimorphism in the shape of the skull occurs in some species (Fig. 12–21), and in males the skull becomes larger and more heavily ridged with advancing age. The dental formula is 3/2, 1/1, 4/4, 1-3/1 = 34-38. The body is covered with fur which is uniformly dark. Weights of otariids range from 60 to 1000 kg.

These seals are generally highly vocal, and utter a great variety of sounds. They tend to be gregarious all year round and are social during the breeding season, when they assemble in large breeding rookeries (see page 335). Propulsion in the water is obtained by powerful downward and backward strokes of the forelimbs; speeds up to 27 km. per hour have been recorded (Scheffer, 1958:13).

Otariids eat mostly squid and small fish that occur in schools, and are highly maneuverable when in pursuit of prey. In the Gulf of California, sea lions frequently assemble around commerical fishing boats, signal their presence by barking, and eat discarded fish.

Family Odobenidae. This is a monotypic family; that is, it contains only one species, *Odobenus rosmarus*, the walrus. This species occurs near shorelines in arctic waters of the Atlantic and Pacific Oceans, but may stray southward to some extent along the coastlines. Odobenids first appear in the late Miocene and, like the otariids, may have had a North Pacific origin.

The walrus is a large pinniped (up to 1270 kg.), with a robust build, a nearly hairless skin, and no external ears. The hind flippers can be brought beneath the body and are used for terrestrial locomotion, which is ponderous and slow. In both sexes the upper canines are modified into long tusks (Fig. 12–

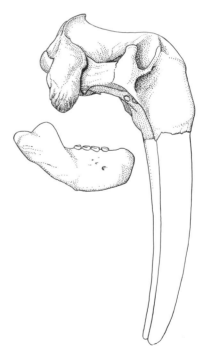

FIGURE 12–23. The skull of a walrus (*Odobenus rosmarus*, Odobenidae); length of skull approximately 355 mm. The tusks are enlarged upper canines.

23), which in the adult lack enamel. There are no lower incisors in adults, and 12 cheek teeth are usually present. The dental formula is 1-2/0, 1/1, 3-4/3-4, 0/0 = 18-24. On the huge mastoid processes attach the powerful neck muscles that pull the head downward.

Walruses feed on mollusks, which they rake from the sea floor by means of the lips and the huge tusks. They have been observed to eat cetaceans (Scheffer, 1958:15, 16), but this must be an extremely unusual food item. Walruses are gregarious and polygynous, and frequently assemble in large groups of more than 1000 individuals. They are migratory to some extent, moving southward in winter. Walruses make a variety of loud noises when out of water, and make a "church-bell" sound and rasps and clicks under water (Schevill et al., 1963; Schevill et al., 1966). The fact that the rasps and clicks are made during swimming suggests their use in echolocation.

Family Phocidae (Fig. 12–24, 12–25, 12–26). These animals, the earless seals, are the most abundant pinnipeds. They occur along most northern (above 30° north latitude) and most southern (below 50° south lati-

tude) coastlines and in some intermediate areas. They appear in the middle Miocene, and presumably originated in the Northern Hemisphere.

The earless seals are more highly specialized for aquatic life than are other pinnipeds. As the vernacular name implies, there is no external ear. The hind flippers are useless on land, but, as a result of lateral undulatory movements of the body, are the primary propulsive organs in the water. The fore flippers are short and well furred. The structure of the cheek teeth is highly variable, but is usually fairly simple. In the crabeater seal (*Lobodon*), however, the cheek teeth have complex cusps (Fig. 12–27). The pelage of most phocids is spotted, banded, or mottled. These seals frequently have extremely heavy layers of subcutaneous blubber that give the bodies smooth contours and, in some cases, a nearly perfect fusiform shape. Most species weigh from about 80 to 450 kg., but male elephant seals (*Mirounga*; Fig. 12–24) occasionally weigh as much as 3600 kg. (7920 lbs.).

Many phocids are monogamous and form small, loose groups in which no social hierarchy is evident; but some, such as the elephant seal, are gregari-

FIGURE 12–24. Northern elephant seal (*Mirounga angustirostris*, Phocidae) on Año Nuevo Island, off the coast of California. (Photograph by Robert T. Orr.)

FIGURE 12–25. Weddell seals (*Leptonychotes weddelli*, Phocidae) near McMurdo Station, Antarctica. (Photograph courtesy of the United States Navy.)

FIGURE 12–26. Harbor seals (*Phoca vitulina*, Phocidae) on Año Nuevo Island, off the coast of California. (Photograph by Robert T. Orr.)

FIGURE 12–27. Medial view of two right lower cheek teeth of the crab-eater seal (*Lobodon carcinophagus*). These complex teeth enable this animal to depend on filter feeding. (After Walker, E. P.: *Mammals of the World*, 2nd ed., The Johns Hopkins Press, 1968.)

ous and polygynous, and have a dominance hierarchy. The monogamous species are quiet, whereas the polygynous species are highly vocal (Evans and Bastian, 1969:437). The sole function of the proboscis of the male elephant seal (Fig. 12–24) is the production of vocal threats (Bartholomew and Collias, 1962).

The usual food of phocids is fish, cephalopods, and other mollusks. The leopard seal (*Hydrurga*), however, is a powerful predator that eats penguins and other kinds of seals. Two species of phocids are filter feeders, and use the complex cheek teeth (Fig. 12–27) to filter crustaceans and other plankton from the water. The Weddell seal (*Leptonychotes weddelli;* Fig. 12–25) is well adapted to life in Antarctic waters and remains beneath the ice all winter; it uses anticlinal, air-filled ice domes as places to breath, and rests on "interior ice shelves" (Perkins, 1945: 278–279). Mortality in some seals is often due to worn teeth and the resulting inability to keep breathing holes open through the ice. Some phocids that rest on ice are able to leap onto ice seven feet above the water.

CHAPTER 13

SUBUNGULATES: ELEPHANTS; HYRAXES; SIRENIANS

If general appearance were used as the single criterion for evaluating relationships, elephants, the largest land mammals, hyraxes, rodent-like creatures, and sirenians, ungainly aquatic mammals, would be judged to be distantly related. But in this case appearances are deceptive, for the fossil record suggests that these groups evolved in Africa from a common ancestral stock related to ungulates. Because of this presumed ungulate ancestry these groups have long been referred to as subungulates.

ELEPHANTS: ORDER PROBOSCIDEA

Through much of the Cenozoic some of the largest and most spectacular herbivores were proboscideans, and in the late Tertiary a varied array of these animals occurred widely in North America, Europe, and Africa. The di-versity of proboscideans was reduced in the Pleistocene, and today but two species represent this remarkable group. Because elephants now often threaten the interests of man, they may be extirpated over wide areas as human populations increase. Elephants occur today only in Africa south of the Sahara Desert (*Loxodonta*), and in parts of southeastern Asia (*Elephas*). Regrettably, we may be witnessing the final stages in the history of one of the most interesting mammalian orders.

The fossil record of proboscideans begins in the late Eocene of Egypt with *Moeritherium*. This tapir-sized animal had a moderately primitive complement of teeth (3/3, 1/0, 3/3, 3/3), but the second incisors above and below were enlarged into short tusks. Proboscideans apparently reached North America in the late Miocene. Late Tertiary proboscideans had brachyodont teeth with few ridges; most or all of the cheek teeth were in place at one time, and both an upper and

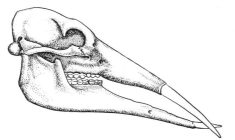

FIGURE 13–1. The skull of *Gomphotherium*, a Miocene proboscidean. Length of skull and tusks roughly 1 m. (After Romer, A. S.: *Vertebrate Paleontology*, 3rd ed., The University of Chicago Press, 1966.)

lower set of tusks were usually present (Fig. 13–1).

Family Elephantidae. This family, to which both living species belong, is represented first by the Pliocene and Pleistocene genus *Stegodon*. Although the skull was not as short as in more advanced elephants and the teeth were brachyodont, the occlusal surfaces of the cheek teeth had laminae and no more than 2/2 cheek teeth were functional at one time. The lower tusks were vestigial, whereas the uppers were long and curved. The Pleistocene woolly mammoth (*Mammuthus primigenius*) was in some ways more specialized than the living elephants. It had a remarkably short, high skull (Fig. 13–2) and long tusks that occasionally crossed; the last molar had up to 30 laminae, more than occur in the living elephants. Entire frozen woolly mammoths have been found in Siberia and Alaska, and many graceful drawings made by Paleolithic man on the walls of caves depict these animals.

The two living proboscideans—the African elephant, *Loxodonta africana* (Fig. 13–3), and the Indian elephant, *Elephas maximus*—are the largest land mammals, reaching weights of 5900 kg. They have a long dextrous proboscis (trunk) with one or two finger-like structures at its tip, large ears, and graviportal limbs. The limb bones are heavy and the proximal segments of the limbs are relatively long; the ulna and tibia are unreduced, and the bones

of the five-toed manus and pes are short, robust, and have an unusual, spreading, digitigrade posture (Fig. 13–4). A heel pad of dense connective tissue braces the toes and largely supports the weight of the animal. As an adaptation allowing the efficient support of great weight, the long axis of the pelvic girdle is nearly at right angles to the vertebral column, and the acetabulum faces ventrally. In addition, when the weight of the body is supported by the limbs, there is little angulation between limb segments; that is to say, each segment is roughly in line with other segments. The gait is unusual. As described by Howell (1944:53), an elephant "relies exclusively upon the walk or its more speedy equivalent, the running walk, which permits it to keep at least two feet always upon the ground. Not only does the weight make it advisable that this be distributed among each of the four feet when the animal in is motion, but the bulk doubtless requires that the equilibrial stresses be shifted as gradually as possible to each foot, rather than more abruptly as in the trot or gallop."

The skull is unusually short and high, perhaps in response to a need for great mechanical advantage for the muscles that attach to the lambdoidal crest and raise the front of the head and the tusks. The skull contains numerous large air cells, particularly in

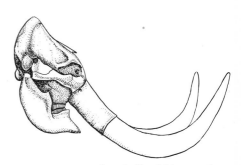

FIGURE 13–2. The skull of *Mammuthus*, a Pleistocene elephantid. Length of skull and tusks roughly 2.8 m. (After Romer, A. S.: *Vertebrate Paleontology*, 3rd ed., The University of Chicago Press, 1966.)

FIGURE 13–3.   African elephants (*Loxodonta africana*) in East Africa. (Photograph by Jean Carror.)

the cranial roof. The highly special-ized dentition consists of the tusks (each a second upper incisor), and six cheek teeth in each half of each jaw. The pattern of cheek tooth replace-ment is remarkable. The cheek teeth erupt in sequence from front to rear, but only a single tooth, or one tooth and a fragment of another, is functional in each half of each jaw at one time. As a tooth becomes seriously worn it is replaced by the next posterior tooth. The last molar does not erupt until the animal is roughly 23 years old, accord-ing to Krumrey and Buss (1968); a specimen judged by these authors to be 60 years old retained a fragment of

the second molar, and the third molar was fully erupted and well worn. The hypsodont cheek teeth are formed of thin laminae (cross ridges), each con-sisting of an enamel band surrounding dentine, with cement filling the spaces between the ridges (Fig. 13–5). The last molar, the tooth that must serve for much of the animal's adult life, has the greatest number of laminae; the pre-molars are considerably smaller, simpler, and less durable than the molars. In old elephants some of the anterior laminae of the third molar may be lost while the remainder of the tooth is still functional. The unique pattern of tooth replacement and the

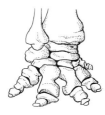

FIGURE 13–4.   The right hind foot of *Masto-don*, a late Tertiary and Pleistocene probosci-dean. (After Romer, A. S.: *Vertebrate Paleon-tology*, 3rd ed., The University of Chicago Press, 1966.)

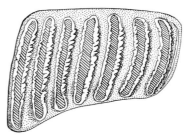

FIGURE 13–5.   The occlusal surface of a molar of the Asiatic elephant (*Elephas maximus*). The ridges of the lamellae are enamel; the crosshatched areas are dentine, and the stippled areas are cement.

complex occlusal surface provide for an enduring dentition and a long life. The life span of elephants is thought by Perry (1954) to be about 70 years.

Elephants occupy forests and semi-open or dense scrub, and are restricted to areas near water. They feed on a variety of trees, shrubs, grasses, and aquatic plants. Elephants characteristically strongly influence their environment. They are gregarious; they repeatedly travel broad trails devoid of vegetation; and each individual eats up to 410 kg. of forage daily. Their great size and strength enable them to "ride down" fairly large trees in order to feed on the leaves. Locally, elephants do great damage to crops. They have been killed for this reason and for sport, and illegal killing of elephants for their ivory is a significant cause of mortality. The large African game preserves, where hunting is prohibited, offer some hope for the survival of wild elephants.

HYRAXES: ORDER HYRACOIDEA

Members of this unusual order are small, rather rodent-like creatures commonly called hyraxes or dassies;

their external appearance (Fig. 13–6) gives little indication of their relationships to the ungulates. This is a small order, with a single Recent family, Procaviidae, and two fossil families of doubtful validity. Recent members include three genera with 11 species. Hyracoids occupy nearly all of Africa except the arid, northwestern part. Hyraxes appear first in lower Oligocene beds in Egypt, from which members of an extinct family and of the living family are known. The relationships of the hyracoids are uncertain, but they perhaps descended from an early "ungulate" stock that was also ancestral to the elephants and sirenians. Some early members of an extinct hyracoid family (Geniohyidae) reached the size of a tapir. The distribution of the structurally more conservative surviving family extended north of its present limits during the Pliocene, when a giant procaviid occurred in western Europe as far north as France.

The roughly rabbit-sized procaviids have short skulls with deep lower jaws (Fig. 13–7). The dental formula is 1/2, 0/0, 4/4, 3/3. The incisors are specialized: the pointed, ever-growing uppers are broadly separated and triangular in cross section, and the flattened posterior surfaces lack enamel; the lowers are chisel-shaped and are

FIGURE 13–6. Rock hyraxes (*Procavia capensis*). (Photograph by Ron Garrison; San Diego Zoo photo.)

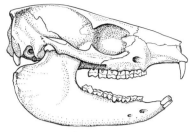

FIGURE 13–7. Skull of a rock hyrax (*Heterohyrax* sp.); length of skull 98 mm. (After Hatt, 1936.)

generally tricuspid. Behind the incisors is a broad diastema, and the cheek teeth are either brachyodont or hypsodont. The molars resemble those of a rhinoceros: the uppers have an ectoloph and two cross lophs, and the lowers have a pair of V-shaped lophs. The body is fairly compact and the tail is tiny. The forefoot has four toes and the hind foot has three, and the feet are mesaxonic (the plane of symmetry goes through the third digit). The digits are united to the bases of the last phalanges (Fig. 13–8), and except for the clawed second digit of the pes, all digits bear flattened nails. The plantigrade feet have specialized elastic pads on the soles that are kept moist by abundant skin glands; in addition, the soles may be "cupped" by specialized muscles and, by functioning as suction cups, can provide for remarkable traction. Although the clavicle is absent, as in cursorial mammals, the centrale of the carpus is present, a feature decidedly not characteristic of runners. The stomach is simple, but digestion is aided by microbiota in the pair of caecae of the colon and in the single iliocolic caecum.

FIGURE 13–8. The sole of the right hind foot of a rock hyrax (*Heterohyrax dorsalis*).

Hyraxes are herbivorous and to some extent insectivorous, and are nimble climbers and jumpers. They occur in a variety of habitats, from forests and scrub country to grassland and lava beds, and up to elevations of over 4000 m. The rock hyraxes live in cliffs, ledges, and talus, and are capable of running rapidly over steep rock faces. These diurnal and crepuscular animals are colonial, and on occasion occur in groups of up to 50 individuals. The body temperature of one species of rock hyrax (*Heterohyrax brucei*) was found by Bartholomew and Rainy (1971) to be quite variable. According to these authors, basking in the sun, a common behavior among rock hyraxes, is probably important because it reduces the metabolic cost of elevating the temperature in the morning from its slightly depressed nocturnal level. Evaporative cooling, aided by profuse sweating of the feet, dissipates metabolic heat at high ambient temperatures. These animals huddle together at low ambient temperatures, a behavior that reduces the rate of heat loss from their bodies. Tree hyraxes are less gregarious, and are largely nocturnal. They climb rapidly over the nearly vertical trunks of trees and can leap from branch to branch. They make distinctive calls at night that, according to Walker (1968:1327), consist of "a series of croaks that gradually mount the scale and end in a loud scream."

SIRENIANS: ORDER SIRENIA

The sirenians—the dugongs, sea cows, and manatee—are the only completely aquatic mammals that are herbivorous, and comprise one of the most anomalous mammalian orders. There are four living species of two genera (*Dugong*, Dugongidae; *Trichechus*, Trichechidae). According to Jones and Johnson (1967:367), sirenians occur in "coastal waters from eastern Africa to Riu Kiu Islands, Indo-Australian Archipelago, western Pacific and Indian oceans; tropical western Africa; coastlines of Western Hemisphere from 30° N to 20° S, Caribbean region, and Amazon and Orinoco drainages in South America; formerly also in Bering Sea." Sirenians probably shared a common ancestry with the proboscideans, and are known from Eocene deposits at such scattered points as Europe, Africa, and the West Indies. This group was once far more diverse and widely distributed than it is today.

Sirenians are large, reaching weights in excess of 600 kg. They are nearly hairless except for bristles on the snout, and have thick, rough or wrinkled skin. The nostrils are valvular, the nasal opening extends posterior to the anterior borders of the orbits, and the nasals are either reduced or absent. The skull is highly specialized and the dentary is deep (Fig. 13–9); the tympanic bone is semicircular and the external auditory meatus is small. The skeleton is dense and heavy—perhaps an increase in specific density is adaptive. Postcranially, sirenians somewhat resemble cetaceans. The five-toed manus is enclosed by skin and forms a flipper-like structure, the pelvis is vestigial, and the tail is a horizontal fluke. There is no clavicle and, unlike cetaceans, the scapula is narrow and blade-like. The teeth are unusual. Functional teeth are present in *Dugong*, but they are large and columnar, lack enamel, and are cement covered. They have open roots, and the occlusal surfaces are wrinkled and bunodont. *Trichechus*, in contrast, has an indefinite, large number of teeth; the teeth are enamel-covered and lack cement, and each has two cross ridges and closed roots. As teeth at the front of the tooth row wear out, they are replaced by the posterior teeth pushing forward. Five or six teeth in each side of each jaw are functional at one time. This unusual style of replacement is basically similar to that of the proboscideans. Horny plates cover the front of the palate and the adjacent surface of the mandible in all genera. The skull of *Trichechus* is curiously modified by elongation of the nasal cavity, and this animal, alone among mammals, has only six cervical vertebrae. Some differences between the two families of sirenians are shown in Table 13–1.

Sirenians are heavy-bodied, slow-moving animals that inhabit coastal seas, large rivers, and lakes, and graze while submerged on aquatic plants. They remain submerged for periods up to about 15 minutes. Some individuals inhabiting the coasts of Florida move into rivers and to springs in the winter, perhaps in an effort to avoid cold water (Layne, 1965:168). Manatees are known to make sounds underwater, but these are probably used for communication rather than for echolocation (Schevill and Watkins, 1965). Man has been responsible for a

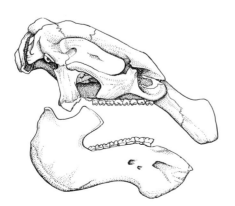

FIGURE 13–9. The skull of a manatee (*Trichechus manatus*). Length of skull 360 mm. (After Hall, E. R., and Kelson, K. R.: *Mammals of North America*, copyright © 1959 by the Ronald Press Company, New York.)

Table 13–1.   Comparison of Sirenian Characteristics.

DUGONGIDAE	TRICHECHIDAE
1. Functional dentition 1/0, 0/0, 0/0, 2-3/2-3	1. No functional incisors; numerous cheek teeth
2. Cheek teeth columnar; no enamel, cement covered; roots single	2. Teeth with cross ridges, covered with enamel; cement absent; roots double
3. Premaxillaries large; nasals absent; nasal cavity short	3. Premaxillaries small; nasals present; nasal cavity long
4. Slender neural spines and ribs	4. Robust neural spines and ribs
5. Flippers lacking nails	5. Flippers with nail in one species
6. Tail notched, like that of whales	6. Tail not notched but spoon-shaped

great restriction of the range of sirenians, and succeeded in exterminating the Steller's sea cow (*Hydrodamalis*) in about 1769, only some 27 years after the animal's discovery (Walker, 1968:1334). This giant sirenian may have reached 4000 kg. in weight, and inhabited parts of the Bering Sea. Present serious declines in the populations of dugongs (*Dugong*) in some areas are due to persistent hunting by man.

CHAPTER 14

UNGULATES

The ungulates are the hoofed mammals, members of the orders Perissodactyla (horses, rhinos, and tapirs) and Artiodactyla (pigs, camels, deer, antelope, cattle, and their kin). The term *ungulate* has no taxonomic status, but refers to this broad group of herbivorous mammals that are more or less specialized for cursorial locomotion.

Cursorial Specialization

Exceptional running ability has developed independently in a number of mammalian groups. It has provided a means of escaping predators (as in some rodents, rabbits, and ungulates) or of capturing prey (as in carnivores). The refinement of cursorial adaptations in ungulates, the most cursorial mammals, was favored by their occupation of the expanding grasslands in the Miocene. For ungulates living in such open country there was probably a high adaptive premium on running ability. Speed became the primary means of avoiding predation, and seasonal movements to seek water or appropriate food probably became an important part of the ungulate mode of life.

Running speed is determined basically by two factors, the length of the stride and the rate of the stride (the number of strides per unit of time), and most important cursorial specializations serve to lengthen the stride or increase its rate. Perhaps the most universal cursorial adaptation that lengthens the stride is lengthening of the limbs. In generalized mammals, or in many powerful diggers, the limbs are fairly short and the segments are all roughly the same length (Fig. 14–1). But in cursorial species the limbs are long, and in the most specialized runners the metacarpals and metatarsals have become greatly elongate, and the manus and pes are the longest segments (Fig. 14–1). The loss or reduction of the clavicle contributes further to the length of the stride. This occurs in carnivores, leporids, and ungulates. With the loss of the clavicle, the scapula and shoulder joint are freed from a bony connection with the sternum, and the scapula can change position to some extent and can rotate about a pivot point roughly at its center. Because the scapula is not anchored to the axial skeleton, as the forelimb

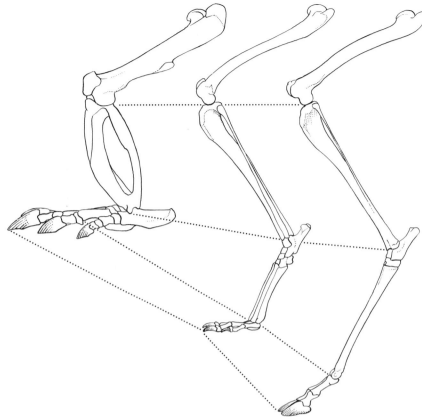

FIGURE 14-1. The hind limbs of three mammals: left, the armadillo (*Dasypus novemcinctus*), a powerful digger with plantigrade feet; middle, the coyote (*Canis latrans*), a good runner with digitigrade feet; right, the pronghorn (*Antilocapra americana*), an extremely speedy runner with unguligrade feet. Note the lengthening of the shank and foot of the coyote and (especially) of the pronghorn; the metatarsals have undergone the greatest lengthening. The limbs are not drawn to scale, but the femur is the same length in each drawing.

reaches forward during the stride the shoulder joint pivots upward and forward, and when the forelimb moves back the shoulder joint swings downward and backward. Hildebrand (1960) estimated that such movements of the scapula added roughly 115 mm. to the length of the stride of the running cheetah. As an additional advantage, when the forefeet strike the ground at the end of a forward bound the impact is cushioned by the muscles that bind the scapula to the body, rather than the shock being transferred directly from the shoulder joint to the axial skeleton via the clavicle. Substantial lengthening of the stride also results from an inchworm-like flexion and extension of the spine (Fig. 14–2).

In small or moderate-sized runners the flexors and extensors of the vertebral column are powerfully developed; the vertebral column extends as the forelimbs reach forward and the hindlimbs are driving against the ground, and it flexes when the front feet move backward while braced against the ground as the hind limbs swing forward. Hildebrand estimated that such movements of the vertebral column could propel the cheetah at nearly 10 km. per hour if the animal had no legs!

The speed of limb movements, and thus the rate of the stride, is similarly increased by a combination of structural modifications. The total speed of the foot, which drives against the ground and propels the animal, de-

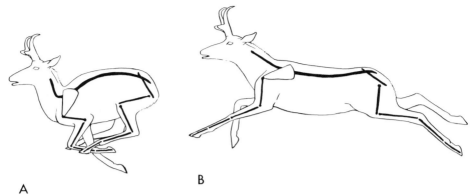

A   B

FIGURE 14–2. Two positions of a running pronghorn, showing the flexion and extension of the vertebral column and the changing position of the scapula. A, the forelimbs have just left the ground; the hind limbs are reaching forward and will contact the ground as the forelimbs swing forward. B, the animal is bounding ahead after the limbs have driven against the ground; the forelimbs are reaching forward.

pends on the speed of movement at each joint of the limb. If another movable joint is added to a limb, the speed of the limb will be increased by the speed of movement at the new joint. The greater the number of joints that move in the same direction simultaneously, the greater the speed of the limbs. For this reason nearly all cursorial mammals have abandoned a plantigrade foot posture in favor of a digitigrade or unguligrade stance. This lifting of the heel from the ground allows another limb joint, that between the metapodials and the phalanges, to contribute to the speed of the limb. In addition, the movable scapula and vertebral column, which help increase the length of the stride, also contribute their motion to the total speed of the feet.

Specializations of the musculature also add importantly to limb speed, and hence to running speed. The trend in many cursorial mammals is toward a lengthening of the tendons of some limb muscles (in association with the elongation of the distal segments of the limbs), and in some cases there has been a proximal migration of the points of insertion of these muscles. Generally, the nearer the insertion of a muscle approaches the joint that it spans and at which it causes motion, the greater advantage for speed it pos-

sesses; such specialized muscles are primarily geared for speed (but not for power). In these animals a complex division of labor among muscles probably allows other muscles, having attachments that confer considerable mechanical advantage for power, to control certain relatively slow movements that get the animal in motion; the less powerful "high-gear" muscles are brought into play during high-speed running.

Speedy limb movements are further facilitated by a reduction of the weight of the distal parts of the limbs and the resultant reduction of the kinetic energy to be overcome at the end of one limb movement and the start of another. Inasmuch as the distal part of the limb moves more rapidly than the proximal part during a stride, reduction of weight of the distal parts is especially advantageous. Several specializations commonly serve this end. The most obvious is the loss of digits. In the most extreme mammalian cases—the front and hind limbs of the horse, and the hind limbs of the pig-footed bandicoot (Fig. 5–17, page 55) and of the red kangaroo (Fig. 5–28) —only one digit that functions during running is retained. The strengthening of the distal joints of the limbs by modifications of bones and ligaments, and the limiting of movements at these joints to a fore and aft plane, obviate the need for muscular bracing and for muscles that produce rotary motions, further reducing distal weight. Also, the heaviest muscles are mostly in the proximal segment of the limb, thus keeping the center of gravity of the limb near the body. The combined effect of these modifications that reduce and redistribute weight is to favor rapid limb movement and to reduce the outlay of energy associated with that movement. (For excellent discussions of cursorial adaptations in mammals see Hildebrand, 1959, 1960, 1965.)

The graceful legs of an antelope, with slim distal segments and the largest muscles bunched near the body (Fig. 14–3), display beautifully many of the cursorial modifications discussed here.

Living ungulates typically have many of the cursorial specializations just discussed. The feet are modified by the loss of toes, by the alteration of the foot posture so that only the tips of the toes touch the ground, and by the development of hoofs. The limbs are usually slender, and the tendency in

the most rapid runners is toward great elongation of the second segments (the forearm of the forelimb and the shank of the hind limb) and distalmost segment (the manus and pes). The distal parts of the ulna and the fibula are strongly reduced in advanced ungulates, and the joints distal to the shoulder and hip joints tend to limit movement to the anteroposterior plane.

In the ungulate ankle joint the calcaneum appears to be pushed aside, so to speak. In mammals in which no drastic reduction of digits has occurred, the distal surface of the astragalus articulates with the navicular and the calcaneum articulates with the cuboid (Fig. 2–23), and the weight of the body is transferred through the digits, the distal carpals, both the astragalus and calcaneum, and the tibia and fibula. In ungulates a different arrangement occurs in association with the reduction of digits: the astragalus rests more or less directly on the distal tarsal bones, which may be highly modified by fusion and loss of elements (Figs. 14–4, 14–5), and the weight of the body is borne by the cen-

FIGURE 14–3. Young Thompson's gazelles (*Gazella thompsonii*); these are among the most abundant of African antelope. The long, slender limbs are typical of the more cursorial ungulates. (Photograph by W. Leslie Robinette.)

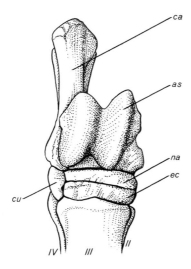

Figure 14-4. The tarsus of the domestic horse (*Equus caballus*). Abbreviations: *as,* astragalus; *ca,* calcaneum; *cu,* cuboid; *ec,* ectocuneiform; *na,* navicular. The metatarsals are numbered.

tral digits (or digit, in the case of the horse), the distal tarsals, and the astragalus. The astragalus thus becomes the main weight-bearing bone of the two proximal tarsals. The calcaneum remains important as a point of insertion for extensors of the foot, but it no longer is a major weight-bearing bone of the tarsus. A similar bypassing of the calcaneum occurs in the cursorial pera-

melid marsupials (see Fig. 5–17, page 55).

Two distinctive ungulate specializations involve connective tissue. The nuchal ligament is a heavy band of elastin (an elastic protein found in vertebrates) that is anchored posteriorly to the tops of the neural spines of some of the anteriormost thoracic vertebrae and attaches anteriorly high on the oc-

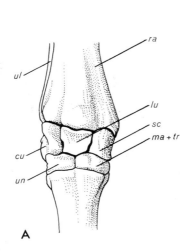

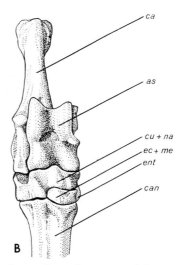

Figure 14-5. The ankle and wrist joints of artiodactyls: A, the right carpus of the mule deer (*Odocoileus hemionus*); B, the right tarsus of the pronghorn (*Antilocapra americana*). Abbreviations: *as,* astragalus; *ca,* calcaneum; *can,* cannon bone (fused third and fourth metatarsals); *cu,* cuneiform of the carpus; *cu + na,* fused cuboid and navicular; *ec + me,* fused ectocuneiform and middle cuneiform; *ent,* internal cuneiform; *lu,* lunar; *ma + tr,* fused magnum and trapezoid; *ra,* radius; *sc,* scaphoid; *ul,* ulna; *un,* unciform.

FIGURE 14–6. A schematic drawing of the nuchal ligament (in black) of an ungulate.

cipital part of the skull (Fig. 14–6). This ligament, especially robust in large, heavy-headed ungulates such as the horse and moose, helps support the head, so that the burden on the muscles that lift the head is greatly lightened. The elasticity of the ligament allows the head to be lowered during

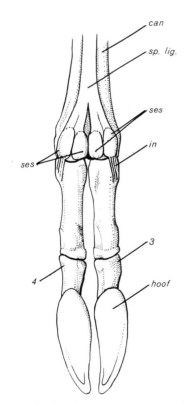

FIGURE 14–7. A posterior view of the left hind foot of an artiodactyl (*Antilocapra americana*), showing the position of the springing ligament. Abbreviations: *can*, cannon bone; *in*, insertion of the springing ligament; *ses*, sesamoid bone; *sp lig*, springing ligament; *3*, third digit (second phalanx); *4*, fourth digit (second phalanx).

eating or drinking. A second specialized ligament, the springing ligament, occurs in the front and hind feet of ungulates, and evolved from muscles that flexed the digits (Camp and Smith, 1942). In the hind foot of the pronghorn (*Antilocapra*), for example, the springing ligament arises from the proximal third of the back of the cannon bone and inserts distally on the sides of the first phalanges of digits three and four (Fig. 14–7). When the foot supports the weight of the body, the phalanges are extended, thereby stretching the springing ligament. But as the foot begins to be relieved of the weight of the body toward the end of the propulsion stroke of the stride, the elastic ligament begins to rebound, and when the foot is leaving the ground the phalanges snap toward the flexed position (Fig. 14–8C). The familiar backward flip of the horse's foot just as it leaves the ground is controlled by the springing ligament. This flip gives a final increase in speed and thrust to the stride, and serves to increase the ungulate's speed afoot without the use of muscular effort.

Feeding Specialization

The herbivorous diet characteristic of most ungulates has favored the development of cheek teeth with large and complex occlusal surfaces that function to finely section plant material as an aid to digestion. Premolars tend to become molariform, and thus to increase the extent of the grinding battery, and the anterior dentition becomes variously modified. In

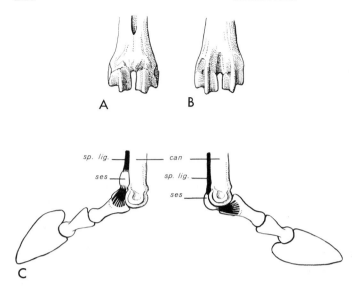

FIGURE 14–8. The left hind foot of the pronghorn (*Antilocapra americana*): A, anterior view of the distal end of the cannon bone; B, posterior view of this bone; C, position of the phalanges when the foot is supporting the weight of the body (right) and the springing ligament (shown in black), is stretched, and the position when the foot leaves the ground and the springing ligament flexes the phalanges (left). Abbreviations are the same as in Figure 14–7.

advanced types, there is a diastema between the anterior dentition and the cheek teeth.

Many ungulates have become large; indeed, the largest members of most mammalian faunas are ungulates. Large size reduces the number of predators to which an animal is vulnerable and is advantageous in terms of temperature regulation and energy requirements. Probably these and other factors have influenced the size of ungulates.

Unusual demands are put on the digestive systems of ungulates by their diet. Vegetation is far less concentrated food than is meat, and is more difficult to digest. In addition, plant material is frequently low in protein. An herbivore must break down the cell wall, a fairly rigid structure formed largely of cellulose, not so much for the energy it yields but to gain access to the proteins within the cell. This breakdown is difficult, however, for mammals lack enzymes that digest cellulose. All herbivores, therefore, must have alimentary canals that are specialized to cope with cellulose by other means than direct enzymatic action. Both perissodactyls (horses, rhinos, and tapirs) and ruminant artiodactyls (camels, deer, antelope, sheep, goats, and cattle) utilize a fermentation process that involves the breakdown of cellulose by the cellulolytic enzymes of microorganisms that live in the alimentary canal.

In perissodactyls, if we can assume that their digestive systems are all similar to the system of the domestic horse, microorganism-aided fermentation takes place in the enlarged colon and large intestine. Protein is digested and absorbed in the relatively small and simple stomach. In ruminants, in contrast, the enlarged and several-chambered stomach harbors microorganisms that break down cellulose. Large food particles float on top of the fluid in the rumen, are passed to the diverticulum (see Fig. 2–12, p. 15), and are then regurgitated and remasticated. In the vernacular, the animals chew their cud. This cycle is repeated until the chemical and mechanical breakdown of the food has reached a point at which the particles sink in the fluid and pass on to the intestine. As a further refinement, in the stomach a complex system of recycling and reconstitution of food constituents ensures that after protein is extracted from food it is used to greatest advantage. When the ruminant's food is high in cellulose and lignin (a substance similar to cellulose that contributes the hard, woody characteristics to plant stalks and roots) it is digested slowly, and the rate of passage of food through the gut is low. The horse's system, however, has no regulation of the rate at which food is processed. Food takes some 30 to 45 hours to pass through the gut of a horse, as compared to 70 to 100 hours to make the corresponding journey in the cow. The central difference, then, is that the digestive system of the horse is less efficient than that of the ruminant, but in compensation the horse eats greater quantities of food; the emphasis in ruminants is on highly efficient digestion and on more selective feeding, but not on high rates of food intake. Given food in short supply, the ruminant will probably survive after the horse has died.

An extremely interesting African grazing succession, strongly influenced by differences between digestive efficiencies and food requirements, has been described by Bell (1971). He studied primarily the most abundant ungulates, the zebra (*Equus burchelli*), a non-ruminant, and the wildebeest (*Connochaetes taurinus*) and Thompson's gazelle (*Gazella thompsonii*), both ruminants. He found that in the Serengeti Plains the zebra was the first of these ungulates to be forced by food shortages to move from the preferred shortgrass area down into the longer, coarser grasses of the lowlands. After the zebras' feeding and trampling activities in the lowlands had removed the coarse upper parts of the grass and had made the lower, more nutritious plant parts more readily available, the wildebeest, a more selective feeder, moved in. By this time the zebras were becoming less able to get sufficient quantities of forage and were moving to new tall-grass pastures. A similar replacement of wildebeest by Thompson's gazelles occurred after the wildebeest had removed still more grass and had made available to the small, highly selective gazelles the fruits and leaves of low-growing forbs. Not only was competition between these abundant ungulates minimized by this grazing pattern, but the activities of the early members of the grazing succession were highly advantageous to the later, more selective members. Bell's study clearly illustrates that differences in the digestive systems of ungulates have pronounced effects on food preferences, migratory patterns, and, in fact, many basic interactions of a grazing ecosystem.

ORDER PERISSODACTYLA

Since Eocene times some of the most specialized cursorial mammals have been perissodactyls. Throughout the early Tertiary these were the most abundant ungulates, but their diversity was reduced in the Oligocene, and with the diversification and "modernization" of the artiodactyls in the Miocene the fortunes of perissodactyls began to decline. The surviving perissodactylan fauna (consisting of six genera and 16 species) is but an insignificant remnant of this once important group, and is vastly overshadowed by an impressive Recent artiodactylan assemblage (consisting of 171 species). Perissodactyls occur today largely in southern areas — Africa, parts of central and southern Asia, and tropical parts of southern North America and northern South America.

Perissodactyls evolved from herbivorous condylarths of the family Phenacodontidae (Fig. 14–9A). The order Condylarthra includes a diverse group of ancient ungulates that occurred from the Cretaceous to the Oligocene, and was probably the basal stock for some 18 mammalian orders (Szalay, 1969), including the Recent orders Artiodactyla, Perissodactyla, Hyracoidea, Proboscidea, and Sirenia. Perissodactyls appeared in the late Paleocene in North America and underwent rapid diversification. Eleven of the 12 families appeared in the Eocene, but in addition to the living families Tapiridae, Rhinocerotidae, and Equidae, only the anomalous extinct family Chalicotheriidae survived into the Pleistocene. The features of sev-

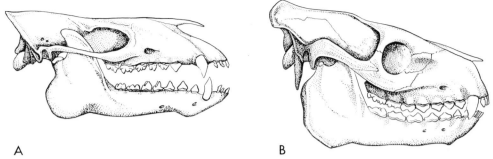

FIGURE 14–9. Skulls of extinct ungulates: A, *Phenacodus*, a Lower Eocene primitive ungulate (Condylarthra); length of skull roughly 230 mm. B, *Oreodon*, a primitive Oligocene ruminant artiodactyl; length of skull roughly 125 mm. (After Romer, A. S.: *Vertebrate Paleontology*, 3rd ed., The University of Chicago Press, 1966.)

eral important perissodactylan families illustrate the considerable structural and functional diversity within the group. The dentition and cranial morphology of perissodactyls developed in response to herbivorous feeding habits. Living perissodactyls have elongate skulls, owing to an enlargement of the facial region to accommodate a full series of large cheek teeth (often hypsodont), and usually have a complete complement of 44 teeth. The teeth are usually lophodont and are either hypsodont in grazing types (all Equidae, and *Ceratotherium* of the Rhinocerotidae) or brachyodont in browsers (all Tapiridae, and *Rhinoceros* and *Didermocerus* of the Rhinocerotidae). Many postcranial specializations further cursorial ability. The clavicle is absent, and usually the manus has three or four digits and the

pes, three digits; but in the equids only one functional digit is retained on each foot (Fig. 14–10C). The feet are mesaxonic; that is, the plane of symmetry of the foot passes through the third digit, whereas this plane passes between digits three and four in the paraxonic foot of artiodactyls.

Family Equidae. (Fig. 14–11). Horses, the most highly cursorial and graceful perissodactyls, now occur wild only in Africa, Arabia, and parts of western and central Asia. There is but one genus with seven Recent species.

Wild horses are in general not as large as domestic breeds. The average weight of a female zebra (*Equus burchelli*) is given by Bell (1971) as 219 kg., but some domestic breeds weigh over 1000 kg. The skull has a fairly level profile, and the rostrum is long

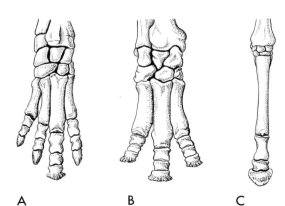

FIGURE 14–10. The front feet of some perissodactyls: A, a tapir (*Tapirus*); B, a rhinoceros (*Rhinoceros*); C, a horse (*Equus*). (After Howell, A. B.: *Speed in Animals*, The University of Chicago Press, 1944.)

FIGURE 14–11. Wild asses (*Equus hemionus*) in the Gobi Desert of Mongolia. (From Allen, G. M.: *Mammals of China and Mongolia*, American Museum of Natural History. Photograph courtesy of the American Museum of Natural History.)

and deep (Fig. 14–12); the dental formula is 3/3, 0-1/0-1, 3-4/3, 3/3 = 36-42. The cheek teeth are hypsodont and have complex patterns on the occlusal surfaces (Fig. 14–13). The limbs are of a highly cursorial type: only the third digit is functional, all but the proximal joints largely restrict movement to one plane, and the foot is greatly elongate. In the tarsus the main weight-bearing bones are the ectocuneiform, navicular, and astragalus; the calcaneum is mostly posterior to the astragalus (Fig. 14–4).

The evolution of horses is well documented by an excellent and largely New World fossil record, and is discussed by Simpson (1951). Equids are first represented by *Hyracotherium* from the late Paleocene of Wyoming (Jepsen, 1969). This primitive type had a generalized skull with 44 teeth (Fig. 14–14A). The upper and lower molars were brachyodont and basically four-cusped. The upper molars bore a protoconule and a metaconule, and the paraconid of the lower molars was reduced (Fig. 14–13A). The premolars were not molariform. The limb structure reflected considerable running ability: the limbs were fairly long and slender; the front foot had four toes and the hind foot had three toes, but the animal was functionally tridactyl. Each digit terminated in a small hoof,

and the foot posture was unguligrade. *Hyracotherium* was the size of a small dog, and presumably browsed on low-growing vegetation in forested or semi-forested areas. Side branches from the main stem of equid evolution developed at various times, but the main evolutionary line can be traced through such intermediate genera as *Merychippus* and *Pliohippus* to the Pleistocene and Recent *Equus*. *Merychippus*, a pony-sized Miocene type, was functionally tridactyl, but retained short lateral digits. The dentary bone was deep, the face was long, and the orbit was fully enclosed. The cheek teeth were high-crowned, were covered with cement, and had an occlusal pattern similar to that of *Equus* (Fig.

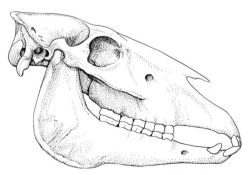

FIGURE 14–12. The skull of the domestic horse (*Equus caballus*). Length of skull 530 mm.

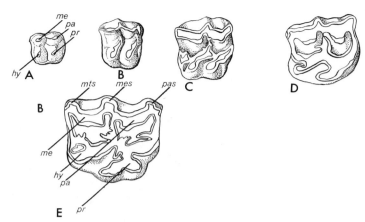

FIGURE 14–13. Right upper molars of equids: A, *Hyracotherium*; B, *Mesohippus*; C, *Parahippus*; D, *Pliohippus*; E, *Equus*. These teeth illustrate stages in the evolution of the equid molars. Abbreviations: *hy*, hypocone; *me*, metacone; *mes*, mesostyle; *mts*, metastyle; *pa*, paracone; *pas*, parastyle; *pr*, protocone. (After Romer, A. S.: *Vertebrate Paleontology,* 3rd ed., The University of Chicago Press, 1966.)

14–13E). *Pliohippus* occurred in the Pliocene; it had the skull features of its progenitor, *Merychippus,* but was more progressive in having higher-crowned teeth and lateral digits reduced to splint-like vestiges. *Equus,* as well as an extinct evolutionary side branch of short-legged South American horses (typified by *Hippidium*), evolved from *Pliohippus. Equus,* the genus to which all living horses belong, differs from *Pliohippus* in greater size and in a more complex crown pattern of the cheek teeth. Major evolutionary trends of the Equidae listed by Colbert (1969:396) include: increase in size; lengthening of legs and feet; reduction of lateral toes and emphasis on the middle toe; molarification of the premolars; increase in height of the crowns of the cheek teeth; lengthening of the facial part of the skull to accommodate the large cheek teeth; and deepening of the maxillary and dentary to accommodate the high-crowned teeth. In addition, the profile of the angular border of the dentary swept progressively further forward, and the origin of the masseter muscles migrated forward. These adaptations increased the force the masseter muscles could exert on the dentary.

Cenozoic changes in climate and in the flora of North America may have had a critical influence on the evolu-

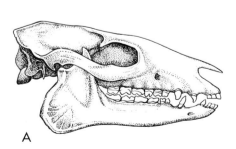
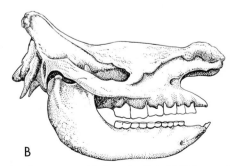

FIGURE 14–14. A, the skull of *Hyracotherium,* the first known equid; length of skull 134 mm. B, the skull of a rhinoceros (*Diceros*); length of skull approximately 610 mm. (After Romer, A. S.: *Vertebrate Paleontology,* 3rd ed., The University of Chicago Press, 1966.)

tion of horses. Especially important was the Miocene development of grasslands over much rolling or nearly level land within the present Great Plains, the Great Basin, and the southwestern deserts. Many of the most progressive equid skull and dental features probably arose in response to the shift to a diet high in grasses. Grass, at least at certain times of the year, has low nutritional value, and must be eaten in large quantities to sustain life. High-crowned, persistently growing teeth were necessary to cope with large amounts of grasses made highly abrasive by silica in the leaves and by particles of soil deposited on leaves by wind and the splash effect of rain. Also of great adaptive value were the highly cursorial limbs with single-toed feet that facilitated rapid and efficient locomotion on the firm, level footing of the grasslands. Cursorial ability was perhaps as advantageous for traveling between widely scattered concentrations of food and distant water holes in semiarid regions as for escaping from predators.

For some unknown reason, horses disappeared from the New World, seemingly their place of origin and primary center of their evolution, before historic times. Although wild horses now occupy only Africa and parts of Asia, within historic times they occurred throughout much of Eurasia. Wild equids inhabit grasslands in areas ranging from tropical to subarctic in climate, and feral domestic horses thrive in North America and in parts of Africa (Thomas, 1971).

Family Tapiridae. Tapirs occupy tropical parts of the New World and the Malayan area. The family includes one living genus and four species. Structurally, tapirs are notably primitive, and according to Romer (1966:269) "are still very close in many respects to the common ancestors of all perissodactyls." "True" tapirs are known first from the Oligocene, but possible ancestral types occurred in the Eocene.

Tapirs are heavy-bodied and weigh up to about 300 kg. The limbs are short, and both the ulna and fibula are large and separate from the radius and tibia respectively; the front feet have four toes (Fig. 14–10A) and vestiges of the fifth (the pollex), and the hind feet have three toes. Tapirs retain a full placental complement of 44 teeth. Three premolars are molariform, and the brachyodont cheek teeth retain a simple pattern of cross lophs. The short proboscis (Fig. 14–15) and reduced nasals are among the few specializations of tapirs.

These animals have probably always occupied moist forests, where their primitive feet serve well on the soft soil, and the teeth are adequate for masticating plant material that is not highly abrasive. Tapirs today inhabit mostly tropical areas, and usually occur near water. They are rapid swimmers, and often take refuge from predators in the water. Tapirs are solitary and nocturnal, and their presence is frequently made known chiefly by their systems of well-worn trails between feeding areas, resting places, and water. Their food is largely succulent plant material including fruit.

Family Rhinocerotidae. This family is represented today by four genera and five species, and is restricted to parts of the tropical and subtropical sections of Africa and southeastern Asia. These ponderous creatures—the armored tanks of the mammal world—are surviving members of the spectacular late Tertiary and Pleistocene ungulate fauna. Although apparently a declining group, rhinos have an illustrious past.

The fossil record of the rhinos and their relatives (superfamily Rhinocerotoidea) is remarkably complex, and parallels that of the horses in documenting the early Tertiary and mid-Tertiary success and the late Tertiary decline of the group. Two genera that illustrate well the diversity of early Tertiary rhinocerotoids are *Hyracodon* and *Baluchitherium*. *Hyracodon* (Hyracondontidae), a small North American Oligocene "running rhinoceros," had slender legs and tridactyl feet, and was probably similar

FIGURE 14–15. South American tapir (*Tapirus terrestris*). (Photograph by Ron Garrison; San Diego Zoo photo.)

in cursorial ability to Oligocene horses. Perhaps as a result of competition with horses, hyracodonts became extinct in late Oligocene. A contemporary of *Hyracodon* in the Oligocene was the Asian form *Baluchitherium* (Rhinocerotidae), the largest known land mammal. This giant was 18 feet high at the shoulder, and the skull (small in proportion to the great size of the rest of the animal) was four feet long. The neck was long, and *Baluchitherium* perhaps browsed on high vegetation in giraffe-like fashion. The limbs were long and graviportal, but the tridactyl feet were unique in that the central digit was greatly enlarged and terminated in a broad hoof, whereas the lateral digits were more strongly reduced than in any other rhinocerotoid. Rhinos died out in the New World in the Pliocene, but remained common and diverse in Eurasia through the Pleistocene. The Pleistocene woolly rhinoceros (*Coelodonta*) was apparently adapted to cold climates. Entire preserved specimens of this rhino have been found in an oil seep in northwestern Spain.

All Recent rhinos are large, stout-bodied herbivores with fairly short, graviportal limbs (Fig. 14–16). Weights range up to about 2800 kg. The front foot has three or four toes (Fig. 14–10B), and the hind foot is tridactyl. The nasal bones are thickened and enlarged, often extend beyond the premaxillaries, and support a horn. Where there are two horns, the posterior one is on the frontals. The horns are of dermal origin and lack a bony core. The occipital part of the skull is unusually high (Fig. 14–14B) and yields good mechanical advantage for neck muscles that insert on the lambdoidal crest and raise the heavy head. The incisors and canines are absent in some rhinos and are reduced in number in others; the dental formula is 0-2/0-1, 0/0-1, 3-4/3-4, 3/3 = 24-34. The cheek teeth have a pattern of cross lophs far simpler than that of equids (Fig. 14–17).

Rhinos inhabit grasslands, savannas, brushlands, forests, and marshes in tropical and subtropical areas. Some species are solitary (*Diceros*), whereas others occur in family groups (*Ceratotherium*) or even in assemblages including up to 24 animals (Heppes,

FIGURE 14–16. A black rhinoceros (*Diceros bicornis*). This species occurs in eastern and southern Africa, and is the most common living species of rhino. (Photograph by Ron Garrison; San Diego Zoo photo.)

1958). Rhinos practice scent marking by establishing dunghills along well-worn trails. A variety of plant material is taken; some species are browsers and some are grazers. Adults are nearly invulnerable to predation, except by man, but young rhinos may be attacked by carnivores as small as spotted hyaenas (Kruuk and Van Lawick, 1968:53). The Asian rhinos (*Rhinoceros* and *Didermocerus*) are probably facing extinction. Because of the supposed medicinal properties of the horn and other parts, Asian rhinos have been hunted persistently for at least 1000 years. According to Walker (1968:1351), there are only some four dozen one-horned rhinos (*Rhinoceros*)

alive; these are restricted to swampy jungles of Java. Regretfully, the survival of the Asian rhinos seems unlikely.

ORDER ARTIODACTYLA

The order Artiodactyla—containing the pigs, peccaries, hippopotami, camels, deer, cows, sheep, goats, and antelope—is by far the most important ungulate group today. Whereas perissodactyls were abundant and reached their greatest diversity in late Eocene, artiodactyls underwent their most important evolution much later, in the Miocene. Since this epoch the perissodactyls have steadily declined, but the artiodactyls have remained diverse and successful. One is tempted to relate the decline of the perissodactyls to the rise of the artiodactyls, and to regard the latter as the more effective competitors. However, although many structural differences between perissodactyls and artiodactyls are apparent, the functional advantages conferred by many features are not easily recognized.

In the order Artiodactyla the structure of the foot is especially diagnostic. The foot is paraxonic; that is, the plane

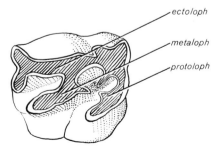

FIGURE 14–17. The right upper molar of a rhinoceros (*Rhinoceros*).

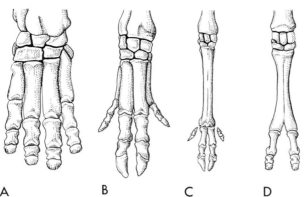

FIGURE 14–18. The right front feet of some artiodactyls: A, hippopotamus (*Hippopotamus*); B, swine (*Sus*); C, elk (*Cervus*); D, camel (*Camelus*). (After Howell, A. B.: *Speed in Animals,* The University of Chicago Press, 1944.)

A   B   C   D

of symmetry passes between digits three and four (Fig. 14–18). The weight of the animal is borne primarily by these digits; the first digit is always absent in living members, and the lateral digits (digits two and five) are always more or less reduced. Four complete and functional digits occur in the families Suidae, Hippopotamidae, and Tragulidae, and in the forelimb of the Tayassuidae (the hind limb has the medial digit suppressed). Two complete toes, with the lateral digits absent (Camelidae, some Bovidae, Antilocapridae, and Giraffidae), or with incomplete remnants of the lateral digits (Cervidae and some Bovidae), occur in the more cursorial families. The cannon bone (a single bone formed by the fusion of the third and fourth metapodials) is present in the families Camelidae, Cervidae, Giraffidae, Antilocapridae, and Bovidae. Typically, the terminal phalanges are encased in pointed hoofs. The astragalus has a "double pulley" arrangement of articular surfaces (Fig. 14–5B) that completely restricts lateral movement. The proximal articulation (with the tibia) and the distal articulation (with the navicular and cuboid, which are fused in many advanced types) of the astragalus are of critical importance in allowing great latitude of flexion and extension of the foot and digits, as is the extension of the articular surfaces and keels on the distal ends of the cannon bones to the an-

terior surfaces (Fig. 14–8A, B). The distinctive artiodactyl astragalus is regarded by some as a primary key to the success of the group. Perhaps more important, however, is the remarkable efficiency of the artiodactyl (ruminant) digestive system (described on pages 226 and 227).

The limbs of artiodactyls, especially the distal segments, are usually elongate and fairly slim. The femur lacks a third trochanter. Whereas this prominence serves as a point of insertion of the gluteal muscles in perissodactyls, in artiodactyls these muscles insert more distally, on the tibia. The distal parts of the ulna and the fibula are usually reduced and may fuse with the radius and tibia respectively; this fusion is associated with the restriction of movement of the limb to one plane. The clavicle is seldom present. The intrinsic muscles of the feet (those that both originate and insert on the feet) are usually absent, being replaced by specialized tendons and ligaments.

The skull usually has a long preorbital section, and a postorbital bar or process is always present. Horns, always of bone or with a bony core, are most often borne on the frontals, which are enlarged at the expense of the parietals. The teeth are brachyodont or hypsodont and vary from 30 to 44 in number. The crown pattern is bunodont or, more often, selenodont. The premolars are not fully molariform, in contrast to the perissodactyl situation, and considerable specialization of the anterior dentition occurs in advanced types.

Although the classification of the Artiodactyla has occupied the attention of a number of the leading paleontologists and mammalogists, it is still not fully resolved (Romer, 1968:209). The system used here is that of Romer (1966), and recognizes the extinct suborder Palaeodonta and the living suborders Suina and Ruminantia, all of which are first known from the Eocene. The Palaeodonta contains the most primitive known artiodactyls from the early Tertiary. Members of this group are small, and some are so primitive in structure as to be nearly unrecognizable as artiodactyls if it were not for the distinctive astragalus.

SUBORDER SUINA

In members of the suborder Suina, which includes pigs, peccaries, and hippopotami, the molars are bunodont, the canines (and in the case of the hippopotami, the incisors also) are tusk-like (Fig. 14–19), and the feet nearly always retain four toes with complete and separate digits. The skull contrasts with those of other artiodactyls in having a posterior extension of the squamosal that meets the exoccipital and conceals the mastoid bone. The stomach is of a non-ruminant type (the animals do not chew their cud) but may have as many as three chambers.

Family Suidae.   Swine are omnivorous, and lack many structural modifications typical of more specialized artiodactyls. The Suidae is an Old World family; the present distribution includes much of Eurasia and the Oriental region, and Africa south of the Sahara. There are eight Recent species of five genera. Suids appeared in the Oligocene. The enteledonts (Enteledontidae), an early branch from the "swine" evolutionary line, were huge pig-like creatures with skulls up to three feet in length.

Most suids resemble the domestic swine (*Sus*). Adults may weigh as much as 275 kg., and typically have thick, sparsely haired skin. The skull is long and low, and usually has a high occipital area and a concave or flat profile. The large canines are ever-growing, and the upper canines form conspicuous tusks that protrude from the lips and curve upward. In *Babyrousa*, an Indonesian member of the Suidae, the upper canines protrude from the top of the snout (Fig. 14–20). Suid molars are bunodont, and the last molars are often elongate, with many cusps and a complexly wrinkled crown surface (Fig. 14–21A). The dental formula is variable even within a species; the total number of teeth ranges from

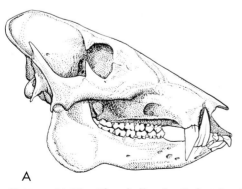
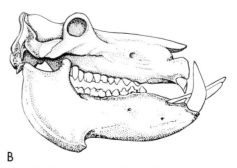

A   B

FIGURE 14–19.   The skulls of artiodactyls of the suborder Suina: A, javelina (*Tayassu tajacu*); length of skull 225 mm. B, hippo (*Hippopotamus*); length of skull 600 mm. (B after Romer, A. S.: *Vertebrate Paleontology*, 3rd ed., The University of Chicago Press, 1966.)

FIGURE 14–20. The babirussa (*Babyrousa babyrussa*), a suid that occurs in the Celebes and several nearby small islands in Indonesia. Note that the tusk-like upper canines emerge from the top of the snout. (San Diego Zoo photo.)

34 to 44. The limbs are usually fairly short, and the four-toed feet never have cannon bones (Fig. 14–18B).

Swine inhabit chiefly forested or brushy areas, but the wart hog (*Phacochoerus*) favors savanna or open grassland and is entirely herbivorous. Most suids are gregarious, and some assemble in groups of up to 50 individuals. Most species eat a broad array of plant food and carrion, and given the opportunity will kill and eat such animals as small rodents and snakes. By comparison with ruminant artiodactyls, cursorial ability in suids is modest. Wart hogs, however, can run at speeds up to 47 km. per hour (Walker, 1968:1362).

Family Tayassuidae. These animals, usually called javelinas or pec-

caries, are restricted to the New World, where they occur from the southwestern United States to central Argentina. There are but two Recent species of one genus (*Tayassu*). The fossil record of tayassuids begins in the Oligocene. Presumably javelinas evolved from primitive Old World pigs, but javelinas are not known from the Old World after the Miocene. Javelinas are more progressive in limb structure than are suids, and are less carnivorous.

Javelinas are much smaller than suids; weights range up to only about 30 kg. The skull has a nearly straight dorsal profile, and the zygomatic arches are unusually robust (Fig. 14–19A). The canines are long and are directed slightly outward, but never

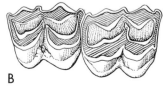

A   B

FIGURE 14–21.   The molars of artiodactyls: A, the second and third right upper molars of the swine (*Sus scrofa*); B, the comparable teeth of the elk (*Cervus elaphus*), with the enamel ridges unshaded, the enamel-lined depressions stippled, and the dentine crosshatched. The molars of *Sus* are bunodont; those of *Cervus* are selenodont.

turn upward, and have sharp medial and lateral edges. These teeth slide against one another, and the anterior surface of the upper, and the posterior surface of the lower, are planed flat by this contact. The molars are roughly square and have four cusps; they lack the complex wrinkled and multi-cusped pattern typical of suids. The dental formula is 2/3, 1/1, 3/3, 3/3 = 38. The feet are slender and appear delicate, and the side toes are small compared to those of suids and usually do not reach the ground. In tayassuids the front feet have four toes and the hind feet have three, and a cannon bone in the hind foot is partially formed by the proximal fusion of the medial metatarsals. Surprisingly, modern javelinas are not as advanced in foot structure as is *Mylohyus*, an extinct Pleistocene javelina, in which the side toes of the forefoot were very strongly reduced and the didactyl hind foot had a fully developed cannon bone.

Javelinas occupy diverse habitats, from deserts and oak-covered foothills in Arizona to dense tropical forests in southern Mexico, Central America, and South America. They are gregarious, and on occasion form groups including several dozen individuals. Javelinas are omnivorous, but seem to rely more heavily on plant material than do suids. The presence of javelinas is often indicated by shallow excavations where roots have been exposed beneath bushes or patches of prickly-pear cactus. Despite their rather chunky build, javelinas are rapid and extremely agile runners in the broken terrain they often inhabit.

Family Hippopotamidae. This family is represented today by the genera *Hippopotamus* and *Choeropsis* (the pigmy hippo), each with one species. The group first appeared in the upper Pliocene in Africa and Asia and occurred widely in the southern parts of the Old World in the Pleistocene. Hippos now occur only in Africa; in North Africa they are restricted to the Nile River drainage, but they occur widely in the southern two-thirds of the continent.

Hippos are bulky, ungraceful creatures with huge heads and short limbs (Fig. 14–22). They are large, weighing up to roughly 3600 kg. in the case of *Hippopotamus* and about 180 kg. in the case of *Choeropsis*. Some of the distinctive features of hippos probably evolved in association with their amphibious mode of life. Specialized skin glands secrete a pink, oily substance that protects the sparsely haired body. The highly specialized skull has ele-

FIGURE 14–22. Hippopotami (*Hippopotamus amphibius*) in East Africa. (Photograph by W. Leslie Robinette.)

vated orbits and enlarged and tusk-like canines and incisors (Fig. 14–19B). The bunodont molars are basically four-cusped; the dental formula is 2-3/1-3, 1/1, 4/4, 3/3 = 38-40. The limbs are robust, and the feet are four-toed (Fig. 14–18A). The foot posture is digitigrade, but only the distal phalanx of each toe touches the ground. The broad foot is braced by a sturdy "heel" pad of connective tissue, and the central digits are nearly horizontal.

Hippos are restricted to the vicinity of water. *Hippopotamus* is gregarious, and groups spend much of the day in the water. When bodies of water are at a premium during the dry season, hippos often concentrate in stagnant ponds that the animals churn into muddy morasses. They are good swimmers and divers, and when submerged are able to walk on the bottoms of rivers or lakes by using a strange slow-motion type of gait. At night hippos may move far inland to feed on vegetation. *Choeropsis* is solitary or occurs in pairs, and inhabits forested areas. Instead of seeking shelter in the water when disturbed, as is characteristic of *Hippopotamus*, *Choeropsis* seeks refuge in dense vegetation.

SUBORDER RUMINANTIA

This suborder includes camels, giraffes, deer, antelope, sheep, goats, and cattle. Members of this most advanced artiodactylan suborder have been in the past, and remain, the dominant artiodactyls. In general, these animals are committed strictly to an herbivorous diet and to highly cursorial locomotion. Ruminants chew their cud; the stomach has three or four chambers (Fig. 2–12, page 15) and supports microorganisms that have cellulolytic enzymes. Ruminants have selenodont molars (Fig. 14–21B), and the anterior dentition is variously specialized by loss or reduction of the upper incisors, by the development of incisiform lower canines, and commonly by the loss of upper canines. The skull differs from those of

members of the suborder Suina in the exposure of the mastoid bone between the squamosal and the exoccipital. Antlers or horns, often large and complex strucures, are present in the most progressive families. In the limbs there is a pronounced trend toward the elongation of the distal segments, the fusion of the carpals and tarsals, and the perfection of the two-toed foot. The ruminants may conveniently be separated into two divisions (infraorders): Tylopoda, the camels and llamas and their extinct relatives; and Pecora, which includes all of the remaining, more progressive ruminants. A diagnostic feature of the pecorans is the fusion of the navicular and cuboid, over which the astragalus is nearly centered (Fig. 14–5B).

An early tylopod family (now extinct), the Merycoidodontidae, usually called oreodonts, deserve mention because they are by far the most abundant mammals in some Oligocene and early Miocene strata in North America. Oreodonts were typically pig-like in general build, with short limbs, digitigrade and four-toed feet, and a continuous tooth row with no loss of teeth (Fig. 14–9B). In contrast to the Suina, however, the cheek teeth were selenodont and became fairly high-crowned in advanced types, suggesting the acquisition of grazing habits. Probably as a result of competition with more advanced artiodactyls, oreodonts declined in the upper Miocene, and disappeared in the Pliocene.

Family Camelidae. These primitive ruminants are restricted to arid and semiarid regions. *Camelus*, with two species, occupies the Old World, and wild populations persist in the Gobi Desert of Asia; *Lama* (Fig. 14–23), with two species, occurs in South America from Peru through Bolivia, Chile, Argentina, and Tierra del Fuego. Camels probably arose in the Old World and migrated in the late Eocene to North America, where their fossil record begins. Of special interest, as an example of a reversal of a well established evolutionary trend, is the development of the camelid foot.

FIGURE 14–23. The vicuña (*Lama vicugna*), a camelid that inhabits the central Andes of South America. (Photograph by Carl B. Koford.)

By the Oligocene, camels were already highly specialized in foot structure; they were nearly unguligrade in foot posture and were didactyl; the distalmost phalanges probably bore hooves. The distinctive distal divergence of the metapodials (Fig. 14–18D), however, was already recognizable. In the Miocene the central metapodials fused to form a cannon bone, but at this same time a retrograde trend toward the secondary development of a digitigrade foot posture began, and from the Pliocene onward camels were digitigrade. Camels are the only living digitigrade ungulates. Because semi-arid conditions developed and became widespread in the Miocene, one is tempted to relate the changes in the camelid foot posture to changing soil conditions. In any case, the peculiar camelid foot clearly provides effective support on soft, sandy soil, into which the feet of "conventional" unguligrade artiodactyls sink deeply. With the establishment of land bridges between North America and Asia, and between North and South America, in the Pleistocene, camels spread to the Old World and to South America.

Although highly specialized in foot structure, camelids are the most primitive living ruminants. These are fairly large mammals, ranging in weight from about 45 to 500 kg., and have long necks and long limbs. The dentition has advanced less toward herbivorous specialization than has that of the Pecora. In camelids only the lateral upper incisor is present, but it is caniniform; the lower canines are retained and are little modified (Fig. 14–24). The lower incisors are inclined forward and occlude with a hardened section of the gums on the premaxillaries. A broad diastema is present, and the premolars are reduced in number (to 3/1-2 in *Camelus* and 2/1 in *Lama*). As in other ruminants, the limbs are long

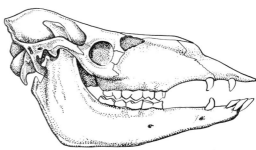

FIGURE 14–24. The skull of an extinct Pleistocene New World camel (*Camelops*); length of skull 565 mm. (After Romer, A. S.: *Vertebrate Paleontology*, 3rd ed., University of Chicago Press, 1966.)

and the ulna and fibula are reduced; in the carpus, the trapezium is absent, and in the tarsus, the mesocuneiform and ectocuneiform are fused. The digitigrade feet are didactyl, but the cannon bone is distinctive in that the distal ends of the metapodials remain separate and flare outward (Fig. 14–18D). The toes are separate, and each is supported by a broad cutaneous pad, which largely encases the second phalanx and serves to increase the surface area of the foot greatly. The short ungual phalanges do not bear hooves, but have nails on the dorsal surfaces.

Camels are remarkably well adapted to arid areas, and their ability to go for long periods in hot weather without water is remarkable (see p. 386). They are grazers, and can survive in regions with only sparse vegetation. The guanaco (*Lama glama*) of South America is gregarious and usually occurs in small groups led by an adult male. Guanacos are fairly speedy runners, but are especially skillful at moving rapidly over extremely rough terrain. Simpson (1965:189) observed that in Argentina guanacos spent considerable time running up and down cliffs and were able to leap delicately up a nearly vertical trail that men could only climb laboriously. He also noted (1965:188) that guanacos are highly vocal. They make a strange yammering noise, and Simpson comments that "there is something distinctly indecent about the noise as it issues from the beast's protrusile and derisive lips."

Family Tragulidae. This family, containing the chevrotains or mouse deer, has only two living genera with four species, but is of interest because these animals probably resemble in many ways the ancestors of the more advanced pecorans. Chevrotains are small, delicate creatures, weighing from 2.3 to 4.6 kg., that occur in tropical forests in central Africa (*Hyemoschus*) and in parts of southeast Asia (*Tragulus;* Fig. 14–25). The tragulid fossil record begins in the Miocene.

Although apparently related to higher pecorans, chevrotains have many primitive features not characteristic of other pecorans. The tragulid skull never bears antlers, but, seemingly in compensation, the upper canines are unusually large, defensive weapons. Otherwise, the dentition resembles that of higher pecorans: the upper incisors are lost, the lower canine is incisiform, and the cheek teeth are selenodont. The limb structure is an unusual mosaic of primitive and advanced features. Although the limbs are long and slender, and the carpus is highly specialized in having the navicular, cuboid, and ectocuneiform fused, the lateral digits of chevrotains are complete, a condition never present in higher pecorans. In addition, although a cannon bone occurs in the hind limb, the metacarpals of the central digits are separate in the African tragulid and are partly fused in the Asian form, whereas the cannon bone is represented by fully fused metapodials in all other pecorans. A further contrast is perhaps associated with differences in bounding ability. In tragulids the articular surfaces at the distal end of the cannon bone do not extend onto its dorsal surface, while in other pecorans these articular surfaces are extended.

Tragulids are secretive, nocturnal creatures that inhabit forests and underbrush and thick growth along watercourses. They escape predators by darting along diminutive trails into dense vegetation. Their food consists of grass, leaves of shrubs and forbs, and some fruit.

Advanced pecorans. The remaining pecorans (the Cervidae, Giraffidae, Antilocapridae, and Bovidae) are advanced artiodactyls and share a series of progressive features. The upper incisors are absent, the upper canine is usually absent, the lower canine is incisiform, and the cheek teeth are selenodont. The dental formula is typically 0/3, 0/1, 3/3, 3/3 = 32. The cannon bone is present in fore and hind limbs, its distal articular surfaces are extensive, and the lateral digits are always incomplete (Fig. 14–18C) and are often lacking. Movement of the foot is

FIGURE 14–25. A chevrotain (*Tragulus napu*, Tragulidae). Note the large upper canine and the lack of antlers or horns. (Photograph by Ron Garrison; San Diego Zoo photo.)

strongly limited to a single plane by the tongue-in-groove contacts between the astragalus and the bones with which it articulates, and by the specialized articular surfaces of the joint between the cannon bone and the first phalanges (Fig. 14–8A, B). Some fusion of carpal elements always occurs and serves further to restrict movement to one plane. The navicular and cuboid are always fused (Fig. 14–5B), and a variety of patterns of fusion of the other elements occurs. The four-chambered stomach is of a ruminant type. Although all pecorans but the tragulids share this basic structural plan, each family has distinctive features usually related to diet and degree of cursorial ability.

Family Cervidae. Members of this family, which includes the deer, elk, caribou, and moose, occur throughout most of the New World, and in the Old World in Europe, Asia, and northwest Africa; they have been introduced widely elsewhere. Living members include some 16 genera and 37 species. Cervids appeared in the early Oligocene in Asia, and reached North America in the early Miocene.

These are the antlered artiodactyls. Antlers are usually present and occur in males only, except in caribou (*Rangifer*), in which both sexes bear antlers. The paired antlers usually arise from a short base on the frontals (the pedicel, Fig. 14–26) and are entirely bony. The antlers are shed annually, usually in late winter, and begin growing again shortly thereafter. During their growth, antlers are covered by fur-covered skin ("velvet"), which is usually shed in autumn when the antlers are fully grown. In some antlered cervids the upper canines are retained but reduced (as in the elk, *Cervus elaphus*). Two cervids with short antlers have enlarged canines (*Cervulus* and *Elaphodus*), and in two deer that have no antlers (*Hydropotes*

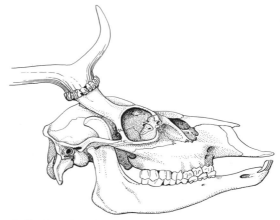

FIGURE 14-26. The skull of a male fallow deer (*Dama dama*, Cervidae). The bony antlers are shed yearly. Length of skull 265 mm.

and *Moschus*) the canines are large sabers. The cheek teeth of cervids are brachyodont, reflecting the browsing habit of these animals. These animals have a wide size range: the musk deer (*Moschus*) weighs but 10 kg., whereas the moose (*Alces*) weighs up to roughly 800 kg. The feet are always four-toed, but the lateral toes are often greatly reduced. Distal to the astragalus and calcaneum, the tarsus is usu-

ally composed of three bones: the fused navicular and cuboid, the fused ecto-cuneiform and mesocuneiform, and the internal cuneiform (Fig. 14–5B).

Cervids occur from the arctic to the tropics. Many cervids are well adapted to boreal regions, and occupy mountainous or subarctic areas with severely cold winters. Effective insulation is provided in many cervids by the long, hollow hairs of the pelage. Some

FIGURE 14-27. A mule deer (*Odocoileus hemionus*, Cervidae) in northern Colorado. This species inhabits much of the western United States. (Photograph by O. D. Markham.)

FIGURE 14–28. A giraffe (*Giraffa camelopardalis*) in African savanna. Note the unusual skin-covered horns, a type unique to giraffes. (Photograph by Ron Garrison; San Diego Zoo photo.)

species are gregarious for much of the year, and may assemble in large herds during the winter and during migratory movements. In the western United States, mule deer (*Odocoileus hemionus;* Fig. 14–27), and in the eastern United States white-tailed deer (*O. virginianus*), are common over wide areas, and are heavily hunted in many states. Some states employ many people to study and manage deer; the expense of this work is generally covered by revenue from the sale of hunting licenses.

Family Giraffidae. This group is represented today by but two genera, *Giraffa* and *Okapia;* each has a single species. The family occurs in much of Africa south of the Sahara Desert.

The robust cheek teeth are brachyodont and are marked with rugosities. Short, bony horns, covered with furred skin, are borne on the front part of the parietals, and a medial protuberance from the posterior ends of the nasals and anterior border of the frontals is also present. Horns occur in both sexes

and are never shed. The lateral digits of the elongate limbs are entirely gone, and the tarsus is highly specialized. Distal to the astragalus and calcaneum, only two tarsal bones are present. One is formed by the fusion of the navicular and cuboid, and the other, by the fusion of the three cuneiform bones. The okapi lacks the extreme elongation of the neck and legs that is typical of giraffes, but has an even more specialized tarsus in which all bones distal to the astragalus and calcaneum are fused. The fossil record of the giraffids begins in the Miocene. The okapi, not known to zoologists until 1900, is remarkable in its close resemblance to primitive upper Miocene and lower Pliocene giraffids long known to paleontologists.

The exceptionally long legs and long neck (Fig. 14–28) enable giraffes to browse on high branches of *Acacia* and other leguminous trees in savanna areas. The okapi inhabits dense tropical forests and eats leaves and fruit. Giraffes, despite their considerable

FIGURE 14–29.   The pronghorn (*Antilocapra americana*), one of the fastest cursorial mammals. (Photograph by F. D. Schmidt; San Diego Zoo photo.)

weight (to nearly 1820 kg. in males), can gallop for short distances at over 55 km. per hour (White, 1948). Relative to lighter, shorter-limbed artiodactyls, the limbs of giraffes are flexed little during each stride, producing a curiously stiff-legged gait. During walking or galloping, the center of gravity of the animal is partly controlled by fore-and-aft movements of the head and neck (Dagg, 1962). When fighting, the long neck allows giraffes to deliver powerful blows with the head.

Family Antilocapridae.   This family is represented today by one species, the pronghorn (*Antilocapra americana;* Fig. 14–29), which occurs from central Canada to north-central Mexico. The fossil record of pronghorns is entirely North American, and begins with *Merycodus* in the middle Pliocene. The horns of this antilocaprid were curiously cervid-like in form: they were forked and had a basal burr.

In contrast to cervids, however, the cores were never shed, and both sexes had horns. In addition, the unusually prominent orbits were situated high and far back in the skull, as in *Antilocapra*. The horns of fossil pronghorns were generally more complex than those of *Antilocapra*, but the limbs and teeth of fossil forms were similar to those of the present species. The antilocaprid fauna of the late Tertiary and Pleistocene was more diverse than it is today. Pleistocene deposits from California, for example, contain two antilocaprid types. *Antilocapra* (probably *A. americana*) is present, together with *Breameryx*, a small, delicate limbed, gazelle-like form that stood but two feet high at the shoulder. Perhaps *Breameryx*, like *Antilocapra*, escaped predators by using its great speed in open country.

The pronghorn is one of the most advanced living artiodactyls, with regard both to dentition and to limb structure.

The dental formula is that typical of pecorans, and, although the animals are largely browsers on low shrubs and forbs, the cheek teeth are high-crowned. The abrasive soil particles adhering to the low-growing vegetation they eat probably make high-crowned teeth advantageous. Perhaps as an adaptation allowing the prong-horn to keep watch for danger while its head is down close to the ground, the orbits are unusually far back in the skull, well behind the level of the last molar (Fig. 14–30). Approximately 62 per cent of the length of the skull is anterior to the orbits, whereas the comparable figures for two cervids, the mule deer (*Odocoileus hemionus*) and the elk (*Cervus elaphus*), are about 52 and 53 per cent respectively. The orbit in the pronghorn is large and protrusive, and its upper rim is above most of the dorsal profile of the skull (Fig. 14–30). The horns are of a unique type. The bony core is covered by a black sheath of agglutinated hairs. The old sheath, beneath which a new sheath is beginning to develop, is shed annually in early winter, and the new

sheath is fully grown by July (Einarsen, 1948; O'Gara, *et al.,* 1971). Whereas the mature sheath is forked, the bony horn core is a single, laterally compressed blade. Both sexes have horns, but those of the females are small and inconspicuous.

The legs are long and slender, and all vestiges of lateral digits are gone. The tarsus distal to the astragalus and calcaneum consists of but three bones (Fig. 14–5B). The posterior part of the bottom of each hoof is protected by a swollen pad of connective tissue. The front hoofs are larger than the rear hoofs and have a larger pad, probably as an adaptation serving to cushion the impact sustained by the forefeet during each stride. Pronghorn hairs contain complex patterns of air cells, and provide remarkable insulation.

Pronghorns inhabit open prairies and deserts that support at least fair densities of low grasses, shrubs, and forbs. The numbers of antelope were seriously reduced during the pioneering period of the western United States, but according to Einarsen (1948: 7), "At no time has the antelope

FIGURE 14–30. The skull of a male prong-horn (*Antilocapra americana*). The sheath of the horn is shed yearly, but the bony core is permanent. Length of skull 292 mm.

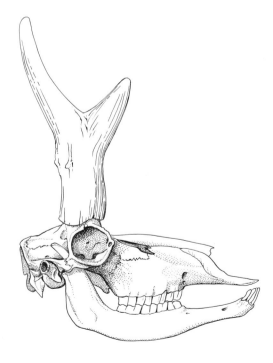

been driven from its original range, being continuously represented by a few individuals in widely scattered sections of the plains country." Today pronghorns are common in several western states, and even the most unobservant tourist, when driving across Wyoming, cannot fail to notice these animals. Pronghorns are gregarious and generally occur in groups. Herds of 50 to 100 individuals are common in the winter. Unlike most mammals, pronghorns do not try to escape danger by seeking shelter. Instead, they depend on their great speed, and generally avoid forested areas or even dense brush. Pronghorns are among the fastest mammals. When running full speed on level footing the pronghorn can attain a speed of about 95 km. per hour for short distances (McLean, 1944; Einarsen, 1948), and their endurance is remarkable. On the short grass prairie of north-central Colorado, I observed pronghorns that had run roughly two miles put on a burst of speed and run away from a closely pursuing light plane that was traveling at 72 km. per hour.

Family Bovidae. The family Bovidae, including the African and Asian antelope, the bison, sheep, goats, and cattle, is the most important and most diverse living group of ungulates. The family includes 44 genera and 111 species, and wild species occur throughout Africa, in much of Europe and Asia, and in most of North America. The domestication of bovids began in Asia roughly 8000 years ago (Darlington, 1957:405), and domestic bovids are nearly as cosmopolitan in distribution as is man.

Bovids were derived from traguloid ancestry in the Old World, and first appeared in the Miocene. Judging from the many kinds of bovids known from the Pliocene, this group underwent a rapid radiation. More Pleistocene than Recent genera of bovids are known (100 *vs.* 44), but toward the end of the Pleistocene most bovids were driven from Europe by the southward advance of cold climate. A few bovids reached the New World in the Pleis-

tocene *via* the Bering Strait land bridge. Because this avenue of dispersal was under the influence of severe boreal climatic conditions at this time, it functioned as a "filter bridge" (Simpson; 1965:88), and only animals adapted to these conditions dispersed across the bridge. As a consequence, the New World received from Asia such bovids as bighorn sheep (*Ovis*), the mountain "goat" (*Oreamnos;* Fig. 14–31), the musk ox (*Ovibos*), and the bison (*Bison*), all animals able to withstand cold. Bovids less able to withstand boreal conditions — the Old World antelopes and the gazelles are prime examples — were forced from the northern parts of Europe and Asia in the Pleistocene to their present strongholds in Africa and Asia, and hence did not disperse across the Bering bridge to North America. An exception is the saiga antelope (*Saiga*), which now inhabits arid parts of Asia but which occurred in Alaska in the Pleistocene. Bison were seemingly extremely abundant members of grassland faunas in the Pleistocene and Recent in North America, where they occurred as far south as El Salvador. Some structural divergence occurred in Pleistocene bison, and in some areas several species may have occupied common ground. Some Pleistocene bison were considerably larger than the present *Bison bison.* Specimens of the Pleistocene species *Bison antiquus* from California indicate that this animal was about seven feet high at the shoulder and had horns that in larger individuals spanned six and a half feet.

Bovids characteristically inhabit grasslands, and the advanced dentition and limbs of bovids probably developed in association with grazing habits. The cheek teeth are high-crowned and the upper canine is reduced or absent. Unlike cervids and antilocaprids, bovids do not have preorbital vacuities in the skull (Fig. 14–32). The lateral digits are reduced or totally absent, the ulna is reduced distally and is fused with the radius, and only a distal nodule remains as a vestige of the fibula. Horns, formed of

FIGURE 14–31. A mountain goat (*Oreamnos americanus*) in the mountains of Idaho. (Photograph by Stewart M. Brandborg.)

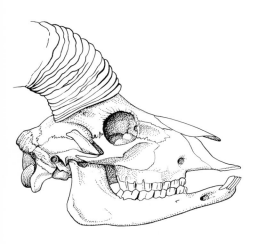

FIGURE 14–32. The skull of a male mouflon sheep (*Ovis musimon*). The horns are never shed, but continue to grow throughout the animal's life. Length of skull 242 mm.

FIGURE 14–33. Several kinds of African antelope, each with a distinctive shape of horn. A, sable antelope (*Hippotragus niger*); B, ellipsen waterbuck (*Kobus ellipsoprymnus*).

(Illustration continued on opposite page.)

a bony core and a keratinized sheath, are present in males of all wild species, and females usually also bear horns. The horns are never shed and in some species grow throughout the life of the animal. Bovid horns are never branched, but are often spectacularly long and form graceful curves or spirals (Fig. 14–33). Males of the Indian four-horned antelope (*Tetracerus quadricornis*) are unique in having four short, daggerlike horns. The horns of bovids are frequently used in fights between males during the breeding season, but in the Grant's gazelle (*Gazella granti*), and in many other species, stylized contests of strength are often substituted for the use of the dangerous horns in encounters between rival males (see p. 338). Some bovids, such as the sable antelope (*Hippotragus niger*) and oryx (*Oryx*), can use their horns as awesome defensive weapons, respected even by lions.

The last great strongholds of bovids are the extensive grasslands and savannas of Africa. Here a diverse bovid fauna occurs, and seemingly every

Figure 14–33 (*continued*).   C, impala (*Aepyceros melampus*); D, eland (*Taurotragus oryx*) with zebras in foreground. (Photograph of *Hippotragus* by N. Watt, courtesy of the Zambia Information Department; other photographs by W. Leslie Robinette.)

conceivable niche that could be filled by bovids has been occupied. Some antelope, such as the bush buck (*Tragelaphus*) and the lechwe (*Kobus*), inhabit river borders and swampy ground, whereas, at the other extreme, the oryx (*Oryx*) and addax (*Addax*) live in arid plains and desert wastes, where they may seldom have access to drinking water. Some African antelope are approaching extinction, but the protection afforded game in some areas may allow the survival of many species of this remarkable group.

CHAPTER 15

ECOLOGY

A modern inclusive definition used by Kendeigh (1961:1) and others describes ecology as "a study of animals and plants in their relations to each other and to their environment." Man's interest in ecology is not new. Paleolithic man doubtless understood some of the relationships between the major game animals and their environments. He could predict where certain species could be found, and he used this knowledge to increase his hunting success and hence his survival. Many thousands of years later, we "civilized" modern men are finally understanding that our very survival may depend on a knowledge of basic ecological relationships.

The value of an ecological approach to the study of mammalogy is difficult to overemphasize. Knowledge of the biology of any species of mammal is clearly incomplete if the relations of the animal to its environment are unknown. An adequate knowledge of the ecology of mammals has been long in emerging, however, not because mammalogists lack interest in ecology but because an understanding of the ecology of any single species demands detailed knowledge of a great many aspects of that species' biology. This problem is even more acute when considering complex interactions among many species in naturally occurring environmental situations.

Because the scope of ecology is extremely broad, only an incomplete and selective coverage is given here. We will concentrate on those ecological relationships or principles well illustrated or best illustrated by mammals.

The environments of animals can be described in terms of physical and biotic factors. Physical factors include temperature, humidity, climatic patterns and precipitation, and soil types; biotic factors are those associated with interactions between organisms.

THE EFFECTS OF PHYSICAL FACTORS OF THE ENVIRONMENT ON HABITAT SELECTION

Heat. Solar radiation provides the energy, in the form of heat and light, on which living organisms depend. The intensity of solar radiation at the earth's surface is influenced largely by the directness with which the sun's rays strike the earth. The angle at which the sun's rays strike the earth

decreases, and climates become progressively cooler, the further areas are
north or south of the equator.

Local conditions may also strongly
affect the amount of heat the surface of
the earth receives. In many mountainous sections of the western United
States steep slopes and precipitous
canyon walls are common. Because the
main axes of most mountain ranges lie
north and south, the drainage systems
are oriented more or less east and west
and the canyon walls face roughly
north or south. In northern latitudes,
no matter what the time of year, the
sun's rays strike a south-facing slope
more directly than a north-facing
slope. South-facing slopes are consequently drier and warmer than are
nearby north-facing slopes. The effects
of slope (steepness of incline) and exposure (the direction the slope faces)
are strongly reflected by the flora, and
the compositions of small mammal
faunas are frequently as conspicuously
different on adjacent opposing slopes
as are the assemblages of plants. In the
precipitous chaparral-covered mountains of southern California, for example, contrasting biotas occupy adjacent north- and south-facing slopes

(Vaughan, 1954). Some species of
mammals that occur on one slope are
absent from the opposing slope (Fig.
15–1).

Vertical temperature gradients are
encountered as one ascends mountain
ranges. The lowering of the temperature with increased elevation is of the
magnitude of approximately 1° C for
every 150 meters. This effect, coupled
with increased precipitation at higher
elevations, shorter growing seasons for
plants, and drastic diurnal-nocturnal
fluctuations in temperature, is associated with a distinct separation of
climatic zones in high mountains
throughout the world. The distributions of some mammals clearly reflect
this zonation. In northern Arizona the
Abert's squirrel (*Sciurus aberti*) occupies the ponderosa pine belt, but is
abruptly replaced by the red squirrel
(*Tamiasciurus hudsonicus*) where
spruce and fir forests appear at higher
elevations. In some areas in the western United States an assemblage of
"desert" mammals, resembling those
typical of arid lands as far south as central Mexico, may occupy the arid or
semi-arid land at the foot of a mountain
range, while the crest of the moun-

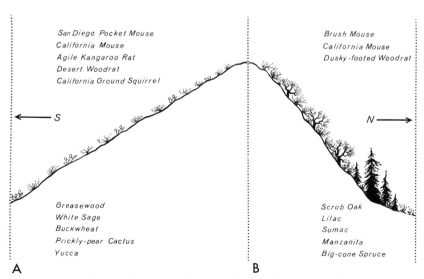

FIGURE 15–1. Assemblages of plants and mammals inhabiting a south-facing slope (A) and a north-facing slope (B) in lower San Antonio Canyon, San Gabriel Mountains, Los Angeles County, California. (Data from Vaughan, 1954.)

tains, a few airline miles away, may support boreal genera that occur as far north as northern Canada or Alaska (Fig. 15–2).

Degrees of tolerance for large and rapid temperature changes differ between species of mammals. Most tropical mammals are adapted to the fairly narrow seasonal and diurnal-nocturnal temperature fluctuations characteristic of the tropics, where freezing temperatures never occur. Some tropical mammals, such as the tree sloths (Bradypodidae), have limited thermal adaptability and are under cold stress at ambient temperatures as high as 22° C (Scholander, 1955), and some species, such as the vampire bat (*Desmodus rotundus*), are sensitive to ambient temperatures above roughly 35° C (Lyman and Wimsatt, 1966). Mammals occupying cold temperate or boreal areas, on the other hand, must have considerable thermal flexibility, and must adapt structurally, physiologically, and behaviorally to the stresses imposed by great seasonal and diel shifts in temperature.

Because the cover provided by such features as vegetation or rocks alters the environment locally, the environment at a terrestrial locality is not uniform, but consists of a complex mosaic of *microenvironments*. As a general rule, few terrestrial mammals can withstand the most extreme temperatures of the (general) habitats they occupy, but are able to select microenvironments in which temperature extremes are moderated or eliminated. A notable example of such a microenvironment is shown in Figure 15–3. Although occupying a temperate region, a pocket gopher may live for much of the year in a "tropical" microenvironment that features even, fairly high temperatures and high humidities. A group of beavers occupying a beaver lodge in the winter are by no means subjected to the extreme air temperatures outside the lodge (Fig. 15–4). Similarly, most species of shrews forage beneath litter, under logs or rocks, or at least beneath dense foliage; not only is their food abundant in such places, but temperature and humidity are moderated by such cover. These animals are actually not adapted to the general

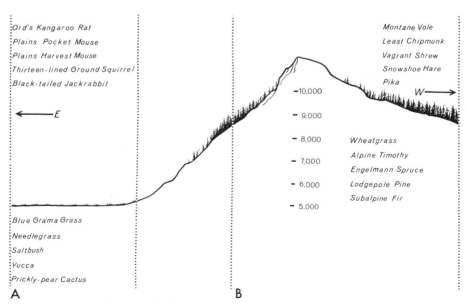

FIGURE 15–2. Assemblages of plants and mammals inhabiting short-grass prairie (A) and subalpine habitats (B) in northern Colorado (Larimer County).

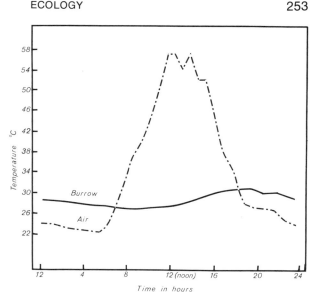

FIGURE 15–3. Temperature fluctuations of the air within a pocket gopher burrow (solid line) and just above the surface of the ground (dashed line). Temperatures were recorded on 24 June 1961, in McLennan County, Texas. (After Kennerly, 1964.)

climatic conditions of the regions they occupy, but are instead adapted to a limited set of conditions which occur in a specific microenvironment.

Water. In mammals, as in other animals, water forms an essential part of protoplasm and body fluids, and the maintenance of water balance is basic to life. The availability of free water and the amount of water in the air affect the habitat selection of mammals.

Fossorial mammals occupy microenvironments generally characterized by high humidities. Under such conditions, pulmocutaneous water loss (water loss from breathing and loss through the skin) is minimal, and fossorial mammals that eat moist food can maintain water balance without drinking water. Careful studies of the microenvironmental conditions of burrows of the plains pocket gopher (Geomys bursarius) in Texas by Kennerly (1964) showed that relative humidities within burrows were usually between 86 and 95%. Humidities within the sealed burrows may be high despite low soil moisture: Kennerly recorded a relative humidity of 95% in a burrow in July when soil from the floor of the burrow contained but 1.0% water. Although the microenvironment within burrows

is such that stresses resulting from high temperatures and low humidities are avoided, other stresses may occur. Concentrations of CO_2 from 10 to 60 times that of atmospheric air were recorded from pocket gopher burrows by Kennerly!

Flooding or high soil moisture may cause seasonal changes in mammal distributions. Ingles (1949) found that pocket gophers (Thomomys monticola) were forced from low-lying ground during spring snowmelt in the Sierra Nevada Mountains of California. The mole rat (Cryptomys hottentotus) in Southern Rhodesia centers its activity during the rainy season in the bases of large termite mounds that rise a few feet above the surrounding grassland and provide relatively dry islands in a sea of waterlogged terrain (Genelly, 1965).

Snow is an environmental feature of great importance in some boreal areas. The continuous snow cover that persists through the winter in many areas is a severe hardship for some large mammals. To most North American artiodactyls, including deer, elk, bighorn sheep, and moose, even moderately deep snow imposes a burden by covering some food and making locomotion difficult. In mountainous areas, deer

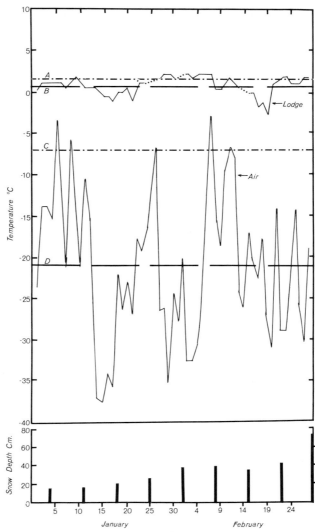

Figure 15–4.   Daily minimum temperatures inside and outside a beaver lodge in Algonquin Park, Ontario, Canada. (After Stephenson, 1969.)

and elk avoid deep snow by abandoning summer ranges and moving to lower elevations. South-facing slopes and windswept ridges, where snow is shallow or periodically absent, are often frequented in preference to deeply snow-covered north-facing slopes and canyon bottoms. In areas with fairly level terrain, deer and moose respond to deep snow by restricting their activities to small areas, called "yards," where they have established trails through the snow. Pro-

longed winters with deep snow commonly cause high mortality among deer and elk. Starvation is the usual cause of death.

Rather than being a source of winter hardship for small mammals, snow is actually a blessing. It forms an insulating mantle that provides a microenvironment at the surface of the ground where activity, and even breeding in some species, continues through the winter. To these small mammals, including shrews (*Sorex*), pocket gophers

(*Thomomys*), voles (*Microtus, Clethrionomys, Phenacomys*) and lemmings (*Lemmus, Dicrostonyx*), the most stressful periods are those in the fall, when intense cold descends but snow has not yet moderated temperatures at the surface of the ground (Formozov, 1946), and in the spring, when rapid melting of a deep snowpack often results in local flooding of much of the ground (Jenkins, 1948; Ingles, 1949; Vaughan, 1969). The snowpack has additional importance to small mammals as a partial protection from some predators; winter may be a period characterized by relatively low mortality among some small mammals. Many boreal mammals are well adapted to coping with a deep snowpack. Snowshoe hares run easily on the surface of the snow, as can most carnivores when the snow is crusted; weasels forage both above and beneath the snowpack, and martens often travel through the trees.

Substrates. Many small mammals seek diurnal refuge in burrows, and many terrestrial mammals of a variety of sizes have specialized modes of locomotion that are effective only on reasonably smooth surfaces. To these mammals, the type or texture of the soil or substrate is a critical environmental feature. Burrowing species may be narrowly restricted to a particular type of soil; for example, some heteromyid rodents that are weak diggers occur only where the soil is sandy. Some scansorial (climbing) species occur only where there are large rocks or cliffs.

Perhaps the most striking examples of mammalian preferences for specific types of substrate are to be found among desert rodents (see Grinnell, 1914, 1933; Hardy, 1945). In most desert localities no single species of rodent occurs on all types of substrate, and some species are tightly restricted to a single type of soil. In Nevada the four species of pocket mice (*Perognathus*) have largely complementary soil preferences (Hall, 1946:358, 364, 371, 376). One (*P. longimembris*) lives "on the firmer soils of the slightly slop-

ing margins of the valleys." A second (*P. formosus*) is "closely confined to slopes where stones from the size of walnuts up to those 8 inches or even more in diameter are scattered over and partly imbeded in the ground. . . ." A third (*P. penicillatus*) occupies "the fine, silty soil of the bottom land. . . ." The most broadly adapted (*P. parvus*) "occurs in a wide variety of habitats as regards soil. . . ." Lines of snap traps set from the desert floor into rocky desert hills will frequently reveal that one assemblage of rodents is associated with the sandy desert floor, another inhabits the gravelly lower slopes of the hills, and another occupies areas marked by boulders and rock outcrops (Fig. 15–5). Although these preferences may in some cases reflect relative digging ability, they appear often to be a reflection of locomotor style and foraging technique. Merriam's kangaroo rat (*Dipodomys merriami*) escapes danger by rapid and erratic hops. It will inhabit several types of soil, but favors fairly open terrain with fine-grained soils where the animal's distinctive style of locomotion can be used effectively. The cactus mouse (*Peromyscus eremicus*), in contrast, is not a speedy runner, but is a capable climber and scrambler. It always occurs where rocks, or in some places cactus or brush, offer immediately accessible retreats from danger.

Large rocks or cliffs are essential environmental features for some mammals. As a few examples, pikas (*Ochotona*) seldom occur away from talus or extensive boulder piles, some wood rats (*Neotoma*) build their houses only in cliffs or steep rock outcrops, and the dwarf shrew (*Sorex nanus*) is apparently restricted to rocky areas in alpine or subalpine situations (Brown, 1967; Marshall and Weisenberger, 1971). Rocks are clearly essential to different species of mammals for different reasons. Neither pikas nor dwarf shrews are accomplished climbers. Pikas seek shelter beneath rocks, and dwarf shrews seemingly forage beneath the rocks and also take shelter there. Some wood rats, in contrast,

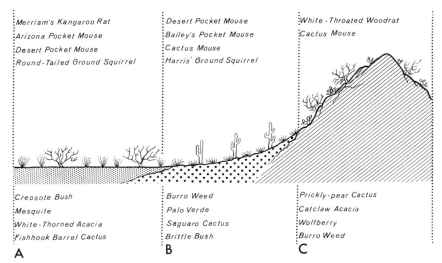

Merriam's Kangaroo Rat
Arizona Pocket Mouse
Desert Pocket Mouse
Round-Tailed Ground Squirrel

Desert Pocket Mouse
Bailey's Pocket Mouse
Cactus Mouse
Harris' Ground Squirrel

White-Throated Woodrat
Cactus Mouse

Creosote Bush
Mesquite
White-Thorned Acacia
Fishhook Barrel Cactus

Burro Weed
Palo Verde
Saguaro Cactus
Brittle Bush

Prickly-pear Cactus
Catclaw Acacia
Wolfberry
Burro Weed

A   B   C

FIGURE 15–5. Assemblages of plants and mammals inhabiting contrasting types of substrate: A, sandy and silty soil; B, gravelly soil; C, rock.

build their houses in fissures or cavities in rocks but forage widely away from the cliffs. To many species of bats, crevices in outcrops or cliffs provide retreats in areas otherwise lacking suitable roosting places. Even some saltatorial species have adopted a rock-dwelling mode of life. The Australian rock wallabies (*Petrogale* and *Peradorcas*) have granular patterns on the soles of the hind feet that increase traction; these animals leap adroitly among rocks and are restricted to rocky areas.

BIOTIC FACTORS IN THE ENVIRONMENT

Food. Food is necessary to animals as a source of energy and for building and maintaining protoplasm, and is therefore one of the most important biotic factors in the environment of a mammal. Mammalian adaptive radiation has involved a progressively broader exploitation of food sources, but although mammals as a group utilize many types of food, a single species usually eats a fairly limited array of foods that it is structurally, physiologically, and behaviorally capable of

utilizing efficiently. Much of mammalian evolution has been "guided" by the necessity of achieving the most favorable possible balance between energy expended in securing and metabolizing food on the one hand and energy gained from the food on the other. The relationships of mammals to their environments must ideally be considered against a background of knowledge of feeding biology.

The most abundant and omnipresent foods for terrestrial mammals are plants and insects. It is not surprising that the three most important mammalian orders in terms of numbers of species—Insectivora, Chiroptera, and Rodentia—depend primarily on these major food sources. Most members of the orders Insectivora and Chiroptera eat insects, and plant material forms an important part of the diet of most rodents. In addition, members of the orders Edentata, Pholidota, and Tubulidentata are primarily insect eaters, whereas some members of the Marsupialia, Primates, and Rodentia, as well as the Lagomorpha, Proboscidea, Hyracoidea, Sirenia, Perissodactyla, and some artiodactyls are herbivores. Although many herbivores are selective in their feeding, a fairly wide variety of foods are generally utilized, and seasonal variations in feeding habits are typical of temperate-zone species. Western wheatgrass (*Agropyron smithii*) forms about 35 per cent of the diet of the plains pocket gopher (*Geomys bursarius*) in July but is not eaten in December (Fig. 15–6). A number of studies have shown that, given a wide variety of plants to choose from, most herbivorous mammals are selective foragers (see, for example, Ward and Keith, 1962; Yoakum, 1958; Zimmerman, 1965); accordingly, an herbivore may show a great preference for one of the least abundant plants in its habitat (Fig. 15–7). Insectivorous mammals perhaps choose prey largely on the basis of size, probably avoiding both those species too large to overcome readily and those too small to yield adequate calories to compensate for the energy used in their capture.

Many mammals prey on higher vertebrates, including reptiles, birds, and mammals. Such carnivores occur in the orders Marsupialia, Insectivora, Chiroptera, Cetacea, Carnivora, and Pinnipedia. The size of a predator obviously determines the range of size of its prey. Many mammalian faunas consist of a wide variety of prey species fed upon by various carnivores; in such faunas each carnivore differs from the others in size and takes a different size class of prey. Many carnivores are highly specialized structurally for killing and eating their prey, and most have behavioral specializations that further their predatory ability. The learning of efficient hunting and killing methods is critical to the survival of carnivores. Prey must be captured without an excessive outlay of energy. Of equal importance, particularly in the cases of the larger carnivores that kill prey as large as or larger than themselves, hunting and killing behavior must be geared to avoiding serious injury to the predator. The high incidence of skeletal damage in the fossil remains of saber-tooth cats (*Smilodon*) suggests that in the Pleistocene, as today, preying on large game was a dangerous undertaking.

Still other mammals are omnivorous and are opportunistic feeders; such mammals eat a wide variety of plant and animal material. Omnivores occur among the orders Marsupialia, Insectivora, Chiroptera, Primates, Rodentia, Carnivora, and Artiodactyla. Omnivores are typically less specialized in structure than are mammals adapted to narrower diets, and within some orders the omnivorous mode of life has been highly successful. Among rodents, for example, the nearly ubiquitous North American genus *Pero-*

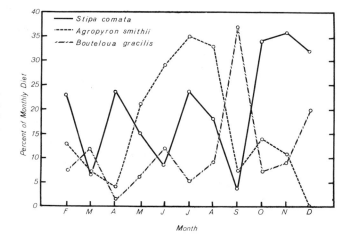

FIGURE 15–6. Seasonal differences in the diet of the plains pocket gopher (*Geomys bursarius*) in eastern Colorado. (After Myers and Vaughan, 1964.)

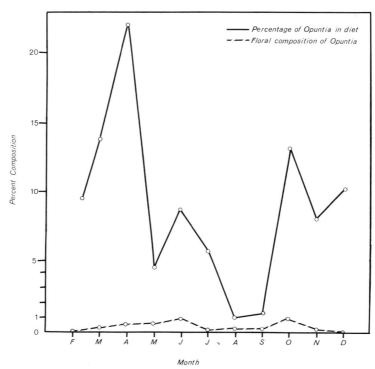

FIGURE 15–7. Seasonal changes in the utilization of prickly-pear cactus (*Opuntia humifusa*) by the plains pocket gopher (*Geomys bursarius*). The percent composition of prickly-pear cactus in the diet varies markedly, whereas the floral composition of prickly-pear (an expression of the percentage of the total coverage of vegetation contributed by *Opuntia*) is nearly constant. (After Myers and Vaughan, 1964.)

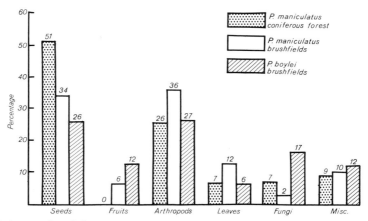

FIGURE 15–8. Foods of deer mice (*Peromyscus*) in different habitats in the northern Sierra Nevadas of California. (After Jameson, 1952.)

myscus includes species seemingly adapted to a wide variety of plant and animal foods (Fig. 15–8). Some terrestrial sciurids are also omnivorous, and in the order Carnivora, such widespread and successful families as Canidae, Ursidae, and Procyonidae have many omnivorous members.

A variety of additional foods are eaten by mammals. Planktonic crustaceans are the major food of the filter-feeding baleen whales and some phocid seals; most odontocete cetaceans eat fish, squid, and, less commonly, a variety of other invertebrates, but the killer whale (*Orcinus*) preys on porpoises, whales, diving birds, and a variety of large sharks and bony fish. Fish, squid, and mollusks are taken by pinnipeds; fish are also eaten by some members of the Carnivora and even by some bats. The leopard seal (*Hydrurga*) of the Antarctic preys on a variety of marine vertebrates and is an important enemy of penguins. A number of marine invertebrates, including sea urchins (*Strongylocentrotus*), provide food for sea otters (*Enhydra*). A few mammals do considerable scavenging; hyaenas (*Crocuta* and *Hyaena*) and jackals (*Canis* spp.) frequently feed on the leftovers of kills of larger carnivores. Nectar and pollen feeding is common among pteropodid and phyllostomatid bats, and one phalangerid marsupial (*Tarsipes*) is highly specialized for a diet that consists partly of nectar. Vampire bats, which have one of the most specialized mammalian feeding techniques, feed entirely on blood. Some carnivores—the coyote is a good example—are highly opportunistic, and under the pressure of hunger may take almost any vulnerable animal, vertebrate or invertebrate, as well as plant material.

Feeding habits are strongly influenced by the different seasonal patterns that characterize different climatic zones. In areas increasingly farther north or south of the tropics, progressively more drastic seasonal differences in climate occur. Attending these differences are equally striking seasonal differences in the availability of food. As a consequence, many mammals of non-tropical areas are more specialized than their tropical relatives in their ability to survive under conditions of seasonal food shortage. Among mammals various patterns of dormancy, food caching, and migration seemingly evolved in association with seasonal food shortages. The long migratory movements of the rorquals (*Balaenoptera* and *Megaptera*) and the gray whale (*Eschrichtius*) are notable examples of behavioral specializations that probably developed in response to "boom or bust" patterns of food availability. The rorquals take advantage of rich warm-season blooms of plankton near the polar ice sheets, and then migrate toward equatorial areas, where they spend the winter and where breeding occurs. These whales and the gray whale fast during the wintering period, living off the thick blubber developed during the summer time of plenty.

Feeding activities of mammals may have pronounced effects on vegetation. Peterson (1955:162, 163) noted that moose (*Alces alces*) were responsible for the local decimation of ground hemlock, quaking aspen, and balsam on Isle Royale, Michigan, and these animals have caused considerable damage to vegetation in Finland (Kangas, 1949). In Rocky Mountain National Park in Colorado, heavy browsing by deer and elk in some areas caused the death of 85 per cent of the sagebrush and 35 per cent of the bitterbrush plants in a five-year period (Ratcliff, 1941), and in California a tract of bitterbrush was killed by overbrowsing by deer within roughly five years (Fischer et al., 1944). Small mammals may also have a marked effect on vegetation. Batzli and Pitelka (1971) found that, during cyclic changes in density, California voles (*Microtus californicus*) had significant effects on preferred food plants. During high mouse densities (160/acre), the mouse's major food plants contributed 85 per cent less volume to the vegetation outside experimental plots from which the mice were excluded than to the vegetation inside the plots. In addition, the fall of seeds of preferred grasses was reduced by 70 per cent on grazed areas.

Some plant-animal associations seem mutually beneficial. Such an interaction occurs between nectar-feeding bats and the tropical or desert plants on which they feed; the bat profits from the food source offered by the nectar and pollen, and the plant is benefited by the efficient pollination by the bats (Alcorn, et al., 1959; Baker and Harris, 1957; Faegri and Van Der Pijl, 1966:111–118). A less clear, but probably mutually beneficial, relationship is that involving the prairie dog (*Cynomys*) and blue grama grass (*Bouteloua gracilis*) on the North American Great Plains. Prairie dogs require open areas with moderately low vegetation, and blue grama has a life-form that frequently provides this sort of habitat. By foraging largely on annual forbs and to a lesser extent on blue grama, prairie dogs may reduce the competition that the grass faces from other plants (Bond, 1945). The blue grama may thus remain dominant and continue to provide the shortgrass habitat most suitable for prairie dogs. Mammals serve as important agents of dispersal for the seeds of many kinds of plants. In Mexico, sprouts from seeds of leguminous trees commonly grow from piles of cow manure deposited far from the plants that produced the seeds. In the western United States, the seeds of prickly-pear cactus (*Opuntia*) are distributed in a similar fashion by skunks and foxes, as are the seeds of manzanita (*Arctostaphylos*) by black bears and coyotes. By dispersing seeds of the Washington palm (*Washingtonia filifera*), coyotes may have helped to spread that plant to suitable sites in the deserts of Southern California (Jaeger, 1950).

Vegetation. Not only are plants important as food for many mammals, but the cover, escape routes, and retreats they provide, as well as the degree to which the plants facilitate or obstruct rapid locomotion, are important aspects of the environments of many terrestrial mammals. Vegetation may also provide environmental features essential to the plant or animal food that a mammal uses. Therefore, plants that are never used as food by a mammal are often as essential a part of this animal's environment as are staple food plants.

A species of mammal is seldom evenly distributed, even within an area of seemingly homogeneous vegetation. On the contrary, the actual distributional patterns of most mammals are discontinuous, indicating that all requirements for the species are not met over broad areas. Close observation of even a limited area usually indicates that there are local changes in the relative densities and the spacing of the plants; further, a given species of mammal is usually restricted to a habitat characterized by plants of a certain *life form*. The size, shape, foliage density, and pattern of branching of a plant determines its life form. Analyses of the environmental requirements of a mammal, therefore, must include considerations of not only the species of plants with which the animal is associated, but (and frequently equally importantly) the life forms of these plants and the "aspect" they give to the habitat. For example, the brush rabbit (*Sylvilagus bachmani*) is a small, short-legged scamperer that occurs only along the Pacific coast of the United States in the chaparral areas. Within these areas, however, it is restricted to dense and continuous patches of evergreen shrubs, and is completely absent from situations where overhead brushy cover is scattered. A brush rabbit will actually run between a man's feet rather than be forced into the open. At the other extreme is the black-tailed jackrabbit (*Lepus californicus*), a long-legged and highly cursorial lagomorph that typically escapes its predators by outrunning them. Although this rabbit is widely distributed within desert and grassland areas, it occurs only in open settings where grasses and shrubs are scattered or low-growing and do not limit rapid running.

PATTERNS OF ECOLOGICAL DISTRIBUTION

Competition and the Ecological Niche. Just as no two species of animals are structurally identical, no two are functionally identical or have exactly the same environmental requirements. The very morphological characters that determine the distinctness of species also determine the distinctness of habitat requirements. Each species requires a specific environment—a particular combination of physical and biotic factors—and each is functionally unique, pursuing a particular mode of life within its environment. This specific environmental setting that a species occupies and the functional role it plays in this habitat constitute the animal's *ecological niche.*

The fundamental niche of Hutchinson (1957) is an abstract formalization of the usual concept of an ecological niche. The fundamental niche is an "*n*-dimensional hypervolume" defined by all of the values limiting the survival of a species and within which every point "corresponds to a state of the environment which would permit the species S to exist indefinitely" (Hutchinson, 1957:417).

Niche segregation among animals has resulted from some of the same evolutionary processes responsible for the origin of species. Competition is one of the most important of these processes. *Competition* occurs when two or more species occupying the same habitat at the same time are utilizing some environmental resource in short supply. Many animals operate during certain periods on tight time and energy budgets; during these periods, energy-consuming interactions between species (such as fighting or threat displays) may threaten survival, and reductions in energy output are highly advantageous. Segregation of animals into niches is one result of natural selection favoring structural, physiological, or behavioral modifications that decrease the energy expended in interspecific competition for such resources as food (Fig. 15–9). In other words, these modifications increase the efficiency with which a species pursues such vital activities as feeding, reproducing, and escaping from predators, and may also restrict the environmental sphere within which the animal functions. These adaptations only survive, however, if they are associated with increased reproductive success. As stated by Mayr (1963): "In order to survive, each species must be a supreme master in its own niche." Interspecific competition, then, is brought to limits that sympa-

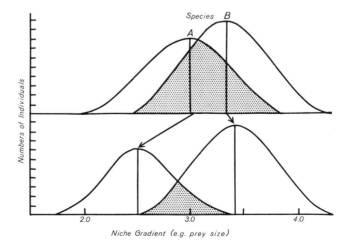

FIGURE 15–9. When two species overlap broadly in some niche characteristic (above), such as choice of prey size, there is reduced survival of the most strongly competing members of each species. An evolutionary trend toward niche divergence typically results (below). (After Whittaker, R. H.: *Communities and Ecosystems,* The Macmillan Company, 1970. Copyright © by R. H. Whittaker, 1970.)

tric species (species whose distributions overlap) can bear by niche segregation.

A full description of an animal's ecological niche must include a consideration of the animal's behavior and biotic interactions, that is, its functional role within its environment. A primary consideration is how an animal seeks its food. Does it burrow to reach a root, as does the pocket gopher, or does it dig roots from above, as does the javelina; does it kill its prey or does it consume prey killed by other animals? The reproductive cycle of the animal is equally important, for the timing of breeding and the seasonal demands put on the habitat due to increased energy requirements during the breeding season, the choice of sites for rearing young, as well as yearly population fluctuations due to reproduction, relate directly to an animal's use of, and its impact on, the environment. The role an animal plays in the story of predator-prey relationships is also of vital importance in understanding the animal's niche. In a terrestrial community the role of an abundant mouse that consumes vegetation is very different from that of a rare carnivore that preys upon this mouse. (Food chains, food webs, and trophic levels are discussed on page 280.)

Similar species may avoid competition in a number of ways. Each can utilize a different microenvironment, or the two species can utilize the same foraging area but with their activity cycles out of phase. Nocturnal cricetid mice and diurnal chipmunks, both omnivores that forage on common ground, virtually never encounter one another, nor do nocturnal insectivorous bats and diurnal insectivorous birds. Asynchronous (out of phase) breeding of sympatric ungulates was found to occur in Southern Rhodesia (Dasmann and Mossman, 1962); this asynchrony offsets to some extent the times at which females of the different species have increased food requirements, and may therefore lessen interspecific competition for food (Table 15–1). In East Africa a yearly pattern of grazing succession by some of the large ungulates results in reduced interspecific competition for forage. Zebras (*Equus burchelli*) and buffalo (*Syncerus caffer*) forage to some extent in the same areas during the early part of the wet season, but for much of the dry season zebras

Table 15–1. Reproductive Seasons of Some Ungulates in Southern Rhodesia, Africa. (After Dasmann and Mossman, 1962.)

	WET SEASON			DRY SEASON							WET SEASON	
SPECIES	Jan.	Feb.	Mar.	Apr.	May	Jun.	Jul.	Aug.	Sept.	Oct.	Nov.	Dec.
Burchell zebra	Y	Y	Y	Y	Y	Y	YP	L	L	Y	Y	YP
Warthog	Y						P	P				B
Giraffe					B	Y	Y	Y	Y			
African buffalo		PY			Y				LY			
Blue wildebeest			Y							P	P	
Waterbuck		YB	YB	Y	L							
Eland									Y			
Kudu		P	BY	Y				LP	L		LY??	
Bushbuck							Y					
Impala	Y	L	L	L	R	R	LP	P	P	P	P	B
Common duiker			Y		Y	Y		YP			PY	
Steenbuck		Y	Y		Y	Y	Y	YP			YP	
Klipspringer								Y				

Key: Y = Young, est. under 1 month of age, observed
    Y = Maximum number of young observed during period
    B = Most births occur during this period
    L = Lactating females observed or collected
    R = Rutting season behavior observed
    P = Pregnant females collected

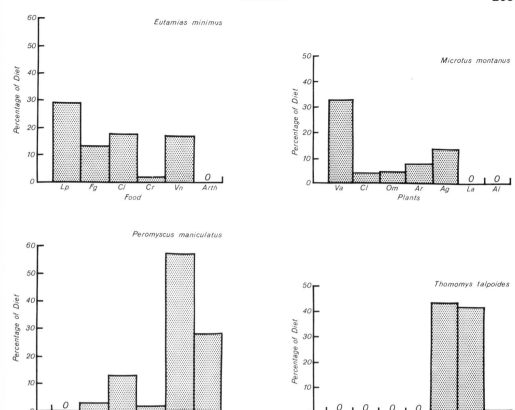

FIGURE 15–10. The diets of four partly or completely herbivorous rodents in part of the summer of 1965 in a subalpine area in Routt County, Colorado. The rodents are the least chipmunk (*Eutamias minimus*), the montane vole (*Microtus montanus*), the deer mouse (*Peromyscus maniculatus*), and the northern pocket gopher (*Thomomys talpoides*). Note the lack of competition among the species for food. Abbreviations: Ag, *Agoseris glauca;* Al, *Achillea lanulosa;* Ar, *Arnica cordifola;* Arth, arthropods; Cl, *Collomia linearis;* Cr, *Carex* sp.; Fg, fungus; La, *Lupinus argenteus;* Lp, *Lewisia pygmaea;* Om, *Oenothera micrantha;* Va, *Vicia americana;* Vn, *Viola nuttali.* (Based on unpublished data.)

utilize areas abandoned earlier by buffalo (Vesey-Fitzgerald, 1960).

Details of food preferences of relatively few mammalian species are well known, but sufficient information is available to indicate that "feeding niche" segregation of similar sympatric species is often striking. As an example, the foraging areas of several small mammals of a subalpine area in Colorado were found to overlap broadly, but the diets of no two species were the same (Fig. 15–10). The great size difference between the sexes of some small predators may reduce competition for food between the sexes by enabling males and females to exploit different sizes of prey. The males of two species of weasel (*Mustela erminea* and *M. frenata*) are roughly twice as heavy as the females, and the skulls of the males are more robust (Hall, 1951:26–28). This situation seems to parallel that described for some birds (Storer, 1955; Selander, 1966), in which pronounced sexual dimorphism allows the sexes to utilize different feeding niches. Where the species of weasel mentioned above occur together, they differ considerably in size. This difference, and the considerable sexual variation in size within each species, not only may reduce interspecific and intraspecific competition for food, but may increase the efficiency of utilization of the food by forcing each predator to specialize on a specific size of prey.

Although the niches of species within a community tend to be complementary, some partial niche overlap occurs, and there is occasionally some interspecific competition between sympatric mammals. Obvious indications of competition in nature are relatively rare, however, and when competition occurs its severity is often difficult to assess. As shown in Figure 15–10, two sympatric herbivores may utilize some of the same foods. Where they occurred in the same habitat, both the northern pocket gopher and the montane vole ate *Agoseris* (mountain-dandelion), but this plant is not equally important to both species. Although it is a major food of the pocket gopher, it is of relatively minor importance in the diet of the montane vole. In addition, *Agoseris* was so abundant where these animals occurred together that they were seemingly not utilizing a resource in short supply. An apparently clearcut case of competition for burrowing sites between two ground squirrels was described by Hansen (1962a), who noted that within the range of the Richardson ground squirrel (*Spermophilus richardsonii*) in Colorado, the golden mantled ground squirrel (*S. lateralis*) was restricted to rocky situations and the Richardson ground squirrel occupied the meadows. Where they do not live in association, however, the golden mantled squirrels occupied both types of habitat. Because the Richardson ground squirrel is extending its range fairly rapidly into areas where golden mantled ground squirrels and other species of ground squirrels occur, competition, and ultimately local competitive displacement of one species by another, is seemingly occurring.

BIOTIC COMMUNITIES

Man has long recognized that animals and plants with similar environmental requirements form recognizable communities. A perceptive description of a community was given by Mobius (1877), who considered an oyster bed as "a community of living beings, a collection of species and a massing of individuals, which find here everything necessary for their growth and continuance..." A community, however, is characterized not only by its unique plant and animal assemblage but by complex interactions between organisms and by the effects the physical environment has on the biota. The term *community* has been used to designate plant-animal assemblages of differing size and importance (Odum, 1971:140). This term can be appropriately used in reference to the biota of a wood rat nest or to that of the extensive deciduous forests of the eastern United States.

Modes of life and structural adaptations of mammals can be fully understood only against a background of knowledge of the communities the mammals occupy. Many of the ungulates important in grassland communities, for example, have adaptations enabling them to feed on grasses and forbs often made abrasive by silica within the plant and by particles of wind-blown sand, to escape predators on terrain largely devoid of cover, and to survive severe weather under conditions of limited cover and limited water. As would be expected, highly cursorial locomotion, high-crowned teeth, and an ability to get sufficient water from vegetation are common adaptations among these mammals. Plant-animal interactions are forces vital to molding the structure of the community. Thus, the life forms of plants of grassland communities are influenced by the grazing and trampling of ungulates, and, in turn, the reproductive success and even the survival of adult herbivores depend on the productivity of the vegetation. In a tropical forest, on the other hand, ungulates are unimportant and scarce, but the most abundant mammals — often bats, rodents, and primates — are arboreal or aerial. Although the tropical community is vastly more complex botanically than the grassland, certain plant-animal interactions are apparent. As an example, the nectar-feeding bats eat largely nectar and pollen, but in the process they act as the most effective pollinators of some plants.

Under some circumstances, as in precipitous mountainous areas where the combined effects of slope, exposure, edaphic (soil) factors, and local patterns of airflow produce remarkably complex distributions of organisms, transitions between communities are abrupt. At the interface where one community adjoins another, as where a grassy meadow meets a coniferous forest, an "edge effect" is produced. At this edge where the communities overlap, a greater diversity and density of animals may occur than within either adjacent community, owing to the increased diversity of vegetation and types of shelter. Carnivores frequently concentrate their efforts on these edge situations, and habitat manipulation by game managers often includes making maximum use of the edge effect.

Sharp dividing lines do not always occur between adjacent communities. A subalpine forest may become progressively more extensively interrupted by open "parks" or alpine meadows until the treeless alpine tundra dominates. The zone of intergradation between communities, whether broad or narrow, is called an *ecotone*. Ecotonal belts between communities

Table 15–2. Some Major Plant Communities and Their Distributions. These Communities are Illustrated in Figures 15–11 through 15–24.

COMMUNITY	DISTRIBUTION
Tropical Rainforest (Fig. 15–11)	South and Central America; Africa; S. E. Asia; East Indies; N. E. Australia
Tropical Deciduous Forest (Fig. 15–12)	Mexico; Central and South America; Africa; S. E. Asia
Temperate Rainforest (Fig. 15–13)	Pacific Coast from northern California to northern Washington; parts of Australia, New Zealand, and Chile
Temperate Deciduous Forest (Fig. 15–14)	Eastern U.S.; parts of Europe and Eastern Asia
Subarctic-Subalpine Coniferous Forest (Fig. 15–15)	Northern North America; Eurasia; high mountains of Europe and North America
Thorn Scrub Forest (Fig. 15–16)	Parts of Mexico; Central and South America, Africa, S. E. Asia
Temperate Woodlands (Fig. 15–17)	Parts of western and southwestern U.S. and Mexico; Mediterranean area and parts of Southern Hemisphere
Temperate Shrublands (Fig. 15–18)	California; Mediterranean area; South Africa; parts of Chile; West and South Australia
Savanna (tropical grasslands) (Figs. 15–19, 15–20)	Parts of Africa, Australia, southern Asia, South America
Temperate Grasslands (Fig. 15–21)	Plains of N.A.; steppes of Eurasia; parts of Africa and South America
Arctic and Alpine (Fig. 15–22)	Tundras north of treeline in North America and Eurasia; some areas in Southern Hemisphere
Deserts (Figs. 15–23, 15–24)	On all continents; widespread in North America, North Africa, and Australia

are broad in some regions, as in western Mexico, where one may travel southward through the region of transition between the Sonoran Desert community and the tropical thorn forest community for many miles. Under such conditions there is a continuous gradient from one major community to another. Despite the difficulty in assigning geographic limits to some communities, the major terrestrial communities are usually readily recognized. Some terrestrial communities and their distributions are shown in Table 15–2, which is based on the community and plant-formation types of Whittaker (1970:52–64). Some of these communities are illustrated in Figures 15–11 to 15–24.

Over broad areas this pattern of gradual clinal changes in the biota in response to gradual climatic and edaphic changes is more typical than is a sudden shift from one community to another. Because the classification of units or areas within such a continuum involves arbitrary choices as to the limits of the units and takes little account of variation within the units, systems of classification of broad environmental units have not been completely satisfactory. Nevertheless, some of them have been widely used. Such a classification is that involving the recognition of terrestrial biomes. According to Odum (1971:378), "The biome is the largest land community unit which it is convenient to recognize." The biome is defined in terms of climate, biota, and substrate. The following biomes are recognized by Odum (1971:379): tundra; northern coniferous forest; temperate deciduous and rain forest; temperate grassland; chaparral; desert; tropical rain forest; tropical deciduous forest; tropical scrub forest; tropical grassland and savanna; and mountains, with com-

(Text continued on page 276.)

FIGURE 15–11. Tropical rain forest near Catemaco, southern Veracruz, Mexico. A, a general view of the forest; B, the understory vegetation beneath a semi-open canopy of taller vegetation. Common mammals of this area include the Mexican mouse-opossum (*Marmosa mexicanus*), opossums (*Didelphis marsupialis* and *Philander opossum*), many kinds of leaf-nosed bats (*Phyllostomatidae*), the howler monkey (*Alouatta villosa*), agouti (*Dasyprocta mexicana*), and coati (*Nasua narica*).

FIGURE 15–12. Tropical deciduous forest near Panuco in southern Sinaloa, Mexico. A, near the end of the dry season (June, 1970); B, in the wet season (August, 1970). Common mammals include the gray mouse-opposum (*Marmosa canescens*), opossum (*Didelphis marsupialis*), many species of leaf-nosed bats (Phyllostomatidae), armadillo (*Dasypus novemcinctus*), spiny pocket mouse (*Liomys pictus*), raccoon (*Procyon lotor*), and coati (*Nasua narica*). (Photographs by Roger B. Smith.)

FIGURE 15–13.   Temperate evergreen rain forest in Olympic National Park, Washington. Mammals typical of this area are the shrew mole (*Neurotrichus gibbsii*), Townsend's mole (*Scapanus townsendii*), mountain beaver (*Aplodontia rufa*), western red-backed mouse (*Clethrionomys occidentalis*), marten (*Martes americana*), black bear (*Ursus americana*), and elk (*Cervus elaphus*). (Photograph by Ray Atkeson.)

FIGURE 15–14. Temperate deciduous forest in Indiana in summer. Mammals typical of this type of community are the short-tailed shrew (*Barina brevicauda*), eastern chipmunk (*Tamias striatus*), gray squirrel (*Sciurus carolinensis*), flying squirrel (*Glaucomys volans*), white-footed mouse (*Peromyscus leucopus*), gray fox (*Urocyon cinereoargenteus*), and white-tailed deer (*Odocoileus virginianus*). (Photograph by U.S. Forest Service.)

FIGURE 15–15.   Subalpine coniferous forest near Rabbit Ears Pass, Routt County, Colorado. The following mammals are common in this community: shrews (*Sorex vagrans* and S. *cinereus*), the red squirrel (*Tamiasciurus hudsonicus*), least chipmunk (*Eutamias minimus*), pocket gopher (*Thomomys talpoides*), montane vole (*Microtus montanus*), red-backed vole (*Clethrionomys gapperi*), beaver (*Castor canadensis*), porcupine (*Erethizon dorsatum*), red fox (*Vulpes vulpes*), mule deer (*Odocoileus hemionus*), and elk (*Cervus elaphus*).

FIGURE 15–16.   Thorn scrub near San Carlos Bay, Sonora, Mexico. Typical mammals are bats (including several members of the Neotropical families Phyllostomatidae and Mormoopidae), the antelope jackrabbit (*Lepus alleni*), Merriam's kangaroo rat (*Dipodomys merriami*), hispid cotton rat (*Sigmodon hispidus*), white-throated woodrat (*Neotoma albigula*), and javelina (*Tayassu tajacu*).

FIGURE 15–17. Temperate woodlands. A, piñon-juniper woodland near Flagstaff, Coconino County, Arizona. Most of the trees are piñons; the tree at the extreme left is a juniper. Common mammals include several bats of the genus *Myotis*, the desert cottontail (*Sylvilagus audubonii*), piñon mouse (*Peromyscus truei*), northern grasshopper mouse (*Onychomys leucogaster*), Stephen's woodrat (*Neotoma stephensi*), gray fox (*Urocyon cinereoargenteus*), bobcat (*Lynx rufus*), and mule deer (*Odocoileus hemionus*). B, oak woodland near Claremont, Los Angeles County, California. A number of vespertilionid bats, the gray squirrel (*Sciurus griseus*), brush mouse (*Peromyscus boylii*), dusky-footed woodrat (*Neotoma fuscipes*), and raccoon (*Procyon lotor*) are common in this habitat.

FIGURE 15–18. Temperate shrubland (chaparral) near San Antonio Canyon, Los Angeles County, California. A shows the steep slopes covered by dense brush; B shows the interlacing stems and branches beneath the chaparral in the right foreground of A. Typical mammals in this area are the western pipistrelle (*Pipistrellus hesperus*), Merriam's chipmunk (*Eutamias merriami*), brush mouse (*Peromyscus boylii*), California mouse (*P. californicus*), dusky-footed woodrat (*Neotoma fuscipes*), gray fox (*Urocyon cinereoargenteus*), bobcat (*Lynx rufus*), and mule deer (*Odocoileus hemionus*).

FIGURE 15–19. Savanna in the Serengeti Plain, Tanzania, Africa. This area supports large numbers of ungulates. Some of the typical kinds are zebra (*Equus burchelli*), buffalo (*Syncerus caffer*), wildebeest (*Connochaetes taurinus*), and Thompson's gazelle (*Gazella thompsonii*). Lions (*Panthera leo*), hyaenas (*Crocuta crocuta*), and African hunting dogs (*Lycaon pictus*) prey on these ungulates. The African savanna supports a richer ungulate fauna than does any other area. (Photograph by W. Leslie Robinette.)

FIGURE 15–20. Savanna (mallee scrub) in the southern part of New South Wales, Australia. The trees are eucalyptus. In this area occur a number of marsupials, including marsupial "mice" (*Sminthopsis crassicaudata* and *Antechinus flavipes*), the vulpine phalanger (*Trichosurus vulpecula*), the "mallee gray" kangaroo (*Macropus giganteus*), and the red kangaroo (*Megaleia rufa*). (Photograph by Diana Harrison.)

FIGURE 15–21. Temperate grassland: shortgrass prairie near Nunn, Weld County, Colorado. Among the common mammals are the white-tailed jackrabbit (*Lepus townsendii*) and black-tailed jackrabbit (*L. californicus*), the thirteen-lined ground squirrel (*Spermophilus tridecemlineatus*), prairie dog (*Cynomys ludovicianus*), Ord's kangaroo rat (*Dipodomys ordii*), northern grasshopper mouse (*Onychomys leucogaster*), coyote (*Canis latrans*), badger (*Taxidea taxus*), pronghorn (*Antilocapra americana*). (Photograph courtesy of Robert E. Bement, Agricultural Research Service.)

FIGURE 15–22. Arctic tundra on the northern slope of the Brooks Range, northern Alaska. Among the common mammals of this area are the arctic shrew (*Sorex arcticus*), collared lemming (*Dicrostonyx groenlandicus*), brown lemming (*Lemmus trimucronatus*), singing vole (*Microtus miurus*), wolf (*Canis lupus*), grizzly bear (*Ursus arctos*), moose (*Alces alces*), caribou (*Rangifer tarandus*), and Dall sheep (*Ovis dalli*). (Photograph by James W. Bee.)

FIGURE 15–23.   High sand dunes in the Kalahari Desert near Gobabeb, South West Africa. Several kinds of mammals that occur in this area are gerbils (*Gerbillus* sp.), ground squirrels (*Xerus* sp.), bat-eared foxes (*Otocyon megalotis*), and gemsbok (*Oryx gazella*). (Photograph by C. K. Brain.)

FIGURE 15–24.   Sandy desert near Lake Victoria, southwestern New South Wales, Australia. Mammals of this community include a marsupial "mouse" (*Sminthopsis crassicaudata*), a kangaroo (*Macropus giganteus*), and a rodent, the Australian kangaroo mouse (*Notomys mitchelli*). (Photograph by Diana Harrison.)

plex zonation of plants and animals. In North America the "life zone" has been used widely in descriptions of the distributions of vertebrates. Life zones were originally described as temperature zones by C. Hart Merriam (1894), but later were used as community zones characterized by assemblages of plants and animals. Merriam's thinking was influenced by his recognition of the sharp elevational zonation of biotas in some mountains of the western United States (Fig. 15–25).

BIOTIC INTERACTIONS

Considerations of interactions between species of animals, between species of plants, or between plants and animals are essential to an understanding of mammalian ecology. A biotic community is a tremendously complex functional unit within which animals live, feed, reproduce and die; it has an evolutionary history and some degree of "dynamic equilibrium."

The role of an organism in a community depends on its interactions with other members of the community and with the physical environment; the fabric of the entire community depends on the combined effects of the interlacing threads of interaction.

Positive Interactions. Interactions between mammals and many other kinds of animals have been observed. Some of these interactions are seemingly of no importance to the mammal involved, and some parasite-mammal interactions are harmful to mammals; on the other hand, mammals derive considerable benefit from other types of interaction. Mammal collectors have frequently noticed that some kinds of pocket mice concentrate their foraging efforts around the prominent mounds built by harvester ants, where the rodents presumably find seeds dropped by the insects. In some cases the activities of mammals improve the habitat for other vertebrates, as in the case of the beaver, which by its dam building creates ponds that often support high densities of trout. Mammals that feed

FIGURE 15–25. Pronounced zonation of vegetation on the San Francisco Peaks in northern Arizona. Observations he made in this area influenced C. Hart Merriam's thinking on life zones. In the foreground, at about 1900 m. elevation, is shortgrass prairie; in the near distance are flats supporting piñon and juniper trees. The mesas and slopes in the middle distance are covered with ponderosa pine; the higher slopes in the distance have spruce and fir forests; and on the treeless and barren tops of the peaks, between roughly 3350 and 3660 m. elevation, grow small alpine forbs and grasses.

to some extent on carrion may be guided to dead animals by concentrations of carrion-feeding birds and by the calls of the birds. Murie (1940) found that such a relationship occurred between coyotes and ravens (*Corvus corax*) in Yellowstone National Park. One of the most remarkable instances of mutually beneficial interactions concerns a bird, the African honey-guide (*Indicator indicator*), and the African honey-badger (*Mellivora capensis*). The bird attracts the badger's attention by raucous chattering and then leads the way to a bee nest. After the nest is torn apart by the mammal (African tribesmen sometimes perform this service), the honey-guide eats fragments of wax, which it has the ability to digest. Two species of African birds, the oxpeckers (*Buphaga*), eat ticks that they remove from big game mammals. The mammals are oblivious of the birds' attentions, but seem to derive benefit from the loss of ticks and from the alarm calls given by the birds when predators approach.

Some species of mammals profit in various ways from the activities of other mammals. Often one species will use shelter provided by another species, as in the case of certain white-footed mice (*Peromyscus*) and shrews (*Notiosorex*) that use wood rat houses for shelter, or of the wart hog (*Phacochoerus*), which seeks refuge in burrows of aardvarks (*Orycteropus*). The burrows of pocket gophers are used by a variety of vertebrates, including amphibians, reptiles, and other mammals (Vaughan, 1961). Scraps from the kills of large carnivores may be eaten by mammalian scavengers, such as hyaenas (*Crocuta* and *Hyaena*), which specialize in this type of feeding. Of particular interest are the close, and seemingly mutually beneficial, associations that occur between two species of mammals. A badger and a coyote have been observed hunting together on many occasions in widely scattered areas (Young and Jackson, 1951:95). Herds of impala (*Aepyceros*) were observed

staying with baboons (*Papio*) through much of the day by De Vore and Hall (1965:48, 49), who judged that the excellent eyesight of the baboons supplemented the acute senses of smell and hearing of the impalas and made the mixed group difficult for a predator to approach undetected. A further advantage was gained by the impalas, for on one occasion these authors observed a large male baboon discouraging three cheetahs that were approaching a mixed group of impalas and baboons.

Territoriality and Home Range. Because individuals of the same species usually have identical niches and are therefore potentially competing for the same environmental resources at the same place and time, intraspecific competition might be expected to be intense. Indeed, this seems to be the case, with the result that individuals of the same species customarily occupy separate, or nearly separate, *home ranges*. Burt (1943) has described the home range of a mammal as "that area traversed by the individual in its normal activities of food gathering, mating, and caring for young." Home ranges may have irregular shapes (Fig. 15–26). Within the home range of some animals is an area that is actively defended against other members of the species. This area, usually not including the peripheral parts of the home range, is called the *territory*, and species that apportion space in this fashion are termed *territorial*. A home range or territory may be occupied by one individual, by a pair, by a family group, or by a social group consisting of a number of families.

To solitary animals, or to members of a group, the occupancy of a home range has several important advantages. Each home range provides all of the necessities of life for an individual or group, permitting self-sufficiency within as small an area as possible; the less extensively the animal must range, the less chance there is for encounters with predators. The home range quickly becomes familiar to the

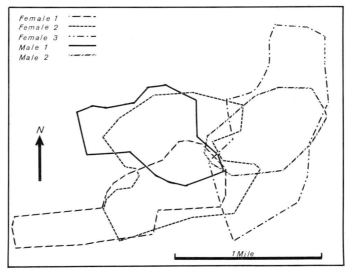

Female 1
Female 2
Female 3
Male 1
Male 2

N

1 Mile

FIGURE 15–26.   Distributions of red fox home ranges at the University of Wisconsin Arboretum. (After Ables, 1969.)

individual, who can then find food and shelter with the least possible expenditure of energy and can escape predators more effectively (because escape routes and retreats are familiar and no time or movement is lost in seeking shelter). Some species, such as rabbits and meadow voles (*Microtus*), maintain trails that serve as routes to food sources and as avenues of escape. Reproductive success may be increased by an animal's knowledge of areas occupied by animals inhabiting adjoining home ranges (in the case of solitary species) or by familiarity with animals sharing his home range (in the case of social species). During early life young can develop under parental care largely free from interference with other individuals of their own species. The spacing of home ranges is often such that the individual or the group is assured a food supply largely untouched by "foreign" members of the species; territorial species tend not to exceed the *carrying capacity* of a habitat (the maximum number of individuals an environment can support).

The sizes of home ranges vary tremendously, from a fraction of an acre in some small rodents and shrews to an area of 100 square miles or more in the

case of some carnivores (Table 15–3). Many mammals within the orders Insectivora, Primates, Rodentia, Lagomorpha, Carnivora, Perissodactyla, and Artiodactyla are known to be territorial. The recognition of territorial boundaries in some species depends on scent marking and other means of territorial marking, and much remarkable behavior is associated with the maintenance of territories (some of this behavior is discussed in Chapter 17). Some territorial species are distributed according to a pattern of home ranges that may persist throughout the lives of many generations. Hansen (1962b) found such a pattern to be typical of northern pocket gophers (*Thomomys talpoides*) in some areas. Each animal occupies an area of raised ground called a mima mound (Fig. 15–27), which is some 10 m. in diameter. The mima mounds are more productive of food than are the relatively narrow intermound areas, which usually have shallow soil. Except in the winter, the intermound areas are used little by pocket gophers, and the chances of survival are slim for an animal that is unable to establish himself in a mima mound. Likewise, wood rat houses may be used over periods of hundreds

Table 15–3. Sizes of Home Ranges of Some Mammals.

SPECIES	HOME RANGE (acres)	SOURCE
White-footed mouse *(Peromyscus)*	.08–10.66	Redman & Sealander, 1958; Blair, 1951
Prairie vole *(Microtus ochrogaster)*	.11 (males); .02 (females)	Harvey & Barbour, 1965
Snowshoe hare *(Lepus americanus)*	14.5	O'Farrell, 1965
Red-backed mouse *(Clethrionomys gapperi)*	.25 (winter only)	Beer, 1961
Least chipmunk *(Eutamias minimus)*	2.1–4.7 (summer only)	Martinsen, 1968
Yellow-pine chipmunk *(Eutamias amoenus)*	3.89 (males); 2.49 (females)	Broadbooks, 1970
Black-tailed deer *(Odocoileus hemionus)*	90 (winter); 180 (summer)	Leopold et al., 1951
White-tail deer *(Odocoileus virginianus)*	126–282	Ruff, 1938
Pronghorn antelope *(Antilocapra americana)*	160–480 (.25–.75 sq. mi.)	Bromley, 1969
Mule deer *(Odocoileus hemionus)*	502–2534 (.78–4 sq. mi.)	Swank, 1958
Red fox *(Vulpes vulpes)*	1280 (2 sq. mi.)	Ables, 1969
Lynx *(Lynx canadensis)*	3840–5120 (6–8 sq. mi.)	Saunders, 1963
Pine marten *(Martes americana)*	4480 (7 sq. mi.)	Marshall, 1951
Russian brown bear *(Ursus arctos)*	6400–8320 (10–13 sq. mi.)	Bourliere, 1956
Mountain lion *(Felis concolor)*	9600–19,200 (males) (15–30 sq. mi.) 3200–16,000 (females) (5–25 sq. mi.)	Hornocker, 1970
Grizzly bear *(Ursus arctos)*	50,240 (1 mother + 3 yearlings) (78.5 sq. mi.)	Murie, 1944
Timber wolf *(Canis lupus)*	23,040 (pack of 2) (36 sq. mi.) 345,600 (pack of 8) (540 sq. mi.)	Stenlund, 1955 Rowan, 1950

FIGURE 15–27. Mima-mounds in Mima Prairie, Thurston County, Washington. These mounds, some 10 m. in diameter, are probably formed by the burrowing activities of pocket gophers. (Photograph by Victor B. Scheffer.)

of years (Wells and Jorgenson, 1964), as indicated by the presence in houses of plants that no longer occur in the area but lived there hundreds of years ago.

Seemingly not all mammals have home ranges or are territorial, but so little is known of the biology of many species that the extent of this phenomenon within the class is unknown. Many whales, porpoises, and dolphins are social and move over wide areas with no evident fidelity to a particular place. Likewise, many African ungulates assemble in herds and seem not to be territorial for much of the year, although temporary "breeding territories" may be defended by males. However, even at times of the year when some animals seem not to be territorial, as when deer and elk concentrate in favored wintering areas, individuals or groups may be faithful to the same area, and may return there year after year.

Feeding Interactions. Just as energy transfers between parts of an organism are vital to life, complex patterns of energy transfer within a biotic community maintain this "superorganism" of interdependent and interacting species. The organisms involved in the transfer of energy within a community—from photosynthetic plants that utilize solar energy and inorganic materials to produce protoplasm, to animals that eat the plants, and thence to animals that eat animals—constitute the *food chain*. Typically, the transfer in a food chain goes from photosynthetic plants (primary producers), to herbivores that eat these plants (primary consumers), to first carnivores (secondary consumers) that eat the herbivores, to secondary or perhaps tertiary carnivores in some extended chains (Fig. 15–28). For the complex sequences of energy transfer that usually occur in nature, often involving predators and primary consumers that figure importantly in more than one food chain, the term *food web* has been used. (The intricacy of a food web is suggested by Figure 15–29.) Animals that occupy comparable functional positions in the food chain, say the position of primary consumers, are at the same *trophic level*. Green plants occupy the first trophic level and are referred to as *autotrophs* (self-feeders). Using mammals as examples, the second trophic level is occupied by herbivorous rodents, rabbits, and ungulates. Small carnivores such as weasels occupy the third trophic level, and large carnivores may be in the third or fourth level.

The food chain is often depicted as a pyramid in an attempt to stress the relationships of *biomass* (the total weight of organisms of a given type in the community), numbers of organisms, and available energy at the different trophic levels. Food chains typically rest on a broad food base of plant material, but energy available to animals in each successively higher trophic level becomes progressively reduced. The reduction results from loss of energy by respiration and by way of organisms that die and are utilized by *decomposers* (organisms that decom-

pose organic material) rather than by animals of the next higher trophic level. Also, energy is lost because of inefficient transfers between levels. Consider, for example, the pyramids of numbers, caloric content, and energy utilization shown in Figure 15–30 and Table 15–4.

The typical relationship of size and abundance of animals in a food chain involves small but numerous primary consumers, larger but much less abundant secondary consumers, and still larger but relatively scarce tertiary consumers. Although the animals at the top of the food chain seem to be in the commanding position of potentially being able to prey on all animals at lower trophic levels without being vulnerable to predation themselves, the top predators occupy precarious positions because they frequently depend on animals from trophic levels with low productivity. Under conditions of food stress, therefore, the fate of the species at the top of the food chain will not be death at the hands of a predator but starvation as a result of low availability of food.

Because of the great loss of energy accompanying food transfer between successive trophic levels, the total

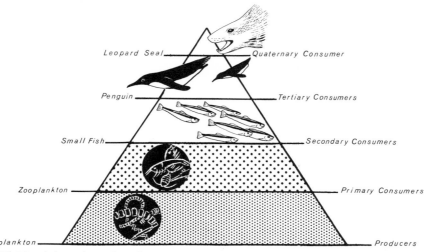

FIGURE 15–28. A hypothetical ecological pyramid based on an Antarctic food chain. The higher the step in the food chain (the higher the trophic level), the larger the individuals and the lower their numbers. Ultimately, gigantic numbers of tiny planktonic plants and animals are necessary to support, through several intermediate steps, one leopard seal.

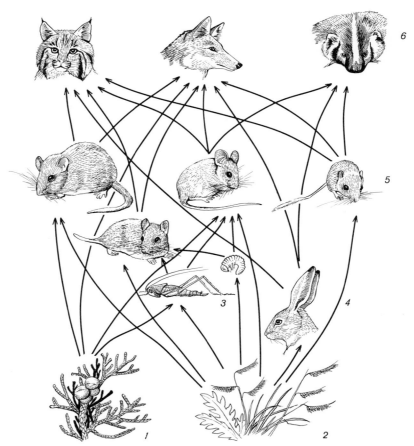

FIGURE 15–29. A simplified food web involving the mammals of a piñon-juniper community in Coconino County, Arizona. The arrows indicate the foods utilized by the mammals. The plants (1, juniper; 2, grasses and forbs) support arthropods (3), rabbits (4), and rodents (5). The rodents (*Neotoma stephensi, Onychomys leucogaster, Peromyscus truei, Perognathus flavus*) and the rabbits (*Lepus californicus* and *Sylvilagus audubonii*) are preyed upon by bobcats, coyotes, and badgers.

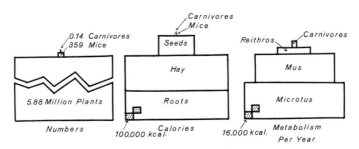

FIGURE 15–30. Pyramids of numbers, calories, and energy utilization for one acre of annual grassland near Berkeley, California. The pyramid showing calories is also approximately to scale for biomass. (After Pearson, 1964.)

Table 15–4. The Standing Crop of Plants, Prey, and Predators on One Acre of California Grassland and the Rate of Use of Vegetation by Rodents and of Prey by Carnivores at Peak Population Levels. (After Pearson, 1964; the Sources of the Figures in this Table are not Listed Here but are Given in Pearson's Paper.)

	STANDING CROP		RATE OF USE PER YEAR	
	kg (dry wt.)	kcal	kcal	% of crop
Roots	2,131	7,269,000		
Hay	2,097	8,141,000		
Seeds	442	1,920,000		
Microtus	1.24	6,402	1,368,750	71
Mus	0.88	4,543	876,000	46
Reithrodontomys	0.084	434	81,650	4
Other prey	0.13	671	27,000	
Carnivores	0.126	650	11,700	97

available energy is largest for consumers at the lower levels. Thus, predators feeding on primary consumers have more energy available to them than do the predators feeding on secondary consumers. Considered in this light, the adaptive importance of filter feeding in some marine mammals becomes clear. The baleen whales and certain plankton-feeding seals exploit tremendously larger sources of energy in primary-consumer plankton than they could if they fed on large fish that are secondary or tertiary predators. Only some 10 to 20 per cent of the energy entering a trophic level can be utilized by the next higher level; this factor limits the length of food chains.

Diagrams of food chains or food webs (Fig. 15–29), although valuable for purposes of illustration, usually of necessity strongly simplify what is really an extremely intricate meshwork of interactions. A broadly adapted carnivore like a coyote, or an omnivore such as the possum (*Didelphis*), may function in all trophic levels above that of the primary producer. The fruit of the prickly-pear cactus or juniper berries may form one meal, while a jackrabbit or deer fawn may be the next. More frequently, seasonal differences in the position of an animal in the food chain may occur. Johnson (1961, 1964) found that the deer mouse (*Peromyscus maniculatus*) in Colorado and Idaho became strongly insec-

tivorous in the summer, and thus functioned during this season as a secondary consumer, whereas it ate largely plant material during the cooler seasons, functioning as a primary consumer at those times. Occasionally in lean times, the ability to utilize various "alternate" foods may be of critical importance to large and powerful species. Even the mountain lion (*Felis concolor*), generally a predator of deer, will stoop under pressure of hunger to such lowly prey as skunks, raccoons, bobcats, or even porcupines (Grinnell et al., 1937:574-575).

FACTORS INFLUENCING POPULATION DENSITIES

The abundance of an animal species at a given time and a given locality depends on (1) the carrying capacity of the habitat and (2) the relationship between the rate at which the animals are added to the population (by reproduction or immigration) and the rate at which they are lost from the population (by death or emigration). Many mammalian populations are in a state of "dynamic equilibrium," and tend to be stabilized within certain limits of density by such interacting processes as competition, reproduction, predation, dispersal, and disease. These factors are usually regarded as density-dependent; that is, they fluctuate in direct relationship to the density of the species involved. The intensity of predation, for example, is generally dependent on the population density of the prey species. A higher proportion of a prey species is frequently taken during high densities of that prey species. Other factors, such as space requirements and weather, are generally density-independent; that is, they do not change in response to changes in population density.

Natality. The number of individuals added to a population through reproduction depends on the *reproductive potential* of a species, which refers to the greatest number of individuals that a pair of animals or a population can produce in a given span of time. The reproductive potential is a function of age and sex ratios, the age at which a female first bears young, mating systems, litter size, and the frequency of litters. Even species of the same genus occupying the same area can have markedly different reproductive potentials as indicated by litter size (Table 15–5). Reproductive performance is also responsive to environmental differences, such as temperature or rainfall, for sharp regional shifts in litter size occur within some species (Table 18–4, p. 360).

The ability of some species to vary reproductive performance in response to environmental conditions or population levels may be of considerable adaptive importance. For example, the reproductive potential of mule deer is lower in poor habitats than in habitats productive of high quality browse; whereas well-nourished does may

breed first at 17 months of age, those that occupy poor ranges may not breed first until as late as 41 months of age (Taber and Dasmann, 1957). The litter size of carnivores is also affected by food supply; Stevenson-Hamilton (1947) reported that the litter size of the African lion dropped when food was scarce. Batzli and Pitelka (1971) found delays in the start of breeding in the California vole (*Microtus californicus*) following times of peak densities; these delays may have resulted from decreased availability of preferred foods during periods of high densities of voles. These authors and other workers (Hoffmann, 1958; Greenwald, 1957) have observed seasonal changes in litter size that are presumably caused by changes in forage quality. In Finnish Lapland, Kalela (1957) found that reproductive maturity was delayed in young male voles (*Clethrionomys rufocanus*) in several localities in a year of high populations, and in one area supporting especially high densities breeding was also delayed in nearly all young females. The result of such delays is usually to reduce the numbers of young produced during that season; also, individuals of late litters may have too little time for maturation before the critical winter period and may suffer unusually high mortality. Breeding of muskrats (*Ondatra zibethica*) in Iowa continued through only part of the usual breeding season in a year of high populations (Errington, 1957), and wild rats (*Rattus norvegicus*) in the city of Baltimore had a markedly low pregnancy rate during population highs (Davis, 1951).

Table 15–5. Differences Between the Reproductive Patterns of Three Species of *Peromyscus* That Are Sympatric in Some Areas. (From McCabe and Blanchard, 1950.)

CHARACTERISTIC	P. maniculatus	P. truei	P. californicus
Number of litters per season	4.00	3.40	3.25
Number of young per litter	5.00	3.43	1.91
Number of offspring per breeding female per season	20.00	11.66	6.21

Significant mortality may occur in some species when adults kill young. In the judgement of Wirtz (1968), bites by adult monk seals (*Monachus schauinslandi*) cause considerable mortality among young seals shortly after they become independent of their mothers.

Survival rates of young also strongly affect population levels. Young are clearly the expendable part of the population and show the greatest fluctuations during population changes. A 74 per cent decline in pocket gopher density in western Colorado in 1958 was associated with an extraordinary drop in survival of young (Hansen and Ward, 1966), and in southern Colorado Hansen (1962b) found that whereas high survival of young pocket gophers was characteristic of periods of high densities, low survival of young was associated with a declining population. Similarly, Krebs (1966) reported better survival for expanding than for declining populations of the California vole.

In mammals the contribution made to a population by reproduction clearly depends on a variety of factors, and is seldom constant within a species from year to year. As mentioned, variation in litter size, numbers of litters and length of breeding season, the age at which young animals breed, and survival of young are all important variables. In addition, the litter size (Table 15–6) and the percentage of females that become pregnant changes with age distribution of a species; the age composition of a population may therefore have a marked effect on the reproductive performance of that population.

Predation. Mortality by predation and other causes is a characteristic of a population that refers to the loss of members by death. Mortality varies with age, and has been carefully studied in some mammals. *Specific mortality* is the number of individuals of a population that have died by the end of a given time span. Specific mortality at given ages can be expressed in a *life*

Table 15–6. Litter Size of Consecutive Litters of the Montane Vole (*Microtus montanus*). Note That Young and Old Animals Have Relatively Small Litters. (After Negus and Pinter, 1965.)

LITTERS	N	MEAN LITTER SIZE	RANGE
1	12	4.2	2–6
2	12	4.7	3–7
3	10	5.0	3–7
4	9	4.2	2–6
5	6	5.8	3–10
6	6	5.5	3–7
7	5	3.4	1–6

table. The life table in Table 15–7 is based on data assembled by Murie (1944) during his study of wolf predation on the Dall sheep (*Ovis dalli*).

Of the mortality factors to which mammals are susceptible, predation looms high in importance. There has been much heated debate on the ability of predators to control or influence densities of prey species or to influence population cycles, and the final word has clearly not been heard. The degrees of impact that predators have on prey have been summed up by Pearson (1971:41): "The effectiveness of predation varies from the relatively ineffective predation of rats on man, in which rats are able occasionally to kill infants or incapacitated adults, through the mink-muskrat system described by Errington (1967), in which mink take a significant proportion of homeless and stressed muskrats, to the almost total effectiveness of carnivore predation on *Microtus* until almost the last one has been killed." A predator-prey relationship must obviously have some stability; as indicated by Lack (1966: 301), "Only those predatory species which have not exterminated their prey survive today, hence we observe in nature only those systems which have proved sufficiently stable to persist, and many others were presumably terminated in the past by extinction." Although the effectiveness of even a single species of predator seems to

Table 15–7. A Life Table for Dall Sheep (*Ovis dalli*) in Mount McKinley
National Park, Alaska. (After Deevey, 1947.)

AGE (years)	AGE AS PER CENT DEVIATION FROM MEAN LENGTH OF LIFE	NUMBER DYING IN AGE INTERVAL PER 1000 BORN	NUMBER SURVIVING AT BEGINNING OF AGE INTERVAL PER 1000 BORN	MORTALITY RATE PER 1000 ALIVE AT BEGINNING OF AGE INTERVAL	EXPECTATION OF LIFE, OR MEAN LIFETIME REMAINING TO THOSE ATTAINING AGE INTERVAL (years)
0–0.5	−100.0	54	1000	54.0	7.06
0.5–1	− 93.0	145	946	153.0	–
1–2	− 85.9	12	801	15.0	7.7
2–3	− 71.8	13	789	16.5	6.8
3–4	− 57.7	12	776	15.5	5.9
4–5	− 43.5	30	764	39.3	5.0
5–6	− 29.5	46	734	62.6	4.2
6–7	− 15.4	48	688	69.9	3.4
7–8	− 1.1	69	640	108.0	2.6
8–9	+ 13.0	132	571	231.0	1.9
9–10	+ 27.0	187	439	426.0	1.3
10–11	+ 41.0	156	252	619.0	0.9
11–12	+ 55.0	90	96	937.0	0.6
12–13	+ 69.0	3	6	500.0	1.2
13–14	+ 84.0	3	3	1000.0	0.7

vary according to the specific situation considered, some general predator-prey relationships that apply to mammals as well as to other animals can be recognized.

The observed responses of a predator to changes in the density of a prey species indicate that predation is density-influenced. The numbers of a preferred prey taken by a carnivore increase as the density of the prey increases (Fig. 15–31), because the greater the number of prey animals per unit area the greater the opportunity for predators to encounter and capture them. This is a *functional response* on the part of the predator (Holling, 1959, 1961). Errington (1937) noted such a response in the predators of muskrats and suggested that the intensity of predation is a function of population levels of the prey. There may also be a *numerical response,* involving an increase in the predator density with a rise in the prey population. The numerical response may be the result of immigration of predators, as in the case of the striking responses to lemming abundance on the part of the pomerine

jaeger (Table 15–8), or may be due to increased breeding success, as in the case of the masked shrew (Fig. 15–32).

The relative populations of predators and their prey have often been regarded as being in dynamic equilibrium. The degree to which a balance between prey populations and predator populations is reached, and the extent to which the relationship is dynamic, varies widely in situations involving mammals. It depends in part

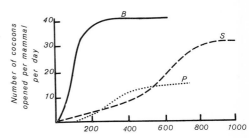

Number of cocoons per acre, in thousands

FIGURE 15–31. Functional responses on the part of three predators, the short-tailed shrew, *Blarina brevicauda* (B), the masked shrew, *Sorex cinereus* (S), and the deer mouse, *Peromyscus maniculatus* (P), to density of prey. The prey are sawfly larvae, which are removed from their cocoons by the predators. (After Holling, 1959.)

on the ratio of predator density to prey density, on the relative sizes of the predator and the prey and the ease with which prey can be captured, and on the degree to which the prey populations are cyclic. Studies by Mech (1966) on Isle Royale in Lake Superior, during the period from 1959 to 1961, indicated that the ratio of moose to wolves was roughly 30 to 1; 20 wolves were supported by approximately 600 moose, the wolves' primary food. The weight differential between an adult moose and an adult wolf was roughly 14 to 1 (approximately 450 kg. to 33 kg.), and the wolves had difficulty killing moose. A strongly contrasting situation was studied in California by Pearson (1971), who found that the

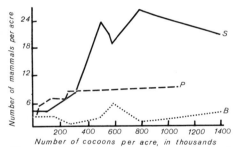

FIGURE 15–32. Numerical responses on the part of the masked shrew (S) and the deer mouse (P) to high densities of sawfly larvae. (After Holling, 1959.)

ratio of numbers of prey to numbers of predators varied from 72 to 1 in 1962, during a period of low vole populations, to 5410 to 1 during a peak in vole numbers (Fig. 15–33). In this case the prey was a cyclic vole (*M. californicus*) with an adult weight of roughly 45 gm. The predators – feral cats, raccoons, gray foxes and skunks – averaged perhaps 2.25 kg. in weight, yielding a rough estimate of prey to predator weight of .02 to 1. The predators could catch voles easily. These examples are based on two very different patterns of predator-prey interaction. The wolf-moose interaction resulted in relatively stable predator and prey densities, whereas the situation involving the California vole was one of great instability. Because of the difficulty with which wolves bring down moose, the pressure they exert on the moose population is highly selective in that primarily young or old animals are taken; adult moose in the prime of their reproductive life are not killed (Mech, 1966). The predators of the vole, on the other hand, show a high preference for this prey and find it easy to catch; their kill is far more nearly random and they are able to kill almost every last vole during times of vole scarcity (Pearson, 1966). As suggested by the above ex-

Table 15–8. Densities of Breeding Pomarine Jaegers (*Stercorarius pomarinus*) and Nesting Success Near Point Barrow, Alaska. Note the Correlation Between High Densities of Lemmings (*Lemmus trimucronatus*) and High Populations of Nesting Jaegers. (After Maher, 1970.)

YEAR	SPRING *Lemmus* DENSITY (no./acre)	NO. OF PAIRS OF JAEGERS	CENSUS AREA (square miles)	DENSITY (pairs/square mile)	MAXIMUM DENSITY (pairs/square mile)	BREEDING SUCCESS (per cent of eggs)
1952	15–20	34	9	3.8	5–6	30–35
1953	70–80	128	7	18.3	25–26	20–25
1954	<1	0	–	–	–	–
1955	1–5	2	15±	0.13	–	0
1956	40–50	114	6	19.0	22–23	4
1957	<1	0	–	–	–	–
1958	<1	0	–	–	–	–
1959	1–5	3	15±	0.20	–	0
1960	70–80	118	5.75	20.5	25	55

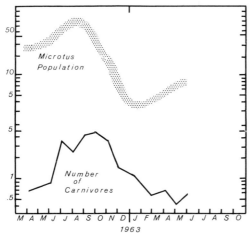

FIGURE 15–33. Densities of California voles (*Microtus californicus*) compared to densities of predators (feral cats, raccoons, foxes, and skunks) during a population cycle of voles. (After Pearson, 1971.)

amples, predator-prey interactions are complex; so many variables are involved that few generalizations relating to such interactions can be universally applied.

The weight of much carefully assembled evidence indicates that some predators tend to remove vulnerable individuals from the prey population. These are individuals that because of inexperience, old age, injury, or sickness are readily captured; or they are animals that are forced by intraspecific competition for space into marginal habitats in which their vulnerability to predators is increased. The excellent work on wolves by Murie (1944), Crisler (1956), and Mech (1966) has shown that vulnerable individuals of the prey species, in these cases Dall sheep, caribou, and moose, were taken more frequently than were healthy, mature individuals. Careful field studies by Hornocker (1970a, 1970b) on the mountain lion in Idaho showed that of 53 lion-killed elk and 46 lion-killed deer, 75 per cent of the elk were young (less than 1.5 years) or old (more than 8.5 years), and 62 per cent of the deer were young or old. These percentages of young and old animals were considerably greater than would be expected if the lions

had killed randomly. On Isle Royale 18 of 51 wolf-killed moose were calves; most of the remainder of the animals killed were from 8 to 15 years old, and 39 per cent of these old animals showed evidence of debilitating conditions (Mech, 1966). The wolves had killed no adults from 1 to 6 years old. Errington (1943, 1946, 1963) found that in Iowa as muskrat populations rose above a "threshold of security" a number of animals were forced into marginal habitats by intraspecific competition for space. This "vulnerable surplus" was preyed upon heavily by mink and red foxes which made a marked functional response to this available food source. This pattern of predation has been called *compensatory predation*.

But can predators control densities of mammals, or are they simply killing individuals that would quickly be removed from the population by other means? The answer clearly depends on the specific predator-prey interaction considered. Murie (1944:230) concluded that, in Alaska, wolf predation on Dall sheep lambs was the most important factor limiting numbers of sheep, and on Isle Royale the wolves seemingly kept moose densities below the level at which food supply would be the limiting factor (Mech, 1966: 167). In Idaho, however, predation by mountain lions had little impact on the populations of deer and elk; here the densities of deer and elk were controlled by the winter food supply (Hornocker, 1970b).

Heavy predation on populations of small mammals has been shown to affect population levels. Data reflecting the severity of predation on lemmings demonstrates how important this source of mortality may be locally. On the coastal plain near Point Barrow, Alaska, during times of high lemming densities, the combined impact of several predators deals a staggering blow to lemming populations and is the primary factor causing the population crash (Pitelka et al., 1955; Pitelka, 1957a). The combined kill of lemmings by the major predators amounted to at least 49/acre during a cycle of abundance (Table 15–9). Studies in California by Pearson (1963, 1964 and 1966) on *Microtus* have demonstrated that predators preying upon these cyclic rodents are unable to control a *rising* prey population, but that "carnivore predation during a crash and especially during the early stages of the subsequent population low determines to a large extent the amplitude and timing of the microtine cycle of abundance." (Pearson, 1971:41)

Disease. Parasitism and disease is known to be a significant cause of mortality (Elton, 1942) among mammals, and may occasionally cause dramatic population crashes, as in the case of a die-off of prairie dogs (*Cynomys gunnisoni*) in Colorado (Lechleitner et al., 1962) caused by bubonic plague. Talbot and Talbot (1963) estimated that 47 per cent of the total mortality suffered by wildebeest was caused by diseases, of which rinderpest seemed most important. The blood parasite *Babesia* is a source of mortality among African lions. Disease in relation to population regulation, however, has been a dif-

ficult factor to assess (Chitty, 1954). Disease as the single cause of death may be relatively unimportant, but it may be important in contributing to the vulnerability of an animal to predation or to stressful environmental conditions. Parasitism has been regarded periodically as an important cause of mortality, but careful observation often indicates that otherwise healthy animals can tolerate a moderately heavy parasite load. Heavy parasitism has been found to accompany a general decline in health in rodents and rabbits during or following times of high density (Batzli and Pitelka, 1971; Erickson, 1944).

Weather. Certain unusual weather-caused conditions, such as flooding, are known to result in significant mortality among mammals. A series of beaver colonies were decimated during a flood in Colorado (Rutherford, 1953), populations of small mammals in Oklahoma were reduced by stream-valley flooding (Blair, 1939), and a 70 per cent decrease in the combined population of golden mice (*Ochrotomys nuttalli*) and cotton mice (*Peromyscus gossypinus*) was caused by a

Table 15–9. Impact by Predators on High Population of Lemmings (*Lemmus trimucronatus*) Near Point Barrow, Alaska. The Data for the Least Weasel are From Thompson, 1955; Those for the Snowy Owl are From Watson, 1958. The Table is After Maher, 1970.

PREDATOR	AGE CLASS	DENSITY (ind./ square mile)	DAILY FOOD CONSUMP-TION (g/ind.)	SEASON'S LEMMING CONSUMPTION (per acre)			
				(per ind.)	25 May to 15 July	16 July to 31 Aug.	TOTAL
Pomarine jaeger	Adult	38	250	338	10	21	31
	Young	38	200	167	–	–	
Snowy owl	Adult	2	250	350	1.3	1.6	3
	Young	7	150	160	–	–	
Least weasel		64	50	100	5	5	10
Glaucous gull		20	250	125	0.7	–	1
Waste					4	–	4
Totals					21	28	49

flood of three weeks' duration in eastern Texas (McCarley, 1959). Small mammals may face considerable cold stress and mortality during autumns when extreme cold descends but an insulating snow cover is late in developing, and rapid snowmelt with resultant flooding is an annual "catastrophe" that faces small mammals in areas with heavy snowfall.

POPULATION CYCLES

Mammalian population cycles are among the most impressive biological phenomena. Striking changes in density occur primarily in temperate, subarctic, and arctic areas, but are not known to occur in tropical or subtropical regions. This difference is probably related to differences in species diversity between these areas. High latitude areas are characterized by biotic assemblages and food webs that are simple relative to those of tropical areas. The typical boreal community has a limited biota and supports few species of vertebrates, but some species may, at least periodically, be remarkably abundant. The simplicity of the northern community is seemingly partly responsible for its instability, for where so few kinds of organisms exist any marked fluctuation

in the density of one species seems to disrupt the entire community. In tropical habitats, by contrast, there is an enormous diversity of plants and animals that support many species of vertebrates. However, few of the many species of vertebrates have high population densities. In the complex tropical community the diversity of carnivores, the intricate patterns of niche displacement and potential competition, and the relatively small percentage of the energy resources available to any one species provide a buffer system against population outbreaks by any species. A complex food web also provides a cushion against drastic population declines. Even in northern Alaska population cycles are more pronounced in coastal areas, where only two microtines occur, than in the foothills, where there are five species. Pitelka (1957b:85) states, "Similarities in their feeding and sheltering activities strongly suggest that where more than one species is important, competition may act to depress their respective populations and hence to depress the likelihood of strong fluctuations." Population fluctuations occur in many temperate and boreal areas, but Aumann (1965) postulated that high densities (over 1000 per acre) of microtines only occur in areas with large amounts of sodium in the soil (Table 15–10). Soils

Table 15–10. Population Densities of Several Species of Microtines and Sodium Levels in the Soils. (After Aumann, 1965.) L Stands for Low, M for Medium, and H for High.

DENSITY (per acre)	SPECIES	REGION	REFERENCE	SODIUM LEVEL
1–20	M. pennsylvanicus	N. Minn.	Beer et al., 1954	L
6–67	M. pennsylvanicus	N. York	Townsend, 1935	L
3000	M. montanus	N.W. U.S.	Spencer, 1958a	H
200–4000	M. montanus	Oregon	Spencer, 1958b	H
25–81	M. californicus	N. Cal.	Greenwald, 1957	M
425	M. californicus	N. Cal.	Lidicker & Anderson, 1962	H
25–145	M. ochrogaster	Kansas	Martin, 1956	L
250–300	M. agrestis	England	Chitty & Chitty, 1962	M
1900	M. arvalis	France	Spitz, 1963	H
1004	M. guentheri	Israel	Bodenheimer, 1949	H
2400	M. sp.	U.S.S.R.	Hamilton, 1937	Unknown
50–100	L. trimucronatus	Alaska	Rausch, 1950	M
200–300	L. lemmus	Sweden	Curry-Lindahl, 1962	M

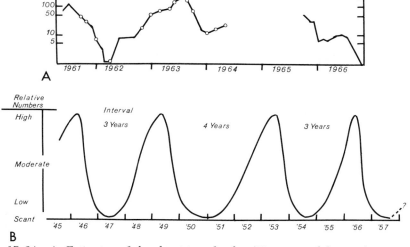

A

B

FIGURE 15-34. A, Estimates of the densities of voles (*Microtus californicus*) in an area near Berkeley, California, during a six-year period. (After Pearson, 1971.) B, Generalized curves (the amplitudes of successive cycles are actually not the same) showing fluctuations in brown lemming (*Lemmus trimucronatus*) populations near Barrow, Alaska. (After Pitelka, 1957b.)

with high sodium levels support plants that provide a ready supply of sodium to microtines, and presumably this availability of sodium is in some way necessary for maintaining high populations and high reproductive rates.

Characteristics of Population Cycles. In areas where well-marked cycles of microtine rodents occur, population peaks may occur at three- to four-year intervals (Fig. 15–34). On the coastal slope of northern Alaska, oscillations of populations of lemmings (*Lemmus trimucronatus*), the chief herbivore, are characterized by: (1) a precipitous drop in density in the late summer and winter following a population peak; (2) a period of one or two years of extremely low populations and localized distribution; (3) an upsurge in the population in the winters following the low population, with peak numbers occurring early in the third or fourth summer (Pitelka, 1957b: 85, 86). The short-term lemming cycles also occur in temperate and boreal parts of the Old World (Elton, 1942; Siivonen, 1954). Longer-term cycles, from eight to twelve years, are known in populations of the snowshoe rabbit and the Canada lynx (Fig. 15–35). Mammalian population cycles occur

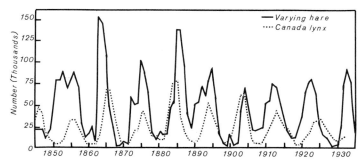

FIGURE 15-35. Cycles of population density of the snowshoe hare (*Lepus americanus*) and the Canadian lynx (*Lynx canadensis*). The figure is based on the numbers of pelts purchased by the Hudson Bay Company. (After MacLulich, 1937.)

in some temperate areas, where perhaps the most striking cycles involve microtines, which in some temperate areas reach amazingly high densities (Table 15–10).

Accompanying population explosions of Norwegian lemmings (*Lemmus lemmus*) are the famous mass wanderings during which lemmings appear far from their preferred habitats. One such movement occurred in Norway in the summer and fall of 1963 and is well described by Clough (1965). The movement began in mid-July, when as many as 40 lemmings passed an observation point per hour; all animals were heading downhill. Lemmings also began to appear on the streets of a nearby town and in pine forests 12 miles down the valley. Clough was struck by the complete intolerance of one lemming for another. Each animal moved alone. Every time lemmings met one another, some aggressive action or strong avoidance behavior occurred, and large, mature females were dominant in all conflicts involving them. The wandering lemmings, however, were chiefly middle-sized or small, often sexually immature individuals. By October lemmings were living in virtually all habitats, including those that were highly unsuitable, and the population was on the decline. The following March Clough did not find a single lemming, nor were any found during his intensive work in May and June in areas where lemmings had been abundant the previous summer. The winter's die-off had seemingly been complete.

Factors Controlling Population Cycles. Although population cycles in mammals have been known for many years, and the patterns of fluctuation have been carefully described for some species, considerable uncertainty and controversy remains as to what factors control cycles. Following are several of the most important theories about the factors controlling animal populations.

Food is of ultimate importance according to some workers. According to Lack (1966), food is the primary factor setting an upper limit on populations, but this limit is rarely reached in many abundant species because the population is kept within bounds by predation or parasitism. Considerable work on big game animals suggests that food shortages, especially during winter periods, are critical in limiting density in some species (see Leopold et al., 1947, 1951). Several workers have upheld the nutrient-recovery hypothesis, which holds that microtine populations drop in response to reductions in food supply, and that a subsequent build-up in density can only occur with a recycling and renewed availability of nutrients in the community (Lack, 1954a, 1954b; Pitelka, 1957a, 1957b; Schultz, 1964). Certainly the winter use of vegetation by lemmings may be extreme in some years. Bee and Hall (1956:86, 87) found that in the winter of 1951–1952, a period of build-up in lemming populations, from 95 to 98 per cent of the grass beneath the winter snowline was eaten. That food actually limits the populations of rodents, however, has not been conclusively demonstrated. Other factors than food shortages seemingly operate to reduce populations before the lack of food becomes limiting (Christian, 1963:328), but the importance of food in controlling cycles has been widely considered, discussed and debated (Chitty, 1960; Kalela, 1962; Krebs, 1964; Pitelka, 1964). Until evidence is available on the changes in the nutritional quality of the plants eaten by lemmings during population oscillations, the importance of food will be difficult to establish. The fact that wild populations of voles (*M. californicus*) fed unlimited food under otherwise natural conditions declined after an initial moderate rise suggested that food may not be of prime importance in limiting their numbers (Krebs and DeLong, 1965). Further, it seems unlikely that food would limit populations of highly motile mammals with highly specialized requirements for daytime retreats, such as bats. In the opinion of Hornocker (1970b), food does not limit populations of the highly territorial mountain lion in parts of Idaho.

Stress is known to have pronounced effects on the mental and physical health of man (witness the high incidence of mental illness and ulcers among people living for long periods under stressful conditions), and is now known to have pronounced physiological effects on some other mammals. Since such pioneer works as those of Selye (1936, 1955), Green and his co-workers (1938, 1939), and Christian (1950) on the physiological responses of mammals to stress and to high populations, much attention has been focused on this subject and on the relationships between endocrine reactions to stress and population cycles in mammals. The "stress syndrome" theory relies basically on the considerable body of evidence amassed in recent years indicating that as populations rise and interactions between individuals of populations become progressively more frequent, adaptive physiological responses tend to reduce the population levels. A consistent pattern seems to be followed. As populations grow, social pressures, such as intraspecific competition and social strife, mount. These increasing pressures provide progressively stronger stimuli to the central nervous system (hypothalamus), which (through neuroendocrine mechanisms) affects the function of the anterior pituitary. The anterior pituitary responds by reducing the output of both growth hormones and gonadotropins (hormones that stimulate the reproductive organs) but by increasing the levels of adrenocorticotropic hormones (hormones that stimulate the adrenal cortex). The alteration of the hormonal balance causes certain behavioral changes, such as unusually high activity and heightened intolerance between individuals. Of greater importance is suppression of normal reproductive function in the adult, decreased

intrauterine and postnatal survival of young, unusual metabolic patterns, and lowered resistance to disease. As a result, mortality increases and reproduction may virtually cease. Increased adrenocortical activity, as indicated by increased adrenal weight, was found to be associated with high populations of Norway rats (Christian and Davis, 1956; Christian, 1959a), voles (Christian and Davis, 1966; Christian, 1959b; Louch, 1958), short-tailed shrews (Christian, 1954), muskrats (Beer and Meyer, 1951), and sika deer (Christian et al., 1960). Some reliable data on wild mammals, however, do not support the stress-syndrome theory. Adrenal glands of Canadian lemmings were found not to enlarge in times of high density (Krebs, 1963), and Clough (1965) observed no difference in adrenal weights or resistance to stress in voles from high populations compared to those from low populations. Much controversy still surrounds the question of the importance of the stress-syndrome in governing population cycles of mammals.

As discussed in the section on predation (p. 288), population cycles of some microtines are thought to be influenced by predation. The population crash that terminates the brown lemming cycle in northern Alaska is brought about by the concentrated efforts of a number of predators (Pitelka, 1957a:88). Pearson (1966, 1971) has presented data suggesting that the amplitude and timing of the microtine cycle is determined by intense predation during and immediately after the population crash. The impact of even a single predator (the pomerine jaeger) on high summer lemming populations has been shown to be heavy (Maher, 1970), and intense winter predation by a weasel (*Mustela nivalis*) may be responsible for nearly wiping out lemmings and delaying the recovery of the population until the weasels themselves decline (Maher, 1967; Thompson, 1955:173). Clearly, predators influence the population cycles of some rodents, but a complete understanding of their effects on cycles of other kinds of mammals awaits further field research.

A theory based on the supposed ability of populations of animals to be self-regulating was developed by Wynne-Edwards (1959, 1960, 1962). According to his theory, individuals of a population are kept dispersed by behavioral means and the population is maintained at near "optimum" density, a density below that at which starvation would limit the population. Wynne-Edwards held that dispersal is a result of behavior, such as territorial displays, that evolved through "group selection." Such selection operates on groups of populations of animals, and favors those populations able to maintain their density near "the level at which food resources are utilized to the fullest extent possible without depletion" (Wynne-Edwards, 1962:132). The fact that natural populations commonly fluctuate widely beyond an "optimum," and that the evolution of behaviors that tend to limit density can be explained reasonably in terms of natural selection operating on the individual rather than on entire populations, has led to strong criticism of this theory (for a careful critique of the ideas of Wynne-Edwards see Lack, 1966:299–312).

CHAPTER 16

ZOOGEOGRAPHY

One of the most familiar kinds of biological information concerns zoogeography. Children learn that lions and zebras live in Africa and not in North America, and that kangaroos are typical only of Australia. This same type of knowledge of the presence or absence of various kinds of animals in different parts of the world is the substance of zoogeography, the study of animal distribution.

Considerations of zoogeography include two major approaches. The first is descriptive and static, and seeks to delineate the distributions of living species. Such information can be gained by field work and careful observation; it can be dealt with directly by using presently available evidence. The second approach is ecological or historical, and attempts to explain the observed distributions. Such inquiry often involves syntheses based on diverse lines of evidence. The ecologist, for example, may try to explain past or present distributions of animals on the basis of their environmental requirements. But scientists studying what Udvardy (1969:6, 7) calls "dynamic zoogeography" ask the most demanding question: How, when, and from where did animals reach the areas they now occupy? Virtually every fauna consists of animals that reached the area at different times, from different regions, and by different means. Our knowledge of the complex history of a fauna depends basically on the completeness of the world-wide fossil record and on our understanding of the geological history of the major land masses. Regrettably, however, our knowledge in these areas is incomplete, and even some of the major questions can only be tentatively answered.

Mammals occupy all continents, from far beyond the arctic circle in the north to the southernmost parts of the continents and large islands in the south. (Antarctica has no land mammals.) In the New World the northernmost lands, the northern coasts of Greenland and of Ellesmere Island, are inhabited by the arctic hare (*Lepus arcticus*), collared lemming (*Dicrostonyx groenlandicus*), wolf (*Canis lupus*), arctic fox (*Alopex lagopus*), polar bear (*Ursus maritimus*), short-tailed weasel (*Mustela erminea*), caribou (*Rangifer tarandus*), and musk-ox (*Ovibos moschatus*). A similar group of mammals, but lacking the musk-ox, lives on the north coast of the Taymyr Peninsula (Soviet Union), the northernmost coast of Asia (Berg, 1950:

19). The southernmost part of Africa has a rich mammalian fauna. Tasmania, the southernmost part of the Australian region, supports two monotremes, many marsupials, several native rodents, and several bats. On Tierra del Fuego, at the southern tip of South America, occur a bat, several rodents, a fox, otters, and a llama. The chiropteran family Vespertilionidae occurs almost everywhere there is land except in arctic areas, and the families Leporidae, Cricetidae, Sciuridae, Canidae, Mustelidae, and Felidae are native to all continents but Australia. All oceans, and seas connected to the oceans, are inhabited by cetaceans, and odontocetes (toothed whales and porpoises) also live in some large rivers and lakes.

DISPERSAL AND FAUNAL INTERCHANGE

Animal Dispersal. Dispersal occurs when an individual or a population moves from its place of origin to a new area. The ability to disperse is as basic as the ability to reproduce, and is as necessary to the survival of a species. A spacing of members of a population such that each individual can satisfy its environmental needs is critical to all organisms. Territoriality is one familiar means by which this spacing is insured, and young of territorial species usually establish home ranges largely separate from those of other individuals, including their parents. The pressures exerted by reproduction and the necessity for the spacing of individuals create a tendency of populations to occupy ever-increasing areas, to colonize formerly unoccupied localities, and to repopulate areas where the animals were previously extirpated. The more widespread a species, the less likely it is to be forced into extinction by local mortality, and as a result natural selection has usually favored those species that have broad distributions. A high adaptive premium is placed on dispersal ability. Udvardy

(1969:12) has stated that "without evolved means of dispersal most animal populations would have succumbed, over a period of time, to the vicissitudes of the environment."

The ability of a population to expand into new areas depends on its innate dispersal ability (which is greater, for example, in fliers than in burrowers), on the breadth of environmental conditions that it can tolerate, and on the presence of barriers. Barriers may be ecological, with environmental conditions under which a species cannot survive, or more simply physical, such as bodies of water, precipitous cliffs or mountains, or rough lava formations. If enough information were available, much of the story of zoogeography could be told by considering the patterns of dispersal of animals as modified by the locations, effectiveness, and longevity of barriers.

Migration and Faunal Interchange. Certain regions have apparently been major centers of origin of mammalian groups. Many families first appear in the fossil record in the Eurasian area, and North America seems also to have been the place of origin for many groups. The present mammalian faunas of regions such as Africa and South America were partly derived from migrations of mammals from northern continents. Despite uncertainty as to the place of origin of many mammalian groups (where a group first appears in the fossil record is generally taken as its place of origin), movements of mammals from place to place are in some cases well documented by the fossil record.

Simpson (1940) recognized several avenues of faunal interchange. The *corridor* is a pathway that offers relatively little resistance to mammalian migration and along which considerable faunal interchange would be expected to occur. Such a continuous corridor now exists across Eurasia; interchange of animals between Europe and Asia is highly probable and has apparently occurred frequently. A *filter route* has the effect of allowing passage of certain animals but stopping others. Selective filtering has occurred at times along the land bridge that has periodically connected Siberia and Alaska. When this bridge was present in the Pleistocene, as an example, conditions were such that only animals adapted to cold climates could migrate between these two continents; mammals intolerant of cold conditions were denied use of this route. Mountain ranges, deserts, or tropical areas may also form filter barriers. The third and most restrictive route is the *sweepstakes route*. This is a pathway that will probably not be crossed by large numbers of any given type of animal, but is a route that an occasional individual may follow. Such a pathway is that between New Guinea and Australia or between Africa and Madagascar. Dispersal *via* a sweepstakes route must occur by swimming or flying, or by such uncertain means as rafting between one continent and another or between islands ("island hopping") on floating vegetation or debris. The probability that an animal will follow a sweepstakes route is extremely low if the route is long, as, for example, from North America to Hawaii, but is increased if an animal is small and can cling to floating material, is aquatic, or can fly. (The only land mammal that reached Hawaii without the help of man was a bat.) Despite the unlikelihood of a mammal's dispersal *via* a sweepstakes route, such dispersal has occurred, and has resulted in the establishment of unusual faunas of the sort that occurred on Madagascar or New Zealand before the coming of European man.

MAMMALS OF THE ZOOGEOGRAPHIC REGIONS

The zoogeographical realms shown in Fig. 16–1, which are the basis for the organization of the following section, were proposed by Wallace (1876) and have been widely used in discussions of zoogeography.

Palearctic Region. This region includes much of the northern part of the Old World, and is the largest of the zoogeographic regions. Included within this vast area are Europe, North Africa, Asia (except the Indian subcontinent and Southeast Asia), and the

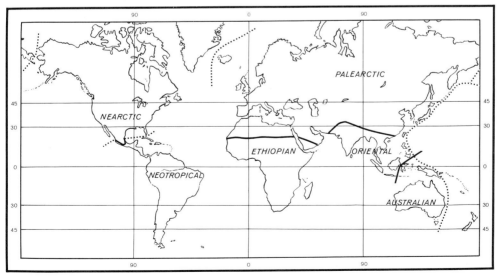

FIGURE 16–1. A map of the world, showing the zoogeographical regions discussed in this chapter.

Near East (Fig. 16–1). The climate is largely temperate, but contrasting conditions exist, from the intense heat of North Africa to the arctic cold of northern Siberia. Broad areas of coniferous forests, comparable in many ways to those of northern North America, are typical of much of the northern Palearctic Region, and deserts are widespread in the south. The Palearctic is separated from the Ethiopian Region by deserts, from the Oriental Region by the Himalayas, and from the Nearctic by the Bering Strait.

The Palearctic mammalian fauna is fairly rich, including some 40 families (Table 16–1). Roughly 78% of the Palearctic families also occur in the Ethiopian Region, and 70% reach the Oriental Region (Fig. 16–2). Although a land bridge between North America and Asia was present in only part of the Tertiary, the Palearctic shares 50% of its mammalian families with the Nearctic. Many genera, and a few species, within the families Soricidae, Vespertilionidae, Cricetidae, Canidae, Ursidae, Mustelidae, Felidae, and Cervidae occur in both regions. Only two small families are restricted to the

Table 16–1. A List of Palearctic Families of Mammals and Their Geographic Origins. Two Families are Endemic.

FAMILIES OF MAMMALS	ENDEMIC FAMILIES	GEOGRAPHIC ORIGIN OF FAMILIES
Erinaceidae		North America
Talpidae		Europe
Soricidae		Europe
Macroscelididae		Africa
Pteropodidae		Europe
Rhinopomatidae		?
Emballonuridae		Europe
Nycteridae		?
Rhinolophidae		Europe
Vespertilionidae		Europe or North America
Molossidae		Europe
Cercopithecidae		Africa
Hominidae		Southern Asia
Ochotonidae		Eurasia
Leporidae		North America
Sciuridae		Europe or North America
Castoridae		North America
Cricetidae		North America, Europe, or Asia
Spalacidae	X	Europe
Rhizomyidae		Europe
Muridae		Europe or Asia
Gliridae		Europe
Seleviniidae	X	?
Zapodidae		Europe
Dipodidae		Asia
Hystricidae		Europe
Canidae		North America or Europe
Ursidae		Europe
Procyonidae		North America
Mustelidae		North America, Europe, or Asia
Viverridae		Europe
Hyaenidae		Asia
Felidae		North America or Europe
Procaviidae		Africa
Equidae		North America
Suidae		Europe
Hippopotamidae		Asia
Camelidae		North America
Cervidae		Asia
Bovidae		Europe

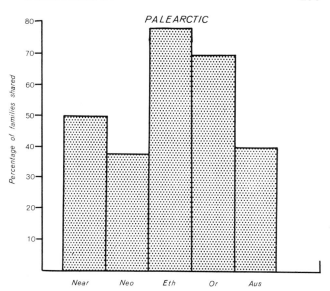

FIGURE 16–2. The percentages of the families that occur in the Palearctic region that are shared by other zoogeographical regions.

Palearctic (Spalacidae and Selviniidae).

Nearctic Region. This area includes nearly all of the New World north of the tropical sections of Mexico (Fig. 16–1). Habitats ranging from semitropical thorn forest to arctic tundra occur within the area. The mammalian fauna is correspondingly diverse, and includes some families that are mostly tropical in distribution (for example, Emballonuridae, Desmodontidae, and Tayassuidae) together with some primarily boreal families (Zapodidae, Castoridae, and Ursidae). Only two Nearctic Families, Aplodontidae and Antilocapridae, are endemic (Table 16–2). (An animal is endemic to an area if it lives nowhere else.) The mammalian fauna of the Nearctic resembles most closely that of the Neotropical (Fig. 16–3).

Neotropical Region. This region features great climatic and biotic diversity and includes all of the New World from tropical Mexico south. Much of the area is tropical or subtropical, and broad areas are covered with spectacular tropical rain forest. Tropical savanna and grasslands occupy much of the southern half of South America, and there are deserts in the south and along the west coast.

The higher parts of the Andes support montane forests and alpine tundra. The South American part of the Neotropics has been isolated from the rest of the world through most of the Cenozoic, but the Isthmus of Panama has provided a connection between South America and North America since the late Pliocene.

This region is second only to the Ethiopian Region in diversity of mammals. The Neotropical supports 46 families of mammals, and has the largest number of endemic families (20; see Table 16–3). Especially characteristic of the Neotropical are marsupials, bats (including three endemic families), primates (two endemic families), edentates (two endemic families), and histricomorph rodents (11 endemic or nearly endemic families). Two species of the genus *Lama* live in South America and are the only New World representatives of the family Camelidae. Wild Old World camelids occur only in the Gobi Desert of Mongolia. Tapirs are restricted to the Neotropical and Oriental Regions. The Neotropical mammalian fauna most strongly resembles that of the Nearctic, but it also shares over 33 per cent of its families with the Palearctic (Fig. 16–4).

Ethiopian Region. This region

Table 16–2.   A List of Nearctic Families of Mammals and Their Geographic Origins.
Two Families are Endemic.

FAMILIES OF MAMMALS	ENDEMIC FAMILIES	GEOGRAPHIC ORIGIN OF FAMILIES
Didelphidae		North America
Talpidae		Europe
Soricidae		Europe
Emballonuridae		Europe
Phyllostomatidae		?
Mormoopidae		?
Desmodontidae		South America (?)
Natalidae		South America (?)
Vespertilionidae		Europe or North America
Molossidae		Europe
Hominidae		Southern Asia
Dasypodidae		South America
Ochotonidae		Eurasia
Leporidae		North America
Aplodontidae	X	North America
Sciuridae		Europe or North America
Geomyidae		North America
Heteromyidae		North America
Castoridae		North America
Cricetidae		North America, Europe, or Asia
Zapodidae		Europe
Erethizontidae		South America
Canidae		North America or Europe
Ursidae		Europe
Procyonidae		North America
Mustelidae		North America, Europe, or Asia
Felidae		North America or Europe
Tayassuidae		Europe
Cervidae		Asia
Antilocapridae	X	North America
Bovidae		Europe

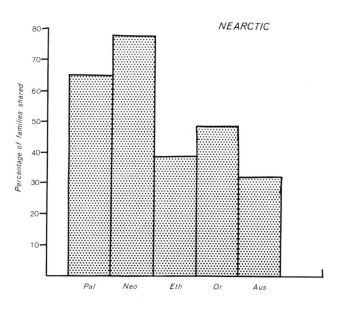

FIGURE 16–3. The percentages of the families that occur in the Nearctic region that are shared by other zoogeographical regions.

Table 16–3. A List of Neotropical Families of Mammals and Their Geographic Origins. Twenty Families are Endemic.

FAMILIES OF MAMMALS	ENDEMIC FAMILIES	GEOGRAPHIC ORIGIN OF FAMILIES
Didelphidae		North America
Caenolestidae	X	South America
Solenodontidae	X	?
Soricidae		Europe
Emballonuridae		Europe
Noctilionidae	X	?
Phyllostomatidae		?
Mormoopidae		?
Desmodontidae		South America (?)
Natalidae		South America (?)
Furipteridae	X	?
Thyropteridae	X	?
Vespertilionidae		Europe or North America
Molossidae		Europe
Cebidae	X	South America
Callithricidae	X	?
Hominidae		Southern Asia
Myrmecophagidae	X	South America
Bradypodidae	X	?
Dasypodidae		South America
Leporidae		North America
Sciuridae		Europe or North America
Geomyidae		North America
Heteromyidae		North America
Cricetidae		North America, Europe, or Asia
Erethizontidae		South America
Caviidae	X	South America
Hydrochoeridae	X	South America
Dinomyidae	X	South America
Heptaxodontidae	X	South America (?)
Dasyproctidae	X	South America
Chinchillidae	X	South America
Capromyidae	X	West Indies (?)
Myocastoridae	X	South America
Octodontidae	X	South America
Abrocomidae	X	South America
Echimyidae	X	South America
Canidae		North America or Europe
Ursidae		Europe
Procyonidae		North America
Mustelidae		North America, Europe, or Asia
Felidae		North America or Europe
Tapiridae		Europe
Tayassuidae		Europe
Camelidae		North America
Cervidae		Asia

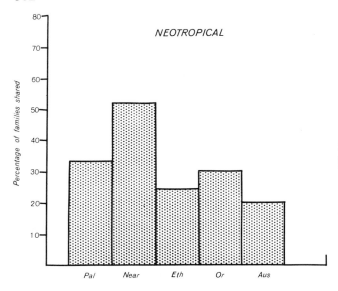

FIGURE 16–4. The percentages of the families that occur in the Neotropical region that are shared by other zoogeographical regions.

takes in Madagascar and Africa north to the Atlas Mountains, the Sahara Desert, and southern Arabia. The area is joined to the Palearctic by a land bridge (now broken by the Suez Canal) in northeastern Egypt. Deserts, tropical savannas, tropical forests, montane forests, and alpine tundra are all represented, and the most extensive tropical savannas in the world occur in Africa.

Next to the Neotropics, the Ethiopian region has the greatest number of endemic families of mammals (Table 16–4). The impressive array of ungulates that inhabits the savannas of Africa is unmatched elsewhere, and Africa is the last stronghold of the families Equidae, Rhinocerotidae, Elephantidae, and Hippopotamidae. Although the only endemic artiodactylan family is the Giraffidae, nearly all of the African genera of antelope (Bovidae) are endemic. The primitive lemuroid primates of Madagascar and the diverse group of cercopithecid primates of Africa are especially typical of the region, and two of the four genera of great apes live in Africa. Apart from South America, Africa is the only area with a fairly diverse histricomorph rodent fauna. Viverrid carnivores reach their greatest diversity in the Ethiopian Region, where about 23 of the 25

genera are endemic. Over 60 per cent of the Ethiopian families of mammals also occur in the Oriental Region (Fig. 16–5).

Oriental Region. Included in this region are India, Indochina, southern China, Malaya, the Philippine Islands, and the islands of Indonesia east to a line (imaginary and controversial) between Borneo and Celebes and between Java and Lombok (Fig. 16–1). The area is dominated by tropical climates and once, before extensive clearing of lands by man, supported almost continuous tropical forests. Deserts occur in the Pakistan area. The Oriental Region is partly isolated from the Palearctic by deserts in the west and by the Himalaya Mountains to the north and northeast.

The mammalian fauna of the Oriental Region resembles most strongly that of the Ethiopian area, with which it shares 78% of its families of mammals (Fig. 16–6). Many (70%) of the Oriental families of mammals also occur in the Palearctic Region. The most distinctive elements of the Oriental mammalian fauna are all of tropical affinities (Table 16–5). Four families of primates occur in this region. Four families of mammals—Tupaiidae (tree shrews), Cynocephalidae (flying

Table 16–4. A List of Ethiopian Families of Mammals and Their Geographic Origins. Fourteen Families are Endemic.

FAMILIES OF MAMMALS	ENDEMIC FAMILIES	GEOGRAPHIC ORIGIN OF FAMILIES
Erinaceidae		North America
Tenrecidae	X	Africa
Chrysochloridae	X	Africa
Soricidae		Europe
Macroscelididae		Africa
Pteropodidae		Europe
Rhinopomatidae		?
Emballonuridae		Europe
Nycteridae		?
Megadermatidae		Europe
Rhinolophidae		Europe
Myzopodidae	X	?
Vespertilionidae		Europe or North America
Molossidae		Europe
Lemuridae	X	Madagascar
Indridae	X	Madagascar
Daubentoniidae	X	Madagascar
Lorisidae		Asia
Cercopithecidae		Africa
Pongidae		Africa
Hominidae		Southern Asia
Manidae		Europe (?)
Leporidae		North America
Sciuridae		Europe or North America
Anomaluridae	X	?
Pedetidae	X	Africa
Cricetidae		North America, Europe, or Asia
Rhizomyidae		Europe
Muridae		Europe or Asia
Gliridae		Europe
Dipodidae		Asia
Hystricidae		Europe
Thryonomyidae	X	Africa
Petromyidae	X	Africa
Bathyergidae	X	Africa
Ctenodactylidae	X	Europe or Asia
Canidae		North America or Europe
Mustelidae		North America, Europe, or Asia
Viverridae		Europe
Hyaenidae		Asia
Felidae		North America or Europe
Orycteropidae	X	Europe or Asia
Elephantidae		Asia
Procaviidae		Africa
Equidae		North America
Rhinocerotidae		Europe
Suidae		Europe
Hippopotamidae		Asia
Tragulidae		Europe
Giraffidae	X	Asia
Bovidae		Europe

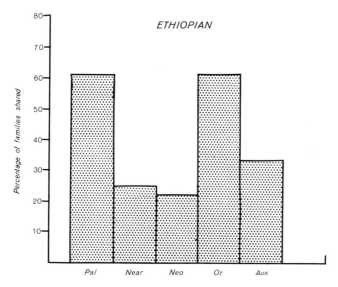

FIGURE 16–5. The percentages of the families that occur in the Ethiopian region that are shared by other zoogeographical regions.

lemurs), Tarsiidae (tarsiers), and Platacanthomyidae (spiny dormice)—are endemic, and each occupies forested tropical areas. Some 15% of the Oriental families occur elsewhere only in the Ethiopian Region (Lorisidae, Pongidae, Manidae, Elephantidae, Rhinocerotidae, and Tragulidae). The presence in both areas of rhinos and elephants, great apes and lorises, and a diversity of viverrid carnivores makes the mammalian faunas of the Oriental and the Ethiopian regions seem much alike, but between these areas there are some striking faunal differences. Lacking in the Oriental Region are lemuroid primates, the distinctive African histricomorph rodents, and the diverse assemblage of antelope so typical of African savannas.

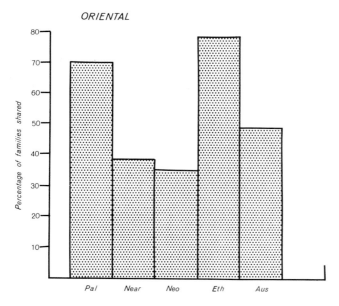

FIGURE 16–6. The percentages of the families that occur in the Oriental region that are shared by other zoogeographical regions.

Table 16–5. A List of Oriental Families of Mammals and Their Geographic Origins. Four Families are Endemic.

FAMILIES OF MAMMALS	ENDEMIC FAMILIES	GEOGRAPHIC ORIGIN OF FAMILIES
Erinaceidae		North America
Talpidae		Europe
Soricidae		Europe
Tupaiidae	X	Southeast Asia
Cynocephalidae	X	?
Pteropodidae		Europe
Rhinopomatidae		?
Emballonuridae		Europe
Nycteridae		?
Megadermatidae		Europe
Rhinolophidae		Europe
Vespertilionidae		Europe or North America
Molossidae		Europe
Lorisidae		Asia
Tarsiidae	X	Europe (?)
Cercopithecidae		Africa
Pongidae		Africa
Hominidae		Southern Asia
Manidae		Europe (?)
Leporidae		North America
Sciuridae		Europe or North America
Cricetidae		North America, Europe, or Asia
Rhizomyidae		Europe
Muridae		Europe or Asia
Platacanthomyidae	X	?
Hystricidae		Europe
Canidae		North America or Europe
Ursidae		Europe
Mustelidae		North America, Europe, or Asia
Viverridae		Europe
Hyaenidae		Asia
Felidae		North America or Europe
Elephantidae		Asia
Equidae		North America
Tapiridae		Europe
Rhinocerotidae		Europe
Suidae		Europe
Tragulidae		Europe
Cervidae		Asia
Bovidae		Europe

Australian Region. This region includes Australia, Tasmania, New Guinea, Celebes, and many of the small islands of Indonesia (New Zealand and the Pacific area are not included). Within the area are islands of varying sizes and degrees of isolation. The island continent of Australia comes closer to New Guinea than to any other large island, but these land masses are separated by the Torres Strait, some 100 miles wide. The northern part of the area, including New Guinea and parts of the east coast of Australia, are covered with tropical forest, but much of Australia is tropical savanna or desert. Some of the most arid deserts in the world occur in the interior of Australia.

The Australian region is famous for its unusual mammalian fauna (see Table 16–6), and to the popular mind Australia itself is an area supporting almost exclusively marsupials. Actually, even Australia has nearly as many placental families (9) as marsupial families (10). Roughly 32% of the families of the Australian Region are mar-

Table 16–6. A List of Australian Families of Mammals and Their Geographic Origins.
Twelve Families are Endemic.

FAMILIES OF MAMMALS	ENDEMIC FAMILIES	GEOGRAPHIC ORIGIN OF FAMILIES
Tachyglossidae	X	Australia
Ornithorhynchidae	X	Australia
Dasyuridae	X	Australia
Notoryctidae	X	?
Peramelidae	X	Australia
Phalangeridae	X	Australia
Petauridae	X	Australia
Burramyidae	X	Australia
Tarsipedidae	X	Australia
Phascolarctidae	X	Australia
Vombatidae	X	Australia
Macropodidae	X	Australia
Soricidae		Europe
Pteropodidae		Europe
Emballonuridae		Europe
Nycteridae		?
Megadermatidae		Europe
Rhinolophidae		Europe
Vespertilionidae		Europe or North America
Molossidae		Europe
Tarsiidae		Europe (?)
Cercopithecidae		Africa
Hominidae		Southern Asia
Sciuridae		Europe or North America
Muridae		Europe or Asia
Hystricidae		Europe
Canidae		North America or Europe
Viverridae		Europe
Suidae		Europe
Cervidae		Asia
Bovidae		Europe

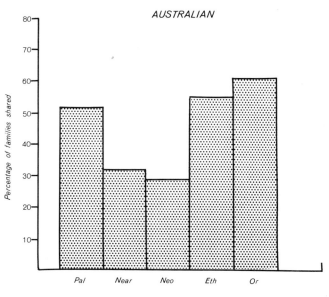

FIGURE 16–7. The percentages of the families that occur in the Australian region that are shared by other zoogeographical regions.

supials, and about 42% (the mono-tremes and marsupials) are endemic. Some 61% of Australia's families of mammals also occur in the Oriental Region (Fig. 16–7), the region from which probably nearly the entire Australian fauna was ultimately derived (the historical biogeography of Australia is discussed on page 314).

Oceanic Region. The oceans of the world comprise the Oceanic Region. Within this region live the whales and porpoises, most of the seals, the sea lions and walruses, and the inhabitants of isolated oceanic islands (usually bats and introduced murid rodents, and other mammals associated with man). The large islands are included within the regions with which their faunas have the most in common. Greenland and Iceland, for instance, are included in the Nearctic and Palearctic Regions, respectively.

HISTORICAL ZOOGEOGRAPHY OF MAMMALS

Mammalian Evolution and Faunal Succession. Since the beginning of their spectacular adaptive radiation in the late Cretaceous and Paleocene, mammals have followed an evolutionary pattern typical of most plant and animal groups. In general, mammals have become progressively more diverse and better able to completely and efficiently exploit the niches available to animals with the basic mammalian structural plan. Along with this trend toward full occupation of the environment has gone a tendency toward increasing specialization. Whereas many Paleocene mammals were generalists and could probably rather inefficiently utilize a wide variety of foods, relatively few living mammals have this mode of life. A modern herbivore, for example, typically eats only a certain type of plant material, but its teeth and digestive system are adapted to efficient utilization of the food. Behavioral adaptations favoring selective foraging have also developed. The "average" modern mammal, compared to his Paleocene or Eocene counterpart, is able to find its food with less expenditure of energy, and to derive more energy from the breakdown of the food.

As mentioned in Chapter 15, biotic communities evolve just as do the interacting organisms that comprise them. Whenever two or more organisms have attempted to play the same role in a community, to occupy the same ecological niche, the unstable situation that developed resulted in faunal change. One organism became master of the niche, and the others either moved away or became extinct; natural selection favored those structural or behavioral modifications that allowed an animal to be the stronger competitor, to most efficiently occupy its niche. Thus, a succession of mammalian faunas have occupied the earth, each better able to efficiently exploit the environmental possibilities of the times than were former faunas (Table 16–7). A number of factors, such as

Table 16–7. Ecological Replacement, as Shown by the Pleistocene Fossil Record, of Older Genera by Younger Genera. (After Hibbard, C. W., et al., in Wright, H. E., and Frey, D. G.: *The Quaternary of the United States*, Princeton University Press, 1965.)

OLDER GENERA	YOUNGER GENERA
Nannippus (3-toed horse) — replaced by	*Equus* (modern horse)
Stegomastodon (mastodon) — replaced by	*Mammuthus,* mammoth (elephant)
Capromeryx (pronghorn) — replaced by	*Antilocapra* (pronghorn)
Hypolagus — replaced by	{ *Lepus* (hare)
	{ *Sylvilagus* (rabbit)
Pliophenacomys — replaced by	*Microtus* (meadow vole)
Arctodus — replaced by	*Ursus* (brown and grizzly bear)

geological changes and changes in weather patterns and vegetation, have importantly "guided" the evolution of mammalian faunas. In addition, a factor of major importance has been the timing of migrations.

Faunal Stratification and Faunal Origins. All large land masses support stratified faunas: not all of the animals have occupied the areas for the same length of time, nor have they all come from the same place. As an example, the mammalian fauna of Africa consists of animals representing families or genera that evolved elsewhere and dispersed to Africa, together with representatives of groups that apparently evolved on the African continent and have occupied the area throughout much of the Cenozoic. Horses had their origin in North America in the Paleocene (Jepsen, 1969) and probably did not reach Africa until the Pleistocene. Today horses (zebras) still form a conspicuous part of the African scene, but are completely absent from the New World. Old World monkeys (Cercopithecidae), on the other hand, probably originated in

North Africa, and have seemingly occupied Africa continuously since the Oligocene. The Old World monkeys clearly represent part of an early African faunal stratum, whereas horses are part of a late one.

Some of the complexities of historical biogeography are well illustrated by considering the histories of the mammalian faunas of several regions.

History of the Nearctic Mammalian Fauna. The mammalian fauna of the Nearctic changed drastically in the early Tertiary, largely due to the evolution of new groups and their dispersal, and to the extinction of old groups; by the Oligocene the most important of today's orders were well established (Fig. 16–8). The aspect of the Nearctic mammalian fauna was strongly influenced by periodic faunal interchange between this region and the Palearctic. The presence of a Bering land bridge is indicated by the fossil record rather than by direct geological evidence. When the fossil record indicates that a genus of mammal occurred in both Asia and North America during a given period, one can assume

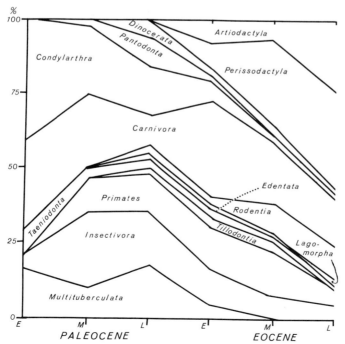

FIGURE 16–8. The changing composition of the mammalian fauna in North America in the early Tertiary. Note that the modern mammalian orders are dominant by the end of the Eocene. (Based upon Figure 24 from *The Geography of Evolution* by George Gaylord Simpson. Copyright © 1965 by the author. By permission of the publisher, Chilton Book Company, Philadelphia.)

FIGURE 16–9. The intensity of interchange of land mammals between Eurasia and North America in most of the
Cenozoic as indicated by the numbers of
closely related mammals on the two continents. (Based upon Figure 28 from *The
Geography of Evolution* by George
Gaylord Simpson. Copyright © 1965 by
the author. By permission of the publisher, Chilton Book Company, Philadelphia.)

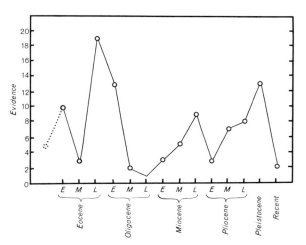

that at that time, or for some period
before it, intercontinental migrations
occurred across the Bering land
bridge.

The intensity of faunal interchange
between the Nearctic and Palearctic
fluctuated during the Cenozoic, with
peaks of migration in the early Eocene,
early Oligocene, late Miocene, and
Pleistocene (Fig. 16–9). No faunal min-
gling seems to have occurred in the
Recent, doubtless because migration
between Siberia and Alaska has been
barred since the Pleistocene by the
Bering Strait. Even during the periods
of most active faunal interchange,
however, limited segments of the
faunas were involved. Degrees of resemblance between faunas of North
America and those of various parts of
Eurasia indicate that mammals that
migrated to North America were from
northeastern Asia, the area closest to
the supposed Bering land bridge. This
filter bridge did not allow passage of
mammals poorly adapted to cool or
cold climates (Table 16–8). There are
many examples of families now established in North America, and still
present in Europe, that had their origins in Europe or Asia and reached
North America *via* the Bering land
bridge. The Talpidae and Soricidae originated in Europe in the Eocene and reached North America in
the Oligocene; the Zapodidae ap-

peared in the Oligocene in Europe,
but did not reach North America until
the Miocene; the Cervidae arose in
Asia in the Oligocene and arrived in
North America in the Miocene. The
rodent families Heteromyidae and
Geomyidae probably evolved in response to the arid and semi-arid conditions that developed in the southern
part of the Nearctic in the middle of
the Tertiary. As might be expected,
these rodents have never reached the
Palearctic. South America was separated by seas from North America (and
all other continents) from about the
early Paleocene until the late Pliocene, when the connection with North
America across the Isthmus of Panama
was reestablished. Invasion of South
America by many North American
mammals in the late Tertiary had striking effects on the fauna of South
America (these will be described
shortly). Aside from some bats,
members of only two families that
originated in South America were able
to become established in the Nearctic.
A porcupine (Erethizontidae) and an
armadillo (Dasypodidae) have become
widely established in the Nearctic,
and during the Pleistocene large relatives of the armadillo, ground sloths
and glyptodonts, were far more widespread in North America than the armadillo is now.

History of the Neotropical Mamma-

Table 16–8. Earliest Fossil Records in North America of Eurasian Immigrants That Crossed the Bering Land Bridge. Note That all of the Mammals are Boreal Types. (After Hibbard, C. W. et al., in Wright, H. E., and Frey, D. G.: *The Quaternary of the United States*, Princeton University Press, 1965.)

GENERA	EARLIEST RECORDS
Ursus, bear	middle Pleistocene (late Blancan)
Mammuthus, mammoth elephant	" " (late Kansan)
Bison, bison	" " (Illinoian)
Smilodon, saber-toothed cat	" " (Irvingtonian)
Gulo, wolverine	" " "
Cervlaces, extinct moose	upper Pleistocene (Rancholabrean)
Rangifer, caribou	" " (Wisconsin)
Oreamnos, mountain goat	" " "
Ovibos, musk-ox	" " "
Ovis, sheep	" " "
Alces, moose	" " "
Bos, yak	" " "
Saiga, asiatic antelope	" " "
Bootherium, extinct bovid	" " "
Symbos, woodland musk-ox	" " "

lian Fauna. Three groups of mammals probably occupied South America before its Paleocene to Pliocene isolation (Table 16–9). In the late Cretaceous and Paleocene, primitive and generalized didelphid marsupials occurred in North America. This marsupial stock probably reached South America before this area's isolation. A group of placentals had also reached South America by this time; these were from an omnivorous placental group that Simpson (1945:105, 216) called ferungulates. These generalized placentals were not committed to a particular evolutionary pathway, but were more or less intermediate between carnivores and condylarths (primitive ungulates). The palaeanodonts, a group of North American mammals that were probably ancestral to the edentates, formed a third group of what Simpson (1965:171, 172) called "ancient immigrants."

All three of these groups underwent impressive adaptive radiations during their Tertiary isolation. The marsupials radiated in carnivorous and insectivorous directions, and some retained omnivorous habits. The marsupial family Borhyaenidae was particularly remarkable, for within this group developed a variety of carnivorous types that were strongly convergent toward placental carnivores. Some borhyaenids reached the size of a bear, and one developed sabers much like those of the saber-tooth cat (Fig. 5–8B; p. 47). Another group that descended from didelphid ancestry, the Caenolestidae, were insect eaters. Populations of this family still persist in South America. Finally, from the ferungulate placental stock, a remarkable series of highly specialized ungulates evolved. Of greatest importance in terms of diversity were the Notoungulata. This order contained various herbivorous species, one of the largest of which was *Toxodon*, a short-legged and rhinoceros-like beast some nine feet in length. Another order, Litopterna, included a number of highly cursorial ungulates. One advanced Miocene genus (*Thoatherium*) had one-toed feet that were not only much more specialized than those of contemporary North American horses, but were even more specialized than the feet of present-day horses (Fig. 16–10).

Another contingent of mammals, called "old island hoppers" by Simpson (1965:171, 172), entered South America from North America at various times in the Tertiary. These mammals

Table 16–9.   Mammalian Faunal Stratification in South America.
(After Simpson, G. G.: *The Geography of Evolution,* Chilton Books, Publishers.
Copyright © 1965 by George Gaylord Simpson.)

TIME OF EMPLACEMENT	FAUNAL STRATUM	GROUPS INTRODUCED	DIFFERENTIATION IN SOUTH AMERICA
Pliocene to Recent	III Late (Island-hoppers and) Immigrants	Deer Camels Peccaries Tapirs Horses Mastodons Cats Weasels Raccoons Bears Dogs Mice Squirrels Rabbits Shrews	Already differentiated before emplacement in South America. Local differentiation of many genera, species, subspecies, no families or higher groups.
Late Eocene to Oligocene	II Old Island-hoppers	Primitive rodents Advanced lemuroids	Caviomorph rodents (many groups) New World monkeys
Around earliest Paleocene	I Ancient Immigrants	Ferungulates (Condylarth-like complex)	Litopterns Notoungulates (many groups) Astrapotheres Pyrotheres
		Palaeanodonts	Xenarthrans Ground sloths Tree sloths Anteaters Armadillos Glyptodonts
		Didelphoids	(Didelphoids continued) Borhyaenoids Caenolestoids

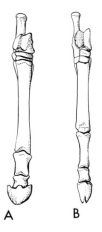

A   B

FIGURE 16-10. A, the hind foot of a modern horse (*Equus*) and B, that of a Miocene South American litoptern (*Thoatherium*). Note that the vestiges of digits two and four are more strongly reduced in the litoptern than in the horse.

perhaps crossed the seaway separating the continents by being rafted between islands. Lemur-like primates, which gave rise to the distinctive New World monkeys and marmosets, probably arrived in the late Eocene or Oligocene; rodents ancestral to the South American histricomorphs seemingly arrived in the late Eocene; and procyonids apparently reached South America in the Miocene. Each of these groups was successful in South America, but the rodents had the most spectacular radiation. About a dozen families, occupying diverse habitats and pursuing a variety of modes of life, evolved from the original Eocene rodent stock (or stocks).

From the late Pliocene through the Recent, faunal interchange between North and South America has occurred either by island hopping (Simpson called animals that followed this route "late island hoppers")—perhaps this occurred as the Isthmus of Panama was being elevated—or by migrations (of Simpson's "late immigrants") across a land bridge. Figure 16-11 indicates some of the interchange of mammals that took place between North and South America. The late Cenozoic history of the mammalian fauna of South America is largely a story of the outcome of interactions between endemic South American groups and the invading Nearctic mammals.

When northern placentals entered

South America there was apparently considerable duplication of functional roles. Intense competition developed between northern placental carnivores and South American marsupial carnivores, and between North American and South American ungulates. Perhaps because of great competitive ability developed by species in the extensive North American area where there was repeated Tertiary faunal interchange with the Palearctic, the northern invaders in many cases supplanted the endemic South American types. The northern placental carnivores were particularly devastating competitors, and forced the South American borhyaenids into extinction by the Pleistocene. Likewise, the North American ungulates completely replaced the distinctive South American notoungulates and their relatives. Perhaps the combined effects of predation by such awesome types as saber-tooth cats and competition from North American ungulates tipped the delicate balance of survival against the South American ungulates. These remarkable mammals disappeared in the Pleistocene.

Some South American groups, however, withstood the onslaught from the north and were even able to move against its flow. Notably successful were the edentates, which migrated northward and occupied much of the Nearctic in the Pleis-

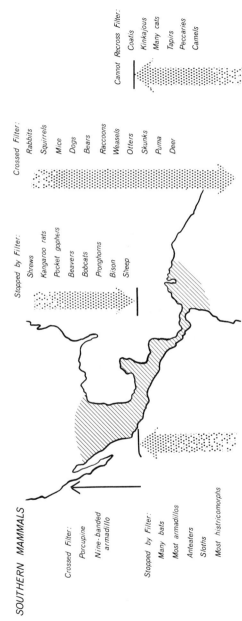

NORTHERN MAMMALS

SOUTHERN MAMMALS

Crossed Filter:

Rabbits
Squirrels
Mice
Dogs
Bears
Raccoons
Weasels
Otters
Skunks
Puma
Deer

Stopped by Filter:

Shrews
Kangaroo rats
Pocket gophers
Beavers
Bobcats
Pronghorns
Bison
Sheep

Cannot Recross Filter:

Coatis
Kinkajous
Many cats
Tapirs
Peccaries
Camels

Crossed Filter:

Porcupine
Nine-banded
armadillo

Stopped by Filter:

Many bats
Most armadillos
Anteaters
Sloths
Most histricomorphs

FIGURE 16–11. A diagrammatic illustration of the effectiveness of Middle America as a filter barrier. The crosshatched area is the filter zone, and the success or failure of animals in crossing the barrier is shown. (Based upon Figure 45 from *The Geography of Evolution* by George Gaylord Simpson. Copyright © 1965 by the author. By permission of the publisher, Chilton Book Company, Philadelphia.)

tocene; some of the large ground sloths did not become extinct in this area until a few thousand years ago. The South American anteaters, tree sloths, and armadillos are still prominent members of the South American fauna. The didelphid marsupials (which actually were "ancient immigrants" from the north) are also still important members of the South American mammalian fauna, and have moved northward to some extent since the Pliocene. Similarly, cebid monkeys, marmosets, and histricomorph rodents are abundant in South America and have moved northward into Central America and Mexico roughly to the limits of tropical rainforest. The rich South American bat fauna includes several families that probably originated in South America. Some of these have expanded their ranges into the southern Nearctic.

The present mammalian fauna of South America, then, consists of an early Tertiary element, an intermediate group that entered the area at various times in the early and middle Tertiary, and a later element that invaded after reestablishment of the Panamanian land bridge (Table 16–9). The camels and tapirs are unique among the later element. These animals have persisted in the Neotropical region, but have become extinct in the Nearctic, whence they originally came. The Isthmus of Panama has obviously been a filter route allowing the passage of some groups and stopping others (Fig. 16–11); this factor has strongly influenced the composition of the South American mammalian fauna.

History of the Australian Mammals. The native mammalian fauna of the Australian region, and especially of Australia itself (Table 16–10), is famous for its uniqueness. Many of the functional roles played elsewhere by rodents or carnivores are pursued in Australia by marsupials that are restricted either to Australia itself or to the Australian zoogeographical region. Simpson (1961) has discussed in detail the history of the Australian mammalian fauna; the following remarks are based largely on his work.

Australia has been isolated by seas from the rest of the world probably since at least the Cretaceous, and apparently the first mammalian stock to reach Australia consisted of small, arboreal, didelphid-like marsupials that reached the area by island-hopping and rafting from the New Guinean (and ultimately Asian) area. That marsupials rather than placentals gained the first foothold in Australia was perhaps strictly a matter of chance, a result of the vagaries of the New Guinean-Australian sweepstakes route of dispersal. From the original marsupial immigrants — conceivably a single pair or a pregnant female — have evolved the diverse Australian marsupial fauna of today, which consists of 10 families and some 68 genera. Monotremes, the other "original" group of Australian mammals, may have evolved on the

Table 16–10. Native Recent Mammals of Australia. (Partly after Simpson, G. G.: *The Geography of Evolution,* Chilton Books, Publishers. Copyright © 1965 by George Gaylord Simpson.)

	FAMILIES		GENERA	
	No.	*Endemic*	*No.*	*Endemic*
Monotremata	2	100%	2	100%
Marsupialia	10	100%	68	100%
Chiroptera	7	0%	21	29%
Rodentia	1	0%	13	77%
Carnivora	1	0%	1	0%
Totals	21	57%	105	82%

island from advanced therapsid reptilian ancestry.

Nine of the 21 families of Australian mammals are placentals (Table 16–10). These groups arrived in Australia at various times, and all can easily be assigned to families that occur in other areas. Some have undergone little change since their arrival in Australia.

Bats are a group that have remained nearly unchanged. Apparently bats entered Australia at various times in the Tertiary. Of the 21 Australian genera of bats, only two are restricted to Australia. Interchanges of bats between New Guinea and Australia, which were perhaps frequent because of bats' ability to fly, have kept the Australian bats from differentiating markedly from those of the Oriental Region.

Rodents are abundant and diverse (about 13 genera) in Australia, and some species have undergone great specialization, but all of these rodents are clearly assignable to the widespread family Muridae. Simpson has divided the Australian murids into four groups, according to their history on this island continent. (1) The *"Rattus* group" consists in part of species commensal with man that were introduced by European man, but also includes species endemic to Australia that perhaps developed from pre-Pleistocene immigrants. (2) The "old Papuan" group contains genera that evolved in Australia from murid ancestors that arrived probably no later than the Pliocene. (3) The "old Australians" are a fairly diverse group of, in some cases, highly specialized rodents (remarkable adaptations to dry climates by two members of this group are discussed on p. 391). Their ancestors were probably the first rodents to reach Australia, and must have arrived in the Miocene. (4) The "hydromyine" group (of the murid subfamily Hydromyinae) consists of two semi-aquatic genera. One apparently evolved in Australia from ancestors that came from New Guinea, and the other came recently from New Guinea.

The family Canidae is represented in Australia by the feral dingo (*Canis dingo*), which was probably brought to Australia by aborigines.

Perhaps the most remarkable feature of the Australian mammalian fauna is the presence of a marsupial assemblage that is fairly balanced in the sense that it fills most of the terrestrial niches. Kangaroos and wallabies take the place of ungulates, dasyurids take the place of shrews and to some extent of rodents, phalangerids take the place of squirrels, and so on. Placental mammals that reached Australia without man's help have in large part either filled adaptive zones that marsupials could not occupy, as in the case of the bats, or have occupied niches that they could perhaps fill more effectively than could marsupials, as in the case of the murid rodents. Such recently man-introduced placentals as dingos, and more recently rabbits and foxes, have had the unfortunate effect of displacing native marsupials or preying heavily on them.

The Unusual Mammalian Fauna of Madagascar. Islands long isolated from continents frequently have unusual mammalian faunas. These faunas may be dominated by a group equally important nowhere else, as in the case of the marsupials of Australia, or they may be extremely poor in mammals, as in the case of New Zealand, where the only native mammals are bats. Madagascar is an interesting example of an area supporting a mammalian fauna with little ordinal diversity, many endemics, and a seemingly incomplete exploitation of habitats.

Madagascar is a large island some 995 miles in length and with a maximum width of 350 miles. It lies 260 miles east of the east coast of Africa, and has probably been isolated from large land masses since the Mesozoic. The island has supported six orders of mammals in Recent times (Table 16–11). Most of the mammals are endemic, and the most highly diversified groups, the lemuroid primates and the tenrecid insectivores, probably reached Madagascar from Africa by island hopping and rafting early in the Tertiary. Many of the Malagasy mammals occupy

Table 16–11. A Comparison of the Mammalian Fauna of the Panama Canal Zone and Madagascar. The Artiodactyl from Madagascar, a Hippopotamus, Is Now Extinct. (After Eisenberg, J. F., and Gould, E.: *The Tenrecs: A Study in Mammalian Behavior and Evolution,* Smithsonian Institution Press, 1970.)

	MARSUPIALIA	INSECTIVORA	CHIROPTERA	PRIMATES	EDENTATA	LAGOMORPHA	RODENTIA	CARNIVORA	PERISSODACTYLA	ARTIODACTYLA
Panama Canal Zone	6	0	40	5	7	1	19	11	1	3
Madagascar	0	10	12	10	0	0	8	6	0	1

niches filled elsewhere by mammals of different taxa (Table 16–12). This phenomenon is called *complementarity* by Darlington (1957:23). The only artiodactyl that was present before man's arrival was the now-extinct hippopotamus (*Hippopotamus lemelii*), and today the introduced river hog (*Potomochoerus*) is the only wild artiodactyl. The ungulate niche has largely gone unfilled, although a group of large Pleistocene lemurs, now extinct, may have been terrestrial herbivores. These lemurids had large skulls that strongly resemble those of some ungulates (Fig. 16–12).

Climate and Mammalian Distribution. Climate has had in the past, and still has, a pronounced effect on animal distribution. The present patterns of distribution of many mammals can be explained in terms of climatic changes in the Pleistocene and Recent.

The Pleistocene was a time of pronounced climatic shifts, and periods of lowered temperatures alternated with periods of relative warmth. Temperatures during the cool periods may have been 4° to 8° C below present temperatures, and temperatures in the intervening warm periods were probably higher than those now. Geological and paleontological evidence both indicate four major episodes of cool climates separated by three warm intervals. Accompanying the periods of cooling, which were apparently worldwide, were a number of spectacular environmental changes. Precipitation increased everywhere, and with increased snowfall continental glaciers developed and pushed southward. At one time in the Pleistocene over 25% of the land surface was covered with glaciers: Eurasia had 3,200,000 square miles of ice; the Nearctic ice sheet covered 4,500,000 square miles, and during its greatest push southward reached what is now Kansas. Glaciers on Mount Kenya in Africa extended about 1700 m. below the present vestigial snow fields (at some 4500 m.), and New Guinea and Madagascar had montane glaciers. The distributions of floras were changed. Boreal vegetational zones were pushed downward on mountains, and coniferous forests spread southward over areas that previously supported less boreal floras. Concurrently, tropical floras receded toward the equator, and deserts became far more restricted than they are today.

The Pleistocene began 2 to 3 million years ago (Evernden et al., 1964) and ended about 10,000 years ago with the extinction in the Nearctic and Palearctic of such common Pleistocene mammals as elephants, camels, woodland musk-ox, ground sloths, horses, and the giant beaver. Following the retreat of the last continental glacier at the close of the Pleistocene was a warm, moist climatic phase (perhaps 8,000 to 6,000 years before the present), followed by a warm, dry phase (about

Table 16–12. Some Major Feeding Niches and the Mammals that Occupy them in Panama and Madagascar. (After Eisenberg, J. F., and Gould, E.: *The Tenrecs: A Study in Mammalian Behavior and Evolution,* Smithsonian Institution Press, 1970.)

	ANT EATERS		PRIMARY INSECTIVORE AND SECONDARY FRUGIVORE		CARNIVORE	
	Arboreal	*Terrestrial*	*Arboreal*	*Terrestrial*	*Arboreal*	*Terrestrial*
PANAMA (Major genera only)	Edentata: Myrmecophagidae *Cyclopes* *Tamandua*	Edentata: Myrmecophagidae *Myrmecophaga*	Marsupialia: Didelphidae *Marmosa* Primates: Callithricidae *Saguinus* Cebidae *Aotes*	Edentata: Dasypodidae *Cabassous* *Dasypus*	Carnivora: Mustelidae *Eira* Felidae *Felis*	Carnivora: Mustelidae *Mustela* *Galictus* Canidae *Urocyon*
MADAGASCAR			Insectivora: Tenrecidae *Echinops* Primates: Lemuridae *Microcebus* *Cheirogaleus* Daubentoniidae *Daubentonia*	Insectivora: Tenrecidae *Centetes* *Hemicentetes* *Oryzorictes* *Geogale*	Carnivora: Viverridae *Cryptoprocta* *Galidia*	*Fossa* *Viverricula*

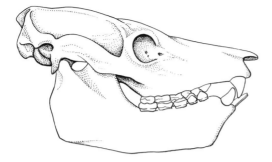

6,000 to 4,000 years before the present), when temperatures were higher than those today and precipitation was lower.

Considerable fossil evidence documents the southward extensions of the ranges of boreal mammals during glacial advances. Remains of the musk-ox (*Ovibos*), arctic shrew (*Sorex arcticus*), and collared lemming (*Dicrostonyx*) have been found well south of their present northern ranges (Fig. 16–13). Two voles that today have separate ranges that extend far north occurred together in the Pleistocene in the area that is now Virginia and Pennsylvania (Hibbard et al., 1965). Abundant evidence verifies the occurrence of northern assemblages of mammals during the Pleistocene as far south in the

United States as Kansas and Oklahoma. There were reciprocal northward movements of subtropical or desert mammals during interglacial times, as indicated by the fossil occurrence of such animals as the hognosed skunk (*Conepatus*) and jaguar (*Panthera onca*) far north of their present ranges. A fossil record of the jaguar, for example, is from northern Nebraska, some 1300 km. north of the animal's present range.

One of the most common and obvious patterns of mammalian distribution—that of the occurrence of isolated or semi-isolated populations of northern mammals on mountain ranges at fairly low latitudes—is the result of Pleistocene southward migrations of boreal faunas. During glacial ad-

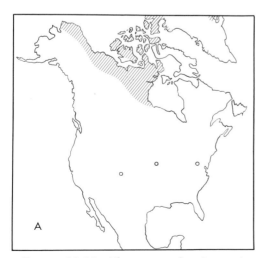

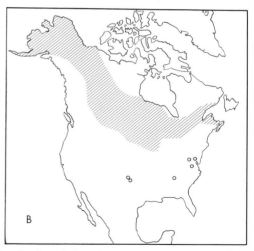

FIGURE 16–13. The present distribution (crosshatched) and the Pleistocene records (circles) of the musk-ox *Ovibos* (A), and the arctic shrew *Sorex arcticus* (B). (After Hibbard, C. W., et al. in Wright, H. E., and Frey, D. G.: *The Quaternary of the United States,* Princeton University Press, 1965.)

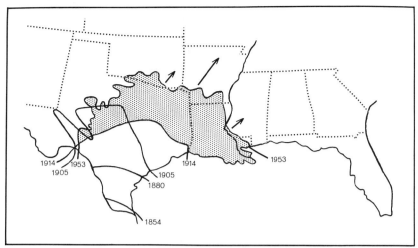

FIGURE 16–14. Recent changes in the distribution of the nine-banded armadillo (*Dasypus novemcinctus*). The stippled area is that occupied between 1914 and 1953. (After Udvardy, M. D. F.: *Dynamic Zoogeography*, Van Nostrand Reinhold Company, 1969.)

vances, assemblages of boreal mammals were widespread in lowlands well south of their present ranges. Concurrent with the movements of these mammals northward during the retreat of cool climates were movements of boreal mammals into montane areas. Here, because of the effect of elevation on climate, cool refuges were available. Many of these montane populations have persisted in "boreal islands" far south of the northern stronghold of their closest relatives, and the zonation of mammalian distributions on some mountain ranges in the southwestern United States has perhaps resulted from Pleistocene faunal movements (Findley, 1969). The White Mountains of east-central Arizona constitute a "boreal island," where several species having northern affinities are isolated from other populations of their species. Populations of least chipmunks (*Eutamias minimus*) and red-backed mice (*Clethrionomys gapperi*) are isolated in this way, and the water shrews (*Sorex palustris*) in the White Mountains are separated by some 300 km. of inhospitable (for this shrew) habitat from the nearest other water shrew populations, which occur in northern New Mexico.

Some relict populations, left behind after northward and eastward retreats of boreal conditions, also occur in some moist lowland localities. One such isolated population of the southern bog lemming (*Synaptomys cooperi*) occurs in a restricted marshy area in western Kansas (Hibbard and Rinker, 1942); another population, that occupies an area of moist habitat only 100 yards wide and about one mile long, is at a fish hatchery in extreme southwestern Nebraska (Jones, 1964: 220, 221).

Man has obviously reduced the ranges of many mammals, but some species are extending their ranges northward today, perhaps in response to the present warm climatic cycle. The armadillo has extended its range from southern Texas into much of central and southern United States in recent years (Fig. 16–14). The cotton rat (*Sigmodon hispidus*) and opossum (*Didelphis marsupialis*) are also moving northward (Hall, 1958). Many ranchers have told me that the raccoon (*Procyon lotor*) has extended its range within the last 50 years from the river valleys in the plains of eastern Colorado deep into the foothills of the Rocky Mountains, where it is common today.

Continental Drift and Mammalian Evolution. In 1915 Wegener proposed the theory of continental drift. He observed that the shapes of the eastern coastlines of the New World could be fit in jigsaw-puzzle fashion against the western coastlines of the Old World, and took this evidence to indicate that the continents of the two hemispheres had originally formed one great land mass. His theory triggered wide controversy, and most New World scientists came to discard Wegener's wild-eyed theory. But scientific opinion was abruptly reversed in the late 1960's when the weight of considerable geological and paleontological evidence became too much for even the most staunch "anti-drifters" to bear. (References on continental drift appear under Dietz and Holden, 1970, and Menard, 1969).

One present view of the history of the major land masses, based on an acceptance of continental drift, holds that some 200 million years ago there was a single great land mass, Pangaea (Wegener's term). This supercontinent was divided by a series of rifts that by the end of the Triassic (some 180 million years ago) had divided Pangaea into a northern land mass, Laurasia, and a southern series of land masses,

collectively called Gondwana. By the end of the Cretaceous (roughly 65 million years ago) South America had moved westward, well away from Africa, which was separated from Laurasia by a narrow sea. The east coast of North America and the west coast of Europe were presumably still in contact, but further drifting of the continents throughout the Cenozoic led to the existing arrangements of the continents.

But what bearing does continental drift have on the zoogeography and evolution of mammals? The Cenozoic radiation of mammals has occupied a shorter span of time than that available to reptiles for their radiation during the age of reptiles (65 compared to 200 million years), but mammals have diversified to a much greater extent, as reflected by a greater number of orders (about 30 orders of mammals to 20 orders of reptiles). Kurten (1969) believes that the greater diversity of mammals is a result of continental drift. Mammals evolved on several land masses under conditions of isolation or semi-isolation, whereas reptiles developed before the continents had moved far apart and therefore devel-

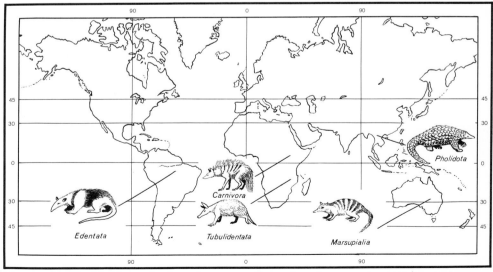

FIGURE 16–15. Members of at least five mammalian orders that occur in southern continents are adapted to eating ants and termites. Edentata—all members of the family Myrmecophagidae; Carnivora —the hyaenid *Proteles;* Tubulidentata—the aardvark, *Orycteropus;* Marsupialia—the numbat, *Myrmecobius;* Pholidota—the pangolins, *Manis.*

oped under conditions allowing freer faunal interchange between evolving stocks. A striking feature of mammalian evolution has been duplication of functional, and to some extent, structural types in separate groups. Examples of such convergent evolution are abundant: the members of several orders specialize in eating ants and termites (Fig. 16–15); the orders Marsupialia, Rodentia, Lagomorpha, Artiodactyla, and Perissodactyla all contain herbivorous, cursorial mammals that pursue basically similar modes of life; and small, terrestrial, insect-eating mammals have developed within at least four orders (Insectivora, Marsupialia, Edentata, and Rodentia). The greatest duplication has occurred in southern land masses, which have been longer and more completely isolated than have the Nearctic and Palearctic areas. Mammalian diversity, then, may be as much a result of the progressive Mesozoic and Cenozoic separation of the continents as of the structural and functional adaptability of the mammals themselves.

CHAPTER 17

BEHAVIOR

The behavior of any animal is of great interest because it is, so to speak, "the proof of the pudding." In the case of the pronghorn, for example, great running speed became part of a unified functional system only because of a complex of behaviors that evolved in association with this ability. The formation of herds, the preference for open situations, the "flashing" of the white rump patch as a danger signal to other antelope, and the remarkable ability to detect enemies at a distance all allow the pronghorn to utilize its great speed effectively to escape from predators. How an animal uses its morphological and physiological equipment is of vital adaptive importance, and forms the substance of behavior.

The behavior of mammals is of particular interest because of its flexibility and variability. Relative to other vertebrates, mammals learn extremely rapidly and can modify behavior on the basis of past experience. This ability, superimposed on a rich array of innate (instinctive and unlearned) responses or behaviors, makes for complex behavioral patterns that often differ widely between species. Remarkably well developed sense organs, coupled with a brain capable of rapid evaluation of complex sensory information, have enlarged the perceptual sphere of mammals and have facilitated the evolution of communication and social behavior.

THE ETHOLOGICAL APPROACH

This chapter will deal largely with *ethology*, the study of behavior in relation to structure and mode of life, or as put by Tinbergen (1963), "the biological study of behavior." One might suppose that behavior could be more readily observed and analyzed than could other aspects of biology, and that detailed behavioral information on many species would have been assembled relatively early. But this is not the case; indeed, little is known of the behavior of many animals that are well known morphologically. A number of recent field studies under natural conditions (such as those of McCullough, 1969; Mech, 1966; Schaller, 1963, 1967; and King, 1955) have provided a foundation of information on the behavior of mammals, but a substantial frame of knowledge is yet to be built on this foundation.

Although mammals are remarkable in their ability to learn and to profit from experience, built-in patterns of behavior form an important part of the behavioral repertoire of mammals. Such innate behavior is not individually variable within a species, but is unlearned and is a part of an individual's heritage that is shared with other members of his species. These built-in behaviors are best regarded as simple sequences of movement elicited by specific stimuli. The term *Erbkoordination,* coined by Conrad Lorenz (1950), seems especially appropriate, and refers to a simple but specific hereditary movement or pattern of coordination. Such a behavior in baby woodrats *(Neotoma)* is the action of turning toward objects that contact the facial vibrissae. This simple turning response probably forms one part of a more complex innate behavioral sequence enabling the baby to regain contact with its mother or its littermates. In the parlance of modern ethologists, the touching of the vibrissae is the *innate releasing mechanism* (IRM) that triggers the innate response. These innate movements are inherited just as are structural features, and were favored by selection because of their adaptive importance.

Clearly, mammals are not completely unique behaviorally: they are set apart from other vertebrates by their superior ability to learn, remember, and innovate, but they resemble other vertebrates in their wide use of innate behaviors.

NON-SOCIAL BEHAVIOR

A number of behaviors relate basically not to social interactions but to such vital activities as feeding and seeking or preparing shelter. Feeding behavior is highly variable and in some species is a non-social activity, and in some species the preparation of shelter involves remarkably complex non-social behavior.

Feeding Behavior. Herbivores utilize food that is unable to escape, and therefore are spared some of the problems that face carnivores. But an herbivore is generally equipped to utilize efficiently only specific types of vegetation, and must often face seasonal shortages of food and must cope with plant material that is difficult to digest.

Some of the most specialized foraging behaviors occur among rodents. Pocket gophers (Geomyidae) dig extensive tunnels, and part of their diet consists of roots they encounter; but much of their food is above-ground vegetation that is usually gathered by brief forays from an open entrance to the otherwise closed burrow. The animal scurries a few inches from the burrow to a plant, clips it hurriedly, and retreats tail first into the burrow as if yanked back forceably by a string tied to the tail. Seldom does the animal venture more than a few inches from the security of its burrow. The closely related kangaroo rats (Dipodomys), by contrast, travel far from their burrows and gather seeds from the soil by using the long claws of the small forefeet. Dry seeds are immediately eaten or taken directly to storage chambers in the burrow. Kangaroo rats can apparently discriminate between dry seeds and those that are moist and would spoil in the burrow. Moist seeds are placed in shallow pits near the burrow, covered with soil, and allowed to dry before being transferred to underground caches (Shaw, 1934). In some desert areas the Merriam's kangaroo rat (Dipodomys merriami) seems to use the shallow pits as "seed traps." In the summer in southern Arizona, where afternoon winds drift seeds into depressions in the soil, these kangaroo rats regularly visit the shallow pits and remove blown-in seeds. The seed-curing behavior may therefore not only provide for a well preserved seed cache, but also form a series of effective seed traps that increase the foraging efficiency of the kangaroo rats in seasons when seeds are not being produced.

Hoarding behavior is highly developed in a number of mammals. Some rodents, some shrews, some carnivores, and the pika (a lagomorph) store food. Many sciurid, cricetid, and heteromyid rodents store food in underground or protected caches. "Scatter hoarding" is practiced by some squirrels, which bury food at scattered points within the home range. The caching behavior of the North American red squirrels (*Tamasciurus hudsonicus*) is particularly notable. These squirrels depend for food on seeds of fir and pine. Cones are cached in holes in large middens formed by the litter of cone fragments that accumulates beneath a squirrel's favorite feeding sites. The middens are frequently 20 to 30 feet in diameter and contain from 2 to 10 bushels of cones, and are in shady situations where the moisture retained within the midden aids in the preservation of the green cones (Finley, 1969). Small numbers of cones are commonly cached in logs or pools of water. The cones are harvested in late summer and fall, and are cut on occasion at the rate of 29 per minute (Shaw, 1936:348). The squirrels are such effective harvesters that one pine in northern California lost 93% of its 926 cones to them (Schubert, 1953). Seeds from the cached cones are eaten during winter, and when snow is deep, access burrows through the snow into the midden are maintained.

The dextrous forefeet of rodents are often used in food handling. The bipedal, squirrel-like feeding posture frees the hands, which grip and manipulate the food while it is gnawed by the incisors. This behavior is seemingly innate, for hand-reared rodents were found to handle food in an essentially adult manner when they were still too poorly coordinated to maintain their balance (Ewer, 1968:32). The small forefeet of kangaroo rats and pocket mice skillfully sift seeds from the soil and put them in the cheek pouches. Even some rodents with forelimbs highly specialized for digging retain considerable manual dexterity. The plains pocket gopher (*Geomys bursarius*), which has powerfully built forelimbs, with large forefeet equipped with long claws, holds food

in typical rodent fashion when it is eating. This fossorial rodent has several specialized feeding behaviors that serve to avoid the ingestion of soil and that demand considerable dexterity. Dirt or water are carefully stripped from leaves by the claws, and unusually wet or sandy food is often held in both forepaws and rapidly shaken (Vaughan, 1966b). Probably the importance of the forelimbs of rodents for food handling has limited the extent to which the limbs have become specialized for digging or locomotion.

The efficient capture of food by predaceous mammals is only possible because of complex behavioral modifications. The capture of flying insects in darkness by bats involves incredibly specialized behaviors (Chapter 21). To all predators, however, the pursuit, capture, and killing of live prey, which in many cases have their own finely tuned behaviors adjusted to defense or to escaping predators, presents a considerable problem.

Except for the use of cooperative action, the canid style of predation is one of the least specialized. Most canids are quite cursorial, and are able to capture prey by virtue of speed or by group effort. A pack of wolves will often bring down prey much larger than a single animal could handle. Wolves develop by experience a remarkable ability to recognize vulnerable prey (Crisler, 1958:106; Murie, 1944:109, 166-174). Mech (1966:121, 124) observed that moose that showed signs of weakness by a lack of defense or by being easily overtaken were attacked, whereas wolves repeatedly abandoned the pursuit of a moose that put up an aggressive defense. Wolves occasionally chase prey for several miles, seemingly testing the animal's stamina. Mech (1966:126-138) found that wolves first slashed at the rump of a moose, perhaps because this is the least dangerous site to attack, and when injuries to this area had partially immobilized the moose the wolves tore at the throat and shoulders and often grasped the nose. Similar styles of killing by wolves are used on deer

(Stenlund, 1955:31), on caribou (Banfield, 1954:47), and on elk (Cowan, 1947:159). Hyaenas often bring down prey by attacking the hind limbs (see Fig. 12–14; p. 203). Canids often kill small prey, such as rabbits, by grabbing the animal across its back and shaking it violently, and this method is used widely by other carnivores.

The cats employ more specialized hunting and killing techniques. Cats are not long distance runners, but usually depend on short rushes directed against surprised prey. The sudden rushes of lions seldom cover more than 50 to 100 yards, and leopards and smaller felids frequently only make several bounds to reach their prey. The cheetah, an exceptional felid, may chase an antelope several hundred yards at speeds up to 112 km./hr. (70 miles per hour!). In order to use this technique effectively, a cat must get close to its prey. The stalking of prey by felids involves a series of beautifully coordinated behaviors described in detail by Leyhausen (1956). When prey is sighted the cat crouches low to the ground, and approaches using the "slink-run" and taking advantage of every object offering concealment. As the distance to the prey is reduced the cat moves more slowly and cautiously. At the last available cover the cat stops and "ambushes." The body is held low and just before the attack the heels are raised from the ground and the weight is shifted forward, just as a sprinter readies himself in the last tense instant before the sound of the gun. The brief rush to the prey ends in a spring; the forefeet clutch the animal but the hind feet remain planted and steady the cat for the possible struggle. Some of the components of this total behavior pattern may be observed in kittens as they stalk an insect or a scrap of paper. The cat makes the kill itself, not by belaboring the prey as do many canids, but either by a powerful bite at the base of the skull or the neck, which crushes the back of the skull or some of the cervical vertebrae and the spinal cord, or by strangulation. The shortening of the felid jaws is a specialization that contributes to the power of the bite. The tiger may kill prey as large as buffalo by gripping the throat and waiting for strangulation (Schaller, 1967). The cheetah also frequently uses this style of killing. Most cats are solitary, and cooperative effort in killing prey is rare. An exception is the African lion, the only truly social felid, which often hunts in groups in which there is some cooperation between individuals, with adult females doing most of the killing. The specialized killing behavior of cats enables a single animal to kill large prey and reduces the risk of injury to the predator.

A wide variety of both placental and marsupial carnivores have been observed to use the neck bite to kill prey. This killing behavior was studied in the house cat by Leyhausen (1956), who presented the predators with normal and headless rats, and with rats with the head fastened to the tail end. The cats aimed their bite at any constriction in the body; with normal prey this obviously results in the neck bite.

Unusual behavior patterns enable some carnivores to break the shells of mollusks and eggs. Viverrids such as the mongoose (Herpestes, Atilax, and Mungos) use the forefeet to throw the objects downward against the ground with great force or to fling them backward between the hind legs (Ewer, 1968:48; Ducker, 1957). The spotted skunk (Spilogale putorius) uses a technique for breaking eggs that involves kicking the egg with a hind leg and sending the egg against an object such as a rock (Van Gelder, 1953). The sea otter (Enhydra lutris) smashes the sturdy shells of mollusks by using a tool: the otter floats on its back with a flat stone on its chest and pounds the mollusk against the stone (Fisher, 1939). An individual was observed to pound mussels (Mytilus) on a stone a total of 2237 times during a feeding period lasting 86 minutes (Hall and Schaller, 1964:290). These otters are clearly selective in their choice of stones, and may use the same one repeatedly.

Shelter Building Behavior. Many mammals have evolved elaborate shelter-building behaviors that aid them in maintaining homeostasis. The nests, burrows, or houses of mammals provide insulation that augments the animal's own pelage and reduces the rate of thermal conductance from the animal to the external environment or vice versa (Fig. 15-4; p. 252). The woodrat *(Neotoma)* collects a variety of materials with which it builds houses or improves the shelter provided by rock crevices or vegetation (Fig. 17-1). Debris gathering behavior seems well developed in all species of *Neotoma*. The feces of carnivores and cattle, bones, shotgun shells, or the snap traps of a mammalogist may be incorporated into the shelter. At the center of the woodrat house, or in a burrow beneath it, is a carefully constructed nest (Fig. 17-1). The nest may be globular or cup-shaped, and is formed from grasses and plant fibers. Many terrestrial rodents construct similar nests beneath logs or rocks or in burrows. Arboreal rodents frequently build nests in the branches of trees or within hollows or holes in trees. Some nest-building behaviors are perhaps common to many rodents, but the choice of nesting site seems to be species-specific. For example, red tree mice (*Phenacomys longicaudus*) of the humid coast belt of Oregon and California build their nests only in Douglas firs (*Pseudotsuga menziesii*), the needles of which provide their primary food (Benson and Howell, 1931).

Fossorial rodents follow rather complex patterns of movement when digging. Probably many of the specific components of the total digging sequence are innate behaviors. Pocket gophers (Geomyidae) and kangaroo rats and pocket mice (Heteromyidae) employ very similar digging styles. The forefeet loosen the soil by powerful downward sweeps, and the hindlimbs kick the accumulated soil backward from beneath the animal. Pocket gophers periodically eject soil from a burrow entrance by pushing it with the chin and forelimbs (Fig. 17-2). Careful studies of the pocket gopher by Kennerly (1971) have shown that the long and complex series of behavior patterns associated with mound building are basically innate, but may be modified by learning. "Autoformulated releasers" probably play important roles in guiding mound building in rodents (and probably many other behavioral sequences in other mammals). Such a releaser is any alteration by an animal of its perceptual environment that acts to release the animal's subsequent behavior. Thus, the mound of earth itself, and changes in the mound due to the pocket gopher's activity, release subsequent behaviors associated with mound building. The animal characteristically alternates direction in pushing soil from the burrow; it pushes a series of 5 to 20 loads to the right, a similar series to the left, and so on. The frequency distribution of directions of pushing soil from the mouth of the burrow to the rim of the mound indicate that efforts are mainly in three directions: the pocket gopher usually pushes soil either directly in front of the mouth of the burrow or at an angle of 90° to either side. This results in the fan-shaped mound so typical of pocket gophers. That learning plays a part in burrowing and mound-building activity is suggested by the fact that young animals are less successful in plugging the openings of burrows than are older animals. Somewhat different styles of digging and mound-building are used by different rodents (see the account of the rodent family Ocotodontidae, page 175); future studies will perhaps indicate the degree to which these differences are determined by innate, behavioral patterns.

COMMUNICATION

Communication obviously involves any interactions that transmit information between at least two animals, and is best developed in social species.

FIGURE 17–1. Above, the house of a Stephen's woodrat (*Neotoma stephensi*) at the base of a juniper tree (*Juniperus* sp.). Note the pile of sticks to the right of the tree and in the crotch between two trunks. Below, the nest of a Stephen's woodrat. This nest, about 180 mm. in diameter and composed of grass and shredded juniper bark, was exposed when a woodrat house was dismantled. (Photographs by David M. Kuch.)

FIGURE 17–2. A pocket gopher (*Thomomys bottae*) pushing soil from its burrow. (Photograph by Robert J. Baker.)

"By far the greatest part of the whole system of communication seems to be devoted to the organization of social behavior...." (Marler, 1965:584). This comment was applied to primates, but may be equally valid for all mammals. Each type of communication—visual, olfactory, auditory, and tactile—will be considered separately, but it should be stressed that usually a complex of several kinds of communication signals passes between animals.

Visual Communication. Highly developed facial musculature, the ability of the body and ears to assume a variety of postures, and the control many mammals have over the local erection of hair allow mammals a breadth of visual communication found among no other animals. Some visual signals are familiar: the dog wags its tail as a sign of friendship, and the cat arches its back, erects its fur, and raises its tail in a defensive threat. Such displays can be observed readily and their functions or messages can often be recognized, but only a start toward an understanding of the broad area of visual communication in mammals has been made.

Facial expressions are of great importance in communication, and natural selection has often favored the development of distinctive facial markings that focus attention on the head (Fig. 17–3). As described by Lorenz (1963), the facial expressions and postures of the ears of dogs signal degrees of aggressiveness or submissiveness. The posture of the head and facial expression of many ungulates provide visual signals to other members of the herd or to territorial or sexual rivals (Fig. 17–4). A tule elk (*Cervus elaphus nannodes*) ready to run from danger elevates its nose and opens its mouth (McCullough, 1969); the Grant's gazelle (*Gazella granti*) holds its head high, elevates its nose, and pricks its ears forward when challenging another male (Estes, 1967). The heads of both of these animals are conspicuous: the elk's because the dark brown head and neck contrast strongly with the pale body, and the gazelle's because of bold black patterns. The intricate facial expressions of primates are frequently made more obvious by distinctive and species-specific patterns of pelage coloration and by brightly colored skin.

The body is used for signaling in

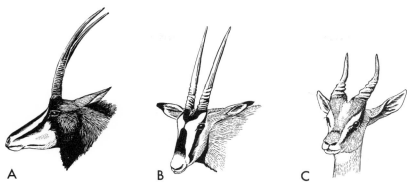

FIGURE 17–3. Facial markings of antelope: A, sable antelope (*Hippotragus niger*); B, oryx (*Oryx beisa*); C, young Thompson's gazelle (*Gazella thompsonii*).

many species. This type of signaling is particularly well developed in ungulates that inhabit open areas and that gain an advantage from coordinated herd action. The Grant's gazelle and the Thompson's gazelle (*G. thompsonii*) of Africa, which have two warning displays (Estes, 1967), twitch the flank skin (conspicuously marked in *G. thompsonii*) just as they begin to run from a predator that has entered the minimum flight distance (the minimum distance at which an approaching enemy causes the animals to run). The most effective display is a stiff legged bounding gait, called "stotting," used as the gazelles begin to run. The conspicuousness of this display is enhanced by the erection and flaring of the hair of the white rump patch. In some monkeys and apes the presentation of the hind quarters as if inviting

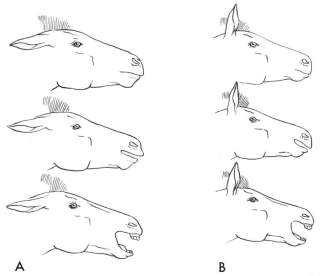

FIGURE 17–4. Facial expressions of horses: series A, three intensities of threat expressions; series B, three successive stages of greeting expressions. (After Trumler, 1959.)

copulation is a social gesture symbolic of friendship, and is accepted by a brief "token" mounting (Heinroth-Berger, 1959). Kangaroos threaten one another by standing bipedally at their maximum height. An understanding of this display helps to explain the extremely aggressive attitude of adult male kangaroos toward people. Because of his bipedal, erect stance, man is constantly assuming a posture interpreted by the kangaroo as a hostile threat.

Olfactory Communication. The great development of the sense of smell in mammals is associated with broad use of olfactory communication. With the exception of anthropoid primates and cetaceans, most mammals can distinguish between individuals of their own kind by smell. A female can discriminate between the smell of her own and another female's young, and in social species an individual can probably identify each member of the group by smell. The perception of predators and of food and water are clearly facilitated by olfaction. In addition, chemical signals from urine, feces, or specialized scent glands may be of great importance.

The effectiveness of reproduction is enhanced by olfactory signals. Males of most species of mammals recognize by smell when a female is in estrus, and in many species copulation will not be attempted until this time. Olfactory communication may even play an important role in the reproductive cycles of some primates. The smell of vaginal secretions of rhesus monkeys (*Macaca rhesus*) in estrus is sexually stimulating to males, and promotes copulation (Michael et al., 1971). Smelling of the genital region by males occurs widely among mammals. A bull elk with a harem of females during the breeding season regularly smells the rump of each female as a means of recognizing those that are entering their fairly brief (17 hour) estrus (McCullough, 1969:87). Mature bull elk during the rut (the breeding season) do not urinate normally; instead, the penis is extended and the animal squirts urine on the belly and the thick

hair on the chest. The smell of the urine may have an important communication function (McCullough, 1969:82, 99, 110). By "self marking" with his metabolic wastes the bull is probably advertising his general physical condition. While in excellent condition his urine advertises this information to rival bulls, but when his condition declines this is also conveyed to rivals *via* the smell of the urine. This type of communication may avoid useless disruption of breeding activity and of the harem by delaying attempts at deposing the harem master until his exhaustion allows a fresh bull to replace him. Perhaps most importantly, this system assures that the breeding is carried on by a succession of fresh bulls that are ready to service cows when they are in estrus.

Many mammals have specialized scent glands that are employed in scent marking (leaving chemical signals on other individuals of their own species or on objects in the environment). These glands may be in such places as the flank, sternum, tarsus or metatarsus, chin, front of the head, preorbital area, or anal areas. Some mammals leave chemical signals produced by anal glands with deposits of feces as markers of their territory. This type of marking is characteristic of the mountain lion (*Felis concolor*), for example. But in her careful discussion of scent marking, Ralls (1971:449) indicates that mammals use scent marking "in any situation where they are both intolerant of and dominant to other members of their species." Scent marking is used in many social situations not involving territoriality, and the most frequent marking is done by dominant animals (Table 17–1). Male lemurs (*Lemur catta*) have "stink fights," in which two males threaten each other by marking their palms and tails with urine. First one "combatant" and then the other marks, and the dominant animal slowly advances and scent marks with his palms branches that his opponent previously marked (Jolley, 1966). Longworthy and Horst (1971) found that dominant males of a

Table 17-1.   The Mean Number of Scent Markings Made During a 10 Minute Period by Maxwell's Duikers (*Cephalophus maxwelli*), Small, Forest-Dwelling Antelope. Males Mark More Than Females, and Dominant (Type A) Females Mark More Than do Subordinate (Type B) Females. (From Ralls, K.: Mammalian Scent Marking. *Science*, 171:443–449. Copyright 1971 by the American Association for the Advancement of Science.)

GROUP MEMBERSHIP	MARKING ACTIVITY WHEN WITH OWN GROUP	MARKING ACTIVITY AFTER PRESENCE OF ADDITIONAL	
		Male	*Female*
Males			
I	6.6	15.2	6.1
II	5.8	10.7	6.2
III	4.4	8.6	4.1
Type A Females			
I	3.5	3.7	18.6
II	3.4	3.1	12.2
III	1.5	0	1.7
Type B Females			
I	0.06	0	0.09
II	0	0.1	0
III	0.04	0	0.03

captive colony of molossid bats (*Molossus ater*) frequently mark subordinate males with secretions from well developed throat glands. The specific messages conveyed to mammals by scent and scent markings are not known, and much study remains to be done in this area of mammalian behavior.

Auditory Communication. The sense of hearing in mammals is acute, and auditory communication is of great importance. Indeed, the sounds of some mammals that are rarely seen are commonly heard. Impressive choruses of howls of the coyote may be heard nightly in some parts of the western United States where the animals themselves are only occasionally seen. The importance of nearly constant auditory communication to a herd animal is difficult for one to imagine. Virtually continuous noises made by the members of a herd serve to integrate the group by keeping individuals apprised of the locations of one another. When he was very close to a herd of tule elk, Mc-Cullough observed (1969:71) "that there is a continuous array of sounds —foot bone creaking, stomach rumbling, teeth grinding, and others." The crunching of vegetation by Mc-Cullough was instantly distinguished by the elk as distinct from similar sounds made by their own feeding activities. In caribou (*Rangifer tarandus*) also, the creaking and snapping of foot bones can be heard for considerable distances and enable scattered members of a herd to remain in auditory contact (Kelsall, 1970).

Vocal communication is widely used by mammals. In man, of course, this type of communication reaches its most complicated development, but even in other primates some type of "language" can be recognized. The Japanese macaque (*Macaca fuscata*) has a repertoire of some 25 sound signals (Mizuhara, 1957). The more basic sounds used by the rhesus monkey may be linked by a series of intermediate sounds, and one basic sound may grade independently into other calls (Rowell, 1962). This yields a remarkably complex and rich vocal repertoire. The functional importance of some sounds can be recognized. The quiet "grunt," for example, is used by many primates to maintain contact with each other (Marler, 1965:568), and vocal sounds are used by many mammals to maintain or reestablish contact with one another. This seems to be the function of howling choruses in canids and the calls of young in a variety of species. The function of the complex vocal noises of cetaceans are as yet poorly understood, but is being intensively studied (see, for example, Dreher, 1966, and Reysenbach de Haan, 1966).

Recent research has shown that ultrasonic pulses play a significant part in several aspects of the social behavior of rodents. Parents respond to the ultrasonic distress vocalizations of helpless young by bringing them back to the nest, and it has been found that a decrease in the acoustical energy of the calls as the young grow older is associated with the development of homoiothermy (and an attending decrease in the vulnerability to cold) by the young (Noirot, 1969). The ultrasonic vocalizations that are uttered by all myomorph rodents that have been studied may be of further importance in inhibiting the aggressiveness of a retrieving adult. Ultrasonic pulses made by rodents during mating may function to reduce aggressive behavior, and some ultrasonic vocalizations may even serve as territorial announcements (Sewell, 1968).

Tactile Communication. Little is known about tactile communication in mammals, but its use is probably widespread. The sexual behavior of many mammals includes precopulatory activities by the male such as laying the chin on the rump, nuzzling the genital area, or touching various parts of the female's body. These behaviors are presumably sexually stimulating to the female or at least cause her to accept mounting by the male. Perhaps tactile stimuli are of greatest importance in connection with sexual activities in most mammals, but in primates they have assumed other roles.

Mutual grooming by primates, which clearly provides tactile stimuli,
serves to cleanse a partner's fur in places inaccessible to self-grooming. But mutual grooming may also have an important social function in promoting social contact and allowing familiarity between individuals. In anthropoid primates embracing and touching of hands or of the genital areas are types of "friendly behavior," and formalized neck biting is a gesture of domination in some species.

TERRITORIAL SYSTEMS AND SOCIAL PATTERNS

The importance of the maintenance of a home range and a territory was mentioned on page 278, but the different types of territorial behavior found among mammals were not discussed. They will be considered briefly here.

Territoriality in Solitary Mammals. Many mammals, including some marsupials, some rodents, some insectivores, and many carnivores, are essentially solitary and territorial. The home range of a pocket gopher (Geomyidae), for example, contains its burrow system, which, except in the breeding season, is occupied by a single animal and is apparently defended against interlopers. Although the burrow system may be extended and some sections may be plugged, most individuals probably occupy the same area throughout most of their lives. These animals may be colonial, but are never social. Concentrations of burrow systems often occur locally, but each burrow system in exclusively "owned." Similarly, in the case of two species of ground squirrel (*Spermophilus mexicanus* and *S. armatus*), the colony consists of a number of solitary individuals living in close proximity (Edwards, 1946; Balph and Stokey, 1963). Leyhausen (1964) regards some carnivores, such as various kinds of felids, as territorial but not truly solitary. Individuals occupying neighboring territories probably meet on occasion and gain some familiarity with one another, and establish a social order featuring mutual respect for one another's territorial rights. For many solitary species, little is known of the brief periods of social life. In these animals the primary function of territoriality is the spacing of individuals and the assurance that each has adequate environmental resources for all vital activities.

The Territorial Family Group or Colony. Perhaps the simplest type of social arrangement is that between members of a territorial family group. A beaver colony is a family group in which mutual tolerance is the rule, and apparently no social integration exists (Tevis, 1950). In wild colonies of rats (*Rattus norvegicus*) a similar colonial system prevails, but a dominance hierarchy gives order to the social system (Steiniger, 1950). Colonies of marmots include a number of females and young and one adult male (Armitage, 1962). Individual adults have separate burrows and pathways to food, but the animals display considerable mutual tolerance.

Among rodents, the most complex social system known is that of a prairie dog (*Cynomys ludovicianus*), which has been carefully studied by King (1955, 1959). Prairie dogs formerly occupied many parts of the western United States, where they occurred in large "towns," often including over 1000 animals and covering many acres. (Now, lamentably, prairie dog towns are rare in many parts of the west, owing partly to intensive, government-sponsored poisoning campaigns.) The functional social units are *coteries*, which generally consist of an adult male, several adult females, and a group of young. No dominance hierarchy exists within the coterie. (A dominance hierarchy is a fairly permanent social system based on dominance. Each individual recognizes his "position;" that is, he recognizes the animals that he can dominate and those that dominate him.) The paths, burrows, and food within the area held by a coterie are shared by its members, but hostility between coteries is the universal pattern. Members of the coterie become familiar with each other in part by grooming, playing, and "kissing" behaviors. During kissing the mouths are open and the incisors are bared. This behavior is seemingly a ritualized method of distinguishing between friend and foe. Faced with the threatening expression of open mouth and bared teeth a trespasser retreats, while a fellow coterie member meets its "friend" and kisses. A two-syllable territorial call is used to proclaim ownership of territory. A repetitive, high-pitched yelp is a warning of danger. During the spring period when females are pregnant or are lactating, the coterie system partially dissolves, and some yearlings and adults establish themselves beyond the territorial limits of their coteries. The personnel of coteries thus may change, but the territory itself is stable. An individual gains several advantages from this social system. Many eyes are watchful for danger, and many voices are ready to sound a warning. The effect of the foraging of the animals is to keep vegetation low over a wide area and to provide terrestrial carnivores with little concealment. Perhaps equally effective in providing for long-term occupancy of an area, the animals are kept spaced so that overuse of food plants is generally avoided.

Among carnivores, social groups consisting of one or more families are formed by many species. The lion "pride," the spotted hyaena "clan," and packs of wolves (*Canis lupus*) or African hunting dogs (*Lycaon pictus*) are examples. Although members of a hyaena clan are intolerant of "foreign" hyaenas, the clan members will often tolerate jackals (*Canis mesomelas*) that are competing with the hyaenas for food (Fig. 17–5). In *Lycaon* an area roughly one kilometer square, surrounding the burrow in which pups are raised, is the defended territory, but the pack hunts over a broad area of some 100 to 200 square kilometers (Ruhme, 1965). Lion prides appear to be territorial, and scent

FIGURE 17–5. Hyaenas *(Crocuta crocuta)* tolerating a jackal *(Canis mesomelas)* feeding with them on a carcass. (Photograph by George B. Schaller.)

marking and roaring are means of advertising territorial occupancy (Schenkal, 1966a).

In some ungulates a small group of one or more families may form the social unit. The warthog *(Phacochoerus africanus)* group does not defend its home range, but male hippopotami *(Hippopotamus amphibius)* that live in "schools" have a hierarchy that determines the animals able to occupy favored sites near the females. Males defend their paths to foraging areas.

A variety of types of social organization occur among primates. In general, the forest-dwelling and arboreal species seem to form small social groups, whereas the baboons, which live in open country where immediate arboreal refuge is not available, aggregate into troops that, in the case of the chacma baboon *(Papio ursinus)*, occasionally contain over 100 individuals (Hall and DeVore, 1965). An order of dominance, with males occupying dominant positions, is typically established in these social groups. Home ranges of primate groups often overlap to some extent, but there is little evidence of strife between groups. In baboons some safety is gained from numbers, for several dominant males may put up a united front that discourages some predators.

The great apes (Pongidae), all of which occupy tropical or subtropical forests, assemble in groups containing one or more families, and the groups are often quite mutually tolerant. This system is perhaps "allowed" by the large amounts of food available and the resulting lack of intense selective pressure favoring the rigid spacing of groups so characteristic of savanna-dwelling baboons (Ewer, 1968:98).

Large Assemblages. Large ungulates that occupy grasslands are clearly visible to predators and are unable to seek concealing shelter. These animals seek safety in their ability to run rapidly and in the formation of large herds. The combined watchfulness of many pairs of eyes makes a large herd less vulnerable than a single individual to an undetected approach. Further, the organized mass movement of the herd makes the job of the predator still more difficult: "The safety of the herd consists of the cohesive mass of animals running in an organized manner. The animals exposed are only those on the outside, and even these are protected by the number of flying hooves and the ebbs and surges within the group. The vast array of movement has a disorienting effect on the observer's vision." (McCullough, 1969: 72). McCullough found that in the tule elk even the speed of the running herd was regulated. Although individuals can run much faster, a herd travels at about 22 to 24 miles per hour, and at these speeds a closely knit herd formation is maintained. Of considerable interest, and perhaps of adaptive importance, is the ability of the elk herd to recognize cripples and to banish them from the herd by aggressive behavior (McCullough, 1969:76, 77). A herd typically has a home range, but this may be abandoned seasonally when herds migrate in response to changes in availability of food and water.

Some herds of large ungulates actually may not be cohesive social units. Large herds of the Burchell's zebra (*Equus burchelli*) are of this type. In this species the basic social unit is a small group of females and their foals overseen by a single stallion (Klingel, 1967). Such groups are stable for long periods of time, not because the stallion keeps the group together but because of social bonds between the females. Young males leave these groups and form bachelor herds. When excellent grazing conditions occur locally, large assemblages of zebras form; but these herds have no social unity, for the bachelor groups and family groups still maintain their integrity and the large herd disperses as forage becomes less plentiful.

Harems and Modified Territorial Behavior During the Breeding Season. Some mammals have breeding cycles that feature harems of females, each maintained by a dominant male. The area occupied by the females is the strongly defended territory, or in some cases the male maintains a "breeding territory" that is defended even when females are not present. The essence of this territorial behavior is that it persists only through the breeding season and apparently fosters breeding largely by mature, vigorous, and aggressive males. McCullough (1969:99) reported that in a herd of elk he studied, only 12 per cent of the bulls—the largest individuals—did 84 per cent of the copulation he observed.

In the pinnipeds that are polygynous—the otariids, some phocids, and the walrus—the males are extremely vocal, are much larger than the females, and maintain breeding territories (Fig. 17–6). In the California sea lion (*Zalophus californianus californianus*) studied by Peterson and Bartholomew (1967), large bulls establish territories at sites adjacent to the water that are favored as hauling out places by females, which arrive at the rookery a few days before they give birth. Non-territorial bulls usually form aggregations apart from the breeding rookery. Because the same females do not continuously occupy a male's territory, and males make no effective effort to herd females into territories, the term "harem" cannot be applied to the females within a territory. Females enter estrus and copulation occurs roughly two weeks after parturition. Fighting between males occurs during the establishment of territories, and males on established territories signal their possession by incessant barking. Little actual fighting occurs after territories are established, but a "boundary ceremony" between males on adjoining territories periodically reaffirms territorial boundaries. These ceremonies involve an initial charge toward one another, followed by open-mouthed head shaking as the animals confront each other at close quarters (Fig. 17–7), and a final standoff in which the bulls stare obliquely at each other. The ceremony is so carefully ritualized that should animals get uncomfortably close to one another they adroitly avoid actual contact. Females are aggressive toward one another through much of the breeding season; again, however, injury is avoided by ritualized aggressive threats. Although males may be on territories in a rookery from May through August, each male maintains a territory for only a week or two; territories are thus occupied by a succession of males.

The northern fur seal (*Callorhinus ursinus*) has a breeding season that extends from June to December, and large numbers of animals assemble on the Pribilof Islands of the North Pacific. Nonbreeding males form bachelor "cohorts" around the edges of the breeding grounds. Territorial males

FIGURE 17–6.   Harems of northern fur seals *(Callorhinus ursinus)* on St. Paul Island in the Bering Sea. (From Orr, R. T.: *Vertebrate Biology*, 3rd ed., W. B. Saunders Company, 1971.)

FIGURE 17–7.   A boundary ceremony between two male California sea lions; the animals are confronting each other and "head shaking." (From Peterson, R. S., and Bartholomew, G. A.: *The natural history and behavior of the California sea lion*, American Society of Mammalogists, 1967.)

herd females into their territories and maintain fairly stable harems (Peterson, 1965).

Many kinds of artiodactyls establish harems during the breeding season. The male impala (*Aepyceros melampus*) maintains a harem of a group of adult females and juveniles and threatens or attacks males that try to mingle with the harem (Schenkel, 1966b), but has no definite territory. A male Grant's gazelle (*Gazella granti*), in contrast, maintains a harem within a fairly stable territory (Walther, 1965). An unusual pattern of "territorial" defense was found by Talbot and Talbot (1963) in the blue wildebeest (*Connochaetes taurinus*). Breeding occurs during mass migrations from open grassland to savanna; nonetheless, the herd is divided into breeding units (Fig. 17–8). Each unit consists of from two to roughly 150 adult females and yearling young. One male may be master of a small harem, but several bulls generally cooperate in defending a large harem. Considerable behavioral flexibility is shown by this species. In areas where wildebeests are not migratory, males defend parts of their home ranges and during the breeding season try to herd as many females as possible into their territories. Oddly enough, in the Ngorongoro Crater both patterns of

harem organization occur (Estes, 1966).

The rutting behavior of the elk is especially well known due to the studies of Darling (1937), Graf (1955), Struhsaker (1967), McCullough (1969), and others. McCullough recognized four main categories of bulls during the breeding season. Primary bulls are powerful, mature individuals that shed the velvet from their antlers early, and are the first to establish harems. Secondary bulls are large individuals that take over the harems by defeating the primary bulls as the latter become exhausted. Tertiary bulls assume control of the harems after the secondary bulls decline. Opportunist bulls are those whose only contact with cows is by chance. When a bull becomes exhausted through constantly keeping cows herded together, driving rival bulls away, and copulating, all while unable to obtain adequate food and rest (Figs. 17–9, 17–10), it is beaten in a fight with a fresh bull, who takes over the harem from the deposed "master."

Of particular interest in the present discussion are ritualized social behaviors evident during the rut. Bulls advertise their location, vigor, and sexual readiness by several acts. The most obvious and characteristic of these is *bugling*, a stirring, high pitched call

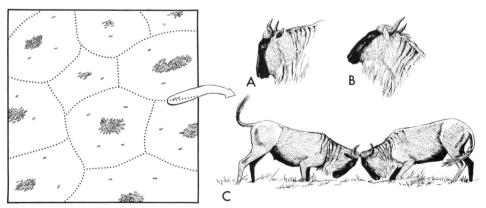

FIGURE 17–8. The diagram on the left shows the tightly packed groups of females and young wildebeest *(Connochaetes taurinus)* protected by bulls. The diagrams on the right show the head down posture (A) typical of females, young, and males not involved in rutting activities, the head up posture (B) of challenging herd bulls, and the kneeling position (C) assumed by sparring bulls. (After Talbot and Talbot, 1963.)

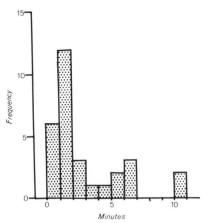
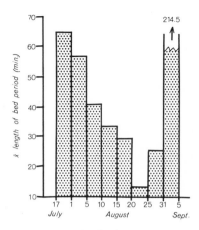

FIGURE 17–9. Left: length of feeding periods of a bull elk (*Cervus elaphus*) at an intense rutting stage. Right: changes in bedding (resting) periods of a bull elk as the rut progressed; this bull was defeated and gave up his harem on 31 August. (After McCullough, D. R.: *The Tule Elk: Behavior and Ecology*, 1969. Originally published by the University of California Press; redrawn by permission of The Regents of the University of California.)

with considerable carrying quality. Another act, performed by many artiodactyls, is grimacing (called *Flehmen* in German), which involves lifting the head, drawing back the lips, and exposing the lower incisors. Knappe 1964) believes that this action does not serve as a social signal, but instead serves to bring Jacobson's organ (p. 18) into play when the bull is "scenting" cows. In addition, a bull frequently thrashes low vegetation and rakes the ground with his antlers while spurting urine onto his venter. Bulls also wallow in boggy places and gouge the soil with the antlers, and frequently rub the antlers and scrape the incisors against trunks of trees. The master bull drives competitors away from the harem by a ritualized charge: with the head and neck extended, the bull jogs stiff-legged at the intruder while grinding the teeth and lifting the upper lip to display the upper canines. If the interloper does not retreat, a series of preliminary behaviors lead to the bulls facing each other and smashing their antlers together. Such a serious encounter typically occurs after the physical condition of the primary bull is on the wane, and is a

serious test of strength, with each contestant attempting to lunge forward and catch its adversary off balance. Such encounters may end indecisively, but if the master bull is clearly defeated his reign as harem master is ended. In the case of one bull observed by McCullough (1969:89), the animal was deposed from the position of harem master one month after shedding the velvet, and exhibited no further rutting behavior; it even ignored cows in estrus.

The matter of fighting among horned or antlered artiodactyls merits further comment. An interesting correlation seems to exist between the degree of specialization or stylization of fighting and the structure of the head weaponry. In elk the pattern of branching of the antlers of mature bulls is such that when antlers interlock during a fight there is little chance that tines will inflict injury. These animals can afford the luxury of forceful contact during tests of strength; injuries seem to occur largely when the antlers are disengaged and one animal is attempting to escape. In the genus *Oryx* the horns are long, unbranched, and pointed, and are extremely dangerous

FIGURE 17–10.   A bull elk during the rut. Top, the bull bugling; the coat of this animal is caked with mud after wallowing. Middle, the bull on the right (the same animal whose feeding and bedding periods are shown in Figure 17–9) is in poor condition, has broken antlers, and is close to the end of his reign as harem master. Bottom, the same bull on 12 September, nearly two weeks after his defeat. Note the scars and broken antlers. (Photographs from McCullough, D. R.: *The Tule Elk: Behavior and Ecology,* 1969. Originally published by the University of California Press; reprinted by permission of The Regents of the University of California.)

Figure 17-11. The neck display of the Grant's gazelle (*Gazella granti*). (From Estes, 1967.)

weapons. Fighting is highly stylized, with the pushing occurring with the foreheads together but with little contact between the horns (Walther, 1958). The test of strength occurs within the limits of a pattern of behaviors that avoids injury from the dangerous horns. Encounters between male Grant's gazelles, which have quite dangerous horns, usually go no further than intimidation by a neck display (Fig. 17-11), whereas in the Thompson's gazelle Estes (1967:189) found that "natural selection has operated on horn configuration and fighting style to produce a relatively safe type of parry-thrust combat, thus obviating the need for a display substitute." In most cases mutual adaptations between head-weapons and fighting style has insured that opponents come together and measure each other's strength without the use of the full offensive or defensive potential of the weaponry. The pattern of rings around the horns of many African antelope may possibly serve as "non-slip" devices that reduce the danger of injury to fighting males. Of considerable interest is the fact that encounters between well-armed carnivores usually feature highly restricted instead of stylized fighting. According to Ewer (1968: 153), this may be because the survival of a carnivore depends on its efficiency as a killer, and any behavioral characteristics that would reduce this efficiency would be disadvantageous. Of interest at this point is the fact that the neck-hold used by a male cat when copulating is similar to the grip used when killing prey. Rather than a different type of grip developing, the killing grip was simply modified for another purpose.

Highly specialized behaviors in a bat (*Saccopteryx bilineata*) have been found by Bradbury (1971) to be associated with territoriality and harem formation. Males establish small territories in roost trees, and the territories are actively defended against "foreign" males. Males have harems of females, but considerable interchange of females occurs between adjacent harems. Males attract females by singing, and the songs (parts of which are within man's audible range) are frequently long and complicated. Several stereotyped displays are used. Males hover before females that have entered their territories, and during hovering the wing gland is often opened. Males also display in another way to females or to other males with which they are involved in territorial disputes. During such displays a male extends one wing (with the digits folded) toward another bat, opens the wing gland, and shakes the wing some five to ten times. Reciprocal displays of this type may occur between males. Inasmuch as the wing gland, which has a distinctive spicy smell, figures prominently in these displays, olfactory signalling may be an important aspect of these behaviors.

Threat and Appeasement. Threat behaviors are among the most familiar activities of mammals. A dog lifts its upper lip to expose the length of its upper canines, a cat opens its mouth and hisses, and some rodents grind their teeth. These actions all signal a readiness to fight or to attack if the antagonist does not retreat or take other appropriate action. A threat is typical of a situation in which conflicting tendencies preclude either an immediate attack or a hasty retreat.

Threats can be simple, as in animals that merely open the mouth widely to display the teeth (Fig. 17–12), or complex, as in some horned artiodactyls in which both distinctive postures and movements are involved, but usually seem to advertise the most important offensive weapons (Fig. 17–13). Visual threats may be made more impressive or startling by audible threats such as explosive hisses or growls. Strictly defensive threats are often used by animals that are under pressure from an aggressor but that would gladly escape. The cat's threat with the back arched and the side of the body confronting an aggressor is such a behavior, and the "oblique stare" used by bull sea lions on adjacent territories is seemingly a ritualized defensive threat. In an extreme defensive posture, many carnivores such as cats and weasels lie on their backs with the teeth and claws ready for action. The animal has retained its intention to defend itself while abandoning any inclination to attack.

Some mammals have carried this type of defensive behavior a step further by discarding the pretense of defense. Such an appeasement posture or behavior is a complete surrender, and contains no elements that are likely to trigger an opponent's aggression. Complete vulnerability is emphasized, and the response on the part of the dominant animal is to cease its hostile activity. Wolves and many other mammals appease their dominant opponents by lying on the back with the vulnerable throat and underside unprotected. In several artiodactyls lying down serves as appeasement (Walther, 1966; Burckhardt, 1958), and a subor-

FIGURE 17–12. Open-mouthed threat by a pouched "mouse" (*Dasyuroides byrnei*, Marsupialia). (Photograph by Jeffrey Hudson.)

FIGURE 17–13. The threat "yawn" of a male baboon *(Papio ursinus)* displays the large canines. The eyes are closed during the threat, and the whitish lids are conspicuous. (Photograph by Irven DeVore.)

dinate black wildebeest *(Connochaetes gnou)* may even roll on its side with its belly towards its superior and the side of its head on the ground (Ewer, 1968: 177). In the Grant's gazelle a lowering of the head, the reverse of the high-headed threat posture, is adopted by a submissive animal (Walther, 1965). In some primates the presenting of the rump as the female does prior to copulation is an appeasement gesture. The brightly colored skin on the rump of some Old World monkeys may serve in part to make the rump conspicuous and thus to make "presenting" appeasement gestures more effective. A "grin" serves as an appeasement in some anthropoid primates—perhaps

this is akin to the cowboy's demand to "smile when you say that, podner."

Appeasement behavior clearly serves several purposes. It allows an animal being defeated in a fight to avoid further injury, and in many cases allows a subordinate animal to avoid a contest. In the cases of highly social species, threat and appeasement behaviors foster the peaceful perpetuation of a dominance hierarchy, and allow animals to be close to one another with a minimum of energy wasted on aggressive interactions. Appeasement behavior may even be important in permitting a subordinate animal to seek social contact without risking attack (Schenkel, 1967). The evolution of appeasement behavior has followed the same course toward ritualization as have other behavioral patterns. Ritualization serves to make signals as simple, obvious, and as unambiguous as possible.

Friendly Behavior. In many social species patterns of friendly behavior, and often close bonds between individuals, help maintain social structure. Grooming is the most common type of friendly behavior. This may involve the grooming of infants by the mother, mutual grooming of the fur by adults as in primates, or grooming as part of courtship behavior. Other friendly or recognition behaviors include smelling of the mouth and the anal and genital region in dogs, the mutual embrace of chimpanzees (*Pan troglodytes*) described by Goodall (1965:471, 472), and the mutual pressing together by duiker antelope (*Cephalophus maxwelli*) of the maxillary scent gland (Ralls, 1971:446). The choral howling so characteristic of such social canids as coyotes, jackals, and wolves may keep members of a group apprised of the locations of other members and may strengthen familiarity and bonds between individuals of the group. The nuzzling and tail wagging of wolves preparing for a hunt may serve to create a common mood in preparation for a cooperative effort (Lorenz, 1963).

Accompanying social behavior and strong friendly ties between individuals is often extreme pugnacity towards "outsiders." An individual recognized as not belonging to the social group may be attacked and driven away. Smell is an important clue used in such recognition (Ewer, 1968; Crowcroft, 1966; Ralls, 1971) but to some species visual signals are also important.

ACTIVITY RHYTHMS

A striking aspect of the behavior of animals is the rhythmic or cyclic pattern of activity. Some species are active at night (*nocturnal*) and some during the day (*diurnal*), and some are active primarily at dawn and dusk (*crepuscular*). The activity periods tend to be at regular intervals — the time of emergence of a particular species of bat may differ by no more than two or three minutes night after night. Animals also exhibit other kinds of cyclic behavior. The timing of reproduction is cyclic, and in some mammals such as bats daily or seasonal shifts occur between highly active and torpid states. Migratory movements are also cyclic. Daily activity rhythms, those based on the 24-hour cycle, are termed *circadian rhythms*, and are better understood than are other types of rhythms.

Circadian rhythms differ markedly between species. Most mammals are nocturnal, but even between two nocturnal species there are contrasts between the patterns of activity (Fig. 17–14). In general, small mammals that are especially vulnerable to predation, such as rodents, tend to be nocturnal (chipmunks and ground squirrels are exceptions), whereas less vulnerable species, such as many ungulates, may be more or less active during the day. The activity cycles of carnivores seem to be geared to the circadian cycles of their prey or to the period when hunting is most rewarding. Martens (*Martes americana*) frequently forage by day, when red squirrels (*Tamasciurus hudsonicus*) are active, whereas coyotes (*Canis latrans*) hunt at dusk and at night, when rabbits and rodents are feeding.

Circadian rhythms doubtlessly are also influenced by interactions between species with similar environmental needs; in some cases competition between species is reduced or eliminated because their activity cycles are out of phase. Two species of fishing bats (*Noctilio*), both of which feed over water, avoid intense competition partly by foraging at different times of the night (Hooper and Brown, 1968). Clearly, the circadian rhythm of an animal is part of its total adaptation to its particular mode of life and envi-

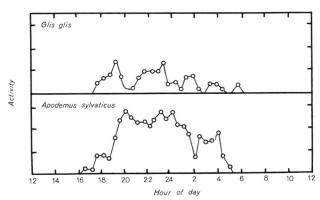

FIGURE 17–14.   The autumn activity cycles of two nocturnal rodents. (After Eibl-Eibesfeldt, 1958.)

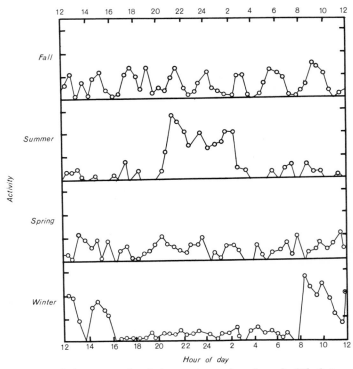

FIGURE 17–15.   Seasonal changes in the daily activity cycles of a vole (*Clethrionomys glareolus*). (After Eibl-Eibesfeldt, 1958.)

ronment, and has evolved just as have morphological characters.

The question of whether circadian cycles are endogenous (internally controlled) or exogenous (ultimately regulated by external stimuli) has occupied the attention of many biologists. Clearly, some strong endogenous control is present in many species. As an example, careful work on the flying squirrel (*Glaucomys volans*) by De Coursey (1961) showed that even under constant environmental conditions, including continuous darkness, flying squirrels maintained regular activity periods that only deviated ±2 minutes from the mean value for activity periods. When a laboratory animal whose circadian cycle is not in phase with the natural 24-hour light-dark cycle is again exposed to normal day and night conditions, however, its cycle rapidly shifts and becomes "synchronized." The cycle becomes adjusted and locked—*entrained*— to the 24-hour cycle (Bruce, 1960). Circadian cycles, and other animal behaviors, are seemingly regulated by intricate, and as yet poorly understood, interactions between endogenous and exogenous factors.

As might be expected if circadian cycles are adaptive, they shift seasonally in some species (Fig. 17–15). Attending the seasonal changes in environmental temperatures are changing metabolic demands put on small mammals, and some shifts in circadian rhythms may allow the animals to avoid activity during times of most intense temperature stress. Studies by Wirtz (1971) have shown that in the deserts of California the antelope ground squirrel (*Ammospermophilus leucurus*) is most active at midday in the winter, whereas in the summer the greatest activity is in the morning and late afternoon; the midday heat is then avoided. The shift from nocturnal activity in the summer to diurnal activity in the winter by a bank vole (Fig. 17–15) probably results in a considerable saving of energy.

Activity cycles are also clearly geared to the basic metabolic demands of animals. In shrews, activity periods are distributed more or less evenly through the day, and the smaller shrews have shorter and more frequent bursts of activity (Crowcroft, 1953). The high metabolic rates of the smaller shrews require frequent periods of feeding and short intervals between feedings. Larger mammals such as rabbits and rats, which have much lower metabolic rates than those of small shrews, can meet their energy needs by feeding at dusk and at night, and some rabbits probably have only two feeding periods per 24-hour cycle.

CHAPTER 18

REPRODUCTION

Compared to the types of reproduction in all other vertebrate classes, the reproductive pattern typical of mammals departs furthest from that of primitive vertebrates. Primitive, ancestral vertebrates were presumably egg layers, and this style of reproduction, or fairly modest variations on this theme, is typical of all classes of vertebrates but the Mammalia. In all mammals but prototherians, young remain during their embryonic and fetal life within the uterus, and here embryonic differentiation of tissues and organs and growth of the fetus occurs. Nourishment and protection for the intrauterine young are provided by the mother, and under most conditions survival rates of the fetuses are high. After birth all young mammals are nourished by milk from the mother, and parental care, or in most cases maternal care, lasts until young are reasonably capable of caring for themselves. The young of some mammals stay with their parents through an additional period of learning that increases the chances for survival of young when they become independent. In sharp contrast, in most non-mammalian vertebrates (the birds are an exception) the young have little or no parental care after hatching, or in the case of ovoviviparous animals, after birth. In mammals the combined effect of the high survival rate of fetuses and extended post-partum care is an increase in the efficiency of reproduction in terms of expenditure of energy per young that reaches reproductive maturity. Most lower vertebrates lay great numbers of eggs at tremendous metabolic expense, and the success of the species depends on the survival of an extremely small percentage of the young; considering any given young of a lower vertebrate, survival is unlikely. In mammals, on the contrary, relatively few young are produced, but the likelihood for survival of any given young is fairly high.

The following sections of this chapter consider unique features and typical major patterns of mammalian reproduction. No attempt has been made to catalogue exhaustively the reproductive cycles of all families of mammals, but tables are included that review features of the reproductive cycles of selected mammals representing most orders. Descriptions of reproduction in many mammalian species are given by Asdell (1964), and the tables are partly based on his work.

THE MAMMALIAN PLACENTA

One of the most distinctive and important structures associated with the reproduction of therian mammals is the placenta. Differences between the major placental types have been used in distinguishing between some of the higher taxonomic categories of mammals (subclasses and infraclasses), and some primary contrasts between reproductive patterns in mammals depend partly on placental differences. Further, the relative competitive abilities of marsupials and eutherians may be strongly influenced by basic differences between the structures and functions of the placentas of these groups.

A functional connection between the embryo and the uterus is necessary in animals in which development of the fetus occurs within the uterus, and in which nutrients for the fetus come directly from the uterus rather than from yolk stored in the ovum. This connecting structure, the placenta, allows for nutritional, respiratory, and excretory interchange of material by diffusion between the embryonic and the maternal circulatory systems, and consists of both embryonic and uterine tissues. The placenta also functions as a barrier that excludes from the embryonic circulation bacteria and many large molecules. In addition, in eutherians the placenta produces certain food materials and synthesizes hormones important for the maintenance of pregnancy. Mammals are not unique in having a placenta, for certain fishes and reptiles establish placenta-like connections allowing diffusion of materials between the vascularized oviduct and the embryo. Among mammals, the major types of placentae differ sharply in structure and in the efficiency with which they facilitate the nourishment of the embryo.

Chorio-Vitelline Placenta. This, the most primitive type of mammalian placenta, occurs in all marsupials except those of the family Peramelidae, the bandicoots. In marsupials with a chorio-vitelline placenta, the yolk sac is greatly enlarged to form a placenta. The blastocyst does not actually implant itself deep in the uterine mucosa, as is the case in eutherians, but merely sinks into a shallow depression made by erosion of the mucosa. The contact is strengthened by the wrinkling of the wall of the blastocyst that lies against the uterus, and this wrinkling serves to increase the absorptive surface of the blastocyst. The embryo is nourished largely by "uterine milk," a nutritive substance secreted by the uterine mucosa and absorbed by the blastocyst. The embryo also derives nourishment from limited diffusion of substances between the maternal blood in the eroded depression in the mucosa and the blood vessels within the large yolk sac of the blastocyst (Fig. 18–1). Seemingly, the mechanical weakness of the fetal-maternal connection, and the inefficiency of the system of nourishment of the fetus, are factors limiting the length of the gestation period, which is remarkably short in some marsupials (Tables 18–1 and 18–2).

Chorio-Allantoic Placenta. This type of placenta occurs in the bandicoots and in all eutherian mammals. Although similar to the eutherian chorio-allantoic placenta in basic structure, the peramelid placenta achieves less effective transfer of substances between the fetal and maternal circulations. In peramelids the allantois is fairly large and becomes highly vascularized; the blastocyst rests against the endometrium on the side where the allantois contacts the chorion. At the point of contact with the blastocyst, the uterus is highly vascularized, and the part of the chorion against the vascularized endometrium is more or less lost. At this point of approximation of the maternal blood stream and the allantois, exchange of materials occurs across the allantoic membranes. Because the peramelid allantois lacks villi and only its corrugations serve to increase its absorptive surface, a limited surface area is available for exchanges of material between the maternal and fetal blood streams;

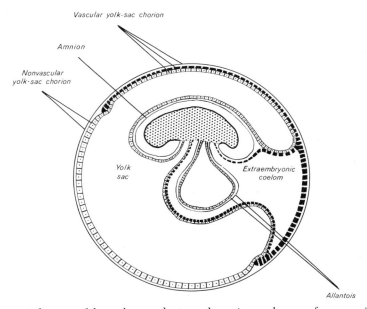

FIGURE 18–1. A diagram of the embryo and extraembryonic membranes of a marsupial *(Didelphis).* (After Torrey, T. W.: *Morphogenesis of the Vertebrates,* 3rd ed., John Wiley and Sons, Inc., 1971. By permission of John Wiley and Sons, Inc.)

Table 18–1. Information on the Breeding Cycles of Several Families of Marsupials. (Data on Suckling Period from Sharman (1970).)

FAMILY	BREEDING SEASON; POLYESTROUS OR MONESTROUS	LITTER SIZE	GESTATION PERIOD IN DAYS	SUCKLING PERIOD IN DAYS
Didelphidae	March – Oct.; monestrous or polyestrous	2–25	13	70–80
Dasyuridae	April – Dec.; monestrous	3–12	8–34	49–150
Peramelidae	March – June; polyestrous	1– 7	15	59
Phalangeridae	all year; 1 litter/2 yrs. or polyestrous	1– 6	17–35	42–165
Macropodidae	all year; monestrous or polyestrous	1– 2	24–43	64–270

Table 18-2. Reproductive Patterns of Some Families of Mammals.
(Data Largely from Asdell, 1964, and Anderson and Jones, 1967.)

FAMILY	LITTER Frequency	Size	BREEDING SEASON	TIME OF BIRTH	GESTATION PERIOD
Monotremata					
Tachyglossidae	1/year	1	winter, early spring		
Ornithorhynchidae	1/year	1–3	later winter, early spring		3–5 weeks
Marsupialia					
Didelphidae	1–2/year	10–25			13 days
Dasyuridae		3–10	May–Oct.	Jan.–Sept.	8–34 days
Notoryctidae	1/year	1–2		Nov.	
Peramelidae	6/year	1–7	March–June		15 days
Phalangeridae	1/2 years	1–3	all year	all year	17–35 days
Phascolomyidae	1/year	1		Jan.–July	
Macropodidae		1–2	Sept.–March		27–40 days
Insectivora					
Erinaceidae	2/year	5	Apr.–Aug.		34–49 days
Talpidae	1/year				35 days
Tenrecidae		2	Oct.–Jan.		56–66 days
Chrysochloridae	1/year	1–2	rainy season		
Soricidae	1/year				13–28 days
Tupaiidae		1–4	all year		56 days
Dermoptera					
Cynocephalidae	1/year	1			60 days
Chiroptera					
Pteropodidae	1/year	1–2	varies with species		
Nycteridae		1	early Dec.		
Megadermatidae	1/year	1	May		
Rhinolophidae		1–2	Oct. or mid-Apr.		
Phyllostomatidae		1–2	spring, summer		highly variable
Desmodontidae	1–2/year	1	all year		
Natalidae	1/year		Jan.–July		
Thyropteridae	1/year	1		Aug.	
Vespertilionidae		1–4	Apr.–May		3 mo.
Molossidae	1/year	1	early spring		90 days +
Primates					
Lemuridae	1/year		Jan.–Feb.	July–Oct.	2–5 mo.
Lorisidae		1–2	all year		6 weeks– 6 mo.
Tarsiidae	1/year	1		Nov.–Feb.	
Cebidae		1	all year		6 mo.
Callithricidae		1–3			140–150 days
Cercopithecidae		1	all year		140–210 days
Pongidae		1	all year		200–260 days
Edentata					
Myrmecophagidae		1			190 days
Bradypodidae		1			4–6 mo.
Dasypodidae	1/year	1–12		Oct.	4–5 mo.
Pholidota					
Manidae		1–2		Jan.–Mar. or Nov.–Dec.	
Lagomorpha					
Ochotonidae	1–2/year	2–5	May–Sept.		30.5 days
Leporidae	several	2–8	all year		26.5–42 days
Rodentia					
Aplodontidae		2–6	late winter, early spring	Feb.–Apr.	28–30 days

Table 18–2. Reproductive Patterns of Some Families of Mammals.
(Data Largely from Asdell, 1964, and Anderson and Jones, 1967.) *(Continued)*

FAMILY	LITTER Frequency	LITTER Size	BREEDING SEASON	TIME OF BIRTH	GESTATION PERIOD
Sciuridae	1–3/year	1–15	Jan.–March, May–Aug.	Apr.–July	22–45 days
Geomyidae		2–6	all year		
Heteromyidae		1–8			24–33 days
Castoridae	1/year	1–6	Jan.–March		42–128 days
Anomaluridae	2/year	1–4			
Cricetidae		1–18		April	20–30 days
Rhizomyidae		3–4		Apr.–May	
Muridae	3–8/year	1–15	all year		20–27 days
Gliridae		2–9	May–June		1 mo.
Zapodidae	1–2/year	3–6	May–Aug.	June–Sept.	18 days
Dipodidae		3	Apr.–Aug.		42 days
Hystricidae	2/year	1–2			6–8 weeks
Erethizontidae	1/year	1–4	Feb. or Nov.–Dec.		2–7 mo.
Caviidae	2/year	1–5			2 mo.
Hydrochoeridae	1/year	3–8			104–111 days
Dasyproctidae		1–4	all year		104 days
Chinchillidae	1–2/year	1–6	March or Oct.		90–120 days
Myocastoridae	2/year	2–8			4 mo.
Echimyidae	2/year	1–5		early spring, late summer	
Thryonomyidae		2–4		June–Aug.	
Petromyidae	1/year	2	hot weather		
Ctenodactylidae		1–4	Jan.; April		
Mysticeti					
Balaenidae	alternate years	1		Jan.–Apr.	12 mo.
Eschrichtiidae	not in successive years	1		Jan.–Feb.	12 mo.
Balaenopteridae	not in successive years	1			10–12 mo.
Odontoceti					
Ziphiidae		1	Feb.	Dec.	10 mo.
Monodontidae	1/2 years	1	spring		14 mo.
Physeteridae	1/4–5 years	1	spring	summer	15.5 mo.
Platanistidae		1	July–Sept.	Apr.–July	8–9 mo.
Phocoenidae	1/year	1	July–Aug.	May–June	11 mo.
Delphinidae	1/3 years– 3/4 years		summer		11–16 mo.
Carnivora					
Canidae	1/year				51–80 days
Ursidae	alternate years	1–4			6 mo.
Procyonidae		1–6			54–77 days
Mustelidae	1/year	1–14			36–350 days
Viverridae	2/year	2–4			49–60 days
Hyaenidae		1–6			3 mo.
Felidae	1/year	1–6			56–100 days
Pinnipedia					
Otariidae		1	varies with species		11 mo.
Odobenidae	1/3 years	1			11–12 mo.
Phocidae		1			276–340 days
Tubulidentata					
Orycteropodidae		1	Apr.–May	winter	7 mo.
Proboscidea					
Elephantidae		1	Jan.–Feb.		18–24 mo.
Hyracoidea					
Procaviidae		1–6			8 mo.

Table 18–2. Reproductive Patterns of Some Families of Mammals.
(Data Largely from Asdell, 1964, and Anderson and Jones, 1967.) *(Continued)*

FAMILY	LITTER Frequency	Size	BREEDING SEASON	TIME OF BIRTH	GESTATION PERIOD
			Sirenia		
Dugongidae		1	winter	winter	11–12 mo.
Trichechidae		1	all year	all year	152 days
			Perissodactyla		
Equidae		1			336–350 days
Tapiridae		1–2	just before rainy season		390 days
Rhinocerotidae		1	July–Oct.		17–19 mo.
			Artiodactyla		
Suidae		2–14	all year	all year	112–150 days
Tayassuidae		2		Apr., Aug., Nov.	112–116 days
Hippopotamidae	1/237 days		all year	all year	201–210 days
Camelidae		1			10–12.5 mo.
Tragulidae		1–2	June–July	close of the rainy season	152–172 days
Cervidae		1–6	late fall, winter		5–10 mo.
Giraffidae		1	all year		14–15 mo.
Antilocapridae		1–2	Aug.–Sept.		230–240 days
Bovidae		1–5			4–11 mo.

supplementary nutrition is supplied by uterine milk. Probably due partly to the lack of villi and the resulting lack of absorptive efficiency of the allantois, the gestation period of peramelids is fairly short, and the suckling period is long (Table 18–1).

In eutherian mammals the chorio-allantoic placenta reaches its most advanced condition with regard to facilitating rapid diffusion of materials between the fetal and uterine blood streams. In eutherians the blastocyst first adheres to the uterus and then sinks into the endometrium. The mechanisms by which implantation occurs are not fully understood. Proteolytic enzymes secreted by the chorion have been thought to erode the endometrium and allow the blastocyst to sink into the cavity thus formed, but little definite evidence supports this view. Studies by Boving (1959) have shown that in the domestic rabbit dissociation of uterine epithelial cells overlying blood vessels facilitates the implantation of the blastocyst. This dissociation is initiated by a local rise in pH caused by a bicarbonate compound produced by the blastocyst, and by a reciprocal reaction on the part of the endometrium to remove the bicarbonate. How widespread this reaction is among mammals, and therefore how important it is as an implantation-furthering device is not known. As implantation proceeds, chorionic villi grow rapidly and push further into the endometrium as local breakdown of uterine tissue continues. The resulting tissue "debris" is often called embryotroph; this nutritive substance is absorbed by the blastocyst and nourishes the embryo until the villi are fully developed and the embryonic vascular system becomes functional. In response to the presence of the blastocyst, the uterus becomes highly vascularized at the site of implantation. When the eutherian placenta is fully formed, the complex and highly vascularized villi provide a remarkably large surface area through which rapid interchange of materials between the maternal and fetal circulations can occur (Fig. 18–2). The extent to which the villi increase the surface area available for diffusion is difficult to imagine; the extent of this increase is suggested by the fact that the total length of the villi in the human placenta is roughly 30 miles (Bodemer, 1968).

Among eutherians the degree to which the maternal and fetal blood streams are separated in the placenta varies widely. Lemurs, some ungulates (suids and equids), and cetaceans have an *epithelio-chorial placenta,* in which the epithelium of the chorion is in contact with the uterine epithelium and the villi rest in pockets in the endometrium. Under these structural conditions, oxygen and nutrients must pass through the walls of the uterine blood vessels and through layers of connective tissue and epithelium before entering the fetal blood stream. In ruminant artiodactyls the uterine epithelium is eroded locally, and contact between the chorionic ectoderm and the vascular uterine connective tissue occurs. This is a *syndesmo-chorial placenta.* In carnivores erosion

of the endometrium is carried further and the epithelium of the chorion is in contact with the endothelial lining of the uterine capillaries. This is called an *endothelio-chorial placenta.* Destruction of the endometrium in some mammals may involve even the endothelium of the uterine blood vessels, allowing blood sinuses to develop in the endometrium; the chorionic villi may then be in direct contact with maternal blood. This *hemo-chorial placenta* occurs in some insectivores, bats, anthropoid primates, and some rodents. In rabbits and some rodents the destruction of placental tissue is so extreme that only the endothelial lining of the blood vessels in the villi separate the fetal blood from the surrounding maternal blood sinuses (Arey, 1965:147). In this case a *hemo-endothelial placenta* results.

The shape of the placenta is governed by the distribution of villi over the chorion. Several different distributions of villi occur in mammals. The lemurs, some artiodactyls, and perissodactyls have a *diffuse placenta;* this type of placenta has a large surface area because the villi occur over the entire chorion. Ruminant artiodactyls have *cotyledonary placentae,* consist-

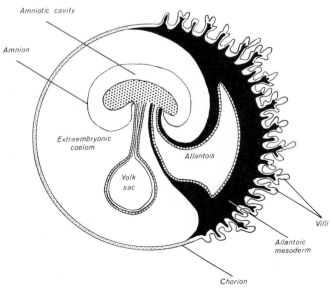

FIGURE 18–2. A diagram of the embryo and extraembryonic membranes of a placental mammal. (After Balinsky, B. I.: *An Introduction to Embryology,* 3rd ed., W. B. Saunders Company, 1970.)

ing of more or less evenly spaced groups of villi scattered over the mostly avillous chorion. Carnivores have a *zonary placenta;* a continuous band of villi encircles the equator of the chorion. In the *discoidal placenta,* villi occupy one or two disc-shaped areas on the chorion; this type occurs in insectivores, bats, some primates, rabbits, and rodents.

At birth the fetal contribution to the placenta is always expelled as part of the "afterbirth," but the maternal part of the placenta may or may not be lost at this time. In mammals with the *epithelio-chorial* type of placenta the villi pull out of the uterine pits in which they fit, none of the endometrium is pulled away, and no bleeding occurs at birth. This placenta is termed *nondeciduous.* In mammals with placentae allowing more intimate approximation of uterine and fetal blood streams, because of the extensive erosion of the uterine tissue and the extensive intermingling of uterine and chorionic tissue, at birth the uterine part of the placenta is torn away, resulting in some bleeding. This type of placenta is *deciduous.* The hemorrhaging after birth is soon stopped by the collapse of the uterus and by contractions of the myometrium, which tend to constrict the blood vessels, and by clotting of blood.

The rate of movement of substances from the maternal to the fetal blood stream in the placenta is of course increased by the reduction of the number of interposed membrane barriers. Because of the difference between the number of such barriers in the man and the pig, for example, sodium is transferred 250 times more efficiently by the placenta of man than by that of the pig (Flexner et al., 1948). The remarkable absorptive ability of the allantoic placentae of such mammals as insectivores, bats, primates, rabbits, and rodents is due largely to the great surface area afforded by the complex system of villi, to the extensive erosion of uterine mucosa and the resulting development of blood sin-

uses into which the villi extend, and to the loss of nearly all of the membranes separating uterine from fetal blood.

THE ESTROUS CYCLE, PREGNANCY, AND PARTURITION

In mammals reproduction is characterized by a series of cyclic events that are under nervous and hormonal control. As with many complex functions of the vertebrate body, the regulation of the reproductive cycle is maintained through reciprocal controls between endocrine organs and their secretions. The events characterizing the stages in the reproductive cycle in mammals are well known, but details of the hormonal regulation of these events are not completely understood. The ovarian cycle results in the development of ova, their release from the ovary, and their passage into the uterus; the uterine cycle involves a series of cyclic changes in the uterus.

The ovarian cycle includes two major phases: (1) the growth of the follicle and the release of the ovum; and (2) the development of the corpus luteum from the ruptured follicle. This cycle seems to be controlled largely by pituitary and ovarian hormones. The pituitary produces FSH (follicle stimulating hormone) and LH (luteinizing hormone), which act together to stimulate growth of the follicle and to initiate the secretion of estrogen by the ovary. Estrogen acts on the pituitary to stimulate increased production of LH, to initiate production of LTH (luteotrophic hormone), and to reduce the secretion of FSH; under the influence of interactions between these hormones ovulation occurs (ova are released from follicles) and the corpus luteum forms from the ruptured follicle. Maintained by LTH from the pituitary, the corpus luteum produces progesterone, which sensitizes the uterus for implantation. If fertilization of ova does not occur, the corpora lutea regress and estrogen and progesterone production is reduced. The pituitary responds by resuming production of FSH, and another ovarian cycle is initiated. The pattern described above, involving spontaneous ovulation, is the typical situation in mammals, but some deviations are known. In the cat and some rodents the follicles develop but ovulation does not occur until after copulation. In rabbits the follicles do not develop fully before copulation and a long estrus may occur; ripening of the follicles and ovulation are initiated by copulation.

As the ovarian cycle proceeds, a series of cyclic changes occurs in the uterus. The uterus consists of an outer peritoneal layer, an intermediate layer of smooth muscle, the *myometrium*, and an inner layer of uterine mucosa, the *endometrium*. Just before ovulation, the endometrium becomes thicker; this uterine stage is called the *proliferation phase*. In many mammals a period of "heat," during which the female is receptive to the male, occurs at the end of the proliferation phase. This time of receptivity immediately prior to ovulation is termed *estrus*. After ovulation the endometrium develops further and becomes highly vascularized; this is the *progestational phase* of the uterine cycle. In most mammals, if the ova are not fertilized, the endometrium shrinks and the vascularization is reduced. In non-primate mammals no extensive bleeding occurs during the regression of the endometrium and the period of receptivity is short; in these animals the uterine cycle is referred to as the *estrous cycle*. Some mammals have repeated estrous cycles during a year and are said to be *polyestrous;* others have a single estrous cycle per year and are *monestrous*. In some species sexual cycles are seemingly influenced by day length. The eyes receive light over a progressively longer period of time each day during the spring and initiate a nervous reflex transmitted by the optic nerves to the brain, where certain centers stimulate the release of gonadotrophic hormones by the pituitary. The ovarian cycle may be triggered in some rodents by vi-

tamins or nutritional factors present in the green vegetation that appears in the spring (Negus and Pinter, 1966). Although considerable information relating to mechanisms controlling reproductive cycles is scattered through the literature, the importance of these mechanisms remains obscure. It seems doubtful that in most mammals any single factor or simple combination of factors is responsible for controlling the estrous cycle.

The ovarian and uterine cycles in primates are different, to some extent, from those in other mammals. In man and most other primates considerable bleeding is typical of the time of endometrial breakdown, ovulation occurs at regular intervals throughout the year, and females are receptive over an extended period. This primate cycle is called the *menstrual cycle*.

When copulation occurs the sperm reach the oviducts within a matter of minutes in some species, and fertilization of the ova usually occurs within 24 hours of ovulation. The zygotes move down the oviducts, aided by contractions of the muscles of the oviducts, and usually reach the uterus and implant within a few days.

The delicate hormonal control of pregnancy is exerted in eutherians by interactions between hormones produced by the pituitary, the ovary, and the uterus. In the early part of pregnancy, chorionic gonadotrophin is of critical importance in preserving the corpora lutea and in preventing regression of the endometrium. This hormone is produced first by the trophoblast during its implantation in the endometrium, and then by the chorion, which develops from the trophoblast. During early pregnancy the corpus luteum, because of its production of progesterone, is important in maintaining pregnancy by keeping the endometrium in a thickened and highly vascularized condition and by altering the ability of the myometrium to perform coordinated contractions that might expel the embryo. In some species progesterone sensitizes the endometrium and increases the efficiency of implantation of the blastocyst. The maintenance of pregnancy in many mammals is not entirely under the control of the corpus luteum; instead, as pregnancy continues, hormones are produced progressively more by the placenta and less by the ovary. In man the placenta is thought to produce chorionic gonadotrophin and estrogens, and is probably the most important source of progesterone. During the latter stages of pregnancy the placenta seems to be a nearly independent endocrine gland, which in man, at least, takes over the functions of the pituitary gland and the corpus luteum. An important hormone of pregnancy, but one whose function is important mainly at parturition, is *relaxin.* This hormone is known to occur in a variety of mammals, and may be universal among mammals. The concentration of relaxin in the blood stream increases toward the end of pregnancy; this hormone causes relaxation of the pelvic ligaments and the pubic symphysis in preparation for parturition. In some mammals, such as pocket gophers (Geomyidae), the connective tissue joining the pubic bones is resorbed at puberty and a gap between the pubic bones remains during the rest of the life of the animal (Hisaw, 1924). Relaxin may be produced by the uterus or by the placenta, and in man is known to be produced by the ovaries during pregnancy (Guyton, 1971:981).

Birth is accomplished by rhythmic and powerful contractions of the uterine myometrium aided by the abdominal muscles. Continued contractions force the placenta from the uterus and the vagina. The uterine contractions are seemingly under the control of interacting hormones. *Oxytocin,* produced by the hypothalamus and stored in the posterior lobe of the pituitary, occurs in increasingly higher concentrations in the maternal blood stream toward the end of pregnancy; oxytocin can initiate contractions of the uterus. Apparently the reduced concentrations of progesterone late in pregnancy are insufficient to block the effects of the increasing levels of oxytocin, with resulting contractions of the myometrium and parturition. The amount of estrogen also increases late in pregnancy, and the sensitizing of the uterus by estrogen shortly before parturition may allow oxytocin to initiate uterine contractions.

The newborn mammal is nourished by milk produced by the female's mammary glands. Under the influence of estrogen and progesterone from the placenta, these glands undergo considerable growth during pregnancy. Milk secretion is stimulated and regulated by *prolactin,* produced by the anterior lobe of the pituitary. Prolactin is secreted in progressively larger amounts during the latter part of pregnancy and after parturition, and when its inhibition by placental hormones is removed after birth, milk secretion begins. Milk production is partly under neural control and continues only as long as the suckling stimulus persists.

MAJOR REPRODUCTIVE PATTERNS

Although several deviations from the usual scheme occur, most mammals follow a similar reproductive pattern with regard to development of the embryo. After ovulation the ovum passes down the oviduct, where it is fertilized. Early cell cleavages occur during the several days occupied by passage of the zygote down the oviduct, and by approximately the time the zygote enters the uterus it has become a hollow ball of cells enclosing a fluid-filled cavity. This stage is called the *blastocyst.* After further enlargement, the blastocyst implants in the endometrium. Implantation occurs between the fifth and fourteenth day after copulation, and the timing of implantation varies little within a species. After implantation the embryo develops a system of membranes and blood vessels in the placenta that allow diffusion of nutrients and waste materials between the uterine and embryonic blood vessels. In eutherian mammals the length of time from fertilization to implantation is considerably shorter than the period between implantation and birth. Typically, fertilization occurs shortly after ovulation, and the development of the embryo from fertilization to birth is an uninterrupted process. Perhaps in response to specialized activity cycles, some mammals have abandoned this usual pattern of continuous development. One departure involves a delay of ovulation and fertilization until long after copulation (delayed fertilization); another is typified by normal fertilization and early cell cleavages but also by an arresting of embryonic development at the blastocyst stage (delayed implantation).

Delayed Fertilization. This unusual pattern of development occurs in a number of bats inhabiting north temperate regions. As early as 1879 Fries recognized that males of some species of the families Rhinolophidae and Vespertilionidae had the unusual ability to store viable sperm through the winter, long after spermatogenesis had ceased; later studies detailed the reproductive cycles of the females of these species (see Guthrie, 1933; Hartman, 1933; Wimsatt, 1944, 1945). These remarkable reproductive cycles are seemingly adaptations to continuous or periodic winter dormancy, and occur in a number of New World and Old World species included in the following genera: *Rhinolophus, Myotis, Pipistrellus, Eptesicus, Nycticeius, Lasiurus, Plecotus, Miniopterus,* and *Antrozous.* Delayed fertilization may be the typical pattern in all but the tropical members of the family Vespertilionidae. Excellent papers by Wimsatt (1944, 1945) and by Pearson and his colleagues (1952) describe delayed fertilization as it occurs in vespertilionids, and the following remarks will be based largely on those studies.

The reproductive cycle of the big-eared bat (*Plecotus townsendii*) in California follows a timetable similar to that of many vespertilionids of temperate zones. The testes descend into the scrotum in the spring. This migration is caused mostly by increased production of testosterone, which is cyclical in bats. The males become reproductively active in August. The testes begin to enlarge in the spring and are largest in September; spermatogenesis occurs mostly in late August and September. The testes regress and spermatogenesis ends before winter, but the caudal epididymides retain motile sperm through February, and the accessory reproductive organs remain enlarged throughout the winter. Young males are not fertile in their first autumn. In the females a single Graafian follicle enlarges in the autumn, but remains in the ovary throughout the winter. A given female may be inseminated repeatedly in the fall and winter, and males frequently copulate with hibernating females, although usually all females are inseminated by the end of November. The most typical vespertilionid pattern is for most copulation to occur before hibernation. The sperm are stored in the uterus, where they remain motile for at least 76 days. In *P. townsendii* ovulation usually occurs in late February or March, either while the females are still at the winter roost or shortly after they leave. In many species inhabiting cold regions, ovulation occurs shortly after the females emerge from the hibernacula. Implantation is nearly always in the right horn of the uterus in *P. townsendii*, but ovulation occurs with equal frequency in each ovary. The gestation period in this species is highly variable, ranging from 56 to 100 days. This variation is probably due to regional differences in ambient temperatures and hence to the different body-temperature routines that occur in bats of widely separated colonies. Periodic torpor or low body temperatures after the beginning of gestation slow the development of the embryo.

Several features of the unique cycle are especially noteworthy: (1) The development of the male reproductive organs is out of phase; that is, the testes have regressed when the caudal epididymides and accessory organs are most enlarged and when breeding activity is at its peak. (2) Males retain viable sperm in the caudal epididymides long after spermatogenesis has ceased. (3) Females do not ovulate until long after they have been inseminated, but are able to store viable sperm for several months. (4) Because of differing metabolic routines in different individuals, the rate of development of the embryo is highly variable.

Delayed fertilization is seemingly a highly advantageous adaptation in mammals with long periods of dormancy. Spermatogenesis, enlargement of reproductive organs, and copulation require considerable energy. In species that practice delayed fertilization these activities occur in the late summer and autumn, when males are in excellent condition and have abundant food, rather than in the spring, when the animals are in their poorest condition and when food (insects) may not yet be abundant. Ovulation and zygote formation occur almost immediately upon emergence from dormancy, rather than being delayed until after males attain breeding condition and copulation occurs. The female can therefore channel more energy into nourishment of the embryo than would be available if copulation were occurring immediately after hibernation. Perhaps the major advantage is that of hastening the time of parturition and allowing the longest possible time for development of young before the winter period of dormancy.

Delayed Implantation. This deviation from the "normal" reproductive pattern occurs in a variety of mammals, representing the orders Chiroptera, Edentata, Carnivora, Pinnipedia, and Artiodactyla (Table 18–3). These mammals obviously do not share a common heritage; in addition, they occupy a wide variety of habitats, and pursue differing modes of life. Delayed implantation in each group, therefore, has probably evolved separately and in

Table 18–3.   Periods During Which Blastocysts Remain Dormant in Some Mammals with Obligate Delayed Implantation. (Data mostly from Daniel (1970); Data on *Artibeus jamaicensis* from Fleming (1971).)

SPECIES	DORMANCY OF BLASTOCYST (in months)
Order Chiroptera	
Equatorial fruit bat (*Eidolon helvum*)	3+
Jamaican fruit bat (*Artibeus jamaicensis*)	2½
Order Edentata	
Nine-banded armadillo (*Dasypus novemcinctus*)	3½–4½
Order Carnivora	
Black bear (*Ursus americanus*)	5–6
Grizzly bear (*U. arctos*)	6+
Polar bear (*U. maritimus*)	8
Marten (*Martes americana*)	8
Fisher (*M. pennanti*)	10–11
Badger (*Taxidea taxus*)	6
River otter (*Lutra canadensis*)	9–11
Mink (*Mustela vison*)	½–1½
Long-tailed weasel (*M. frenata*)	7
Order Pinnipedia	
Alaskan fur seal (*Callorhinus ursinus*)	3½–4
Harbor seal (*Phoca vitulina*)	2–3
Grey seal (*Halichoerus grypus*)	5–6
Walrus (*Odobenus rosmarus*)	3–4
Order Artiodactyla	
Roe deer (*Capreolus capreolus*)	4–5

response to different selective pressures. Delayed implantation is either *obligate*, and constitutes a consistent part of the reproductive cycle, or is *facultative*, and provides for a delay of implantation on occasions when an animal is nursing a large litter. A good discussion of delayed implantation is given by Daniel (1970).

In mammals with obligate delayed implantation ovulation, fertilization and early cleavages up to the blastocyst stage occur normally, but further development of the blastocyst is arrested and it does not implant in the uterine endometrium. The blastocyst remains dormant in the uterus for periods of from 12 days to 11 months (Table 18–3). Little growth of the blastocyst occurs during its dormancy, which begins generally when the embryo consists of from roughly 100 to 400 cells. Western forms of the spotted skunk (*Spilogale putorius*) studied by Mead (1968) follow a reproductive pattern fairly typical of mammals with delayed implantation. Males become fertile in the summer, and copulation and fertilization of the ova occur in September. The zygote undergoes normal cleavage but stops at the blastocyst stage; the blastocysts float freely in the uterus for 180 to 200 days. The gestation period is from 210 to 230 days, and the young are usually born in May. During the period of dormancy, each blastocyst is covered by a thick and durable *zona pellucida*, a non-cellular protective layer.

The adaptive advantage of delayed implantation is not understood for all species. In his study of a population of Jamaican fruit bats (*Artibeus jamaicensis*), Fleming (1971) showed that implantation is delayed from August until mid-November, and that the delay allows young to be born in early spring, when fruit is abundant. In this species development of the blastocyst does not stop during diapause, but cell division continues at an extremely reduced rate. Delayed implantation also occurs in the Neotropical phyllostomatid *Macrotus waterhousii* (Bradshaw, 1962; Wimsatt, 1969) and in the Euro-

pean vespertilionid *Miniopterus schreibersii*. In these cases the delayed implantation may confer some of the same advantages as those resulting from delayed fertilization in vespertilionids of temperate zones.

Facultative delayed implantation occurs in some species in which the female is inseminated soon after the birth of a litter. This type of delay is known in some marsupials, some insectivores, and some rodents. In certain rodents that have post-partum estrus, implantation of blastocysts is delayed when the female is suckling a large litter.

Our understanding of the factors controlling normal blastocyst development or dormancy of the blastocyst in eutherian mammals is incomplete. Present evidence suggests that estrogen causes the uterine endometrium to form proteins essential for rapid growth of the blastocyst, and that a deficiency of these proteins results in dormancy of the blastocyst (Daniel, 1970). One protein seemingly responsible for regulation of the differentiation and growth of the blastocyst was named "blastokinin" (Krishnan and Daniel, 1967). Experimentally administered doses of estrogen and/or progesterone have been used in an attempt to initiate growth of the dormant blastocyst in mammals with the obligate type of delayed implantation. These procedures have not been successful in renewing growth of the blastocyst, indicating that in such animals some blocking of the action of estrogen in the endometrium must occur (Daniel, 1970). McLaren (1970) has proposed that during lactation in mice (*Mus*) implantation is delayed by an initial inability of the blastocyst to "hatch" from the zona pellucida, which must be shed before implantation can occur.

Delayed implantation is an important part of the reproductive cycles of many marsupials. In most marsupials the suckling of the young in the pouch inhibits estrus and ovulation; but in some kangaroos and wallabies (Macropodidae) a type of delay occurs that is termed *embryonic diapause* by Sharman (1970). In most macropodids for which embryonic diapause is known, the mother undergoes post-partum estrus; copulation occurs and the ovum is fertilized early in the life of the young she carries in the pouch. The suckling of the pouch young initiates neural and hormonal responses that arrest the activity of the corpus luteum and induce dormancy of the blastocyst; cell division in the blastocyst ceases and it does not implant. In contrast to the dormant eutherian blastocyst, which consists of an inner cell mass that gives rise to the embryo and a hollow sphere of cells that gives rise to extraembryonic membranes, the marsupial blastocyst consists of only 70 to 100 cells that form a single spherical layer of cells of one type (protoderm). The marsupial blastocyst is surrounded by protective coverings consisting of an albumen layer and a shell membrane. When the young leaves the pouch, development of the corpus luteum and growth of the blastocyst resume, the blastocyst implants, and rapid growth of the embryo resumes.

In marsupials the young suckle after they leave the pouch for a period roughly comparable to the suckling period in eutherian mammals of similar size, but the intrauterine period for the marsupial fetus is often short (Table 18–1). As a result, a newborn young may be attached to a nipple and suckling while a much older young is returning periodically to suckle from a separate nipple. In both marsupials and eutherians the composition of the milk changes during pregnancy. In marsupials the milk secreted early in the suckling period contains little or no fat, whereas milk secreted later in the period may contain as much as 20% fat. During "double suckling" in kangaroos a remarkable thing occurs: separate mammary glands concurrently produce vastly different milks, although both glands are seemingly under the same hormonal influences. Compared to the low-fat-content milk produced by the gland supporting the pouch young, the milk produced by

the gland supporting the advanced young has three times as much fat. The physiological basis for this remarkable arrangement is not known.

The available evidence suggests that in marsupials no extraovarian hormones are secreted during pregnancy. The placenta, important as an endocrine organ in eutherians, does not serve this function in marsupials. The reproductive physiology of marsupials is reviewed by Sharman (1970), who believes that the differences between marsupial and eutherian reproduction point to a separate evolution of viviparity in these two groups after they diverged from a common oviparous ancestral stock.

REPRODUCTIVE CYCLES OF MAMMALS

Tremendous variation occurs among the reproductive cycles of different species of mammals, and this variation is reflected by differences among the lengths of the gestation periods in different mammals (Table 18–2). The duration of gestation depends in part on the size of the animal, on whether or not delayed fertilization or delayed development occurs, and on the rate of intrauterine development of young. The longest gestation periods are those of elephants (up to 22 months). But gestation periods cannot be predicted on the basis of size of the mammal alone. The blue whale (*Balaenoptera musculus*), for example, is the largest living animal, and probably the largest animal that has ever lived, but its gestation period is only 11 months. For some unknown reason, growth of the whale fetus is amazingly fast. As might be expected, the gestation periods in species with delays in fertilization or development are characteristically long. The gestation period of the fisher (*Martes pennanti*), a small carnivore with delayed implantation, is roughly the same length as that of the blue whale. At the other extreme, many marsupials have remarkably short gestation periods (Table

18–1). This is a result of the unique marsupial reproductive pattern typified by brief intrauterine development of young, birth while the young are still poorly developed, and a long period of suckling while the young are in the pouch. In the common opossum (*Didelphis marsupialis*) the gestation period is but 12 or 13 days, considerably less than that of some tiny shrews (Table 18–2).

The frequency of breeding and the size of the litters in any mammalian species are adaptive features that have doubtless evolved in response to a great number of factors. Among these are longevity of individuals, duration of suckling period of young, time during which the young are dependent on the parents, frequency of mortality of young, annual activity cycles of adults, and such environmental factors as availability of food supply and severity of seasonal changes in temperature and precipitation. Under the influences of these and other factors, a reproductive pattern has evolved in each species that is geared to the greatest possible success in rearing the young. According to Lack (1948, 1954), litter size is a result of natural selection that tends to favor the most consistently successful litter size in terms of survival of young. Williams (1967) offered a refinement of Lack's principle that takes into account the total reproductive performance of adult animals. Be-

cause the metabolic cost of raising large and well nourished litters is paid by a lowering of future reproduction, in any population litter size will represent the best reproductive investment for the environmental situation under which the population is operating. This "best investment," as represented by litter size, differs from area to area, even within a species (Spencer and Steinhoff, 1968). Within a species litter size generally becomes larger at northern latitudes and at higher elevations because the severe winters and brief growing seasons in these areas limit the number of litters an animal can have during its lifetime (Table 18–4). The most adaptive pattern in boreal areas, then, is one involving a few large litters; in less severe climates more but smaller litters are produced. In some tropical areas breeding continues through much of the year, but the pronounced wet and dry season in many tropical regions partially restricts breeding to periods of high productivity of food. Pat statements regarding timing of breeding can seldom be made, however, for considerable interspecific variation in reproductive patterns can occur in mammals of a single locality. In the Panama Canal Zone most rodents breed throughout the year, but in some species breeding is restricted to certain seasons (Fleming, 1970); and in a subalpine locality in Colorado some ro-

Table 18–4. Geographic Variation in Litter Size in a Deer Mouse (*Peromyscus maniculatus*) as Indicated by Embryo Counts. (Data for California is from Jameson (1953:48); that for Larimer Co., Colorado, from Spencer and Steinhoff (1968:283); and that for Routt Co., Colorado, from Vaughan (1969:60).)

AREA	ELEVATION IN FEET	NUMBER OF FEMALES	MEAN LITTER SIZE
Plains:			
Larimer Co., Colorado	5100–5300	56	4.0
Mountains:			
Plumas Co., California	3500–5000	96	4.6
Larimer Co., Colorado	5500–6500	37	4.4
Larimer Co., Colorado	8000–11,000	47	5.4
Routt Co., Colorado	10,500	111	5.6

dents are polyestrous and breed over a several-month period, whereas others are strictly monestrous (Vaughan, 1969).

Especially critical is the seasonal timing of the gestation period, the suckling period of the young, and, in many species, the period during which young are becoming independent of the parents. During the intrauterine development of the young and the period of lactation, unusually heavy metabolic demands are put on the female; also, the survival of the young depends on their ability to obtain sufficient food when they are becoming independent. The reproductive patterns of most mammals are timed so that these critical parts of the cycle occur during times of high productivity of food. Most mammals of temperate or boreal areas bear young in the spring or summer, when food is abundant and optimal weather conditions for survival of young occur. Because the summer snow-free period is short in some northern or mountain regions, the young of many small mammals are frequently born soon after snowmelt, and the periods of lactation, weaning, and early independence of young occur during the brief period of rapid growth and flowering of the plants and the time of maximum activity of insects. For some hibernating species the summer period in boreal areas seems barely long enough for growth sufficient to prepare young for dormancy. Some populations of least chipmunks (*Eutamias minimus*) are probably under such environmental pressure. Spring populations of this species in the high mountains of northern Colorado often contain some non-reproductive individuals. These animals are considerably lighter in weight than the average members of the population and are probably individuals born the previous summer that were unable to gain sufficient growth and develop sufficient fat to provide energy for both hibernation and rapid gonadal development prior to emergence from hibernation (Vaughan, 1969:60-62).

In deserts, similarly, the reproductive cycles of small mammals are timed to take advantage of periods of plant growth. In the deserts of Arizona, for example, a burst of growth of small forbs occurs in the spring, after the winter precipitation, and another period of plant growth occurs in late summer, in response to the summer thunderstorms. Most litters of kangaroo rats (*Dipodomys*) and pocket mice (*Perognathus*) are born in these periods. This pattern of restricted breeding seasons is repeated in some tropical areas. In some parts of the Neotropical Region parturition is sharply limited to certain times of the year, seemingly in response to a yearly climatic cycle featuring an extremely wet season and a dry season. In Panama the two periods of natality of the Jamaican fruit bat (*Artibeus jamaicensis*) are at times when fruit is available (Fleming, 1971). In western Mexico four insectivorous bats of the family Mormoopidae have their single yearly young within a brief period in the early part of July at the time when the rains begin and insects suddenly become abundant (Bateman and Vaughan, 1972).

Reproductive rates vary markedly between species and are dependent on many factors such as litter size, duration of the suckling period, and the time required for young to reach sexual maturity. As might be expected, the reproductive rate is low in large animals that have long gestation periods and a single young, that suckle their young for long periods, and that have young that take many years to reach sexual maturity. Because of the lengths of the periods of gestation and suckling in the elephant, for example, a female may bear a young once every four years or even less often, and an average female may have only four young in her lifetime. By comparison, some small rodents have amazingly high reproductive potentials because of polyestrous reproductive cycles and rapid growth and maturation of the young. The montane vole (*Microtus montanus*) of the western United

States is a good example of a small rodent with a high reproductive rate. This vole is polyestrous and often breeds throughout the late spring, summer, and early autumn. The gestation period is only 21 days, and a post-partum estrus occurs. A female may have three of four litters in the breeding season, the litter size averaging four to six. Young females can breed at 21 days of age, and males are fertile at roughly twice this age. In a summer period, therefore, a female can bear 20 young, and, as a conservative estimate, these young might produce an additional 20 young before winter. Thus, a roughly 20-fold increase from the original pair of mice could occur. In the European field vole (*M. agrestis*) breeding may continue through all but the mid-winter months (Baker and Ranson, 1933), and the California vole (*M. californicus*) may breed nearly all year (Greenwald, 1956). It is not surprising that occasionally, under conditions that favor high survival of young, tremendously high population densities of microtine rodents occur.

GROWTH OF YOUNG

Growth and development of young are remarkably rapid in some mammals. The young least shrew (*Cryptotis parva*) roughly doubles its birth weight at the end of four days (Conaway, 1958). Young evening bats (*Nycticeius humeralis*), which weigh 2 gm.

at birth, roughly double their weight by 18 days of age, and by this time the wings more than double their length (Jones, 1967). These bats can fly when 18 days old, but adult weight is not attained until roughly 60 days. Similar growth rates occur in the cave myotis (*Myotis velifer*; Kunz, 1970); such rates are probably typical of many small bats. Growth rates are also quite rapid in rodents, but rates may differ sharply between closely related animals. One-half of adult body weight was attained by several species of *Peromyscus* (white-footed mice) at from 18 to 49 days after birth (Table 18–5). Characteristically, the early growth of small mammals is rapid, but the rate declines shortly after weaning. As an example, in three species of kangaroo rats that are weaned between 15 and 25 days of age, young attained roughly half the adult weight within 30 days, but full adult weight is not reached in two of these species until from 150 to 180 days after birth (Fig. 18–3).

Some of the most astounding growth rates known in mammals have been recorded for the southern elephant seal (*Mirounga leonina*). This is a huge animal, the males reaching weights of over 3000 kg. (more than three tons). An average female weighs 46 kg. at birth. The weight doubles by 11 days of age and quadruples by 21 days (Laws, 1953:33). The young Weddell seal (*Leptonychotes weddelli*) doubles its weight within two weeks after birth (Bertram, 1940:32). The rapid growth

Table 18–5. Rate of Postnatal Growth of Deer Mice (*Peromyscus*) as Indicated by Percentages of Mature Weight and Age at One-Half Growth. (After Layne, J. N., in King, J. A.: Biology of *Peromyscus* (Rodentia), Spec. Publ. No. 2, American Society of Mammalogists, 1968.)

SPECIES	AGE IN WEEKS						ONE-HALF GROWTH IN DAYS
	1	*2*	*4*	*6*	*8*	*10*	
P. maniculatus	24	38	60	81	92	99	23
P. truei	19	30	51	72	85	89	27
P. megalops	9	14	27	41	51	75	48
P. floridanus	19	27	51	68	74	78	28
P. californicus	23	34	55	73	85	90	25

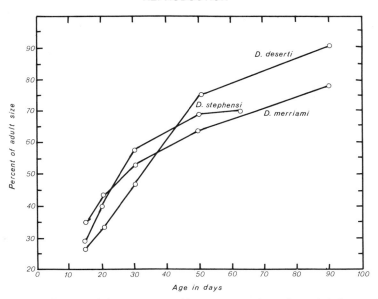

FIGURE 18–3. Growth rates of three species of kangaroo rats *(Dipodomys)*. (After Lackey, 1967.)

of pinnipeds is facilitated by high-energy milk that is up to 53% fat (Amoroso and Matthews, 1951). The period of suckling is usually long in pinnipeds, lasting for a year and a half in the walrus *(Odobenus;* Chapskiy, 1936:120) and for a year in some California sea lions *(Zalophus californianus;* Peterson and Bartholomew, 1967:44). According to Scheffer (1958: 25), the long suckling period and rich milk are adaptations allowing considerable growth and fat deposition before the young animals must face the stresses associated with their first winter at sea.

Some newborn mammals are helpless and poorly developed (altricial), whereas others are well developed and capable of taking care of themselves to some extent (precocial). Most mammals are of the former type, and many are born hairless, with the eyes closed and the external ear opening sealed. Many small rodents are not covered with fur until at least one week of age, and the eyes open between the 10th and the 20th day. Newborn cricetid rodents are typically uncoordinated and lack locomotor ability, and such ability is not developed to any extent until the eyes open. Newborn carnivores, although usually fully furred, are helpless, and the eyes and ears are not open. At the other extreme, some mammals are remarkably well developed and alert at birth or soon thereafter. Perhaps the prime examples of such young are those of perissodactyls and artiodactyls. Newborn of these animals are fully furred, their eyes are open, and they are soon able to run. Within four days after birth, although still somewhat unsteady on their feet, young pronghorns *(Antilocapra)* are able to outrun a man, and at one week of age young pronghorns can outdistance the average dog (Einarsen, 1948:109).

CHAPTER 19

METABOLISM AND TEMPERATURE REGULATION IN MAMMALS

Major barriers to mammalian distribution are easily recognized. Bodies of water, arid lands, or mountains may be absolute barriers to dispersal, depending on the environmental tolerances of the specific mammals considered. Equally limiting, however, are environmental temperatures; the distributions of some mammals — Neotropical sloths, for example — might be described most precisely by reference to the extreme temperatures and the seasonal patterns of temperature change that can be tolerated. Air temperatures from $-50°$ to $50°\,C$ may be encountered at various times and places on the earth, but at best mammals can only survive body temperatures of approximately $45°$ to $0°\,C$, and can be normally active only within the range of body temperatures between roughly $45°$ and perhaps $30°\,C$. In mammals, interspecific differences in the ability to withstand temperature extremes occur even between closely related species (Fig. 19–1), and it is not surprising that no one species is adapted to facing the full range of temperature extremes known for mammals as a group. Just as some mammals are adapted to a few food sources or to a restricted type of habitat, some can live only within a narrow range of temperatures. Knowledge of metabolism and temperature regulation in mammals is essential to an understanding of their ability to adapt to the great variety of ecological settings they occupy.

Most animals are *ectothermic*. In these animals body temperature is regulated by heat gained from the environment rather than by heat produced by the animals' own metabolic processes. Mammals and birds are unusual in being *endothermic;* their body temperatures are controlled largely by metabolic activity and by modifications that carefully regulate the rate of heat exchange with the environment.

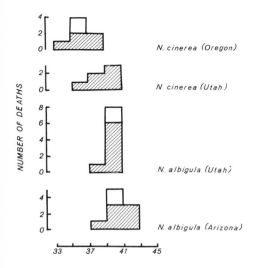

FIGURE 19–1. Lethal ambient temperatures for four populations of wood rats (*Neotoma*). Figures are based on four-hour exposures; hatched areas indicate numbers of deaths. (After Brown, 1968.)

The most obvious advantage of endothermy is the freedom it allows from dependence on environmental temperatures. Mammals and birds, whether terrestrial or aquatic, remain active and typically retain a remarkably constant body temperature under an imposing variety of temperature extremes, ranging from intense desert heat to extreme arctic cold. This type of thermal regulation allows these endotherms to have broad ecological and geographical distributions. Some mammals and birds maintain constant body temperatures year round, even in areas with drastic seasonal changes in temperature. Of advantage to some mammals and birds, however, is the ability to maintain a constant body temperature at some times but to allow the body temperature to fluctuate at other times. Such animals are said to be *heterothermic*. This pattern of regulation is of critical adaptive importance to some animals by facilitating metabolic economy particularly during energetically stressful periods.

Endotherms have gained an advantage over ectotherms by developing partial independence of environmental temperatures, but this advantage is maintained at considerable metabolic expense. Some 80 to 90% of the oxidative energy produced by endotherms is used for maintaining thermal homeostasis. To maintain a constant body temperature when the ambient temperature is below the body temperature, an animal must keep a balance between heat lost to the environment and heat produced by metabolism. When the environmental temperature exceeds the body temperature, heat gained from the environment must be dissipated by some cooling device. Both heat production and heat dissipation demand some outlay of energy, and in some extreme environments the energy demands of endothermy are extremely high.

A major adaptive trend in some mammals is toward the reduction of the metabolic cost of endothermy. This is often done by decreasing *thermal conductance*. This is expressed as the metabolic cost (in cubic centimeters of oxygen per gram of body weight) for a given time interval per °C difference between the body temperature and the environmental temperature. (The units of this quantity are thus cm.^{3} O_2/gm./hr./°C.) In some mammals in extremely cold environments, the difference between body temperature and environmental temperature may be 70° C or more; because the rate of loss or gain of heat in a body is proportional to the difference between the body's temperature and that of the environment, reduction of loss of heat by the lowering of thermal conductance is essential in these mammals. Insulation, in the form of fur or blubber (or feathers in birds), is the primary means of reducing thermal conductance. Control of the blood supply to peripheral parts of the body is also important.

Each endotherm has a *thermal neutral zone* within which little or no energy is expended on temperature regulation. The thermal neutral zone is "the range of temperatures over which a homoiotherm can vary its thermal conductance in an energetically inexpensive manner and on a short time scale" and keep a constant body temperature (Bartholomew, 1968). Within this zone the fluffing or compressing of the fur, local vascular changes, or shifts in posture suffice to maintain thermal homeostasis. At the lower limit of the thermal neutral zone is the *lower critical temperature*, the point below which the balance between metabolic heat production and heat loss to the environment cannot be maintained by metabolically inexpensive variations in thermal conductance. Below the lower critical temperature, at which the rate of heat flow is minimal, oxidative metabolism must be increased to keep the body temperature constant. Obviously, if a constant body temperature is to be maintained over a wide variety of ambient temperatures, adjustments of both thermal conductance (through changes in insulation) and heat production (through metabolic changes) are necessary. The *upper critical temperature* is the point above which a constant body temperature can only be maintained by an increase in metabolic work above the resting level to dissipate heat. This temperature is far less variable than is the lower critical temperature, but is of great importance to desert mammals. These mammals usually do not have

access to water and must strictly minimize water loss. Animals faced with temperatures above the upper critical temperature dissipate heat by evaporative cooling, which involves considerable water loss. Because such loss in desert species is extremely disadvantageous, these animals generally avoid temperatures above the upper critical temperature. In some mammals the upper critical temperature is unusually high or may even be difficult to detect, as in the anomalous case of the African rock hyrax (*Heterohyrax brucei*), studied by Bartholomew and Rainy (1971). The body temperatures of experimental hyraxes rose as the environmental temperatures rose, and at an environmental temperature of 42.5° C, and body temperatures in the vicinity of 41° C (105° F), the rate of oxygen consumption was actually lower than at temperatures nearer the mean body temperature of animals not under heat stress (36.4° C). Figure 19–2 summarizes the usual relationship between oxygen consumption and environmental temperatures in a mammal.

The following considerations of the reactions of different mammals to conditions of thermal stress provide a basis for a general understanding of metabolism and temperature regulation in mammals.

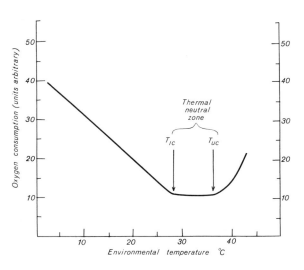

FIGURE 19–2. Oxygen consumption in relation to environmental temperatures in a hypothetical mammal. Abbreviations: T_{lc}, lower critical temperature; T_{uc}, upper critical temperature. (After Bartholomew, G. A., in Gordon, M. S.: *Animal Function: Principles and Adaptations*, The Macmillan Company, 1968.)

HOMOIOTHERMY AND COLD STRESS

Homoiothermy in Terrestrial Mammals. A primary adaptation of mammals inhabiting cold terrestrial environments is the development of highly effective insulation. This insulation may be so remarkably effective that the thermal neutral zone of an animal may extend downward to −40° C, as in the case of the arctic fox (*Alopex lagopus*). The length of the woolly underfur and the longer guard hairs varies seasonally; the summer pelage, which is acquired in the spring, has reduced insulating qualities, but the longer winter coat, which replaces the summer pelage in the fall, has great insulating ability and allows the fox to keep a constant body temperature in extreme cold with relatively little metabolic drain. This reduction in energy loss may be of critical importance to arctic foxes, for they are at the top of the food chain (see page 281), and are in an energetically precarious position because they depend on such prey species as lemmings, which fluctuate widely in density and in availability.

Additional metabolic savings may be gained by reductions in peripheral circulation. During cold stress, extremities such as legs and ears, which dissipate heat rapidly because of their relatively poor insulation, receive a reduced blood supply and are allowed to become cool, thereby reducing the temperature differential between these parts and the environment. Such regional heterothermy is common among mammals of cold areas, and differences between temperatures of different parts of the body are surprisingly high. In an eskimo dog with a deep body temperature of 38° C, for example, the temperature of the foot pads may be 0° C, that of the dorsum of the foot, 8° C, and that of the carpal area of the forelimb, 14° C (Irving, 1966).

The hollow hair of some large ungulates is remarkable insulation, and allows the animals to modify their activities relatively little in the winter. Pronghorns (*Antilocapra americana*) commonly remain in open and often windswept situations in temperatures far below 0° C. The metabolic saving in the winter resulting from decreased thermal conductance is of great importance in some species that must endure not only extreme cold but reduced availability of food as well.

As might be expected, behavior plays a part in reducing cold stress. Many small mammals curl up in a ball or hunch the body so that the overall shape is nearly spherical. By bringing the ratio of the outer surface to the volume to a minimum, the most advantageous shape in terms of retention of heat is obtained, and lightly insulated surfaces (feet, face, and parts of the venter) are well protected. Even the tree-roosting red bat (*Lasiurus borealis*), which seemingly goes into short-term hibernation during winter cold spells in the central United States, assumes a posture with the furred tail membrane covering the venter and the head tucked downward; the total form approaches the shape of a sphere (Davis, 1970). The response of seeking shelter may also be of great importance. An animal burrowed deeply into the snow faces an ambient temperature of roughly 0° C, whereas the ambient temperature above the snow might be many degrees below zero. Wintering herds of deer and elk often frequent ridge tops or south-facing slopes where the cold of the night is quickly moderated by the first rays of the sun, and during the Alaskan winter, moose abandon many basins into which cold air drains from the surrounding mountains.

Size in Relation to Homoiothermy. Small size is disadvantageous in terms of heat conservation, but favors heat dissipation. In general, the smaller the animal, the greater the surface area relative to volume; the surface area is proportional to the square of the body length and the volume is proportional to the cube of the length, given similar, nearly spherical shapes. Assuming that body weight is proportional to length, the surface-to-volume ratio, then, varies as the two-thirds power of the weight. The empirical relationship between body temperature (T_B), lower critical temperature (T_{LC}), and body weight (W) in mammals is represented by the expression $T_B - T_{LC} = 4W^{0.25}$. "Because T_B is essentially independent of body weight (W) in mammals, as weight decreases T_{LC} approaches T_B." (Bartholomew, 1968:322). The calculated lower critical temperature for a mouse weighing 20 gm. is 29° C, a temperature considerably higher than that usually encountered by nocturnal mammals. Basal metabolic rate (the metabolic rate necessary for simply the maintenance of life in a resting organism), lower critical temperature, and thermal conductance all vary inversely with body size, and all are intimately related. The metabolic rate, as measured in oxygen consumption per gram of body weight per hour, climbs so precipitously with decreasing body weight that a mammal weighing less than about 3.5 gm. would be unable to eat sufficient food to sustain activity. This is shown in Figure 19–3, which also illustrates the fact that rates of oxygen consumption differ strikingly even between small mammals: the tiny masked shrew consumes oxygen

at a rate over four times that of the larger deer mouse. Carrying the comparisons further, the metabolic rate of a horse is only approximately one tenth that of a mouse (Krebs, 1950). Small mammals must face an additional acute problem related to insulation, for they are limited as to the length that hairs of the pelage can attain, and the pelage is correspondingly limited in insulative effectiveness.

Most small mammals (such as mice and shrews) that inhabit cold areas and are active throughout the winter remain beneath the snow pack, thereby avoiding the intense cold. These animals are often active at the soil-snow interface, in the zone of "depth hoar." This is a stratum of loose snow that develops beneath a fairly deep snow pack and through which small animals can readily travel. These mammals forage in winter in a subnivean environment (an environment beneath the snow) with a constant temperature near 0° C, and when resting presumably seek refuge in nests, which provide insulation that augments the animal's pelage. Small mammals that are intermittently active above the snow in the winter typically maintain thermal homeostasis by increasing heat

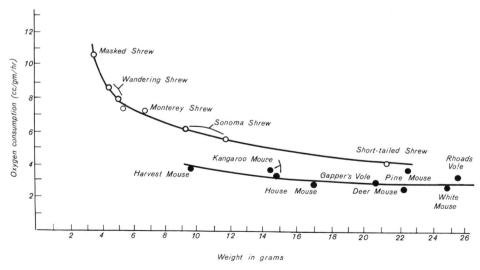

FIGURE 19–3. Oxygen consumption in relation to body weight in some small mammals. (After Pearson, 1948.)

production *via* increases in oxidative metabolism. The metabolic cost to small mammals of activity in the open in winter is extremely high. Field workers soon learn that these animals, when captured in live traps in cold weather, will die in a very short time if not supplied with ample food and nesting material.

At the other end of the size scale, it might be expected that exceptionally large mammals would commonly be faced with the problem of heat dissipation because of a more limited surface area relative to volume. This seems to be the case. In whales, for example, the flippers and flukes, and at least parts of the skin, are important as heat radiating surfaces, even when these animals are immersed in cold water.

Homoiothermy in Aquatic and Semi-Aquatic Mammals. Temperature regulation is a demanding problem to mammals that inhabit cold water. The rate of heat loss by an endotherm in water is some 10 to 100 times as great as the rate of loss in air of the same temperature (Kanwisher and Sundnes, 1966:398). Arctic and antarctic waters are near 0° C year round, and northern lakes and rivers approach this temperature in the winter. Consequently, a temperature differential of some 35° C between deep body temperature and ambient temperature is usual in mammals swimming in these waters. Despite the thermal inhospitability of this environment, cold waters are temporarily or permanently inhabited by a variety of mammals, some of which are highly specialized for coping with cold. The muskrat (*Ondatra*), the beaver (*Castor*), some shrews (*Sorex*), the river otter (*Lutra*), and the mink (*Mustela*) spend considerable time in cold water. In these animals heat is lost to the water mainly by the foot pads, the nose, and other bare surfaces, for most of the body is insulated by a layer of air that is entrapped by the fur. Some nearly permanent inhabitants of cold marine waters, such as otariid seals, similarly use entrapped air as insulation. The cetaceans, phocids (earless

seals), and walruses (*Odobenus*), however, lack appreciable fur, and the surfaces of their bodies are in contact with water that may in extreme cases be 40° C below the deep body temperature. The means by which the latter animals maintain a constant body temperature under such demanding conditions are of considerable interest.

These cold-water, hairless marine mammals have thick layers of subcutaneous blubber that form an insulating envelope around the deep, vital parts of the body. A substantial amount of the weight of a marine mammal may be contributed by the blubber. For example, blubber constitutes 25% of the weight of the Weddell seal (*Leptonychotes weddelli;* Bruce, 1915:3). In the small (75 kg.) harbor porpoise (*Phocoena*) 40 to 45% of the weight is blubber, and only 20 to 25% is muscle (Kanwisher and Sundnes, 1966:405). Studies of seals by Irving and Hart (1957) have shown that the skin temperature varies directly with the water temperature down to 0° C. In seals, the cooled surface of the body and the thick blubber are seemingly an effective insulation, as indicated by the fact that the lower critical temperature of some seals is 0° C. The skin of seals has a well developed vascular supply, and the temperature gradient between the deep parts of the body and the skin is controlled largely by changes in the blood supply to the skin.

Some of the most extreme thermal demands faced by endotherms are those met by cetaceans. Whales and porpoises live their entire lives in the water, and some species continuously occupy water at or near the freezing point. All cetaceans have insulating layers of blubber, but an extreme situation is faced by a small porpoise that must maintain a deep body temperature some 40° C higher than that of the sea, from which it is insulated by only 2 cm. of blubber. No inflexible pattern of thermoregulation is adequate even in inhabitants of the sea, which offers a relatively constant thermal environment. Some cetaceans migrate seasonally from cold waters to warm tropical seas. Because of the high thermal conductivity of water, skin temperatures generally equal water temperatures, and variations in water and skin temperatures of roughly 20 to 30° C may occur seasonally. The temperature of the body core, however, remains constant, and insulation requirements, therefore, may vary fivefold. Heat production by mammals varies tremendously as a result of changes in metabolic level. Metabolic activity and heat production increase roughly ten times in an animal going from a resting state to one of maximum exertion. It has been estimated that a cetacean at rest in cold water needs roughly 25 to 50 times the insulation it needs when swimming at high speed in tropical waters (Kanwisher and Sundnes, 1966: 399).

Gigantic differences in the ability of cetaceans to keep warm result from differences in size and in thickness of blubber. The biggest whale is some 10,000 times as heavy as the smallest porpoise, has roughly a 10 times more advantageous mass-to-surface ratio with regard to heat retention, and has a much thicker shell of blubber. Because of these differences, the whale has approximately a 100 fold advantage over the small porpoise in its ability to keep warm. The very factors working in favor of heat retention in the large cetaceans, however, are obviously disadvantageous under conditions of great activity or warm waters. Because of the vast bulk of large cetaceans, dissipation of heat is an acute problem. As an example of the slowness of diffusion of heat in large cetaceans, the deep muscle temperature of a dead and eviscerated fin whale (*Balaenoptera*) dropped only 1° C in twenty-eight hours (Kanwisher and Leivestad, 1957). Clearly, cetaceans must have considerable "thermal versatility." How is this versatility achieved?

As is usual in considerations of biological problems, no single answer is appropriate, nor has sufficient research been done on the problem to suggest even most of the probable answers. Al-though much remains to be learned, several points seem well established. First, metabolic rates of cetaceans differ markedly in different species. The small porpoises have much greater basal metabolic rates than do large whales, as could be expected because of the inverse relationship between basal metabolic rate and size, and the former thus have a much more rapid rate of heat production even when resting. Metabolic rate seems to be in part an adaptive feature, for certain animals that have difficulty keeping warm, such as small porpoises, have even higher metabolic rates than would be expected on the basis of size alone. Second, blood flow through the well developed vascular system in the flippers, dorsal fin, and flukes of cetaceans allows these structures to function effectively as heat dissipators under conditions of heat stress. The flow can apparently be shut down during cold stress, allowing for a minimum of heat loss from these surfaces. Third, a remarkable series of vascular specializations allow for great variations in the thermal resistance offered by the blubber. A system of countercurrent heat exchange in the vascular network supplying the blubber minimizes heat loss to the blubber and skin, and hence to the environment. This system, utilized by many mammals, involves arterioles and venules that lie against one another, often in a complex network. Heat diffuses from the arterioles to the venules and serves to heat the venous blood before it enters the body core; much of the heat of the arterial blood is thus returned to the body core before it is lost to the environment from such poorly insulated surfaces as bare skin or appendages. In cetaceans a second venous system in the blubber bypasses the countercurrent system during heat stress and allows considerable heat loss to the environment when heat dissipation is of prime importance. Similar countercurrent and bypass systems occur in the flippers and fins. The extremities and much of the surface of the body can thus serve to dissipate heat, or can be maintained under an altered vascular supply that provides for maximal heat retention. The longitudinal folds of blubber on the throat of the rorqual (*Balaenoptera*) probably function partly as a cooling device (Gilmore, 1961) by providing increased surface area for heat dissipation. The highly vascularized skin at the bottom of these grooves can be exposed to the water. Morrison (1962) found that these grooved anterior parts of the humpback whale (*Megaptera*) were slightly cooler than were other parts of the body, suggesting their importance in heat dissipation.

The great quantities of blubber on large whales (up to 20 cm. in thickness) are seemingly not primarily useful as insulation; these animals could probably maintain a constant deep body temperature, without increased heat production, with much less insulation. These fat deposits may be useful primarily as food stores that can support an animal during periods of migration and fasting. The consumption of only half of a whale's blubber could support the animal at a basal metabolic rate for from four to six months (Parry, 1949).

REACTIONS OF MAMMALS TO HEAT STRESS

Some of the most severe problems in thermoregulation are those faced by mammals living in hot regions. In many low-latitude deserts, daytime surface and air temperatures in the summer rise well above the body temperatures of most mammals. Under such conditions heat from the environment is absorbed, while at the same time the animals themselves are producing considerable metabolic heat. In order to maintain thermal homeostasis these animals must avoid as much as possible the absorption of heat from the environment, dissipate such heat as is absorbed, and lose endogenous heat. Unless the body temperature is elevated, these heat transfers must occur against a thermal gradient, from the relatively cool animal to the relatively hot environment. Such heat transfers invariably involve evaporative cooling, a luxury that most desert organisms cannot afford since they live in a region where water is in critically short supply. Nonetheless, even extremely hot and arid deserts are occupied by mammals, and some kinds, notably rodents, are quite common in such areas. A variety of physiological, anatomical, and behavioral adaptations have allowed mammals to inhabit these seemingly inhospitable regions.

Avoidance of High Temperatures. Most desert animals are never subjected to the extremely high diurnal temperatures, nor are they able to survive them; their success is based on the ability to avoid extremely high temperatures rather than to cope with them. Perhaps the saving grace of the desert is the typically great daily and seasonal fluctuation in temperature. Temperatures frequently drop markedly at night, and winters are usually cool or cold. As a result, soil temperatures well below the surface are never high (Fig. 19–4), even in the summer, and to this refuge of coolness and relatively high humidity nearly all desert rodents retreat during the day. All but a very few desert rodents are strictly nocturnal, and all are more or less fossorial; these animals are active above ground in the part of the diel cycle when temperatures are lowest and humidities are highest. The studies of McNab (1966) and MacMillen and Lee (1970) suggest that most fossorial and nocturnal desert rodents have metabolic rates below those predictable on the basis of body size. The low rates are associated with lowered metabolic heat production while the animals are in the humid burrows during the daytime summer heat. This lowered heat production probably precludes the use of wasteful (in terms of water loss) evaporative cooling in order to dissipate heat while the rodents are below ground.

Most desert carnivores are also largely nocturnal, some are fossorial, and all seek shelter during the hottest

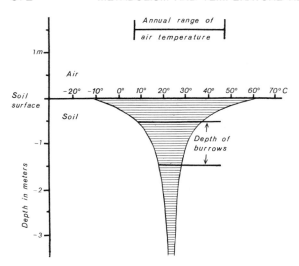

FIGURE 19–4. Annual range of sub-soil temperatures in a desert in Arizona. Note that most rodent burrows are at depths at which heat or cold stress is never encountered. (After Misonne, X.: Analyse zoogéographique des mammifères de l'Iran. Bruxelles, Inst. Royal des Sci. Nat. de Belgique, Mémoires, 2me ser. 59, 1959, 157 pp.)

part of the day. Several other means of avoiding the extremes of daytime heat are used by some mammals of the deserts of the southwestern United States. The white-throated woodrat (*Neotoma albigula*) remains during the day in burrows insulated by piles of sticks and other debris. Frequently these houses are built beneath the shade of low-growing vegetation or under mesquite. In the burrow beneath a woodrat house near Yuma, Arizona, the temperature fluctuated but 2.5° C throughout a 24-hour cycle and never exceeded 34.4° C; temperatures outside the house varied 13.7° C and reached 41.5° C (Brown, 1968:24). Jackrabbits (*Lepus californicus* and *L. alleni*) stay in "forms" in the shade of bushes in the day, and are reluctant to run far in the open in midday. Bighorn sheep (*Ovis canadensis*) and javelinas (*Tayassu tajacu*) often take shelter in rock grottos or in the shade of steep rock outcrops.

Large Mammals and Heat Stress. A few mammals are able to tolerate exposure to extreme desert heat. Large ungulates are obviously unable to use the types of shelter available to small mammals, and often occupy open areas where there is little shelter. These mammals must be able to withstand hours of exposure to environmental temperatures in excess of their body temperatures.

Large size itself is advantageous to mammals that must tolerate high temperatures. Because of the weight-surface ratio discussed earlier, the larger the animal the greater will be its ability to withstand exposure to high temperatures due to a relatively reduced surface area for heat gain. Stated differently, large animals have greater thermal inertia than do small ones. Of additional importance, just as insulation in the form of thick pelage slows the loss of body heat in low ambient temperatures, fur provides a partial heat barrier that slows the penetration of heat to the body surface when temperatures are high. Another advantage of fur under some circumstances is that it reduces water loss through the skin, an extremely critical advantage in arid environments. Although both large size and fairly thick fur are important aids to avoiding rapid heating in hot areas, under conditions of intense heat other factors usually help tip the delicate thermal balance away from lethally high body temperatures.

Studies of temperature regulation in the camel by Schmidt-Nielsen (1959) have revealed a carefully regulated and highly adaptive diel cycle of changes in body temperature. Camels in the Sahara desert in the winter, when cool temperatures (from roughly 0° to 20° C) prevailed, had fairly constant body temperatures that varied between 36° and 38° C. The fluctuations in body temperatures were not random, but followed the same pattern day after day, regardless of weather. In the summer variations in body temperature were considerably greater; generally animals had temperatures in the morning between 34° and 35° C, and the body temperature reached a peak of approximately 40° C late in the day (Fig. 19–5). The camels were seemingly able to regulate their temperatures, but did so only above or below these extremes; when body temperatures reached 40.7° C, evaporative cooling in the form of sweating was used to dissipate heat, and body temperatures never exceeded 40.7° C. Thus, the camel accepted a heat load during the day that sharply elevated its temperature; but during the relative coolness of the desert night the heat stored during the day was passively dissipated, and the body temperature was allowed to drop to some extent (perhaps, according to Schmidt-Nielsen, to a critical lower limit below which metabolism is disrupted). Schmidt-Nielsen

(1964:44) estimated that for a camel to dissipate by evaporative cooling the heat load accepted during a hot day would require the expenditure of some 5 liters of water. In an animal that does not have frequent access to water such daily water loss would lead to fairly rapid dehydration. An additional advantage of high body temperature during the day results from the narrowing of the gap between environmental and body temperature; the smaller this temperature differential, the lower the rate of heat flow from the environment to the body.

As an adaptation to intense heat, similar patterns of temperature fluctuation occur in the oryx (*Oryx*), the eland (*Taurotragus*), and the gazelle (*Gazella*), African antelope that occur in desert or savanna areas. The oryx frequently occurs in situations where no shade is available, and it does not seem to seek shelter in the day. The ability of this animal to withstand a diurnal heat load is exceptional. Under experimental conditions the oryx could withstand exposure to an ambient temperature of 45° C (113° F!) for 12 hours (Taylor, 1969:91). During this period the body temperature rose above 45° C and was sustained at this

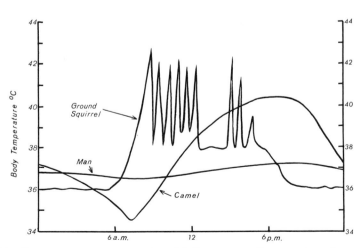

FIGURE 19–5. Diagrammatic representation of daily patterns of temperature change in three mammals subjected to desert heat. Note that the antelope ground squirrel (*Ammospermophilus leucurus*) goes through a series of heating-cooling cycles during the day. (Mostly after Bartholomew, G. A.: *Symposia of the Society for Experimental Biology*, No. 18, Academic Press, Inc., 1964.)

level for up to eight hours with no injury to the animal. Hence, rather than gaining heat from the environment, the oryx was actually losing heat. Such high body temperatures would kill most mammals fairly quickly, but circulatory specializations apparently allow the oryx to survive such extreme "overheating." The brain, seemingly the most heat sensitive organ, is provided in this animal with a specialized countercurrent cooling system of its own in the cavernous sinus beneath the brain (Taylor, 1969:92). The external carotid artery, on its way to the brain, divides into many branches in this sinus, and these branches are in close proximity to veins returning from the nasal passages. These veins carry relatively cool blood, because evaporative cooling of the nasal mucosa tends to cool the blood supplying these surfaces. Heat exchange between this cool venous blood and the blood in the carotid artery network assures that the blood supply of the brain is cooler than that of most of the rest of the body. Measurements of another kind of antelope, a gazelle, have shown that the brain may be as much as 2.9° C cooler than the arterial blood leaving the heart.

Tolerance to high body temperatures may require specializations (as yet unknown) of various enzyme systems, such as those of the muscles. Many of the physiological reactions known to occur in animals that are not subject to heat stress do not proceed properly at elevated temperatures; also, many enzymes are unstable or denatured at high temperatures.

Some reduction of heat load occurs in the eland and the oryx by metabolic adjustment that reduces internal heat production; this is most striking in the oryx. A dehydrated oryx reduces its metabolic rate at high temperatures sufficiently to reduce evaporative water loss to 17% below that of an individual with free access to water. In the oryx, as in any species of mammal, temperature regulation is controlled by a variety of interrelated adaptations. The following features are known to

contribute to heat tolerance in the oryx: diurnal storage of heat and reduced or negligible heat flow from the environment at high body temperatures; decreased metabolism at high temperatures; cooling of blood in arteries supplying the brain; pale-colored pelage that reflects considerable heat; and panting to produce evaporative cooling under extreme conditions.

Many mammals respond to heat stress by resorting to evaporative cooling produced by sweating or by panting. Panting is perhaps the more efficient system for evaporative cooling and is of primary importance in many mammals lacking continuous access to water. In the antelope mentioned above, water economy is usually of vital importance under natural conditions; these animals use evaporative cooling (usually panting) only when the body temperature is above roughly 41° C. Although panting is a very familiar reaction to heat stress in many mammals, not until recently has it been well understood.

Typical panting, involving rapid and shallow respiration, is used entirely for heat dissipation and is an effective aid to temperature regulation. Laboratory studies of dogs, for example, indicated a tolerance of an ambient temperature of 43° C for at least seven hours (Robinson and Lee, 1941). Panting utilizes evaporative cooling of the mouth, tongue, and, probably most importantly, the nasal mucosa (Schmidt-Nielsen et al, 1970). In the dog, and in many other mammals with long snouts and excellent olfactory ability, the turbinal bones of the nasal cavity are intricately rolled and provide a great surface area for olfactory epithelium in the nasal mucosa. This large, moist surface area provides an ideal situation for evaporative dissipation of heat. That the tongue is probably also important as a site of heat loss during panting is indicated by the fact that the blood flow to the tongue increases sharply at the onset of panting, and during heat stress increases six times over normal. High respiratory rates are typical of panting. The resting respiratory rate of a dog is roughly 30 per minute, but this rate rises abruptly, with virtually no intermediate rate, to over 300 per minute during panting. As demonstrated by recent laboratory studies (Schmidt-Nielsen, 1970), air movement during panting in the dog is largely unidirectional: most of the air passes in the nose and out the mouth. This type of flow achieves the maximum effectiveness of evaporative cooling. This system has some advantages over cooling effected by sweating. There is little loss of salt during panting, whereas salt loss during sweating (except probably in donkeys and camels) is always appreciable. In addition, adequate ventilation of evaporative surfaces always occurs during panting; in cool, still air, however, sweating is seemingly not equally efficient. One disadvantage of panting is that the increased activity increases metabolism, thereby contributing more heat to be dissipated. Studies of respiratory frequency in dogs (Crawford, 1962) have indicated that these animals are panting at the resonant frequency of oscillation of the diaphragm (the natural frequency of vibration of this structure) and may therefore economize on energy output. Considering water loss relative to total body surface area of a mammal, the amounts of water loss in sweating and panting are probably similar. Both panting and sweating are obviously not effective means of cooling in high humidities.

Desert Ground Squirrels. Especially remarkable in their ability to be active in desert heat are the small desert ground squirrels. In the heat of the day these are the only small mammals that are conspicuously active. In the case of the antelope ground squirrel (*Ammospermophilus leucurus*), hyperactive is a more appropriate descriptive term. This small (roughly 90 g.) rodent appears extremely nervous, and dashes from enterprise to enterprise, whether in the cool of winter or the searing heat of summer. It is often active when surface soil temperatures are 65° C or more. Studies of the antelope ground squirrel by Hudson (1962) have shown that, compared to most mammals, the thermoneutral zone of this squirrel is high, between roughly 33° C and 41° C. Throughout its thermoneutral zone the body temperature is maintained at least slightly above the air temperature, allowing some dissipation of body heat to the environment even at these high temperatures. This squirrel is able to store heat, and operates in a seemingly normal manner at body temperatures above 43.5° C. Because of the high temperatures at and slightly above ground level, the zone in which this squirrel is usually active, its body absorbs heat rapidly and the animal is forced periodically to unload heat; to do so it retreats periodically to the dense shade of a bush or rock, or into its burrow, where the body loses heat to the relatively cool substrate (Fig. 19–5). Hudson observed that under laboratory conditions an antelope ground squirrel could reduce its body temperature from approximately 41° C to 38° C within three minutes when transferred from high ambient temperatures to a chamber with an ambient temperature of 25° C. As a last resort when subjected to continued heat stress, the antelope ground squirrel salivates copiously and spreads the saliva over the head, where evaporative cooling occurs. The Mohave ground squirrel (*Spermophilus mohavensis*) is also able to tolerate high body temperatures. Ground squirrels that inhabit deserts in both the Old World and New World probably have a tolerance of high temperatures similar to that of the antelope ground squirrel.

Marsupials. Some of the larger kangaroos (*Macropus*), the quokka (*Setonix brachyurus*), a rabbit-sized macropodid, and the Tasmanian devil (*Sarcophilus harrisii*), a raccoon-sized dasyurid, are excellent temperature regulators in the face of heat stress. Panting, resulting in evaporative cooling by the rapid passage of air over the moist surfaces of the mouth and the tongue, is probably a primary means of avoiding heat stress in some large kangaroos. The quokka is an exceptionally good heat regulator. Cooling in this animal may be accomplished by sweating, by panting, or by increased salivation and the licking of appendages, but the relative importance of each of these devices is unknown. In the Tasmanian devil sweating is seemingly the major mode of temperature regulation. Some physiological and behavioral adaptations of marsupials to desert environments are discussed by Schmidt-Nielsen (1964:193-203).

Recent studies by MacMillen and Nelson (1969) and by Dawson (1970) have shown that marsupials have body temperatures equivalent in magnitude to those of placentals but have reduced metabolic rates that are about two thirds those of placentals of comparable sizes.

HYPOTHERMIA AND METABOLIC ECONOMY

Hypothermia (the lowering of body temperature) is critically important in the lives of some small mammals. These animals are able to exploit efficiently such abundant food sources as tiny seeds and small insects, and can use a nearly limitless variety of shelters; but in terms of metabolism, small size is an extremely costly luxury. The weight-to-surface ratio of small mammals favors rapid dissipation of heat, and a small mammal must have a high metabolic rate sustained by frequent feeding to maintain thermal homeostasis. To a shrew that maintains a fairly constant body temperature, for example, a continuously available and rich food source is a necessity. Some areas are highly productive of adequate food for small mammals in some seasons, but are relatively unproductive at other times. Winters in the north and dry seasons in the deserts are typically times of potential food shortage for some small mammals, and are also periods when conditions of temperature or moisture may limit their activity. It is not surprising,

therefore, that some small species have evolved means of surviving periods of food shortage and temperature stress and of taking advantage of times of moderate temperatures and high productivity of food. Many small (up to woodchuck size) mammals periodically conserve energy by allowing the body temperature to drop to near that of the environment. This is not a primitive feature, a manifestation of some ancestral inability to sustain a steady temperature at all times, but is instead a highly adaptive ability. The body temperature in heterothermic mammals may vary widely, but within certain limits, depending on the species, is still under control. This adaptive hypothermia may well have been a factor important in furthering the success of the two largest mammalian orders, the Rodentia and the Chiroptera. Many small bats would be unable to forage only at night and fast throughout the day if they could not conserve energy in the day by hypothermia, and seasonally hostile areas would not be inhabited by some small rodents if these animals retained constant thermal homeostasis.

From evidence assembled in roughly the last 20 years, a rather complex picture of heterothermy in small mammals has emerged, but it is clear that the metabolic economy gained by hypothermia is of importance to many small mammals daily or seasonally. Periodic torpor occurs in a monotreme (*Tachyglossus aculeatus*), in marsupials (Bartholomew and Hudson, 1962), in some members of the orders Insectivora, Chiroptera, Primates, and in many rodents.

A number of physiological changes, all favoring metabolic economy, occur during adaptive hypothermia (Bartholomew, 1968:347, 348): body temperature drops to a level about 1° C above that of the environment; oxygen consumption decreases, in some cases to near 5% of the basal metabolic rate; breathing rates may decline to one per minute or less; the animal sinks into a state of torpor, during which it is less responsive than during the deepest sleep; the heart rate is sharply reduced; and spontaneous arousal, by means of increased heat production and conservation, can occur. These physiological responses are similar in many mammals, but the duration of torpor, the tolerance to low body temperatures, and the environmental factors initiating torpor differ broadly between species.

Hypothermia in Rodents. Diverse patterns of hypothermia in response to a considerable spectrum of environmental stresses occur in various rodents. These patterns have been carefully studied in some species.

Periods of torpor are seemingly characteristic of the life cycles of some rodents inhabiting hot regions. In the cactus mouse (*Peromyscus eremicus*), an inhabitant of the deserts of the southwestern United States and northern Mexico, torpor may occur in both summer and winter (MacMillen, 1965). Cactus mice remain in their burrows for several weeks during the driest part of the summer, and laboratory animals entered torpor in the summer in response to a reduced food supply or, in some cases, to restricted access to water. In the winter labora-

tory animals were torpid by day and active by night when their food was in short supply, and they were able to become torpid at ambient temperatures below 30° C (Fig. 19–6). The cactus mouse has a narrow thermoneutral zone (28° to 35° C) and a low basal metabolic rate for a mammal its size; these features are seemingly typical of small mammals able to enter torpor under moderate temperatures. According to MacMillen, the summer torpor in the cactus mouse is a device for reducing the use of food and water and for surviving periods of water and food shortage on the surface. The California pocket mouse (*Perognathus californicus*), an inhabitant of seasonally dry chaparral areas, undergoes a daily cycle of diurnal torpor in the laboratory when its food supply is reduced, and maintains a delicate balance between food supply and metabolic economy (Tucker, 1962, 1963). Its periods of torpor are adjusted so that the shorter the food supply, the longer the daily torpor becomes. Both the cactus mouse and the California pocket mouse are adapted to moderately high-temperature torpor; neither can arouse, and the animals will die, when the body tem-

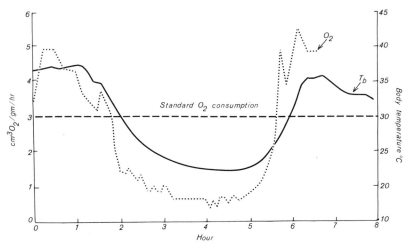

FIGURE 19–6. Pattern of changes in oxygen consumption and body temperature in the cactus mouse *(Peromyscus eremicus)* during entry and arousal from torpor at an ambient temperature of 19.5° C. The straight dashed line shows the standard rate of oxygen consumption in active mice at an ambient temperature of 20° C. The dotted line shows oxygen consumption and the solid line gives body temperature. The cycle of torpor was initiated by deprivation of food and water. (After MacMillen, 1965.)

perature goes below about 15° C. Such dormancy, usually in response to heat, drought, food shortages, or combinations of these stresses, is usually called *estivation*. The Mohave ground squirrel (*Spermophilus mohavensis*), which occupies part of the Mohave desert in California, remains in its burrow from August to March. Laboratory studies indicate that during this period the animals are intermittently torpid for periods of from several hours to several days (Bartholomew and Hudson, 1961). This squirrel is able to elevate its temperature from 20° to 30° C in from 20 to 35 minutes, and even during its active period in spring and summer has an unusually variable body temperature (from 31 to 41.5° C). Among small mammals occupying hot areas, estivation is obviously an important and widespread device for avoiding hot-season stresses. In addition, periods of torpor in some species may strongly reduce competition for limited food between closely related and sympatric rodents during periods of food shortage.

Hibernation is perhaps the most familiar pattern of heterothermy, and usually involves long periods of dormancy in response to winter cold. Mammals that hibernate typically become extremely fat prior to hibernation, and entry into hibernation is usually marked first by brief and then by progressively longer periods of hypothermia. Most hibernators use stored fat as the energy source that carries them through the winter. Some rodents, however, store food in preparation for winter. The European hamster (*Mesocricetus auratus*), for example, responds to cold by storing large amounts of food; this food is eaten during intervals of arousal from hibernation during the winter and is essential to survival through this period (Lyman, 1954). Animals that remain in hibernation for long periods are typically emaciated after hibernation, but usually gain weight rapidly after emergence.

The duration of hibernation varies widely, and in some species late summer periods of torpor (estivation) grade imperceptibly into winter hibernation. Within the same genus sharply contrasting patterns of dormancy may occur. Some members of populations of the California ground squirrel (*Spermophilus beecheyi*) in some parts of California hibernate through part of the winter, but some individuals do not hibernate; there is no prolonged and well marked period of hibernation involving the entire population (Fitch, 1948; Linsdale, 1946; Tomich, 1962). At the other extreme, the Richardson ground squirrel (*S. richardsonii*) in basins at high elevations in northern Colorado often begins hibernation in late August and does not appear above the ground until May. Within the genus *Spermophilus* a large number of variations on the estivation-hibernation theme occur; owing in large part to this thermoregulatory flexibility, this genus occurs in a variety of habitats from the arctic to the deserts.

Hibernating mammals are usually protected from lethally low body temperatures by automatic responses that initiate arousal. During arousal the metabolic rate rises rapidly, some species shiver, and heat production in deposits of brown fat (such as those in the interscapular areas of bats) is of considerable importance. Because of the anteriorly situated deposits of brown fat and the circulatory patterns that shunt blood to anterior parts during arousal, in some species the posterior part of the body at this time may be as much as 15° C cooler than the heart and head. The time required for a hibernating animal to elevate its body temperature to the "active" level differs among species, and also depends on the starting temperature. This time of arousal in some bats is slightly over 30 minutes, but some larger mammals require 90 minutes or more. In the two largest hibernators, the marmot (*Marmota caligata*) and the echidna (*Tachyglossus aculeatus*), arousal takes considerable time. In the marmot arousal is caused by rapid heat production, and takes from two to three hours. Heat production increases slowly during the period of arousal in the echidna, however, and arousal occupies roughly 20 hours.

TEMPERATURE REGULATION IN BATS

Recent studies of a variety of bats have clarified and, at the same time, complicated the picture of temperature regulation in these animals. Among different species contrasting reactions to temperature changes occur, and within the Chiroptera most mammalian styles of temperature regulation are represented.

Seemingly the larger megachiropterans are homoiotherms. Those that have been studied are able to maintain body temperatures within fairly narrow limits (35° to 40° C) over a range of ambient temperatures from roughly 40° to 0° C. No diurnal torpor occurs in these bats, which are usually quite active during the day in their communal roosts. *Pteropus poliocephalus* and *P. scapulatus* react to cold stress by shivering and by enveloping their bodies with the wings; the wings serve as blankets that provide considerable insulation for the body (Bartholomew et al., 1964). Many megachiropterans roost in trees where they are periodically exposed to direct sunlight. Several devices for lowering body temperature were observed by Bartholomew and his co-workers in animals under heat stress. Vasodilation occurred in surfaces such as the scrotum, wing membranes, and ears; these naked surfaces are seemingly efficient heat dissipators. Other reactions to high temperatures were extension of the wings, fanning of the wings, and panting. Under intense heat stress, the animals salivated copiously and licked their bodies. A different thermoregulatory pattern occurs in some megachiropterans; Bartholomew et al. (1970) have shown that some of the smaller species in New Guinea are capable of becoming torpid.

Compared to megachiropterans, microchiropteran bats are highly variable in their responses to temperature extremes. Among tropical or subtropical microchiropterans two extreme patterns of response have been found. The Australian species *Macroderma gigas* (Megadermatidae) is able to maintain a stable body temperature in the face of ambient temperatures as low as 0° C, and many of the reactions to temperature extremes in this bat are similar to those of megachiropterans (Leitner and Nelson, 1967). At the other extreme, the Neotropical species *Desmodus rotundus*, the vampire bat, seems unable to regulate its body temperature (Lyman and Wimsatt, 1966). The inability of this species to dissipate heat causes death at ambient temperatures as low as 33° C. Nor can this species withstand low temperatures: after an initial short-lived attempt at maintaining the body temperature by increasing metabolism, the body temperature drops and body temperatures between 17° C and 27° C are often lethal. Vampires have no ability to rewarm their bodies if the ambient temperature is not raised. The style of temperature regulation of most tropical bats probably lies somewhere between these extremes.

Many tropical microchiropteran bats from the Old World and from the Neotropics have similar diel activity cycles. These bats are active at night and inactive during the day, and this activity cycle is reflected by their temperature cycles. In the Neotropical phyllostomatid bats that have been studied, body temperatures are from 37° to 39° C during the night, and are two or three degrees lower during the day (Morrison and McNab, 1967). In general, these bats are able to maintain high body temperatures despite moderately low ambient temperatures; only one species cannot sustain "normal" body temperatures at ambient temperatures below 12° C. Some Old World members of the families Rhinopomatidae, Emballonuridae, Megadermatidae, and Rhinolophidae have responses to temperature extremes similar to those of phyllostomatids

(Kulzer, 1965). But broadly speaking, cold stress can usually be tolerated by tropical bats for only fairly short periods, after which the body temperature falls uncontrollably. Body temperatures below 20° C are often fatal.

The Molossidae, a widespread but largely tropical and subtropical group, seem to have a pattern of temperature regulation intermediate between that of the tropical families and the pattern typical of microchiropterans inhabiting temperate areas. The strictly tropical molossids have some diel variation in body temperature, but cannot cope with low temperatures, and die when the body temperature goes below 20° C. Molossids that inhabit temperate areas, however, can tolerate body temperatures as low as 10° C. In southern California in the winter the western mastiff bat (*Eumops perotis*) becomes torpid during the day, when its body temperature drops to within 1° to 2° C of the ambient temperature at temperatures from 9° to 28° C. The metabolic rate is spontaneously elevated in the afternoon and evening, and the bats are active at night (Leitner, 1966).

Adaptive hypothermia, often involving (at different seasons) both daily torpor and hibernation, occurs in many vespertilionids of north-temperate areas, and seems to be the key to the survival of some species in cool or cold regions. During the summer some microchiropterans of temperate zones undergo daily torpor ("*Tagesschlaflethargie;*" Eisentraut, 1934) at low ambient temperatures, but under certain conditions these animals are able to maintain high body temperatures even when resting. Daily torpor greatly reduces the total metabolic output; consequently, these bats can survive with less food than would be necessary if they maintained a constant temperature. Some bats abandon homoiothermy well before winter. Fat deposition is known to occur in late summer or early fall in some species of vespertilionids that hibernate (Krzanowski, 1961; Baker et al., 1968; Weber and Findley, 1970; Ewing et al., 1970), and three species of *Myotis*

studied by O'Farrell and Studier (1970) became nonhomoiothermic during this time. In *M. thysanodes* the metabolic rate for homoiothermic individuals at an ambient temperature of 20.5° C is 6.93 cm.^{3} O$_2$/gm./hr.; the rate drops to 0.59 cm.^{3} O$_2$/gm./hr. in nonhomoiothermic bats (O'Farrell and Studier, 1970). This drop in metabolic rate results in the saving of 2.81 kcal./day as a bat becomes nonhomoiothermic. Fat is deposited in preparation for hibernation at the rate of 0.17 gm./day in the period of maximum fat accumulation, which requires 1.60 kcal./day. This energy is available primarily because of the late summer-autumn shift to daily hypothermia (Krzanowski, 1961; Ewing et al., 1970). Some birds accumulate fat in preparation for migration by greatly increasing food intake, but this increased intake does not occur during the period of fat deposition in *Myotis lucifugus* and *M. thysanodes*.

Winter hibernation in bats differs from short-term torpor largely in the length of dormancy and in the levels to which the metabolic rate and temperature drop. The duration of hibernation for bats differs widely between species and within a species depending on the area. In the northeastern United States *M. lucifugus* remains in hibernation for six or seven months, from September or October to April or May (Davis and Hitchcock, 1965). Periods of hibernation for bats in warmer areas are probably considerably shorter. Ewing et al. (1970) used amounts of fat accumulated by bats in the autumn as a basis for estimating durations of hibernation. The estimated lengths of hibernation for several New Mexican bats were as follows: *M. lucifugus*, 165 days; *M. yumanensis*, 192 days; *M. thysanodes*, 163 days. Because no allowance was made for metabolic drain occasioned by intermittent periods of activity, these estimates are probably too high. Hock (1951) estimated that the metabolic rate of hibernating bats at ambient temperatures not much above freezing was 0.1 cm.^{3} O$_2$/gm./hr., which is 257 to 385 times lower than the rate for a flying bat estimated by Studier and Howell (1969). At ambient temperatures near 5° C, bats in deep hibernation maintain their body temperatures about 1° C above ambient temperature. These bats are responsive to certain stimuli, and will begin arousal when handled or when subjected to unusual air movement. As a defense against freezing to death, bats spontaneously raise their metabolic rates at dangerously low ambient temperatures (below roughly 5° C) and either arouse fully or regulate their temperatures and remain in hibernation. At least some members of the micro- chiropteran families Rhinolophidae, Vespertilionidae, and Molossidae are known to hibernate.

An excellent review of thermoregulation and metabolism in bats is given by Lyman (1970). According to this author, "there seems to be a progression in the temperature regulation of bats from a reasonably well regulated homeothermism of the large Megachiroptera of the tropics, through the less adequate temperature regulation of tropical Microchiroptera, to a rather special form of hibernation in the microchiropterans of the temperate zone."

CHAPTER 20

WATER REGULATION IN MAMMALS

Roughly 35 per cent of the earth's land surface is desert, where water is the primary limiting factor for plant and animal life. These desert areas are characterized by intense daytime heat in the summer, intense solar radiation by day and maximal heat loss by night resulting in great daily changes in temperature (commonly up to 30° C), extremely low humidity through most of the year, and small amounts of precipitation, often at irregular intervals. To an animal abroad on the desert on a summer day, the searingly dry winds and the radiation and reflection of heat from the hot and pale-colored soil add to the harshness of the environment. Few equally hostile environments occur on earth, and to the casual observer the desert gives the impression of overwhelming sterility. This impression is deceptive, however, for in reality the desert supports a great variety of animal life. The abundance of mammal life on the desert and the severity of this environment are well described by Hall (1946:1,2): In the morning "scores of burrow openings around sandy dunes attest the density

of population of small mammals — a density equaled in few other habitats — and inspection discloses that in nearly every burrow, a short distance back from the entrance the occupant has snugly packed a plug of moist sand to shut him away from the dangers of day. Before a person's curiosity is half satisfied about the burrows and the dozens of stories told by the tracks, the sun is up — and with it the wind, the wind that obliterates every telltale mark and burrow opening, leaving only smooth sand in their places. Little by little the heat returns." Each desert mammal has evolved means of coping with the extreme conditions of this environment. Of these conditions none is more acute than the lack of water.

Water is absolutely essential to life; to all mammals life depends on the maintenance of an internal *water balance* (water balance occurs when intake, through drinking, eating, and the production of metabolic water equals the output through the skin, respiratory passages and surfaces, feces, and urine) within fairly narrow limits. Most mammals are under intense discomfort when water loss reduces their body weight by as little as 10 or 15%, and death occurs in many mammals when such loss reduces the body weight by 20%. Loss of water occurs rapidly on the desert: water loss in man on a hot summer day in the southwestern deserts of the United States has been recorded as 1.41% of body weight per hour; comparable figures for the donkey and dog were 1.24 and 2.62 respectively (Schmidt-Nielsen, 1964:27). Deprived of drinking water, a man or a dog can survive only a day or two of exposure to the desert in the summer. Nonetheless, some small desert rodents live for long periods on diets of dry seeds and no drinking water. Similarly, all of the mammals in some desert areas must maintain water balance with only occasional access to water. Although much remains to be learned about mammalian adaptations for water conservation in arid environments, excellent studies in this field have provided a solid base of knowledge.

A number of different solutions to the problem of maintaining water balance without drinking water are used by desert mammals. These solutions depend on the size of the animal, the timing of activity cycles, the foods eaten, and a variety of behavioral, structural, and physiological features. It is highly unlikely that any two desert mammals have solved this problem in the same way. The following discussions will not cover the subject of water conservation in mammals exhaustively, but will consider the adaptations that permit several kinds of mammals to maintain water balance in dry environments. A good discussion of water regulation in desert mammals is presented by Schmidt-Nielsen (1964).

MAMMALS DEPENDENT ON "WET" FOOD

A number of mammals that occupy deserts or semi-arid areas are no better adapted to surviving without considerable moisture in their diets than are mammals of fairly moist areas. Even in some areas with fairly high precipitation small mammals do not have regular access to drinking water and, as in the case of some desert rodents, satisfy their water requirements by eating moist food.

Succulent desert plants provide water for some desert rodents. One such rodent is the white-throated woodrat (Neotoma albigula), which occupies the hot deserts of the southwestern United States and northern and central Mexico. Although this woodrat often inhabits extremely barren areas, its distribution is limited to that of the succulent plants from which it obtains water. Paradoxically, this desert rodent is dependent on large amounts of water for the maintenance of water balance; but it has an important adaptation that allows it to use cactus as a water source. This animal has the ability to cope with oxalic acid ($C_2H_2O_4$), which occurs in large amounts in prickly pear and cholla cactus (Opuntia). Oxalic acid is a highly toxic compound to most mammals. The white-throated woodrat, however, is able to metabolize oxalic acid, and eats quantities of cactus that contain sufficient oxalic acid to kill other mammals of equal size (Schmidt-Nielsen, 1964:146-149). MacMillen (1964a) found that the desert woodrat (N. lepida) and the cactus mouse (Peromyscus eremicus) also utilize large quantities of Opuntia as a source of both food and water; apparently these species have also evolved a means of metabolizing oxalic acid and excreting the by-products.

The ability to obtain water from cacti and to deal with oxalic acid metabolically is not limited to the rodents mentioned above, all of which belong to the family Cricetidae, but also occurs in the rodent family Geomyidae, the pocket gophers. The northern pocket gopher (Thomomys talpoides), where it occurs on fairly dry shortgrass prairies of Colorado, obtains water by eating prickly pear cactus (Vaughan, 1967). During the dry midwinter period this plant comprised 79% of the diet of these gophers in one area. In arid parts of Arizona and Colorado, the valley pocket gopher (*T. bottae*) has also been observed to make heavy seasonal use of prickly pear cactus. Anyone who has had contact with prickly pear cacti must admit that the behavioral ability of small mammals to cope with the spiny armor of these plants is as impressive as the physiological ability to deal with oxalic acid.

Other plants provide water for rodents inhabiting arid regions. As an example, the juniper (*Juniperus*), which contains about 70 per cent water, provides water during all seasons for the Stephen's woodrat (*Neotoma stephensi*), an inhabitant of semi-arid areas in Arizona and New Mexico (Hanson, 1971). The desert ground squirrels (*Spermophilus* and *Ammospermophilus*) also obtain moisture from their food, which is largely green vegetation and insects.

Some desert rodents obtain water from succulent plants that contain high salt concentrations; these animals have kidneys that are able to produce highly concentrated urine (urine that has little water relative to the contained solutes). The North African sand rat (*Psammomys obesus*, Cricetidae) is such an animal. This small rodent obtains water from the fleshy leaves of halophytic plants (plants that grow in salty soil), which grow along dry river beds in the desert (Schmidt-Nielsen, 1964:183). These leaves are 80 to 90 per cent water, but contain higher concentrations of salt than seawater does, and also have large amounts of oxalic acid. In order to utilize this water source, the sand rat must produce urine with extremely high concentrations of salt and must be able to metabolize large quantities of oxalic acid. An Australian hopping mouse, *Notomys cervinus* (Muridae), has a remarkably well developed ability to concentrate electrolytes in its urine, and probably also uses the succulent but highly saline leaves of halophytic plants as a water source (MacMillen and Lee, 1969). The tammar wallaby is known to be able to drink seawater

(Kinnear et al., 1968). A similar ability occurs in the western harvest mouse (*Reithrodontomys megalotis*), a rodent that commonly inhabits semi-arid or arid areas. This animal can produce highly concentrated urine, and is able to survive short periods of water deprivation. Some populations of harvest mice inhabit salt marshes, areas regarded as "physiological deserts" because of the physiological problems of utilizing the water from the highly saline sap of the plants or from seawater. MacMillen (1964b) thought that these populations of harvest mice might obtain water by drinking seawater or eating halophytes, but more recent work by Coulombe (1970) suggests that daily torpor and water from dew or fog precipitation may be of equal importance. The fact remains, however, that this rodent can drink seawater and maintain weight on this water source.

Most deserts support a number of carnivorous or insectivorous mammals whose moisture requirements are seemingly met by the water in their food. The grasshopper mouse (*Onychomys*), a small rodent widely distributed in the deserts and semi-arid sections of the western United States and Mexico, is almost exclusively insectivorous at some times of the year. This mouse has thrived in the laboratory on an entirely meat diet, with no drinking water (Schmidt-Nielsen, 1964:185). In the North American deserts kit foxes (*Vulpes macrotis*), badgers (*Taxidea taxus*) and coyotes (*Canis latrans*) must also be able to derive sufficient water from their meat diets, for these animals often live in areas remote from drinking water. Schmidt-Nielsen (1964:126, 127) found that the desert hedgehog (*Hemiechinus auritus*, an insectivore) and the fennec (*Fennecus zerda*, a fox), both inhabitants of North African deserts, could get adequate water from a predominantly carnivorous diet, as could the mulgara (*Dasycercus cristicauda*), an Australian dasyurid marsupial (Schmidt-Nielsen and Newsome, 1962).

Few large ungulates inhabit barren

deserts where no drinking water or cover is available. One notable exception is the oryx or gemsbok (*Oryx*, Fig. 20–1), a large antelope that occurs in arid and semi-arid sections of Africa, and even penetrates the borders of the Sahara Desert. The oryx has an amazing ability to withstand intense desert heat (as described in Chapter 19, p. 373); perhaps more remarkable is the animal's lack of dependence on drinking water. Careful studies by Taylor (1969) have shown that the water needs of the oryx are satisfied by its food, which consists of leaves of grasses and shrubs that by day may contain as little as 1% water. After nightfall, as temperatures drop and the humidity rises, these parched leaves absorb moisture from the air, and probably contain approximately 30% water during much of the night. By feeding at night, therefore, the oryx can manage a nightly intake of some 5 liters of water with its forage. This is at best a minimal amount of water for a roughly 200 kg. mammal living in shelterless desert, and is sufficient for the oryx only because of a combination of mechanisms that favor water conserva-

tion. Some of the major means for reducing water loss in a dehydrated oryx include: (1) voluntary hypothermia and the sparing use of evaporative cooling except under extreme conditions, when panting but not sweating is used; (2) reduction of evaporative water loss by lowering of the metabolic rate at high ambient temperatures during the day; (3) reduced permeability of the skin resulting in reduced water loss by diffusion; (4) reduced respiratory rates and greater extraction of oxygen from inspired air during the night, and hence reduced water loss *via* expired air; (5) lowering of metabolism and respiratory water loss by more than 30% during the cool night. Taylor also studied the eland (*Taurotragus*) and found that its diet of moist acacia leaves, together with physiological specializations to reduce water loss similar to those in the oryx, allowed this largest of African antelope to be independent of drinking water. Our knowledge of the water needs of wild ungulates is rudimentary at best; adaptations that reduce water requirements probably occur in many other ungulates that occupy dry areas.

FIGURE 20–1. Gemsbok (*Oryx gazella*) in the Kalahari Desert of Africa. These animals are able to go for long periods without drinking. (Photograph by Fritz C. Eloff.)

PERIODIC DRINKERS

In many arid or semi-arid regions, scattered water holes or widely separated rivers offer water to mammals that can move long distances. The extensive grasslands of Africa form such an area, as did the North American Great Plains before the coming of white man. Most large mammals in such areas probably drink every day or two in hot weather, and seemingly are unable to survive for long periods without drinking. A few ungulates, however, occupy an intermediate position with regard to water needs. Although they can go for long periods without drinking, these mammals are not independent of drinking water, as are some desert rodents, and must drink water periodically. Such a mammal is the camel.

Our present knowledge of the water metabolism of the camel is largely a result of the work of Schmidt-Nielsen and his associates (1956a, 1956b, 1957a, 1957b). Their work, done in the northwestern part of the Sahara desert on local domestic camels (*Camelus dromedarius*), substantiated the popular idea that camels can tolerate long periods without drinking water; but more important, Schmidt-Nielsen and his group explained the adaptations allowing this tolerance. The ability of their experimental animals to tolerate dehydration was remarkable. One camel went without water for 17 days in the winter on a diet of dry food; during this period it lost 16.2% of its body weight. In some areas camels that foraged on native vegetation in the winter were never watered. Two camels kept without water for 7 days in the heat of the summer lost slightly over 25% of their original body weights. All of these animals drank tremendous amounts of water after their periods of dehydration, and none showed ill effects.

The camel economizes on water in several ways. Because the body temperature of the camel drops sharply at night and, at the other extreme, the camel can tolerate considerable hyperthermia, the day is largely over before the body temperature rises to levels at which evaporative cooling, in the form of sweating, must occur. Thus, relative to man under similar conditions, for example, very little moisture is expended each day in cooling the camel (Fig. 20–2). As in the oryx, excess heat gained by day is lost passively at night. Further water economy results from the ability of the kidneys to concentrate urine, and from the absorption of water from fecal material. But despite these important water-saving devices, the camel loses water steadily through evaporation and in the urine and feces, and perhaps most remarkable is its ability to tolerate tremendous losses of weight (up to over 27% of body weight) during intervals of dehydration.

Apparently the degrees of water loss from various parts of the body differ between man and the camel. When a man on the desert has lost water equal to about 12% of the body weight, the blood plasma has lost sufficient water

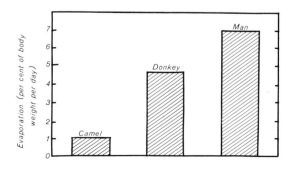

FIGURE 20–2. Amounts of water used for temperature regulation when the subjects were exposed to the sun in the Sahara Desert in June. The camel and the donkey were tested during periods when they were deprived of water. (After Schmidt-Nielsen et al., 1956.)

Table 20–1.   Water Consumption by a Donkey Deprived of Water in June in a Desert. Water Consumed within a Two-Hour Period is Shown in Parentheses; Water Taken within Five Minutes is Shown without Parentheses. (From Schmidt-Nielsen, K.: *Desert Animals, Physiological Problems of Heat and Water,* Oxford University Press, 1964.)

	WEIGHT LOSS (in kg)				DRINKING	
Days Without Water	*Weight Before Dehydr.*	*Dehydr. Weight*	*kg*	*Per Cent of Body Weight*	*Liters*	*Per Cent of Dehydr. Body Weight*
4	107.0	74.5	32.5	30.4	20.5 (23.5)	27.2 (31.5)
4	104.1	73.0	31.1	29.9	20.3 (26.9)	27.8 (36.8)

to become viscous; as a result, the heart has difficulty moving the blood and the rate of the blood circulation to the skin and the rest of the body decreases. This leads to a marked reduction in the rate of dissipation of metabolic heat, to a sudden rise in body temperature, and to rapid death. Even in a camel that has been dehydrated to the extent of losing 20% of its body weight, however, close to the normal plasma water content is retained, whereas large amounts of water are lost from interstitial fluids and from intracellular protoplasm. In a camel deprived of water for eight days, the loss of volume due to water loss was 38% for interstitial fluids, and 24% for intracellular water. However, only a 10% reduction in water in the plasma occurred (Schmidt-Nielsen, 1964:65). Although the camel becomes strikingly dehydrated and emaciated during periods without water, the blood apparently retains its fluidity and its ability to contribute to heat dissipation without straining the circulatory system. The donkey (*Equus asinus*) was also studied by Schmidt-Nielsen (1964:81-93) and proved to be as capable of tolerating weight loss due to dehydration as was the camel, but the donkey had a rate of water expenditure some 2½ times that of the camel and could therefore not be independent of water for intervals of more than a few days. Both the camel and the donkey (Table 20–1) have an amazing ability to

recoup their water losses rapidly. A camel that had lost 25% of its body weight recovered its original weight within 10 minutes by drinking, and a similarly dehydrated donkey was able to drink back its weight in but two minutes. The amount of water drunk in a brief interval is nearly unbelievable. Within a few minutes a dehydrated male camel studied by Schmidt-Nielsen drank 104 liters of water (weighing 228 pounds)!

Perhaps other mammals inhabiting dry areas have adaptations for water conservation similar to those of the camel. Studies of South American members of the family Camelidae might indicate that they share some of their Old World relative's abilities to cope with dry conditions.

Many desert-dwelling marsupials have specializations that minimize excretory water loss. For example, water conservation in the red kangaroo (*Megaleia rufa*), an inhabitant of arid parts of central Australia, is partially accomplished by a high degree of concentration of urine by the kidneys. (The more concentrated the urine, the less water lost with excretory wastes.) The ability to concentrate urine in this animal is roughly equal to that in some North American desert rodents (*Ammospermophilus leucurus* and *Dipodomys merriami*). The urine-concentrating ability of the tammar wallaby allows this animal to drink sea water (Kinnear et al., 1968). The mulgara (*Dasycercus cristicauda*), a small carnivorous da-syurid that inhabits sandhill deserts in central Australia, needs no drinking water and avoids heat stress by taking refuge in burrows. Excretion of the large amounts of urea contributed by the carnivorous diet of this animal is accomplished with little water loss by urine concentration, and the moisture in the mulgara's food is sufficient to maintain water balance. It is of considerable interest that this marsupial is almost identical in its ability to concentrate urine to the grasshopper mouse (*Onychomys torridus*), an insectivorous-carnivorous rodent that inhabits deserts of the southwestern United States.

NON-DRINKERS
DEPENDENT ON DRY FOOD

The most impressive feats of water conservation are those performed by granivorous desert rodents (seed-eating rodents). These small mammals are often common in barren sand dunes or rocky hills where drinking water is rarely available, and their diets through much of the year include only dry seeds that are picked from the surface of the ground or are sifted from the soil. Some New World kangaroo rats and pocket mice (Heteromyidae), some jerboas (Dipodidae) and gerbils (Cricetidae, Gerbillinae) of the Old World, and some Australian mice (Muridae) are capable of living on dry diets. Recent work by Abbott (1971) has shown that the cactus mouse (*Peromyscus crinitus*), a cricetid rodent, can also live on a dry diet. As might be expected, the centers of distribution of these rodents are all desert regions. The following discussion is based largely on the work of Schmidt-Nielsen and his associates, whose publications dealing with water conservation in Merriam's kangaroo rat are cited in the bibliography.

In the kangaroo rat a precarious water balance is maintained by a combination of physiological and behavioral adaptations. In all animals intake of water must equal or surpass loss of water if water balance is to be maintained. Inasmuch as kangaroo rats do not drink water, water can only be obtained from food, either from free water in the food or from water formed by oxidation of food. Because the moisture content of the kangaroo rat's food is generally low, their chief source of water is probably "oxidation water." Channels of water loss from the body are evaporation and loss in urine and feces.

Water intake in the kangaroo rat can be accounted for fairly easily. The seeds eaten by this animal are high in carbohydrates, which yield a large amount of oxidation water. As an example, for every 100 gm. of dry pearled barley metabolized, a total of 53.7 gm. of oxidation water is produced (Schmidt-Nielsen, 1964:173). This may be augmented under natural conditions by free water in the food. Seeds scattered on or slightly beneath the surface of the ground absorb moisture at night as the temperature drops and the humidity rises, and seeds stored in humid burrows may contain as much as 30% free water. If kangaroo rats show a preference for seeds that have been stored in burrows for a time, rather than for those freshly collected, the amount of moisture taken in with the food could be appreciable. This preference has not yet been demonstrated, but might be expected.

Water loss in kangaroo rats is minimized by several devices. Evaporative water loss in kangaroo rats is unusually low, due in part to a nearly total absence of evaporation from the skin. But equally important is the reduction of respiratory water loss. This economy results from the nasal passage operating as a countercurrent heat exchanger (a feature characteristic of many birds and mammals; Schmidt-Nielsen et al., 1970) with alternating flow in opposite directions in a single tube rather than with steady flow in opposite directions in adjacent tubes. Inspired air drawn through the nasal passage causes evaporative cooling of the moist nasal mucosa. This air then becomes warmed and saturated with water in the lungs.

During expiration this humid air passes back into the nasal passage and over the cool nasal mucosa, where the air is cooled and loses some of its moisture by condensation on the mucosa. This pattern, repeated with every respiratory cycle, results in cool expired air that is more than 10° C below the body temperature. Although this expired air is saturated with moisture, because it is far cooler than the air in the lungs it contains considerably less water than does saturated air at the higher lung temperature. (In man, by comparison, the temperature of expired air is nearly that of the body, resulting in high rates of pulmonary evaporation.) An additional reduction of evaporative water loss occurs when the animal is in its burrow, in which the air is usually fairly humid.

MacMillen (1972) has discussed the relationship between evaporative water loss and metabolic water production in desert rodents when these animals are active above ground on cool evenings. Although under conditions of thermal neutrality evaporative water loss surpassed metabolic water production in several species of rodents, when they were subjected to ambient temperatures below their thermoneutral zones (temperatures of 20° C to 0° C, a range commonly encountered in deserts at night) these animals produced more metabolic water than they lost by evaporation (Table 20–2). Two factors favor the maintenance of water balance under cool conditions. First, as has been shown by Schmidt-Nielsen et al. (1970), the water-conserving effectiveness of the nasal countercurrent system in rodents varies inversely with temperature; thus, evaporative water loss due to respiration decreases with decreasing temperature. Secondly, as the ambient temperature drops progressively further below the zone of thermal neutrality there is a progressive rise in the metabolic rate (this is especially pronounced in small rodents with high surface-volume ratios and poor insulation); with higher metabolic rates, the production of metabolic water increases. In light of these relationships, it is MacMillen's opinion that while active on the surface at night most small desert rodents produce more water metabolically than they lose through pulmocutaneous evaporation.

An important reduction in water loss results from the great ability of the kidneys to concentrate urine. The kangaroo rat's urine is roughly five

Table 20–2. Relationships Between Metabolic Water Production (MWP) and Evaporative Water Loss (EWL) in Several Nocturnal Rodents Under a Range of Ambient Temperatures (T$_A$) and Low Relative Humidities (R.H.%). Underlined Measurements Were Taken of Animals in Thermal Neutrality. Abbreviations of generic names: *P., Peromyscus; L., Leggadina; N., Notomys.* (From MacMillen, R. E., in Maloiy, G. M. O.: *Comparative Physiology of Desert Animals,* Academic Press, New York, 1972. Reprinted with permission of the Zoological Society of London.)

T$_A$ °C	P. eremicus R.H.,%	P. eremicus MWP/EWL	L. hermannsburgensis R.H.,%	L. hermannsburgensis MWP/EWL	N. alexis R.H.,%	N. alexis MWP/EWL	N. cervinus R.H.,%	N. cervinus MWP/EWL
5		1.05						
10		1.26	39.9	1.07	25.8	1.14	30.3	1.00
15		1.08						
20		0.88	14.6	0.97	12.4	0.90	13.9	0.89
25		0.69						
28			9.7	0.56	6.9	0.61	6.0	0.84
30		0.43						
33			11.6	0.34	5.1	0.47	5.5	0.39
35		0.40	6.7	0.36	5.2	0.44	6.9	0.32
37		0.19	9.3	0.31	9.4	0.26	8.0	0.27

times more concentrated than that of man; in excreting comparable amounts of urea, the kangaroo rat uses one-fifth as much water as does man. The concentration of dissolved compounds in the urine of kangaroo rats may be roughly twice that of seawater, and under laboratory conditions these animals have maintained water balance by drinking seawater. The urine of man, on the other hand, has a concentration of dissolved compounds lower than that of sea water. Thus, when man drinks seawater the excretion of the dissolved salts requires the withdrawal of water from the body tissues, resulting in dehydration; this is made more acute by diarrhea caused by the magnesium and sulfate in the seawater.

Fecal water loss is also low in the kangaroo rat. In eliminating wastes from comparable amounts of food, the laboratory white rat (*Rattus*) used five times as much water for the formation of feces as did the kangaroo rat (Table 20–3). This saving in the kangaroo rat is partly the result of more efficient utilization of food and the accompanying reduction in the production of fecal material, but is also due to an exceptional ability to withdraw water from intestinal contents.

Schmidt-Nielsen calculated that in the kangaroo rat water intake equalled water losses at atmospheric humidities higher than 20% relative humidity at 25° C. More advantageous conditions of humidity, in terms of water loss, are faced by kangaroo rats in their burrows and when foraging during the most humid parts of even summer nights.

Although the margin for survival is slim, it appears that the kangaroo rat's remarkable ability to subsist on dry food and no water can be satisfactorily explained on the basis of the several behavioral and physiological specializations that aid in conserving water.

The ability to be independent of exogenous water is by no means unique to a few rodents. Each major desert region in the world is well populated by rodents, and seemingly each of these regions supports some highly specialized species adapted to dry diets. Such specializations occur in at least some members of the North American heteromyid genera *Perognathus*, *Dipodomys*, and *Microdipodops*. Probably all species of heteromyids that are restricted to deserts can survive on dry diets; not all inhabit deserts, however, and not all are equally well adapted to dry conditions. For example, *Dipodomys agilis*, an inhabitant of coastal southern California and northwestern Baja California, Mexico, occurs largely in coastal sage scrub, where conditions of extreme aridity occur only in midsummer. This kangaroo rat is marginal in its ability to subsist on a dry diet and cannot survive indefinitely without exogenous water (MacMillen, 1964; Carpenter, 1966). Among desert rodents of the Old World, two jerboas (*Jaculus jaculus* and *Dipus aegypticus*) and several gerbils of two genera (*Meriones* and *Gerbillus*) can live on dry food, and some surpass kangaroo rats in their ability to concentrate urine (Table 20–4). Adaptations to dry diets have doubtlessly evolved independently in

Table 20–3.   The Amount of Water Lost in the Feces of the Kangaroo Rat (*Dipodomys merriami*) and the White Rat (*Rattus norvegicus*).
(After Schmidt-Nielsen, K.: *Desert Animals, Physiological Problems of Heat and Water*, Oxford University Press, 1964.)

	FECES, GRAMS OF DRY MATTER/100 GM. FOOD	WATER, MG./GM. DRY FECAL MATTER	WATER LOST WITH FECES PER 100 G. BARLEY EATEN
Kangaroo rat	3.04	834	2.53 g
White rat	6.04	2246	13.6 g

several rodent families (Heteromyidae, Cricetidae, Dipodidae, Muridae). Striking convergent evolution in these families has led to saltatorial adaptations in some members of each family as well as to similar specializations favoring water conservation. The most concentrated urine yet measured is that of a murid Australian hopping mouse (*Notomys alexis*). This rodent occupies some of the most arid deserts in the world, regions where 10 years may pass between rains. This species and two other desert-inhabiting murids of Australia were studied by MacMillen and Lee (1967, 1969), who found that all three species could live on dry seeds with no drinking water. Compared to the kangaroo rat, these murids had higher rates of pulmocutaneous water loss and lost more water in the feces, but in general they had a

greater ability to concentrate urine (Table 20–4).

Two species of spiny mice (*Acomys*, Muridae) studied by Shkolnik and Borut (1969) in the desert of Israel are remarkable in their unusual pattern of adaptation to arid conditions. These animals have highly specialized kidneys that can concentrate urine to a greater degree than can the kangaroo rat kidney; but the spiny mice have an evaporative water loss two to three times as great as that in Merriam's kangaroo rat. Probably primarily due to high water loss through the skin, spiny mice are unable to subsist on a diet of dry seeds. Apparently the high cutaneous water loss is important as a means of dissipating heat in a hot climate, and the great ability of the kidney to concentrate urine, coupled with a diet high in land snails (which have a

Table 20–4. Concentration of Urea in the Urine, Water Loss from Lungs and Skin, and Percentage (by weight) of Water in the Feces of Some Desert Rodents. (Data from MacMillen and Lee, 1967.)

	URINE	WATER LOSS	
SPECIES	Maximum Urea Concentration (mmole/liter)	Pulmocutaneous (mg. H_2O cm.^{3} O_2)	Feces (% H_2O)
Sciuridae			
Ammospermophilus leucurus (North America)	2860	0.53	
Heteromyidae			
Dipodomys merriami (North America)	3840	0.54	45.2
Dipodomys spectabilis (North America)	2710	0.57	
Cricetidae			
Gerbillus gerbillus (Egypt)	3410		
Dipodidae			
Jaculus jaculus (Northern Africa)	4320		
Muridae			
Notomys alexis (Australia)	5430	0.91	48.8
Notomys cervinus (Australia)	3140	0.76	51.8
Leggadina hermannsburgensis (Australia)	3920	1.15	50.4

high water content), compensates for the extravagant use of water in thermoregulation.

The kidneys of some bats are specialized to concentrate urine, but these animals are seemingly not independent of drinking water. Carpenter (1969) found that two insectivorous bats that inhabit deserts produced concentrated urine, but that their need for water was increased by high evaporative water losses during flight and when the animals were not torpid. He estimated that the bats lost approximately 3.09% of their body weight through evaporation per hour of flight. After a careful consideration of the bats' water budgets, Carpenter concluded that the bats were not independent of drinking water but that their ability to fly long distances to drink water enabled them to maintain water balance in desert areas. A marine fish- and crustacean-eating bat (*Pizonyx vivesi*, Vespertilionidae) that inhabits desert islands and desert coasts of the Gulf of California has the ability to concentrate urine to the extent that it can utilize seawater as a water source (Carpenter, 1968). Because of high evaporative water losses, particularly during flight, the water gained from this bat's food probably is not sufficient to meet its water requirements, and presumably it must drink seawater. The fact that members of the genus *Myotis* (which is a close relative of *Pizonyx*) have been observed drinking seawater (Dalquest, 1948) suggests that the ability to use seawater as a source of water may be widespread among vespertilionid bats.

ECOLOGICAL CONSIDERATIONS

The importance of knowledge of interspecific differences in water metabolism to an understanding of the roles of mammals in the ecosystems of arid regions has only recently been appreciated. The ecological displacement of some species of rodents results in part from their differing means of satisfying water requirements. In a study of a semi-desert rodent fauna in California, MacMillen (1964a) found that water metabolism differed significantly between species. Observations in other areas indicate that different preferences with respect to water sources reduce competition between some sympatric and congeneric species. Where *Neotoma stephensi* and *N. albigula* occur together along the Mogollon Rim of Arizona, for example, the former uses juniper as its source of water, whereas the latter uses cactus; competition for food and for nest sites

Table 20–5. Temperature and Relative Humidity (R.H.) of Roosting Sites and Body Weight Loss During the Period From 8 A.M. to 8 P.M. for Bats (*Myotis*) Either Caged Singly or in Groups of Four. The Mean and Range are Given; Sample Sizes Appear in Parentheses. (From Studier et al., 1970.)

SPECIES	AMBIENT TEMP. (°C)	AMBIENT R.H. (PER CENT)	PER CENT WEIGHT LOSS INDIVIDUALS	PER CENT WEIGHT LOSS GROUPS
M. lucifugus (site 1)	26.8 15.6–31.1	23 18–31	10.5(5) 7.7–12.8	9.9(4) 7.7–11.5
M. lucifugus (site 4)	26.1 15.6–30.4	32 27–40	10.9(7) 8.8–13.0	11.2(5) 9.0–11.5
M. thysanodes	26.8 15.6–31.1	23 18–31	15.8(8) 9.0–21.8	10.9(3) 10.1–11.5
M. velifer	22.0 20.7–23.3	64 53–96	8.2(4) 5.6–9.8	8.4(2) 8.1–8.8

between these wood rats seems slight in this area (Hanson, 1971). In some deserts the patterns of seasonal dormancy in small mammals may be devices that favor water economy by permitting these mammals to avoid periods of intense heat and low humidity. Future research on water metabolism in mammals may contribute importantly to our knowledge of many aspects of mammalian ecology.

The choice of microhabitat by some mammals may be strongly influenced by humidity. Although the skin of mammals does not usually allow rapid water loss, some small mammals lose water from the lungs and the skin together at such a rate that they become dehydrated if unable to retire periodically to humid microenvironments. Small mammals such as shrews, partly because of a high surface-to-mass ratio and a concomitant high metabolic rate, have high pulmocutaneous water losses and must live in areas with high humidities or remain within fairly humid microenvironments. Pulmocutaneous water loss is also high in some bats exposed to moderately high ambient tempera-

Table 20–6. The Average Percentage of Body Weight Lost by Four Species of Bats (*Myotis*) Deprived of Water Until They Were Under Stress From Water Loss. (From Studier et al., 1970.)

SPECIES	SAMPLE SIZE	PER CENT OF WEIGHT LOSS
M. lucifugus	12	32.3
M. thysanodes	8	31.7
M. yumanensis	3	31.6
M. velifer	4	22.8

tures, and the choice of roosting sites as well as the geographic distributions of some bats may be limited by the animals' inability to avoid daily dehydration. Clustering during roosting, a common behavior in some bats, reduced pulmocutaneous water loss markedly in one species (*Myotis thysanodes*) studied by Studier et al. (1970), as shown by Table 20–5. It was further found that several species of *Myotis* have an ability to tolerate considerable weight loss caused largely by pulmocutaneous water loss (Table 20–6); but because of high rates of such loss two species were unable to survive two days without access to water.

CHAPTER 21

ACOUSTICAL ORIENTATION AND SPECIALIZATIONS OF THE EAR

Because the mammals most familiar to us depend largely on vision for perceiving their environments, it is surprising to note that about 20 per cent of the known species of mammals use acoustical orientation as their primary means, or at least as an important secondary means, of "viewing" their surroundings. Most bats, some members of the orders Insectivora and Pinnipedia, and probably all odontocete cetaceans depend on acoustical orientation. Future research may demonstrate that the use of such orientation is even more widespread among mammals.

ECHOLOCATION IN BATS

Foraging insectivorous bats and bats flying within deep caverns face seemingly insurmountable problems: they must perceive tiny prey and obstacles under conditions of nearly complete darkness. The remarkable nocturnal performance of bats indicates that these problems of perception have been effectively solved. Bats are able to capture insects with great efficiency and speed in darkness. The pursuit and capture of an insect take a mustached bat (*Pteronotus psilotis;* Fig. 21–1) but 1/4 to 1/3 of a second (Novick, 1970). Bats deep in caverns, flying in complete darkness, can not only detect the walls of the cavern but avoid collisions with hundreds of other bats that are circling and maneuvering abruptly.

Bats habitually fly in darkness when vision is clearly of little use; as a consequence, bats (at least microchiropterans) have largely abandoned vision in favor of acoustical orientation. Ultrasonic pulses are emitted by the bat, and the echoes of these pulses reflected by objects allow the bat acoustically to "see" in the dark. Even the

394

FIGURE 21–1.   Left, a moustached bat (*Pteronotus* sp.), in a tropical forest in Mexico, with an insect it has just captured. Right, a big brown bat (*Eptesicus fuscus*). The animal is emitting ultrasonic pulses through the open mouth. (Photographs by J. Scott Altenbach.)

"nature" of objects can be analyzed to some extent.

When considering echolocation in bats we are dealing almost entirely with members of the suborder Microchiroptera, all of which use echolocation. Most members of the suborder Megachiroptera use visual perception instead of echolocation; but how this serves at low levels of illumination is not known. Among megachiropterans, only members of the genus *Rousettus* are known to echolocate; their technique is unique, however, in that the pulses are audible and are made by tongue clicking (Mohres and Kulzer, 1956; Kulzer, 1956, 1958; Novick, 1958).

Although microchiropterans clearly depend on echolocation for perceiving their environments in darkness, the eyes are present in all bats and sight has by no means been totally abandoned. Many phyllostomatids (leaf-nosed bats), for example, have large eyes and obviously make use of them. Olfaction may be of great importance for detecting food in bats that eat fruit, nectar, and pollen or small vertebrates. In species with well developed visual and olfactory capabilities, echolocation is perhaps primarily used for perceiving nearby obstacles.

An accurate picture of the bat's use of acoustical orientation was long in emerging. As early as 1793, Lazaro Spallanzani performed experiments that suggested that bats use acoustical rather than visual perception when avoiding obstacles and when feeding. Not until the early 1940's, however, was the use of echolocation by bats conclusively demonstrated by the careful laboratory experiments of Griffin and Galambos (1940, 1941, 1942), and by the observations of Dijkgraaf (1943, 1946). Continued research and the use of refined electronic equipment have contributed to our present detailed, but as yet incomplete, knowledge of echolocation in bats.

Because citation of the many studies that have contributed to our knowledge of echolocation would unduly disrupt the continuity of the following discussions, very few citations of original literature are given here; but these references should give a student some points of entry into the extensive literature on the subject.

Evolution of Echolocation. In his excellent studies of vocalization and communication in bats, Gould (1970, 1971) hypothesized that the sonar pulses of bats may have been derived originally from vocalizations that served to establish or maintain spacing or contact between bats. The repetitive communication sounds used by infant bats, and similar pulses perhaps used originally during flight to maintain adequate spacing of foraging individuals, may have secondarily become important in connection with detecting prey and avoiding obstacles. According to Gould (1971:311): "The prominence with which continuous, graded signals pervade the lives of such social and nocturnal mammals as bats suggests that echolocation is an inextricable and integral part of a communication system." This author suggests that some of the vocalizations used by early bats may have been inherited from their insectivore ancestors, in which auditory communication was perhaps as important as it has been shown to be in some living insectivores (Gould, 1969; Eisenberg and Gould, 1970).

Adaptive Importance of Echolocation. Probably ever since their appearance in the Paleocene or late Cretaceous, bats have "owned" the aerial adaptive zone during the nocturnal segment of the diel cycle. No birds can match the nocturnal insect-catching ability of bats, nor have birds exploited nocturnal fruit eating or nectar feeding. In many tropical areas the most important flying predators of small vertebrates are seemingly bats, not birds. Bats and birds, the only two flying vertebrates of the Cenozoic, may, early in the era, have divided the diel cycle: birds dominated the aerial zone during daylight, whereas bats dominated this zone during darkness. In tropical areas, which characteristically support a great diversity of both bats and birds, the changing of the guard at sunset is dramatic. Bird activity and bird songs suddenly begin to diminish as the sun

disappears; at the same time bats begin to appear and become increasingly more evident as darkness descends. When the twilight glow in the west is gone most birds are inactive and silent, but the cries of bats regularly penetrate the cacophonous blending of frog and insect noises. This chiropteran domination of the nocturnal air is seemingly due largely to their ability to use echolocation. The perfection of highly maneuverable flight and echolocation in bats may have occurred more or less concurrently during their early evolution, and together these abilities were probably responsible for the spectacular nocturnal success of bats.

Production of Ultrasonic Pulses. The pulses used by microchiropterans for echolocation are produced by the larynx. The cricothyroid muscles (muscles that tense the vocal cords) were shown by Novick (1955) to be essential for normal emission of pulses. These are mostly of frequencies well above those that man can hear. Man can detect frequencies up to roughly 20 kHz (20,000 cycles per second). There are audible components (the so-called "ticklaute") in the vocalizations of many bats, and some sack-winged bats (Emballonuridae) of the Neotropics emit pulses (of 12 kHz) that are audible to man. The pulses, however, are only low-intensity components of cries with higher harmonics (harmonics are frequencies that are integral multiples of the fundamental frequency).

The pulses emitted by bats are usually of high intensity (great loudness), and are emitted through either the mouth or the nose. In most vespertilionids, molossids, and noctilionids the pulses are of the highest intensities recorded for bats, and come from the mouth, which is kept open during flight (Fig. 21–1). The rhinolophids (horseshoe bats) emit high intensity pulses through the nostrils. The Nycteridae, Megadermatidae, and Phyllostomatidae, on the other hand, produce relatively low intensity pulses that have resulted in these groups being called the whispering bats (Griffin,

1958:232-251). Whereas the bats that emit intense pulses catch flying insects, the whispering bats feed largely on fruit, nectar, small vertebrates, insects on the ground or on vegetation, or combinations of these foods. Some authors (for example, Griffin, 1958: 251; Novick, 1970:39,40) have suggested that these soft pulses are well adapted to close-range perception of surfaces, such as rocks or tree trunks, or complex tropical environments with interlacing vines and branches and stratified foliage. Perhaps an advantage is gained by not receiving echoes from distant objects and therefore limiting the complexity of the information from an already complex perceptual situation.

Just how loud are the pulses emitted by bats? Inasmuch as we cannot hear them, their loudness must be expressed in terms of an energy unit called a dyne (a dyne is the force necessary to accelerate one gram of mass one centimeter per second per second). Under ideal circumstances, man's threshold of hearing is roughly 0.0002 dynes per square centimeter. When recorded 5 cm. from a bat's mouth, the least intense pulses of whispering bats are approximately 1 dyne per square centimeter, whereas the loudest pulses of other bats are near 200 dynes per square centimeter. Such loud pulses are comparable to the painfully intense noise made by nearby jet engines.

The strange faces of whispering bats, always a source of amazement to those unfamiliar with bats, may have an important function in connection with echolocation. Mohres (1953) showed that the horseshoe-like structure surrounding the nostrils of a rhinolophid serves as a horn, and "beams" the pulses directly forward from the head. This diminutive megaphone effectively beams the short-wavelength pulses emitted by these bats; these 80 to 100 kHz pulses have wavelengths of only 3 or 4 mm. In addition, because the nostrils are situated almost exactly one-half wavelength apart, the pulses emitted through the nostrils undergo interference and reinforcement of a sort that tends to beam the pulses. The bizarre facial patterns of other whispering bats may well function similarly to direct pulses in such a way that some species can scan their environment with a concentrated beam of ultrasonic pulses much as we probe the darkness with a flashlight beam. The nose-leaf of phyllostomatids (Fig. 21-2) may further serve to shield the ears of the bats from the pulses as they are emitted.

Sensitivity of a Bat to its Outgoing Pulses. Several structural specializations reduce the sensitivity of bats to their own outgoing pulses. Vibrations are transmitted mechanically in mammals by the ear ossicles (malleus, incus, and stapes) from the tympanic membrane to the oval window of the inner ear. Two muscles of the middle ear of mammals serve to dampen the ability of the ossicles to transmit vibrations when animals are subjected to unusually loud sounds or when they are vocalizing. In bats these two muscles of the middle ear—the tensor tympani, which changes the tension on the tympanic membrane, and the stapedius, which changes the angle at which the stapes contacts the oval window—contract immediately prior to the emission of each pulse and probably reduce the bat's sensitivity to pulses. The contraction and relaxation of the muscles are usually synchronized with the emission of pulses and the reception of echoes, respectively. At high rates of emission, however, the muscles probably remain contracted.

An additional structural refinement is the insulation of the bones that house the middle and inner ear from the rest of the skull. This bony otic capsule does not contact other bones of the skull (Fig. 21–3), and is insulated from the skull by blood-filled sinuses or fatty tissue. During the emission of pulses the conduction of vibrations from the larynx and the respiratory passages to the inner ear is thus greatly reduced.

The Advantage of High Frequencies. Because high frequencies are

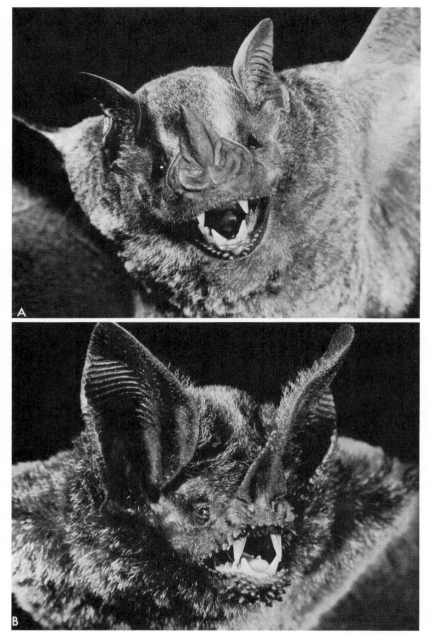

FIGURE 21–2.   Faces of phyllostomatid bats: A, Jamaican fruit-eating bat (*Artibeus jamaicensis*), B, fringe-lipped bat (*Trachops cirrhosis*), a carnivorous-omnivorous species. (Photographs by N. Smythe and F. Bonaccorso.)

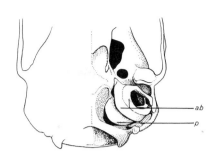

FIGURE 21–3. Ventral view of the posterior part of the skull of a bat (*Myotis volans*), showing the looseness of attachment of the periotic bone (*p*) and auditory bulla (*ab*) to the skull.

more severely attenuated (weakened in intensity) by air than are low frequencies, one might wonder why echolocation in bats is based almost entirely on high frequencies. Interference from such sounds as insect chirps and frog calls is doubtless partially avoided by the use of high frequencies, but this factor may not be of primary importance. Perhaps of greater importance is the relationship between the size of prey and the wavelength of echolocation pulses. The higher the frequency of a sound, the shorter its wavelength. Frequencies of roughly 30 kHz have wavelengths of approximately 11.5 mm., roughly the size of a small moth; this balance between size of object and wavelength is ideal, for objects with sizes of approximately one wavelength reflect sound particularly well. Some species of bats are able to detect wires with as small a diameter as 0.08 mm. (some one-thirtieth of a wavelength), but in general the wavelengths of the pulses emitted by bats are in the range that is most efficient for the detection of small- to medium-sized insects.

Pulses Used For Echolocation. The usual frequencies of bat-produced ultrasonic pulses span the range from approximately 120 kHz to 30 kHz, and the characteristics of the pulses differ widely among species. Novick (1971: 199) recognized four major types of frequency pattern. (1) In bats of such families as Natalidae, Vespertilionidae, and Molossidae, pulses are frequency modulated (FM). Each pulse sweeps roughly an octave downward, usually from levels of 60 to 80 kHz to about 30 or 40 kHz. (2) Members of the Rhinolophidae and Mormoopidae emit pulses mostly of constant frequency (CF) that terminate in a short FM sweep. (3) The pulses of the Nycteridae and Megadermatidae are entirely CF, but are complicated by several harmonics. (4) The Emballonuridae and Phyllostomatidae use complicated FM pulses with several harmonics. All bats that have been carefully studied are able to produce a wide variety of frequencies.

Pulse duration (Table 21–1) and the rate of emission varies widely among different bats, but many follow a basic, widely used pattern. This pattern involves a pulse rate of approximately 10 per second during "searching" flight. The rate is raised to 25 to 50 per second when prey is located, and to 200 or more per second during the final pursuit of the insect. The pulse duration typical of some bats is remarkably short; in *Megaderma lyra* (Megadermatidae) it is down to 0.72 msec. (a millisecond is 1/1000 of a second) during searching flight. By contrast, *Rhinolophus ferrum-equinum* (Rhinolophidae) emits pulses of from 50 to 65 msec. duration under similar conditions. Different types of pulses would be expected to be correlated with different feeding habits, patterns of communication, styles of flight, and habitat preferences. The following considerations of the actual types of echolocation used by different species when approaching perches or prey illustrate some such correlations.

Echolocation and Target-Directed Flight. To a bat approaching a target,

Table 21-1. Pulse Duration Under a Variety of Conditions in Several Species of Bats. (After Novick, 1971.)

BAT	PULSE DURATION	CONDITIONS	REFERENCES
Megaderma lyra	0.54–1.2 msec. (limits) 0.72–1.8 msec. (limits)	Approaching goal or obstacle Searching in laboratory	Mohres and Neuweiler, 1966
Rhinolophus ferrum-equinum	50–65 msec. Decreasing to 10 msec.	Take-off and flight in lab Approaching a landing	Schnitzler, 1968
Rhinolophus euryale	30–45 msec. Decreasing to 7 msec.	Take-off and flight in lab Approaching a landing	Schnitzler, 1968
Noctilio leporinus	7.4 msec. (5.9–9.4 msec.) 14.3 msec. (11.1–16.7 msec.) 13.8 msec. (10.7–16.0 msec.) Decreasing to 1 msec.	Searching flight in outdoor cage Searching flight in wild Searching flight in wild End of prey location	Suthers, 1965
Pteronotus parnellii	14–26 msec. Rising from 20–21 msec. to 28–37 msec. Decreasing to 6.8 msec.	Searching flight in lab Detecting a fruit fly End of insect pursuit	Novick and Vaisnys, 1964
Pteronotus psilotis	2.9–4.8 msec. Decreasing linearly to 0.6–1.0 msec.	Searching flight in lab Fruit fly pursuit and capture	Novick, 1965
Vampyrum spectrum	1.5–1.8 msec. 0.5–1.5 msec. About 0.5 msec.	Searching flight Approach Terminal portion of flight	Bradbury, 1970
Myotis lucifugus	2–3 msec. Decreasing to 0.3–0.5 msec.	Searching flight in lab At end of insect pursuit	Griffin, 1962
Plecotus townsendii	2–5 msec. Decreasing to 0.3–0.5 msec.	Searching flight in lab At end of insect pursuit	Grinnell, 1963
Eptesicus fuscus	2–4 msec. 10–15 msec. 0.25–0.5 msec.	Flying in laboratory Flying in open at 10 m. altitude Late terminal phase	Griffin, 1958
Lasiurus borealis	2.4–3.0 msec. 0.3–0.5 msec.	Pursuit of mealworms in lab Late terminal phase	Webster and Brazier, 1968

several types of information are of critical importance. The distance of the target, its position, the direction of its movement (if it is an insect), its size and shape, and to some extent its nature (whether furry or scaly, for example) are essential to appropriate reactions on the part of the bat. In the case of a bat closing in on an insect, the difficulty of obtaining this information is compounded by the fact that neither the bat nor the insect is stationary, and the means by which the bat is seeking the information is based on pulses that take some time to travel from the bat's larynx to the target and back to the bat's ear. Also, the "processing" of the information—the traveling of the information along sensory pathways to the central nervous system, the bridging of the appropriate synapses, and the resultant initiation of action by pulses relayed along motor nerves—takes time. Nonetheless, the very survival of insectivorous bats depends on their ability to locate, pursue, and capture insects that they perceive by means of echolocation.

The search-and-capture sequence described by Novick (1963b, 1965, 1970, 1971) for a Neotropical moustached bat (*Pteronotus psilotis*) illustrates one major style of echolocation. As *P. psilotis* flies, it emits about 18 pulses per second, each with a duration of approximately 4 msec. Each pulse is followed by a silent interval of some 55 msec., during which the bat moves approximately 100 mm. Sound travels about 345 mm./msec. in air, and a 4 msec. pulse has a length of about 1380 mm. With this pulse duration, the total time required for a pulse to travel to an object and for its echo to return to the bat is such that echoes from objects closer than about 700 mm. will be perceived by the bats as overlapping with an outgoing pulse. With a search-pulse duration of 4 msec., therefore, the world of *P. psilotis* is probably divided into two parts, depending on distance from the bat. "Distant" objects either do not produce detectable echo, or produce an echo that does not overlap an outgoing pulse. "Close" objects are those no more than roughly 600 or 700 mm. away, that produce echoes that overlap an outgoing pulse. When the bat detects pulse overlap caused by an echo from an insect it adjusts its flight path to intercept the prey; at the same time it raises the rate of emission of pulses and reduces the pulse duration. The approach phase of the chase takes perhaps 350 msec., and during this period the pulse-repetition rate accelerates to roughly 100 per second. The terminal phase of the chase may occupy 150 to 214 msec., by which time the pulse-repetition rate has risen to at least 170 per second. Throughout the approach phase and terminal phase the pulses are shortened linearly (they become progressively shorter as the distance is decreased). In *P. psilotis* the echo from the insect during the approach phase overlaps the outgoing pulse by some 1.2 msec., and during the terminal phase by 0.8 msec. After the capture of the insect, when all objects are again so distant that no pulse overlap occurs, the bat resumes its 18-per-second search pulses. Pulse overlap, then, occurs throughout the pursuit of an insect in *P. psilotis* and in other mormoopids (leaf-chinned bats) studied (Fig. 21–4).

Remarkably long pulses, of 50 to 65 msec. duration, are used by the greater horseshoe bat, *Rhinolophus ferrum-equinum*, during searching flight. The pulse duration lengthens at from 5.5 to 2.7 msec. from the target, and then shortens linearly as the target is approached. The briefest pulses, approximately 10 msec., are two or three times as long as the longest pulses of vespertilionids (common, mouse-eared bats). Broad overlap of incoming echoes and outgoing pulses occurs throughout the approach in this bat and in a closely related species, and such overlap may be characteristic of all rhinolophids (horseshoe bats). However, pulse overlap is not characteristic of all patterns of echolocation.

In the little brown bat, *Myotis lucifugus*, and in all other vespertilionids studied, pulse overlap is avoided. In

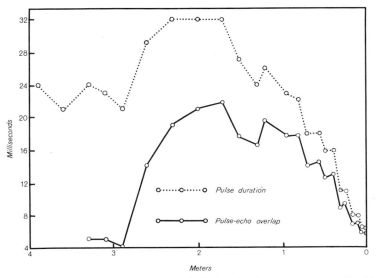

FIGURE 21–4.   An example of the pursuit of a fruit fly by a bat (*Pteronotus parnellii*, Mormoopidae). The pulse duration and calculated pulse overlap are plotted against distance from the insect target. (After Novick and Vaisnys, 1964.)

M. lucifugus the searching pulse rate is roughly 15/sec. (Griffin, 1958:360) and the pulse duration is about 2.6 msec. (Webster and Brazier, 1968). A target is detected at some two meters, but the early approach phase begins at approximately 720 mm. (Novick, 1971: 206). As a bat approaches its target it shortens its pulses by about 1 msec. per 260 mm. of approach; immediately before contact with the insect, pulse duration may be as short as 0.25 msec. and the pulse rate may approach 200/sec. Pulse overlap is avoided by shortening the outgoing pulses before pulse-echo overlap occurs, and then by a linear shortening of the pulses as the bat closes in on the target. Present evidence suggests that avoidance of pulse overlap is probably also typical of bats of the families Megadermatidae, Phyllostomatidae, and Noctilionidae.

Complexities and Speculations. Whereas vespertilionids, among others, begin to shorten their pulses during the early part of the target-directed flight, some of the long-pulsed bats (rhinolophids and *Pteronotus parnelli*) lengthen their pulses at this time. In all bats the pulses

are then shortened as the target is approached (Table 21–1). This unusual pattern on the part of the long-pulsed bats has been discussed by Novick and Vaisnys (1964) and Novick (1971). The lengthening of the pulses at this stage of the approach to a target probably serves to segregate components of the pulse that occur at different times and serve different functions. Perhaps of prime importance is adequate time before the beginning of the echo of the FM sweep for processing of the CF echo by the central nervous system. This processing of the CF echo may involve the bat's use of the Doppler shift as a means of determining its own speed relative to that of its target. (Because of the Doppler shift, the echo from an approaching target would appear to the bat to be of a higher frequency than would an echo from a target moving away.)

Novick suggested that three types of evaluation might be made by a long-pulsed bat during the approach phase: (1) an evaluation of the relative target movement on the basis of an assessment of the frequency of the CF echo; (2) an evaluation of the distance of the target and the speed with which it is being approached on the basis of the interval between the end of the emission of a pulse and the end of its echo (the shorter this time the closer the target); (3) an evaluation of the size and direction of travel of the target on the basis of assessment of the FM echo. The evaluation of the Doppler shift and the interval between the emission of a pulse and the reception of its echo may be continued throughout the pursuit of an insect, and would continue to yield information on the position and movement of the prey.

Observations of the performance of bats during target-directed flights have consistently indicated that these animals can evaluate distance with considerable precision. A series of careful laboratory tests by Simmons (1971) has demonstrated how well bats discriminate between the nearer and farther of two targets. In his experiments the big brown bat (*Eptesicus fuscus*) was able to discriminate a range difference of 12 mm. at a distance of 60 cm., and the spear-nosed bat (*Phyllostomus hastatus*) could discriminate a difference of 13 mm. at the same distance. These bats clearly performed as though they were able to use the echo arrival time to determine target distance.

Various structural and physiological factors set limits on pulse duration, particularly on the brevity of a pulse. The periods of recovery (refractory periods) of nerve-muscle systems that are necessary between successive muscle contractions, delays associated with reaction time, and time required by the central nervous system for processing information contained in the echoes are all limiting factors. The timing of pulses doubtless partially determines the bat's ability to get appropriate acoustical information. In general, during the approach phase bats wait roughly 25 msec. after the termination of the echo of a pulse before they emit the next pulse. This time is presumably necessary for the evaluation of the echo for information about the target. In the terminal phase of the pursuit,

when the bat is closing in on its target, many kinds of bats maintain interpulse intervals of some 2.5 to 5.0 msec. The silent interpulse intervals may be important in allowing bats to receive the FM echoes clearly.

The production of ultrasonic pulses of high intensity probably requires considerable energy. The pulse duration and rate of pulse emission, therefore, may in part be determined by the need to conserve energy (Novick, 1971:209). One might expect a bat to use as few pulses or pulses as short as possible within the limits set by the necessity to perceive targets clearly.

The available evidence indicates that particular styles of echolocation may be species-specific characters. Pulse duration during searching flight may be related to the flight speed of a bat and to its maneuverability. Fast, high-flying, narrow-winged, and frequently small-eared bats with limited maneuverability seem to produce intense, fairly long pulses that facilitate the perception of distant objects in time for appropriate maneuvers. The molossids (free-tailed bats, page 111) probably fit into this pattern. On the other hand, maneuverable, slow-flying, broad-winged, and big-eared bats fly close to complex foregrounds, such as vegetation, and produce short and relatively soft pulses. These bats concentrate on nearby features and avoid the complexity of echoes from more distant objects. An understanding of the extent to which these generalizations are broadly applicable, and the recognition of correlations between specific patterns of echolocation and such features as wing design, size and shape of the ears, and facial features associated with directing pulses, await further research.

Avoidance of Bats by Moths. Studies by Roeder and his colleagues (Roeder and Treat, 1961; Dunning and Roeder, 1965; Roeder, 1965) have revealed a remarkable series of adaptations by certain moths in response to predation by bats. Nocturnal moths of the families Noctuidae, Geometridae, and Arctiidae have an ear on each side

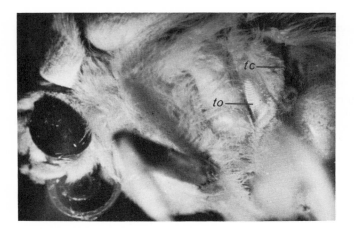

FIGURE 21–5. Photograph of an arctiid moth, showing the tympanic cavity (*tc*) and timbal organ (*to*). (Photograph by K. D. Roeder.)

of the rear part of the thorax (Fig. 21–5). Each ear is a small cavity within which is a transparent membrane; the ears are sensitive to a wide range of frequencies and allow the insects to detect the ultrasonic pulses of foraging bats. Upon detecting the approach of a bat, the moths alter their level flight and adopt various erratic flight patterns. A wide array of loops, dives, and abrupt turns have been photographed (an example is shown in Figure 21–6). Some members of these families of moths have carried the business of evasion of bats to an even greater extreme. These moths have a noise-making organ on each side of the thorax (Fig. 21–5). When the moths are disturbed these organs produce trains of clicks with prominent ultrasonic components. Under laboratory conditions, flying bats about to capture mealworms tossed into the air regularly turned away from their targets when confronted with recorded trains of moth-produced pulses (Table 21–2).

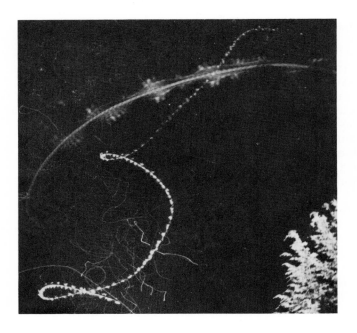

FIGURE 21–6. The evasive tactics of a moth approached by a bat. (Photograph by F. A. Webster.)

Table 21–2. The Responses of Bats to the Simultaneous Presentation of Mealworms and the Sounds of Bats or Moths. (After Dunning, D. C., and Roeder, K. D.: Moth Sounds and the Insect-Catching Behavior of Bats. *Science,* 147:173–174, 1965. Copyright 1965 by the American Association for the Advancement of Science.)

| | | NUMBER OF | | |
BAT NO.	NO. OF TOSSES	Contacts (%)	Dodges (%)	Attempts (%)
		Bat sounds presented		
3	67	88	7	5
4	150	79	8	12
5	92	65	27	5
Total	309	77	14	8
		Mouth sounds presented		
3	95	14	86	3
4	249	14	83	3
5	121	11	87	2
Total	465	13	85	2
		No sounds presented		
3	141	99	0	1
4	373	98	0	1
5	167	97	1	1
Total	681	98	1	1

These pulses apparently protect moths from bats, but why they cause bats to abandon the chase is not known (Dunning and Roeder, 1965). Perhaps they are announcing their identity as bad-tasting prey. Some arctiid moths, and perhaps some noctuids, are unpalatable to bats, and their identification by a bat that had had previous experience with them might be aided by the moth's ultrasonic clicks.

ECHOLOCATION IN CETACEANS

Just as bats must cope with darkness, cetaceans frequently must be able to perceive their underwater environment under conditions that render vision difficult if not useless. In some waters inhabited by cetaceans, suspended material such as soil particles or plankton limits visibility to a few feet or even a few inches. Water transmits light rather poorly, and even under ideal conditions visibility under

water is limited. Also, some cetaceans forage at considerable depths where but a trace of light remains. It is not surprising, then, that some cetaceans have developed echolocation.

A growing body of evidence suggests that all odontocetes (toothed whales and dolphins) use echolocation for obstacle avoidance and for the detection of prey. Mysticetes (the baleen whales), however, are not known to echolocate. A tremendous variety of sounds, some having a fascinating musical quality, are made by both mysticetes and odontocetes; the wailing, creaking, and squealing noises of cetaceans have become commonplace to men operating sonar equipment at sea. Payne (1970) has recorded the remarkable "songs" of the humpback whale (*Megaptera*). The following discussion will primarily consider pulses used by cetaceans in connection with echolocation.

Insulation of the Cetacean Ear from the Skull. Each ear of a cetacean functions as a separate hydrophone, allowing the animals to localize the source of sound by discriminating (as we do) between the times at which a sound is received by each ear. The pressure that sound transmitted through water exerts on the bones of the entire skull causes vibrations to be transmitted by the skull. When the bone that houses the middle and inner ear is attached rigidly to the skull, as it is in most mammals, vibrations from water are transmitted through the bones of the skull and reach the ear from various directions. As a consequence, when a mammal with this type of skull is submerged it is unable to localize accurately the source of a sound. Because the localization of sound is of great importance to cetaceans that use echolocation, these animals have evolved structural features that serve to insulate the bone surrounding the middle and inner ear (tympanic bulla) from the rest of the skull.

Several structural features provide for isolation of the tympanic bulla. First, the tympanic bullae are not fused to the skull in any cetacean, and in the specialized porpoises and dolphins the bullae are separated by an appreciable gap from adjacent bones of the skull. In addition, the bullae are insulated by an extensive system of air sinuses that are unique to cetaceans. These air sinuses surround the bullae and extend forward into the enlarged pterygoid fossae (Fig. 11–5; p. 181), and each sinus is connected by the eustachian tube to the cavity of the middle ear. The sinuses are filled with a oil-mucus emulsion, foamed with air, and are surrounded by fibrous connective tissue and venous networks. These sinuses can apparently retain air even when subjected to pressures of 100 atmospheres (1,470 lb./sq. in.), pressures higher than those to which cetaceans are subjected during deep dives. Purves (1966:360) used a foam in which gelatin was substituted for mucus and oil and found that at such pressures the air became dispersed in the mixture as tiny bubbles. The foam in the air sinuses apparently forms a layer around the bullae that retains remarkably constant sound-reflecting and sound-insulating qualities through a wide range of pressures.

Disagreement remains as to how sounds reach the cetacean middle ear. Fraser and Purves (1960) believe the route to be through the external auditory meatus (which may be highly reduced or even covered with skin in odontocetes), the tympanic ligament, and the ear ossicles. But a divergent view is held by Norris (1964, 1968), who regards the extremely thin back part of the lower jaw of delphinids as an acoustical window. Norris holds that sound passes into the skin and blubber overlying the dentary, through the thin part of this bone, which at its thinnest may be only 0.1 mm. thick, to the intramandibular fat body. This fat body leads directly to the wall of the auditory bulla, into which the sound presumably passes. Weight is given to the Norris hypothesis by experiments done by Yanagisawa et al. (1966), who found that the jaw is the most acoustically sensitive area of the dolphin's head.

Patterns of Echolocation in Cetaceans. Since the account by Schevill and Lawrence (1949) of the underwater noises made by the white whale (*Delphinapterus*), considerable research has been done on the vocalizations of cetaceans. Much research has dealt with one of the common dolphins, *Tursiops truncatus* (Kellogg et al., 1953; Wood, 1954; Schevill and Lawrence, 1956; Norris, 1961; Lilly, 1962, 1963). *Tursiops* is able to detect obstacles and to recognize food by means of echolocation, and in the use of short pulses its sonar system resembles that of bats. *Tursiops* is capable of producing a great variety of sounds, but of primary importance for echolocation are the trains of clicks that it emits. The clicks are audible to man, but cover a wide spectrum of frequencies, some of which are ultrasonic. The rate of emission of clicks varies among different odontocetes: the killer whale (*Orcinus orca*) has a slow repetition rate of from 6 to 18 clicks per second (Schevill and Watkins, 1966), whereas the Amazon dolphin (*Inia geoffrensis*) emits click trains at rates from 30 to 80 per second (Schevill and Watkins, 1962; Caldwell, 1966). The pulse rate rises as a porpoise approaches a target, and *T. truncatus* can distinguish between a piece of fish and a substitute water-filled capsule with a similar shape (Norris et al., 1961), or even between sheets of the same metal but of different thicknesses (Evans and Powell, 1967).

Despite our rather detailed knowledge of the kinds of sounds made by cetaceans, where and how the sounds are produced is not definitely known (see Norris, 1969:404-407). The larynx of odontocetes lacks vocal cords, but has well developed muscles and a complicated structure, and is thought by some to be the site of sound production (Purves, 1966). The nasal plugs (muscular valves at the blowhole that close the nares) and the membranes against which they rest have been thought to produce sound, as have the lips of the blowhole (Norris and Prescott, 1961). Because the echolocation sounds produced by delphinids seemingly come from above the margins of the jaws, some of the structures associated with the blowhole, mentioned above, may produce sound (Norris et al., 1961).

The sperm whale (*Physeter catodon*) presents an unusually interesting case of echolocation among cetaceans because it uses a unique mode of foraging that is probably made possible by echo ranging. (No actual proof that sperm whales echolocate is available, but this ability has been inferred on the basis of data from other odontocetes.) The click of a sperm whale is known to consist of a series of pulses (Backus and Schevill, 1966). The click lasts roughly 24 msec. and is composed of up to nine separate pulses. These vary in duration from 2 to 0.1 msec., and the interpulse intervals are some 2 to 4 msec. The clicks are repeated at rates of from less than one click per second to 40 per second. Of particular interest is the fact that the sperm whale feeds largely on squid that it takes at depths (down to at least 1000 m.) at which prey is scarce and light is virtually absent. It appears, therefore, that the sperm whale is able to utilize deep-sea foraging largely because it is able to use echo scanning to locate food under conditions that require efficient long-range echolocation. Backus and Schevill (1966:525) estimated that sperm whales, by the use of echo scanning, can detect prey up to about 400 m. away.

Some small cetaceans that inhabit turbid water have tiny eyes and presumably are dependent on echolocation. One of the most highly specialized of these is the blind river dolphin (*Platanista gangetica*), an inhabitant of the muddy and murky waters of the Ganges, Indus, and Brahmaputra river systems of India and Pakistan. This unusual dolphin habitually swims on its side with the ventralmost flipper either touching the bottom or moving within 2 or 3 cm. of it (Fig. 11-10; p. 189). *Platanista* has greatly reduced eyes that are barely visible externally. The lens is absent, but the retina apparently retains the ability to detect light, although doubtlessly no image can be formed. The tiny eye opening is surrounded by a sphincter muscle, and another muscle functions to open the sphincter. Blind river dolphins in captivity continuously produced series of pulses at rates of from 20 to 50 per second, primarily in the frequency range between 15 and 60 kHz, and the animals have a remarkable ability to direct their pulses into a narrow beam (Herald et al., 1969).

Studies by Evans and his associates (Evans et al., 1964) have demonstrated that the shape of the skull of *Platanista* effects the directional beaming of pulses (these are emitted through the blowhole). The pulses are reflected by the concave front of the skull and are focused by the "melon," a lens-shaped fatty structure that gives a domed profile to the forehead of many odontocetes. The skull of *Platanista* is grotesquely modified by large flanges from the maxillary bones (Fig. 21-7). These prominent flanges, rounded on

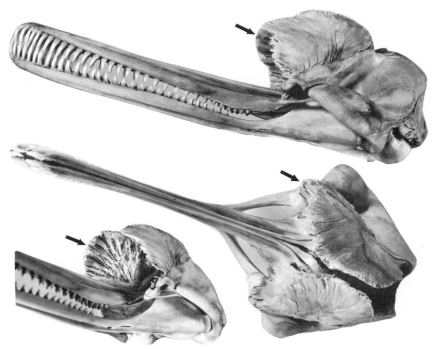

FIGURE 21–7. The skull of the blind river dolphin (*Platanista gangetica*), showing the highly developed extensions of the maxillary bones. (From Herald, E. S., et al.: Blind river dolphin: first side-swimming cetacean, *Science*, 166:1408–1410, 1969. Copyright 1969 by the American Association for the Advancement of Science.)

the outside but with an intricate, radial pattern of latticework on the inside, probably serve as acoustical baffles that, with the melon, concentrate the pulses into a narrow beam (Herald et al., 1969). Observations of swimming dolphins, and reception of these pulses with a hydrophone and amplifier system, showed that pulses were indeed effectively beamed. When a dolphin's snout was directed as little as 10° on either side of the receiver the intensity of the pulses dropped markedly, and a far greater drop occurred when the angle was greater than 40°. As the dolphin swims on its side it moves its head constantly in a sweeping action close to the bottom. One is tempted to speculate that this action serves to scan the area ahead of the porpoise with a beam of pulses, a system similar in some ways to that of the horseshoe bat, which also uses beamed pulses. Such scanning in the

dolphin might serve effectively both in determining bottom contours and in finding food.

ECHOLOCATION BY INSECTIVORES, PINNIPEDS AND OTHER MAMMALS

Echolocation is known to occur in four orders of mammals (Insectivora, Chiroptera, Cetacea, and Pinnipedia) and may have evolved independently in each of these groups. Highly suggestive evidence also points toward the occurrence of echolocation in some other orders.

Many authors have noted high pitched sounds made by insectivores when they explore unfamiliar surroundings or objects (for example, Swinhoe, 1870; Komarek, 1932; Reed, 1944; Crowcroft, 1957). In a series of carefully controlled laboratory experiments, Gould et al. (1964) demonstrated that three species of *Sorex* could echolocate. These shrews searched around an elevated disc, found a lower platform, and jumped to it, without the use of tactile, visual, or olfactory senses. While the shrews searched their environments they emitted pulses with frequencies between 30 and 60 kHz; the pulse duration was about 5 to 33+ msec. The familiar short-tailed shrew of the eastern United States, *Blarina brevicauda*, produced similar pulses, and all shrews studied produced pulses with the larynx. Laboratory trials revealed that tenrecs (Tenrecidae) from Madagascar could also echolocate (Gould, 1965). These primitive insectivores produced pulses by clicking the tongue, and the pulses were of frequencies audible to man (from 5 to 17 kHz). Further research is necessary to determine how ubiquitous echolocation is within the Insectivora.

Pinnipeds produce a variety of underwater sounds (Ray and Schevill, 1965; Schevill and Watkins, 1965), some of which are of a type that might be used for echolocation. The Weddell seal (*Leptonychotes weddelli*) of Antarctic waters, for example, emits "chirps" with frequencies up to 30 kHz. These chirps are produced consistently as the seals swim between air holes, and Watkins and Schevill (1968) suggested that they may have a navigational value. The California sea lion (*Zalophus californianus*) is probably able to echolocate (Evans and Haugen, 1963; Poulter, 1963; Schusterman and Feinstein, 1965), and makes extensive use of vocal signals for communication both in the water and on land. As in the case of the insectivores, our knowledge of echolocation in pinnipeds is limited.

Diverse lines of evidence indicate that some form of echolocation may occur in still other mammals. Flying lemurs (order Dermoptera) move in a series of jumps interspersed with pauses, and emit series of pulses (Burton, 1949; Tate, 1947); these could be used in echolocation. *Antechinus*, a small, nocturnal, tiny-eyed Australian marsupial that is convergent toward shrews, may possibly echolocate (Gould, 1964). Two rodents, the fat dormouse (*Glis*, Gliridae) and the golden hamster (*Mesocricetus*, Cricetidae), are able to locate perches without the use of tactile, visual, or olfactory senses, and can presumably echolocate (Kahmann and Ostermann, 1951). Young mice are known to emit ultrasonic pulses (Zippelius and Schleidt, 1956), but the pulses are important in communication (see page 332) and their use for echolocation has not been demonstrated. Griffin (1958: 297–322) discusses the ability of some blind people to echolocate. One is tempted to speculate that a latent ability to echolocate is common to many mammals.

THE SPECIALIZED EAR OF THE KANGAROO RAT

One of the most remarkable cases of specialization of the ear is that of the kangaroo rat (*Dipodomys*). The enormous auditory bullae of these animals, which have a total volume greater than that of the braincase, were a source of wonder to mammalogists for years. Recently, partly due to the studies of Webster (1961, 1963, 1966, 1968a, 1968b), the functional importance of this and other auditory specializations of the kangaroo rat have been clarified.

Before discussing the inflated bullae, let us consider some other important specializations. First, the malleus lacks the anterior and lateral ligaments that in most mammals brace this bone (Fig. 2–10; p. 14), and therefore rotates unusually freely. In addition, the manubrium of the malleus, which rests against the tympanic membrane, is greatly lengthened; it thus serves as a lever arm that transforms the relatively weak vibrations of the tympanic membrane into more powerful movements transmitted to the incus and stapes. Further, the tympanic membrane is exceptionally large, and the footplate of the stapes, which rests against the oval window of the inner ear, is small. This forms a piston system of sorts. Because the force per unit area is increased in proportion to the difference in surface areas of the two structures, relatively weak pressure on the large tympanic membrane is transformed into relatively great pressure on the fluid within the inner ear *via* the small footplate of the stapes. There is, then, great amplification of force by the combined means of the piston system and the long lever arm of the malleus. The degree of this amplification is expressed by the *transformer ratio*, which in Merriam's kangaroo rat is extremely high, about 97:1. (This ratio in man is roughly 18:1.)

The inflation of the auditory bullae of kangaroo rats (Fig. 10–11; p. 156) results in a great increase in the volume of the air-filled chambers surrounding the middle ear. This increase in volume reduces the resistance the enclosed air offers to the compression caused by inward movement of the tympanic membrane. Consequently, the damping effect on vibrations of the tympanic membrane is diminished. The advantage gained during transmission of low-frequency sounds is especially great, for these sounds cause relatively great movement of the membrane. Laboratory experiments, involving the filling of the bullae of kangaroo rats with plasticine, demonstrated that the transmission ability of the tympano-ossicular system in individuals lacking use of the enlarged middle ear chambers is seriously lowered. Both experimental and control animals, however, were most sensitive to low-intensity sounds with frequencies between 1 and 3 kHz. In an attempt to simulate natural predator-prey confrontations, Webster tested the reactions of kangaroo rats to two predators adapted to hunting in darkness. He used the sidewinder rattlesnake (*Crotalus cerastes*), which detects its prey by using heat sensitive facial "pits," and the barn owl (*Tyto alba*), which can use hearing to locate the position of prey. By using delicate recording equipment, Webster found that the wings of owls produce low frequency (below 3 kHz) whirring sounds as the birds swoop toward prey, and a rattlesnake produces a short burst of low frequency sound (perhaps when scales rub against the substrate) just before it strikes. Under experimental conditions, when a kangaroo rat heard these low intensity and low frequency sounds produced by attacking predators, it made a sudden vertical leap and avoided capture. In contrast, individuals with artificially reduced middle ear volumes, and presumably reduced ability to hear faint sounds, could not evade capture. Clearly, the beautifully coordinated specializations of the tympano-ossicular system and the auditory bullae are highly adaptive: in darkness kangaroo rats can detect the faint sounds of attacking predators, make evasive leaps, and go on to reproduce their sharp-eared kind.

Kangaroo rats are not alone in having enlarged bullae. Many other rodents, including other heteromyids, some South American histricomorphs, gerbils (Cricetidae), jerboas (Dipodidae), the springhaas (Pedetidae), and the Australian hopping mice (*Notomys*, Muridae), have this specialization, as do elephant shrews (Macroscelididae). Whether or not the adaptive importance of enlarged bullae is the same among all these mammals is not known. Many kinds of mammals that live in deserts have auditory bullae that are relatively larger than those of relatives occupying cooler and more humid areas. Sound is known to be absorbed to different degrees under different conditions of temperature and humidity (Knudson, 1931, 1935), and the enlargement of the bullae may compensate for the poor sound-carrying qualities of warm, dry desert air.

CHAPTER 22

COMMENTS ON MAMMALS AND MAN

Man has long been interested in his fellow mammals and has long exploited them. More than a million years ago *Australopithecus* was killing and eating baboons and antelope, and the use of mammals for food remains characteristic of most cultures of men today. Many kinds of mammals have been domesticated by man, and some are taught to work for him: the trained Indian elephant lifts and drags teak logs in the tropical forests of Ceylon, where the periodically saturated soil limits the usefulness of vehicles; dogs help some African natives capture antelope and other game; and trained rhesus monkeys pick coconuts from tall trees nearly inaccesable to man and drop them to their masters. Even the vicious and rebellious camel has been trained to carry man and his burdens. The raising of various kinds of mammals—from homozygous strains of mice to be used for medical research to stocky beef cattle—are important enterprises today. The very distribution of early man was probably influenced by his ability to kill his fellow mammals, for the skins and furs of mammals may have enabled primitive man, probably endowed with a hopelessly inadequate insulation of his

own, to penetrate cool or cold regions.

Wild mammals are still under pressure from man in nearly all parts of the world. In many primitive areas men either are basically hunters and partly depend on mammals for food, or at least hunt to supplement limited food supplies. Even in the United States most boys who live on the farm or away from towns regularly hunt small game, and in Mexico the "veinte-dos" (.22 rifle) is used year round in many areas for hunting anything from squirrels to deer. In the United States trapping for furs has long been important. Not only was the exploration of the western United States largely accomplished by trappers seeking new territory, but in more recent years trapping has been an important source of seasonal income to some. In the early 1920's Henry Duval, trapping marten in the vicinity of Mineral King in the Sierra Nevada of California, took as many as 96 marten in a winter (Grinnell, et al., 1937:205). This probably brought Duval over $1,500 for his winter's work (the average price of a marten fur in this period was $15.71), more than the average worker made in a full year. In 1929, a top year with respect to demand for furs, the return to North American trappers was estimated at 60 million dollars (Hamilton, 1939:393). Fur farming has provided stiff competition to the trapper in the United States, but trapping, particularly for muskrat, continues in some areas. To many boys the excitement of running a trapline is as attractive as is the monetary return.

In some states big game hunting is a major source of income. In Colorado in 1968, for example, expenditures of deer and elk hunters were estimated at over 60 million dollars. Many states realize considerable returns from the sale of deer licenses alone (Table 22–1). Perhaps equally important is the recreational value of hunting to the individual. Only during the hunting season do some men get away from their work in the city to experience the sights, smells, and sounds of primitive country, and many men spend months anticipating and planning the yearly deer hunt.

But wild animals may also be costly. Pocket gophers, rabbits, meadow voles, ground squirrels, and even deer and elk may cause local damage to crops or rangeland, and efforts to combat these losses are frequently expensive. In addition, the Federal Government supports considerable research on economically important mammals, and finances the local control of some mammals. In 1971 Federal animal control programs cost U.S. taxpayers over $8 million.

Unhappily, the long-term exploitation of mammals by man has had a tremendous impact. In about the last 400 years some 36 species of mammals have become extinct, and today over 120 species are threatened by extinction. Some species were disposed of remarkably summarily. The Steller's sea cow was pushed to extinction some 27 years after its first discovery by white man. Sea otters, which were hunted at least as early as 1786, were ruthlessly killed for their valuable fur; probably more than 200,000 were killed between 1786 and 1868 (Evermann, 1923:524). By 1900 the animals were rare over much of their range, and they were seemingly lucky to have survived until protected by legislation in the early 1900's. Not so lucky was the grizzly bear in California. In the 1890's grizzlies still persisted in the San Grabriel Mountains near Los Angeles, but the last known southern California grizzly was killed in 1916, and the last verified occurrence in California was in 1922, in the foothills of the Sierra Nevadas in central California (Grinnell, 1937:93). Only roughly 60 years were required to bring the grizzly in California from fair abundance to extinction. In Mexico, a population of grizzlies that survived in a small mountain range in central Chihuahua in 1957 was probably wiped out by about 1963, the very year when funds were raised by the World Wildlife Fund to set aside a refuge for the

Table 22–1. Revenue Received by Several States from the Sale of Hunting Licenses for Deer. (Data from Material Supplied by These States.)

STATE	YEAR	NUMBER OF DEER LICENSES SOLD (RESIDENT AND NON-RESIDENT)	REVENUE FROM SALES
Arizona	1969	101,337	$ 790,855.00
Colorado	1969	161,919	3,172,527.00
Idaho	1970	168,624°	420,759.00
Michigan	1970		3,767,702.00
Nebraska	1970	36,139	389,510.00
New York	1970	670,239°	2,803,792.65
Utah	1970	92,021	1,320,830.75
Wyoming	1968	100,961°	1,786,830.00

°Includes deer-bear permits.

animals. The quagga, a zebra that inhabited southern Africa, was extirpated in the wild about 1860, and another type of zebra was exterminated by roughly 1910. The Arabian oryx, well on its way toward extinction, has been hunted in recent years with machine guns mounted on jeeps—surely a manly style of hunting. The impact of man has nowhere been more strongly felt than by some marine mammals. Due to persistent hunting by man the blue whale, the largest animal of all times, has been reduced to a total population of probably no more than 200 individuals. Other activities of man have left their mark: DDT has seemingly been responsible for decimating bat populations in some areas, and a freeway in California has transected the unusually small range of the handsome Morro Bay kangaroo rat. The list of mammals extirpated or endangered by man is depressingly long.

A broad view should be taken, however, and one must be realistic. Local extirpation of some mammals is regrettable but unavoidable. The clearing of land for agriculture by the starving people in Asia cannot be stopped because wildlife habitat is being destroyed, nor can the hunting of tribes or populations whose very survival depends on hunting success be seriously limited. Clearly grizzlies and man, at least at his present population densities, could not co-exist in California. But the loss of the grizzly in Mexico could perhaps have been avoided, and most biologists would regret further restriction of the already vastly reduced range of this animal. We must expect restrictions of the ranges of many mammals; but must we tolerate the extinction of more species? Under favorable conditions a species may reoccupy formerly abandoned areas; by no means, however, can an extinct species be brought back.

In the United States the continued survival of a reasonably balanced mammalian fauna, that is to say, a fauna that has not been stripped of all large predators and other mammals that man fancies are threatening his interests or that he enjoys killing, will probably only be possible if man's attitude toward wildlife changes. We will not long be able to afford the luxury of putting a .22 rifle in the hands of any boy and letting him roam the countryside "plinking" at all wild mammals or birds that offer attractive targets. Nor will some species, such as the prairie dog, be able indefinitely to stand the pressure from men with scope-sighted rifles who must "keep in practice" at times of the year when game animals cannot be hunted. This is not to say that the hunting of game animals must be drastically curtailed or stopped. Most states are doing a capable job of managing their game populations. The survival of most game species is not threatened, although the numbers of hunters will probably need to be regulated in most areas in the future by a permit system. But if mountain lions are to survive in mountainous parts of the western United States, if coyotes are to remain common enough for boys camping out over a weekend to hear choruses of howling, and if wolves, wolverines, and marten are to persist in wilderness areas and be occasionally glimpsed by hikers, man's attitude toward wildlife must change.

Man must learn to value wildlife not only for its monetary value but for its great aesthetic and recreational worth. He must be committed to the idea that animals form an essential part of a biotic richness developed over millions of years, and that they are a resource worthy of perpetuation. If the present decimation of the earth's wildlife is to be curtailed, at least the following steps must be taken:

1. Man's population growth must be halted and his need for space and resources stabilized.
2. Large tracts of land undisturbed by man must be maintained for wildlife.
3. Man's exploitation of many species (whales are notable examples) must be drastically reduced.
4. A broad understanding of the

meaning of such biotic interactions as predation must underlie an interest in preserving balanced faunas.

5. Control of animals threatening crops and livestock must be local, with no attempt to exterminate a species over wide areas with little respect to the damage it is doing.

6. The use of biocides must be carefully controlled.

7. Man must accept some types of economic losses and inconvenience caused by wildlife, and must feel that these are more than compensated for by his enjoyment of a rich and balanced biota.

There is some justification for "guarded" optimism. Mammals are clearly adaptable, and balanced mammalian faunas can thrive in close proximity to man. I recently spent a day in the chaparral-covered foothills of the San Gabriel Mountains some 40 miles east of Los Angeles. Tracts of houses, nurseries, and remnants of once-productive orange groves were only two miles away and all but the closest lowlands were obscured by a curtain of smog. But as I followed trails and firebreaks I had not hiked in 20 years, I saw considerable evidence of mammals. I jumped deer from tracts of chaparral, saw the huge stick-pile houses of the dusky footed woodrat, noted scattered tracks of bobcats and gray foxes in arroyos, and in the dust of the trails saw numerous tracks of coyotes and raccoons. Piles of droppings indicated that, as they had done 20 years before, coyotes were eating apricots and other items available in the orchards and gardens of the flatlands. During this day I was greatly encouraged by the impression that in 20 years, during which "civilization" had advanced to the very base of the foothills, the populations of chaparral-dwelling mammals have remained nearly unchanged. Because the fire roads and firebreaks are closed to non-government vehicles, the foothills are reasonably undisturbed by man; virtually no one hikes up the steep arroyos or through the dense tracts of brush. Perhaps there is room for hope. The lowland, checkerboarded by housing developments and criss-crossed with freeways, is one world; the precipitous, brush covered foothills and mountains are another.

Because the survival of the wildlife of the world, and indeed the survival of mankind itself, is in our hands, one cannot help but fervently hope that the word *sapiens* (meaning wise) becomes a justly-earned part of the name *Homo sapiens.*

SELECTED REFERENCES
ON MAMMALS

GENERAL

Asdell, S. A. 1964. Patterns of mammalian reproduction, 2nd ed. Cornell Univ. Press, Ithaca, New York. 670 pp.

Romer, A. S. 1966. Vertebrate paleontology, 3rd ed. Univ. Chicago Press, Chicago. 468 pp.

Simpson, G. G. 1945. The principles of classification and a classification of mammals. Bull. Amer. Mus. Nat. Hist., 85:1–350.

Van Gelder, R. G. 1969. Biology of mammals. Charles Scribner's Sons, New York. 197 pp.

Young, J. Z. 1957. The life of mammals. Clarendon Press, Oxford. 820 pp.

BIOLOGY OF MAMMALS: SPECIFIC GROUPS OF MAMMALS

Andersen, H. T. (ed.). 1969. The biology of marine mammals. Academic Press, New York. 511 pp.

DeVore, I. (ed.). 1965. Primate behavior. Holt, Rinehart and Winston, New York. 654 pp.

Ewer, R. F. 1968. Ethology of mammals. Plenum Press, New York. 418 pp.

Jay, P. C. (ed.). 1968. Primates, studies in adaptability and variability. Holt, Rinehart and Winston, New York. 529 pp.

Norris, K. S. (ed.). 1966. Whales, dolphins and porpoises. Univ. California Press, Berkeley. 789 pp.

Scheffer, V. B. 1958. Seals, sea lions and walruses. Stanford Univ. Press, Stanford, California. 179 pp.

Slaughter, B. H., and D. W. Walton (eds.). 1970. About bats. Southern Methodist Univ. Press, Dallas. 339 pp.

Slijper, E. J. 1962. Whales. Basic Books, Inc., New York. 475 pp.

Wimsatt, W. A. (ed.). 1970. Biology of bats. Academic Press, New York. 406 pp.

MAMMALIAN FAUNAS

World

Littlewood, C. 1970. World's vanishing animals: the mammals. Arco Publ. Co., New York. 63 pp.

Walker, E. P. 1968. Mammals of the world, 2nd ed. 2 vols. Johns Hopkins Press, Baltimore.

New World

Burt, W. H., and R. P. Grossenheider. 1952. A field guide to the mammals. Houghton Mifflin Co., Boston. 200 pp.

Cabrera, A. 1940. Historia natural ediar. Mamíferos Sud-Americanos. Co. Argentina Edit., Soc. Resp. Ltda., Buenos Aires. 370 pp.

_____. 1957 (1958). Catalágo de los mamíferos de America del Sur. Revista Mus. Argentino Cien. Nat., Cien. Zool., Vol. *4(1)*:1–307.

Hall, E. R., and K. R. Kelson. 1959. The mammals of North America. The Ronald Press Co., New York. 2 vols.

Leopold, A. S. 1959. Wildlife of Mexico. Univ. California Press, Berkeley. 568 pp.

Tomich, P. Q. 1969. Mammals of Hawaii—a synopsis and notational bibliography. Bernice P. Bishop Mus. Spec. Publ. 57, Bishop Mus. Press, Honolulu. 238 pp.

Old World

Allen, G. M. 1938. The mammals of China and Mongolia. Amer. Mus. Nat. Hist., New York. 620 pp.

_____. 1940. The mammals of China and Mongolia. Part 2. Amer. Mus. Nat. Hist., New York. pp. 621–1350.

Chasen, F. N. 1940. A handlist of Malaysian mammals. Bull. Raffles Mus., *15*:1–209.

Dorst, J. 1970. Field guide to the larger mammals of Africa. Houghton Mifflin Co., Boston. 287 pp.

Ellerman, J. R., and T. C. S. Morrison-Scott. 1951. Checklist of Palearctic and Indian mammals, 1758 to 1946. British Mus. Nat. Hist. 810 pp.

_____., _____., and R. W. Hayman. 1953. Southern African mammals, 1758 to 1951: a reclassification. British Mus. Nat. Hist. 363 pp.

Jones, F. W. 1968. The mammals of South Australia. Handbook of the flora and fauna of South Australia, issued by the South Australian Branch of the British Guide Handbooks Committee. Photolitho Reprint, Adelaide. 458 pp.

Kingdon, J. 1971. East African mammals. Academic Press, New York. Vol. 1, 446 pp. (More volumes of this series will be published.)

Kuroda, N. 1940. A monograph of the Japanese mammals. The Sansiedo Co., Ltd., Tokyo and Osaka. 311 pp.

Laurie, E. M. O., and J. E. Hill. 1954. List of land mammals of New Guinea, Celebes and adjacent islands, 1758–1952. British Mus. Nat. Hist. 175 pp.

Medway, Lord. 1969. The wild mammals of Malaya and offshore islands including Singapore. Oxford Univ. Press, London. 127 pp.

Miller, G. S. 1912. Catalogue of the mammals of western Europe. British Mus. Nat. Hist. 1019 pp.

Ognev, S. I. 1947. The mammals of Russia (U.S.S.R.) and adjacent countries, Vol. 5. Moscow. 809 pp. (Also see other volumes in this series.)

Ride, W. D. L. 1970. A guide to the mammals of Australia. Oxford Univ. Press, London. 249 pp.

Roberts, A. 1951. The mammals of South Africa. Central News Agency, South Africa. 700 pp.

Scheffer, V. B., and D. W. Rice. 1963. A list of the marine mammals of the world. U.S. Fish and Wildlife Serv., Spec. Sci. Rept. Fisheries, *431*:1–12.

Tate, G. H. H. 1947. Mammals of Eastern Asia. Macmillan Co., New York. 366 pp.

Troughton, E. 1947. Furred animals of Australia. Charles Scribner's Sons, New York. 374 pp.

Van Den Brink, F. H. 1968. A field guide to the mammals of Britain and Europe. Houghton Mifflin Co., Boston.

"PRACTICAL" MAMMALOGY

Anderson, R. M. 1944. Methods of collecting and preparing vertebrate animals. National Museum of Canada, Bull. No. 69, pp. 1–162 (second edition).

Borror, D. J., and D. M. DeLong. 1964. An introduction to the study of insects, 3rd ed. Holt, Rinehart and Winston, New York. 812 pp. (Helpful in some food habits studies.)

Cook, C. W., H. H. Biswell, R. T. Clark, E. H. Ried, L. A. Stoddart, and M. L. Upchurch. 1962. Basic problems and techniques in range research. National Academy of Sciences—National Research Council, Publ. No. 890. 341 pp.

Jaques, H. E. 1947. How to know insects. Wm. G. Brown Co. Publishers, Dubuque, Iowa. 205 pp. (Helpful in some food habits studies.)

Martin, A. C., H. S. Zim, and A. L. Nelson. 1951. American wildlife and plants—a guide to wildlife food habits: the use of trees, shrubs, weeds, and herbs by birds and mammals of the United States. McGraw-Hill Book Co., New York. 500 pp.

Mosby, H. S., and O. H. Hewitt (eds.). 1965. Wildlife investigational techniques. The Wildlife Society, 3900 Wisconsin Ave. NW, Washington, D.C. 419 pp.

Murie, O. J. 1954. A field guide to animal tracks. Houghton Mifflin Co., Boston. 374 pp.

Spence, L. E., Jr. 1963. Study of identifying characteristics of mammalian hair. Wildlife Disease Research Laboratory, Wyoming Game and Fish Commission, Cheyenne, Wyoming (Pittman-Robertson Project FW-3-R-10). 121 pp.

JOURNALS

American Midland Naturalist
Ecology
Evolution
Journal of Animal Ecology
Journal of Mammalogy
Journal of Wildlife Management
Mammalia
Säugetierekundliche Mitteilungen
Zeitschrift für Säugetierekunde

BIBLIOGRAPHY

Note: This list includes, in addition to publications cited in the text, a few additional important references by authors who have publications cited.

Abbott, K. D. 1971. Water economy of the canyon mouse, *Peromyscus crinitus stephensi*. Comp. Biochem. Physiol., *38A*:37–52.

Ables, E. D. 1969. Home range studies of red foxes *(Vulpes vulpes)*. J. Mamm., *50*:108–120.

Alcorn, S. M., S. E. McGregor, G. D. Butler, Jr., and E. B. Kurtz, Jr. 1959. Pollination requirements of the saguaro *(Carnegiea gigantea)*. Cactus and Succ. J. Amer., *31*:39–41.

Allee, W. C., A. E. Emerson, O. Park, T. Park, and K. P. Schmidt. 1949. Principles of animal ecology. W. B. Saunders Co., Philadelphia. 837 pp.

Allen, G. M. 1940. The mammals of China and Mongolia, Part 2. Amer. Mus. Nat. Hist., New York. pp. 621–1350.

Allen, J. A. 1924. Carnivora collected by the American Museum Congo Expedition. Bull. Amer. Mus. Nat. Hist., *47*:73–281.

Altman, P. L., and D. S. Dittmer. 1964. Biology data book. Federation Amer. Soc. Exp. Biol., Washington, D.C. 633 pp.

Amoroso, E. C., and J. H. Matthews. 1951. The growth of the grey seal *(Halichoerus grypus)* from birth to weaning. J. Anat., *85*:426–428.

Andersen, H. T. (ed.). 1969. The biology of marine mammals. Academic Press, New York. 511 pp.

Anderson, J. W. 1954. The production of ultrasonic sounds by laboratory rats and other mammals. Science, *119*:808–809.

Anderson, S. 1967. Introduction to the rodents, pp. 206–209. *In* S. Anderson and J. K. Jones, Jr. (eds.), Recent mammals of the world. The Ronald Press Co., New York.

———., and J. K. Jones, Jr. (eds.). 1967. Recent mammals of the world—a synopsis of families. The Ronald Press Co., New York. 453 pp.

Arey, L. B. 1965. Developmental anatomy. W. B. Saunders Co., Philadelphia. 695 pp.

Armitage, K. B. 1962. Social behaviour of a colony of the yellow-bellied marmot *(Marmota flaviventris)*. Anim. Behav., *10*:319–331.

Asdell, S. A. 1964. Patterns of mammalian reproduction, 2nd ed. Cornell Univ. Press, Ithaca, New York. 670 pp.

Augee, M. L., and E. J. M. Ealey. 1968. Torpor in the echidna, *Tachyglossus aculeatus*. J. Mamm., *49*:446–454.

Aumann, G. D. 1965. Microtine abundance and soil sodium levels. J. Mamm., *46*:594–612.

Backhouse, K. M. 1954. The grey seal. Univ. Durham Coll. Med. Gaz., *48(2)*:9–16.

Backus, R. H., and W. E. Shevill. 1966. *Physeter* clicks, pp. 510–528. *In* K. S. Norris (ed.), Whales, dolphins and porpoises. Univ. California Press, Berkeley, California.

Baker, H. G., and B. J. Harris. 1957. The pollination of *Parkia* by bats and its attendant evolutionary problems. Evol., *11*:449–460.

Baker, J. R., and R. M. Ranson. 1933. Factors affecting the breeding of the field mouse *(Microtus agrestis)*. Proc. Roy. Soc. London, *113B*:486–495.

Baker, W. W., S. G. Marshall, and V. B. Baker. 1968. Autumn fat deposition in the evening bat *(Nyc-ticieus humeralis)*. J. Mamm., 49:314–317.

Balinsky, B. I. 1970. An introduction to embryology. W. B. Saunders Co., Philadelphia. 725 pp.

Balph, D. E., and A. W. Stokes. 1963. On the ethology of a population of Uinta ground squirrels. Amer. Midl. Nat., 69:106–126.

Banfield, A. W. F. 1954. Preliminary investigation of the barren ground caribou. Part II. Life history, ecology and utilization. Can. Wildl. Serv., Wildl. Manage. Bull. ser. 1., no. 10B. 112 pp.

Bartholomew, G. A. 1968. Body temperature and energy metabolism, pp. 290–354. *In* M. S. Gordon, G. A. Bartholomew, A. D. Grinnell, C. B. Jorgensen, and F. N. White (eds.), Animal function: principles and adaptations. Macmillan Co., New York.

———., and T. J. Cade. 1957. Temperature regulation, hibernation, and estivation in the little pocket mouse, *Perognathus longimembris*. J. Mamm., 38:60–72.

———., and N. E. Collias. 1962. The role of vocalization in the social behavior of the northern ele-phant seal. Anim. Behav., 10:7–14.

———., W. R. Dawson, and R. C. Lasiewski. 1970. Thermoregulation and heterothermy in some of the smaller flying foxes (Megachiroptera) of New Guinea. Z. Vergl. Physiol., 70:196–209.

———., and J. W. Hudson. 1960. Aestivation in the Mohave ground squirrel, *Citellus mojavensis*. Bull. Mus. Comp. Zool., 124:193–208.

———.,———. 1961. Desert ground squirrels. Sci. Amer., 205(5):107–116.

———.,———. 1962. Hibernation, estivation, temperature regulation, evaporative water loss, and heart rate of the pigmy opossum, *Cercartetus nanus*. Physiol. Zool., 29:26–40.

———., P. Leitner, and J. E. Nelson. 1964. Body temperature, oxygen consumption, and heart rate in three species of Australian flying foxes. Physiol. Zool., 37:179–198.

———., and M. Rainy. 1971. Regulation of body temperature in the rock hyrax, *Heterohyrax brucei*. J. Mamm., 52:81–95.

Bateman, G. C., and T. A. Vaughan. 1972. Foraging behavior of mormoopid bats. In manuscript.

Bateson, G. 1966. Problems in cetacean and other mammalian communication, pp. 569–579. *In* K. S. Norris (ed.), Whales, dolphins and porpoises. Univ. California Press, Berkeley.

———., and B. Gilbert. 1966. Whaler's Cove dolphin community: an interim report, p. 5. The Oceanic Institute, Makpuu Point, Waimanalo, Oahu, Hawaii.

Batzli, G. O., and F. H. Pitelka. 1971. Condition and diet of cycling populations of the California vole, *Microtus californicus*. J. Mamm., 52:141–163.

Bee, J. W., and E. R. Hall. 1956. Mammals of Northern Alaska. Univ. Kansas Publ., Mus. Nat. Hist., Misc. Publ. No. 8. 309 pp.

Beer, J. R. 1961. Seasonal reproduction in the meadow vole. J. Mamm., 42:483–489.

———., R. Lukens, and D. Olson. 1954. Small mammal populations on the islands of Basswood Lake, Minn. Ecology, 35:437–445.

———., and R. K. Meyer. 1951. Seasonal changes in the endocrine organs and behavior patterns of the muskrat. J. Mamm., 32:173–191.

Bell, R. H. V. 1970. The use of the herb layer by grazing ungulates in the Serengeti, pp. 111–125. *In* Adam Watson (ed.), Animal populations in relation to their food resources. Blackwell Scientific Publ., Oxford. 477 pp.

———. 1971. A grazing ecosystem in the Serengeti. Sci. Amer., 225(1):86–93.

Benson, S. B. 1933. Concealing coloration among some desert rodents of the southwestern United States. Univ. California Publ. Zool., 40:1–70.

———., and A. E. Borell. 1931. Notes on the life history of the red tree mouse, *Phenacomys longicaudus*. J. Mamm., 12:226–233.

Berg, L. S. 1950. Natural regions of the U.S.S.R. Macmillan Co., New York.

Bertram, G. C. L. 1940. The biology of the Weddell and crabeater seals, with a study of the compara-tive behavior of the Pinnipedia. Brit. Mus. (Nat. Hist.) Sci. Repts. Brit. Graham Land Exped. 1934–1937, 1:1–139.

Blair, W. F. 1939. Some observed effects of stream-valley flooding on mammalian populations in east-ern Oklahoma. J. Mamm., 20:304–306.

———. 1951. Evolutionary significance of geographic variation in population density. Texas J. Sci., 1:53–57.

Bloedel, P. 1955. Hunting methods of fish-eating bats, particularly *Noctilio leporinus*. J. Mamm., 36:390–399.

Bodemer, C. W. 1968. Modern embryology. Holt, Rinehart and Winston, Inc., New York. 475 pp.

Bodenheimer, F. S. 1949. Problems of vole populations in the Middle East. Report on the population dynamics of the Levant vole (*Microtus quentheri* D.). Azriel Print. Works, Jerusalem, 77 pp.

Bond, R. M. 1945. Range rodents and plant succession. Trans. N. Amer. Wildl. Conf., 10:229–234.

Bourliere, F. 1956. The natural history of mammals. Alfred A. Knopf, Inc., New York. 364 pp.

Boving, B. G. 1959. Implantation. Ann. N.Y. Acad. Sci., 75:700–725.

Bradbury, J. W. 1970. Target discrimination by echolocating bat *Vampyrum spectrum*. J. Exp. Zool., 173:23–46.

———. 1971. Personal communication.

Bradshaw, G. V. R. 1962. Reproductive cycle of the California leaf-nosed bat, *Macrotus californicus*. Science, *136*:645–646.

Breadon, G. 1932. The flying fox in the Punjab. J. Bombay Nat. Hist. Soc., *35*:670.

Broadbooks, H. E. Home ranges and territorial behavior of the yellow-pine chipmunk, *Eutamias amoenus*. J. Mamm., *51*:310–326.

Bromley, P. T. 1969. Territoriality in pronghorn bucks on the National Bison Range, Moiese, Montana. J. Mamm., *50*:81–89.

Brown, J. H. 1968. Adaptation to environmental temperature in two species of woodrats, *Neotoma cinerea* and *N. albigula*. Misc. Publ. Mus. Zool., Univ. Mich., *135*:1–48.

Brown, L. N. 1967. Ecological distribution of six species of shrews and comparison of sampling methods in the central Rocky Mountains. J. Mamm., *48*:617–623.

Bruce, V. G. 1960. Environmental entrainment of circadian rhythms. Cold Spr. Harb. Symp. Quant. Biol., *25*:29–48.

Bruce, W. S. 1915. Measurements and weights of antarctic seals (part 11, pp. 159–174, 2 pls.). *In* Report on the scientific results of the voyage of S. Y. "Scotia" during the years 1902, 1903, and 1904 . . . Edinburgh, Scottish Oceanogr. Lab.

Bullard, E. 1969. The origin of the oceans. Sci. Amer., *221*:66–75.

Burckhardt, D. 1958. Kindliches Verhalten als Ausdrucksvewegung im Fortpflanzungszeremoniell einiger Wiederkauer. Rev. suisse Zool., *65*:311–316.

Burrell, H. 1927. The platypus. Angus and Robertson Ltd., Sydney.

Burt, W. H. 1940. Territorial behavior and populations of some small mammals in southern Michigan. Misc. Publ. Mus. Zool., Univ. Michigan, *45*:1–58.

_____. 1943. Territoriality and home range concepts as applied to mammals. J. Mamm., *24*:346–352.

Burton, M. 1949. Wildlife of the world. Long Acre, London. 384 pp.

Busnel, R. G. (ed.). 1966. Animal sonar systems, biology and bionics, Tomes I and II. Laboratoire de Physiologie Acoustique, Paris.

Cadenat, J. 1959. Bull. Inst. Franc. Afrique Noire, *21*:1137–1143.

Caldwell, D. K., J. H. Prescott, and M. C. Caldwell. 1966. Bull. So. California Acad. Sci., *65*:245–248.

Caldwell, M. C., and D. K. Caldwell. 1970. Further studies on audible vocalizations of the Amazon freshwater dolphin, *Inia geoffrensis*. Contrib. Sci., *187*:1–5.

Camp, C. L., and N. S. Smith. 1942. Phylogeny and functions of the digital ligaments of the horse. Mam. Univ. California, *13*:69–124.

Campbell, C. B. G. 1966. Taxonomic status of tree shrews. Science, *153*:436.

Carpenter, R. E. 1966. A comparison of thermoregulation and water metabolism in the kangaroo rats *Dipodomys agilis* and *Dipodomys merriami*. Univ. California Publ. Zool., 78:1–36.

_____. 1968. Salt and water metabolism in the marine fish-eating bat, *Pizonyx vivesi*. Comp. Biochem. Physiol., *24*:951–964.

_____. 1969. Structure and function of the kidney and the water balance of desert bats. Physiol. Zool., *42*:288–302.

_____., and J. B. Graham. 1967. Physiological responses to temperature in the long-nosed bat, *Leptonycteris sanborni*. Comp. Biochem. Physiol., *22*:709–722.

Chapskiy, K. K. 1936. The walrus of the Kara Sea. Trans. Arct. Inst., Leningrad, Tom. 67. 111 pp. (In Russian; resumé in English, pp. 112–124.)

Chitty, D. 1954. Tuberculosis among wild voles with a discussion of other pathological conditions among certain mammals and birds. Ecology, 35:227–237.

_____. 1960. Population processes in the vole and their relevance to general theory. Can. J. Zool., 38:99–113.

_____., and H. Chitty. 1962. Population trends among voles at Lake Vyrnwy, 1932–1960. Symposium Theriologicum, Proc. Inter. Symp. on Methods of Mammalogical Investigations, 67–75.

Christian, J. J. 1950. The adreno-pituitary system and population cycles in mammals. J. Mamm., *31*:247–259.

_____. 1954. The relation of the adrenal cortex to population size in rodents, Doctoral dissertation, Johns Hopkins School of Hygiene and Public Health, Baltimore.

_____. 1959a. Control of population growth in rodents by interplay between population density and endocrine physiology. Wildl. Dis., *1*:1–38.

_____. 1959b. The roles of endocrine and behavioral factors in the growth of mammalian populations, pp. 71–97. *In* A. Gorbman (ed.), Comparative endocrinology. Columbia Univ. Symposium, New York.

_____. 1963. Endocrine adaptive mechanisms and the physiologic regulation of population growth, pp. 189–353. *In* W. V. Mayer and R. G. Van Gelder (eds.), Physiological mammalogy. Academic Press, New York.

_____., and D. E. Davis. 1956. The relationship between adrenal weight and population status in Norway rats. J. Mamm., *37*:475–486.

————, ————. 1966. Adrenal glands in female voles (*Microtus pennsylvanicus*) as related to repro-
        duction and population size. J. Mamm., *47*:1–18.
————, V. Flyger, and D. E. Davis. 1960. Factors in mass mortality of a herd of sika deer. Chesapeake
        Sci., *1*:79–95.
Clark, W. E. LeGros. 1959. The antecedents of man. Edinburgh Univ. Press, Edinburgh. 374 pp.
Clemens, W. A. 1968. Origin and early evolution of marsupials. Evolution, *22*:1–18.
Clough, G. C. 1965. Lemmings and population problems. Amer. Sci., *53*:199–212.
Colbert, E. H. 1948. The mammal-like reptile *Lycaenops*. Bull. Amer. Mus. Nat. Hist., *89*:353–404.
————. 1949. The ancestors of mammals. Sci. Amer., *180*:40–43.
————. 1961. Evolution of the vertebrates. Science Editions, Inc., New York. 479 pp.
Cole, R. W. 1970. Pharyngeal and lingual adaptations in the beaver. J. Mamm., *51*:424–425.
Conaway, C. H. 1958. Maintenance, reproduction and growth of the least shrew in captivity. J.
        Mamm., *39*:507–512.
Coulombe, H. N. 1970. The role of succulent halophytes in water balances of salt marsh rodents.
        Oecologia (Berl.) *4*:223–247.
Cowan, I. M. 1947. The timber wolf in the Rocky Mountain national parks of Canada. Can. J. Res.,
        25:139–174.
Crawford, E. C., Jr. 1962. Mechanical aspects of panting in dogs. J. Appl. Physiol., *17*:249–251.
Crisler, L. 1956. Observations of wolves hunting caribou. J. Mamm., *37*:337–346.
————. 1958. Arctic wild. Harper and Bros., New York. 301 pp.
Crowcroft, P. 1953. The daily cycle of activity in British shrews. Proc. Zool. Soc. London, *123*:715–
        729.
————. 1957. The life of the shrew. Max Reinhart, London. 166 pp.
————. 1966. Mice all over. G. T. Foulis & Co., London.
Curry-Lindahl, K. 1962. The irruption of the Norway lemmings in Sweden during 1960. J. Mamm.,
        43:171–184.

Dagg, A. I. 1962. The role of the neck in the movements of the giraffe. J. Mamm., *43*:88–97.
Dalquest, W. W. 1948. The mammals of Washington. Univ. Kansas Publ., Mus. Nat. Hist., *2*:1–444.
Daniel, J. C., Jr. 1970. Dormant embryos of mammals. Bio. Sci., *20(7)*:411–415.
Darling, F. F. 1937. A herd of red deer. Oxford Univ. Press, London. 215 pp.
Darlington, P. J. 1957. Zoogeography: the geographical distribution of animals. John Wiley & Sons,
        Inc., New York 675 pp.
Dasmann, R. F., and A. S. Mossmann. 1962. Population studies of impala in Southern Rhodesia. J.
        Mamm., *43*:375–395.
David, A. 1968. Can young bats communicate with their parents at a distance? J. Bombay Nat. Hist.
        Soc., *65*:210.
Davis, D. E. 1951. The relation between level of population and pregnancy of Norway rats. Ecology,
        32:459–461.
Davis, R. B., C. F. Herreid, Jr., and H. L. Short. 1962. Mexican free-tailed bats in Texas. Ecol. Monogr.,
        32:311–346.
Davis, W. H. 1970. Hibernation: ecology and physiological ecology, pp. 265–300. *In* W. A. Wimsatt
        (ed.), Biology of bats. Academic Press, New York.
————, and H. B. Hitchcock. 1965. Biology and migration of the bat, *Myotis lucifugus,* in New
        England. J. Mamm., *46*:296–313.
————, and W. Z. Lidicker, Jr. 1956. Winter range of the red bat. J. Mamm., *37*:280–281.
Dawson, M. R. 1958. Later Tertiary Leporidae of North America. Univ. Kansas Paleont. Cont., Ver-
        tebrata, Art. 6, pp. 1–75.
————. 1967a. Fossil history of the families of Recent mammals, pp. 12–53. *In* S. Anderson and J. K.
        Jones, Jr. (eds.), Recent mammals of the world. The Ronald Press Co., New York.
————. 1967b. Lagomorph history and stratigraphic record. Essays in Paleontology and Stratigraphy,
        Raymond C. Moore Commemorative Volume, Univ. Kansas, Department of Geology Spec. Publ.
        2, pp. 287–316.
Dawson, T. J., and A. J. Hulbert. 1970. Standard metabolism, body temperature, and surface areas of
        Australian marsupials. Amer. J. Physiol., *218*:1233–1238.
DeCoursey, P. 1960. Phase control of activity in a rodent. Cold Spr. Harb. Symp. Quant. Biol.,
        25:49–55.
————. 1961. Effect of light on the circadian activity rhythm of the flying squirrel, *Glaucomys volans*.
        Z. Vergl. Physiol., *44*:331–354.
Deevey, E. S., Jr. 1947. Life tables for natural populations of animals. Quart. Rev. Biol., *22*:283–314.
Degerbøl, M., and P. Freuchen. 1935. Mammals, vol. 2, p. 278. *In* Report of the fifth Thule expedition,
        1921-1924. Copenhagen, Nordisk Forlag. Part I. Systematic notes, by Degerbøl. Part II. Field
        notes and biological observations, by Freuchen.
DeVore, I. (ed.). 1965. Primate behavior. Holt, Rinehart and Winston, New York. 654 pp.

_____., and K. R. L. Hall. 1965. Baboon ecology, pp. 20–52. *In* I. DeVore (ed.), Primate behavior. Holt, Rinehart and Winston, New York.

Dietz, R. S., and J. C. Holden. 1970. The breakup of Pangaea. Sci. Amer., *223(4)*:30–41.

Dijkgraaf, S. 1943. Over een merkwaardige functie wan den gehoorzin bij vleermuizen. Verslagen Nederlandsche Akademie wan Wetenschappen Afd. Naturkunde, 52:622–627.

_____. 1946. Die Sinneswelt der Fledermäuse. Experientia, 2:438–448.

_____. 1960. Spallanzani's unpublished experiments on the sensory basis of object perception in bats. Isis, *51*:9–20.

Dobzhansky, T. 1950. Mendelian populations and their evolution. Amer. Nat., *84*:401–418.

Dreher, J. J. 1966. Cetacean communication; small-group experiments, pp. 529–543. *In* K. S. Norris (ed.), Whales, dolphins and porpoises. Univ. California Press, Berkeley.

Dücker, G. 1957. Fard- und Helligkeitssehen und Instinkte bei Viverriden und Feliden. Zool. Beitr., Berl., 3:25–99.

Dunning, D. C., and K. D. Roeder. 1965. Moth sounds and the insect-catching behavior of bats. Science, *147*:173–174.

Edwards, R. L. 1946. Some notes on the life history of the Mexican ground squirrel in Texas. J. Mamm., 27:105–115.

Eibl-Eibesfeldt, I. 1958. Das Verhalten der Nagetiere. Handb. Zool. Berlin, 8:1–88.

Einarsen, A. S. 1948. The pronghorn antelope. Wildl. Manage. Inst., Washington, D.C. 238 pp.

Eisenberg, J. F., and E. Gould. 1970. The tenrecs: a study in mammalian behavior and evolution. Smithson. Contrib. Zool., 27:1–138.

Eisentraut, M. 1934. Der Winterschlaf der Fledermäuse mit besonderer Berücksichtigung der Warmeregulation. Z. Morphol. Oekol. Tiere, 29:231–267.

_____. 1957. Aus dem Leben der Fledermäuse und Flughunde. Jena: Veb. Gustav Fischer Verlag. 175 pp.

_____. 1960. Heat regulation in primitive mammals and in tropical species. Bull. Mus. Comp. Zool., Harvard Univ., *124*:31–43.

Ellerman, J. R. 1940. The families and genera of living rodents. Vol. I. British Mus. Nat. Hist. 689 pp.

_____. 1941. *Ibid.* Vol. II. 690 pp.

_____. 1949. *Ibid.* Vol. III. 210 pp.

Eloff, F. C. 1967. Personal communication.

Elsner, R. 1965. Hvalradets Skrifter Norske Videnskaps-Akad. Oslo, 48:24.

_____. 1969. Cardiovascular adjustments to diving, pp. 117–143. *In* H. T. Andersen (ed.), The biology of marine mammals. Academic Press, New York.

Elton, C. 1942. Voles, mice and lemmings. Clarendon, Oxford.

Erickson, A. B. 1944. Helminth infections in relation to population fluctuations in showshoe hares. J. Wildl. Manage., 8:134–153.

Errington, P. L. 1937. What is the meaning of predation? Smithsonian Rep. for *1936*:243–252.

_____. 1943. An analysis of mink predation upon muskrats in North-central United States. Agric. Exp. Sta. Iowa State Coll. Res. Bull., *320*:797–924.

_____. 1946. Predation and vertebrate populations. Quart. Rev. Biol., *21*:144–177, 221–245.

_____. 1957. Of populations, cycles and unknowns. Cold. Spr. Harb. Symp. Quant. Biol., 22:287–300.

_____. 1963. Muskrat populations. Iowa State Univ. Press, Ames, Iowa.

_____. 1967. Of predation and life. Iowa State Univ. Press, Ames, Iowa. 277 pp.

Estes, R. D. 1966. Behavior and life history of the wildebeest (*Connochaetes taurinus* Burchell). Nature, Lond., *212*:999–1000.

_____. 1967. The comparative behavior of Grant's and Thompson's gazelles. J. Mamm., 48:189–209.

_____., and J. Goddard. 1970. Prey selection and hunting behavior of the African wild dog. J. Wildl. Manage., *31(1)*:52–70.

Evans, F. G. 1942. The osteology and relationships of elephant shrews (Macroscelididae). Bull. Amer. Mus. Nat. Hist., *80(4)*:85–125.

Evans, W. E., and J. Bastian. 1969. Marine mammal communication: social and ecological factors, pp. 425–475. *In* H. T. Andersen (ed.), The biology of marine mammals. Academic Press, New York.

_____., and R. M. Haugen. 1963. An experimental study of the echolocation ability of a California sea lion, *Zalophus californianus* (Lesson). Bull. So. Calif. Acad. Sci., 62:165–175.

_____., and B. A. Powell. 1967. Proc. symp. bionic models of animal sonar systems, Frascati, Italy, 1966, pp. 363–398. Labor. d'Acoustique Animal, Jouy-en-Josas, France.

_____., W. W. Sutherland, and R. G. Beil. 1964. pp. 353–372, vol 1. *In* W. N. Tavolga (ed.), Marine bioacoustics. Pergamon Press, Oxford.

Evermann, B. W. 1923. The conservation of marine life of the Pacific. Sci. Mon., *16*:521–538.

Evernden, J. F., D. E. Savage, G. H. Curtis, and G. T. Jones. 1964. Potassium-argon dates and the Cenozoic mammalian chronology of North America. Amer. J. Sci., *262*:145–198.

Ewer, R. F. 1968. Ethology of mammals. Plenum Press, New York. 418 pp.

Ewing, W. G., E. H. Studier, and M. J. O'Farrell. 1970. Autumn fat deposition and gross body composition in three species of *Myotis*. Comp. Biochem. Physiol., 36:119–129.

Faegri, K., and L. Van Der Pijl. 1966. The principles of pollination ecology. Pergamon Press, New York.

Fields, R. W. 1957. Histricomorph rodents from late Miocene of Colombia, South America. Univ. California Publ. Geol. Sci., 32:273–404.

Findley, J. S. 1967. Insectivores and dermopterans, pp. 87–108. In S. Anderson and J. K. Jones (eds.), Recent mammals of the world. The Ronald Press Co., New York.

_____. 1969. Biogeography of Southwestern boreal and desert mammals, pp. 113–128. In J. K. Jones, Jr. (ed.), Contributions in mammalogy. Univ. Kansas, Mus. Nat. Hist., Misc. Publ. No. 51.

Finley, R. B., Jr. 1969. Cone caches and middens of *Tamasciurus* in the Rocky Mountain region, pp. 233–273. In J. K. Jones, Jr. (ed.), Contributions in mammalogy. Univ. Kansas, Mus. Nat. Hist., Misc. Publ. No. 51.

Fink, B. D. 1959. California Fish and Game, 45(3):216–217.

Fischer, G. A., J. C. Davis, F. Iverson, and F. P. Cronmiller. 1944. The winter range of the Interstate Deer Herd. U. S. Dept. Agric., Forest Serv., Region 5:1–20 (mimeo).

Fisher, E. M. 1939. Habits of the southern sea otter. J. Mamm., 20:21–36.

Fitch, H. S. 1948. Ecology of the California ground squirrel on grazing lands. Amer. Midl. Nat., 39:513–596.

_____, R. Goodrum, and C. Newman. 1952. The armadillo in the southeastern United States. J. Mamm., 33:21–37.

Fleming, T. H. 1970. Notes on the rodent faunas of two Panamanian forests. J. Mamm., 51:473–490.

_____. 1971. *Artibeus jamaicensis*: delayed embryonic development in a Neotropical bat. Science, 171:402–404.

Flexner, L. B., D. B. Cowie, L. M. Hellman, W. S. Wilde, and G. J. Vosburgh. 1948. The permeability of the human placenta to sodium in normal and abnormal pregnancies and the supply of sodium to the human fetus as determined with radioactive sodium. Am. J. Obst. & Gyn., 55:469–480.

Formozov, A. N. 1946. The covering of snow as an integral factor of the environment and its importance in the ecology of mammals and birds. Material for Fauna and Flora of the USSR, New Series Zool., 5:1–141.

_____. 1966. Adaptive modifications of behavior in mammals of the Eurasian steppes. J. Mamm., 47(2):208–222.

Fox, R. C. 1964. The adductor muscles of the jaw in some primitive reptiles. Univ. Kansas Publ., Mus. Nat. Hist., 12:657–680.

Fraser, F. C., and P. E. Purves. 1955. The blow of whales. Nature, 176:1221–1222.

_____, _____. 1960a. Anatomy and function of the cetacean ear. Proc. Roy. Soc. (London), B, 152:62–77.

_____, _____. 1960b. Hearing in cetaceans. Bull. Brit. Mus. Nat. Hist. Zool., 7:1–140.

Fries, S. 1879. Über die Fortpflanzung der einheimischen Chiropteren. Zoologischer Anzeiger, Bd. 2:355–357.

Frith, H. J., and S. H. Calaby. 1969. Kangaroos. F. W. Cheshire, Melbourne. 209 pp.

Galambos, R., and D. R. Griffin. 1942. Obstacle avoidance by flying bats; the cries of bats. J. Exp. Zool., 89:475–490.

Gawn, R. L. W. 1948. Aspects of the locomotion of whales. Nature, 161:44.

Genelly, R. E. 1965. Ecology of the common mole-rat (*Cryptomys hottentotus*) in Rhodesia. J. Mamm., 46:647–665.

Gill, E. D. 1957. The stratigraphical occurrence and paleontology of some Australian Tertiary marsupials. Mem. Nat. Mus., Victoria, 21:135–203.

Gilmore, R. M. 1961. Whales, porpoises, and the U. S. Navy. Norsk Hvalfangst-tid., 3:1–9.

Goodall, J. 1965. Chimpanzees of the Gombe stream reserve, pp. 425–473. In I. DeVore (ed.), Primate behavior. Holt, Rinehart and Winston, New York.

Goodwin, C. G. 1954. Mammals of the air, land, and waters of the world, Book 1, pp. 1–680 (vol. 1) and pp. 681–874 (vol. 2). In F. Drimmer (ed.), The animal kingdom. Doubleday & Co., Inc., Garden City, New York.

Gould, E. 1955. The feeding efficiency of insectivorous bats. J. Mamm., 36:399–407.

_____. 1965. Evidence for echolocation in the Tenrecidae of Madagascar. Proc. Amer. Phil. Soc., 109(6):352–360.

_____. 1969. Communication in three genera of shrews (Soricidae): *Suncus*, *Blarina*, and *Cryptotis*. Communications in behavioral biology, Part A, 3(1):11–31.

_____. 1970. Echolocation and communication in bats, pp. 144–161. *In* B. H. Slaughter and D. W. Walton (eds.), About bats. Southern Methodist Univ. Press, Dallas.

_____. 1971. Studies of maternal-infant communication and development of vocalization in the bats *Myotis* and *Eptesicus*. Communications in behavioral biology, Part A, 5(5):263–313.

_____., and F. J. Eisenberg. 1966. Notes on the biology of the Tenrecidae. J. Mamm., 47:660–686.

_____., N. C. Negus, and A. Novick. 1964. Evidence for echolocation in shrews. J. Exp. Zool., 156:19–38.

Graf, W. 1955. The Roosevelt elk. Port Angeles, Wash.: Port Angeles Evening News. 105 pp.

Grand, T. I., and R. Lorenz. 1968. Functional analysis of the hip joint in *Tarsius bancanus* (Horsefield, 1821) and *Tarsius syrichta* (Linnaeus, 1758). Folia Primat., 9:161–181.

Green, R. G., and C. L. Larson. 1938. A description of shock disease in the snowshoe hare. Am. J. Hyg., 28:190–212.

_____., _____., and J. F. Bell. 1939. Shock disease as the cause of the periodic decimation of the snowshoe hare. Am. J. Hyg., 30B:83–102.

Greenwald, G. S. 1956. The reproductive cycle of the field mouse, *Microtus californicus*. J. Mamm., 37:213–222.

_____. 1957. Reproduction in a coastal California population of the field mouse *Microtus californicus*. Univ. California Publ. Zool., 54:421–446.

Griffin, D. R. 1951. Audible and ultrasonic sounds of bats. Experientia, 7:448–453.

_____. 1953. Bat sounds under natural conditions with evidence for echolocation of insect prey. J. Exp. Zool., 36:399–407.

_____. 1958. Listening in the dark. Yale Univ. Press, New Haven, Connecticut. 413 pp.

_____. 1962. Comparative studies of the orientation sounds of bats. Symp. Zool. Soc. London, 7:61–72.

_____. 1970. Migrations and homing of bats, pp. 233–264. *In* W. A. Wimsatt (ed.), Biology of bats. Academic Press, New York.

_____., D. Dunning, D. A. Cahlander, and F. A. Webster. 1962. Correlated orientation sounds and ear movements of horseshoe bats (Part I). Nature, 196:1185–1186.

_____., and R. Galambos. 1940. Obstacle avoidance by flying bats. Anat. Rec., 78:95.

_____., _____. 1941. The sensory basis of obstacle avoidance by flying bats. J. Exp. Zool., 86:481–506.

_____., and H. B. Hitchcock. 1965. Probable 24-year longevity records for *Myotis lucifugus*. J. Mamm., 46:332.

_____., and A. Novick. 1955. Acoustic orientation of Neotropical bats. J. Exp. Zool., 130:251–300.

_____., F. A. Webster, and C. R. Michael. 1960. The echolocation of flying insects by bats. Anim. Behav., 8:141–154.

Grinnell, A. D. 1963a. The neurophysiology of audition in bats: intensity and frequency parameters. J. Physiol., 167:38–66.

_____. 1963b. The neurophysiology of audition in bats: temporal parameters. J. Physiol., 167:67–96.

Grinnell, J. 1914a. An account of the mammals and birds of the lower Colorado Valley with especial reference to the distributional problems presented. Univ. California Publ. Zool., 12:51–294.

_____. 1914b. Barriers to distribution as regards to birds and mammals. Amer. Nat., 48:248–254.

_____. 1914c. The Colorado River as a hindrance to the dispersal of species. Univ. California Publ. Zool., 12:100–107.

_____. 1922. A geographical study of the kangaroo rats of California. Univ. California Publ. Zool., 24:1–124.

_____. 1926. Geography and evolution in the pocket gopher. Univ. California Chron., 30:429–450.

_____. 1933. Review of the Recent mammal fauna of California. Univ. California Publ. Zool., 40:71–234.

_____., J. S. Dixon, and J. M. Linsdale. 1937. Fur-bearing mammals of California. 2 vols. Univ. California Press, Berkeley. 777 pp.

_____., and T. I. Storer. 1924. Animal life in the Yosemite. Univ. California Press, Berkeley. 752 pp.

Grummon, R. A., and A. Novick. 1963. Obstacle avoidance in the bat *Macrotus mexicanus*. Physiol. Zool., 36:361–369.

Guilday, J. E. 1958. The prehistoric distribution of the opossum. J. Mamm., 39:39–43.

Guthrie, M. J. 1933. The reproductive cycles of some cave bats. J. Mamm., 14:199–216.

Guyton, A. C. 1971. Textbook of medical physiology, 4th ed. W. B. Saunders Co., Philadelphia. 1032 pp.

Hall, E. R. 1946. Mammals of Nevada. Univ. California Press, Berkeley. 710 pp.

_____. 1951. American weasels. Univ. Kansas Publ., Mus. Nat. Hist., Vol 4:1–466.

_____. 1958. Introduction, Part. II, pp 371–373. *In* C. L. Hubbs (ed.), Zoogeography. Amer. Assoc. Adv. Sci. Publ., 51.

_____., and W. W. Dalquest. 1963. The mammals of Veracruz. Univ. Kansas Publ., Mus. Nat. Hist., 14:165–362.

_____, and K. R. Kelson. 1959. The mammals of North America. The Ronald Press Co., New York. 2 vols.

Hall, K. R. L. 1965. Behaviour and ecology of the Wild Patas Monkeys, *Erythrocebus patas*, in Uganda. J. Zool. Soc. London, *148*:15–87.

_____. 1968. Behaviour and ecology of the Wild Patas monkey, pp. 32–119. *In* P. C. Jay (ed.), Primates, studies in adaptation and variability. Holt, Rinehart and Winston, New York.

_____, and I. DeVore. 1965. Baboon social behavior, pp. 53–110. *In* I. DeVore (ed.), Primate behavior. Holt, Rinehart and Winston, New York.

_____, and G. B. Schaller. 1964. Tool-using behavior of the California sea otter. J. Mamm., *45*:287–298.

Hamilton, W. J., Jr. 1937. The biology of microtine cycles. J. Agric. Res., *54*:779–790.

_____. 1939. American mammals. McGraw-Hill Book Co., Inc. 434 pp.

Hansen, R. M. 1962a. Dispersal of Richardson ground squirrel in Colorado. Amer. Midland Nat., *68*:58–66.

_____. 1962b. Movements and survival of *Thomomys talpoides* in a mima-mound habitat. Ecology, *43*:151–154.

_____, and A. L. Ward. 1966. Some relations of pocket gophers to rangelands on Grand Mesa, Colorado. Colo. Agric. Exp. Sta., Tech. Bull., 88. 20 pp.

Hanson, D. D. 1971. The food habits and energy dynamics of *Neotoma stephensi*. M. S. Thesis. Northern Arizona Univ., Flagstaff, Arizona.

Hardy, R. 1945. The influence of types of soil upon the local distribution of some mammals in southwestern Utah. Ecol. Monogr., *15*:71–108.

Hart, F. M., and J. A. King. 1966. Distress vocalizations of young in two subspecies of *Peromyscus*. J. Mamm., *47*:287–293.

Hart, J. S. 1956. Seasonal changes in insulation of the fur. Can. J. Zool., *34*:53–57.

Hartman, C. G. 1933. On the survival of spermatozoa in the female genital tract of the bat. Quart. Rev. Biol., 8:185–193.

Harvey, M. J., and R. W. Barbour. 1965. Home ranges of *Microtus ochrogaster* as determined by a modified minimum area method. J. Mamm., *46*:398–402.

Hatt, R. T. 1932. The vertebral columns of ricochetal rodents. Bull. Amer. Mus. Nat. Hist., *58*:599–738.

_____. 1934. The pangolins and aard-varks collected by the American Museum Congo Expedition. Bull Amer. Mus. Nat. Hist., *66*:643–672.

_____. 1936. Hyraxes collected by the American Museum Congo Expedition. Bull. Amer. Mus. Nat. Hist., *72*:117–141.

Hayward, J. S., and C. P. Lyman. 1967. Non-shivering heat production during arousal from hibernation and evidence for the contribution of brown fat, pp. 346–355. *In* K. C. Fisher et al. (eds.), Mammalian hibernation III. Oliver & Boyd, Edinburgh and London.

_____, _____, and C. R. Taylor. 1965. The possible role of brown fat as a source of heat during arousal from hibernation. Ann. N.Y. Acad. Sci., *131*:441–446.

Heezen, B. C. 1957. Whales entangled in deep-sea cables. Deep Sea Res., *4*:105–115.

Heinroth-Berger, K. 1959. Beobachtungen an handaufgezogenen Mantelpavianen (*Papio hamadryas* L.). Z. Tierpsychol., *16*:706–732.

Henshaw, R. E., and G. E. Folk, Jr. 1966. Relation of thermoregulation to seasonally changing microclimate in two species of bats (*Myotis lucifugus* and *M. sodalis*). Physiol. Zool., *39*:223–236.

Henson, O. W., Jr. 1961. Some morphological and functional aspects of certain structures of the middle ear in bats and insectivores. Univ. Kansas Sci. Bull., *42*:151–225.

_____. 1965. The activity and function of the middle-ear muscles in echo-locating bats. J. Physiol., *180*:871–887.

Heppes, J. B. 1958. The white rhinoceros in Uganda. Afr. Wildlife, *12*:273–280.

Herald, E. S., R. L. Brownell, Jr., F. L. Frye, E. J. Morris, W. E. Evans, and A. B. Scott. 1969. Blind river dolphin: first side-swimming cetacean. Science, *166*:1408–1410.

Herreid, C. F., II. 1963. Temperature regulation and metabolism in Mexican free-tail bats. Science, *142*:1573–1574.

_____. 1967. Temperature regulation, temperature preference and tolerance, and metabolism of young and adult free-tail bats. Physiol. Zool., *40*:1–22.

Hertel, A. 1969. Hydrodynamics of swimming and wave-riding dolphins, pp. 31–63. *In* H. T. Andersen (ed.), The biology of marine mammals. Academic Press, New York.

Hesse, R., W. C. Allee, and K. P. Schmidt. 1951. Ecological animal geography, 2nd ed. John Wiley & Sons, New York.

Hibbard, C. W., D. E. Ray, D. E. Savage, D. W. Taylor, and J. E. Guilday. 1965. Quaternary mammals of North America, pp. 509–525. *In* H. E. Wright, Jr. and D. G. Frey (eds.), The Quaternary of the United States. Princeton Univ. Press, Princeton, New Jersey.

_____, and G. C. Rinker. 1942. A new bog-lemming (*Synaptomys*) from Meade County, Kansas. Univ. Kansas Sci. Bull., *28*:25–35.

Hildebrand, M. 1959. Motions of the running cheetah and horse. J. Mamm., *40*:481–496.

_____. 1960. How animals run. Sci. Amer., *202(5)*:148–156.

_____. 1962. Walking, running, and jumping. Amer. Zoologist, 2:151–155.

_____. 1965. Symmetrical gaits of horses. Science, 150:701–708.

Hill, J. E., and T. D. Carter. 1941. The mammals of Angola, Africa. Bull. Amer. Mus. Nat. Hist., 78:1–211.

Hisaw, F. L. 1924. The absorption of the pubic symphysis of the pocket gopher, *Geomys bursarius* (Shaw). Amer. Nat., 58:93–96.

Hock, R. J. 1951. The metabolic rates and body temperatures of bats. Biol. Bull., 101:289–299.

Hoese, H. D. 1971. Dolphin feeding out of water in a salt marsh. J. Mamm., 52:222–223.

Hoffmann, R. S. 1958. The role of reproduction and mortality in population fluctuations of voles (*Microtus*). Ecol. Monogr., 28:79–109.

Holling, C. S. 1959. The components of predation as revealed by a study of small mammal predation of the European pine sawfly. Can. Entomol., 91:293–320.

_____. 1961. Principles of insect predation. Ann. Rev. Entomol., 6:163–182.

Hooper, E. T. 1952. A systematic review of the harvest mice (genus *Reithrodontomys*) of Latin America. Misc. Publ. Mus. Zool., Univ. Michigan, 77:1–255.

_____. 1968. Anatomy of middle-ear walls and cavities in nine species of microtine rodents. Univ. Michigan Occ. Papers, 657:1–28.

_____., and J. H. Brown. 1968. Foraging and breeding in two sympatric species of Neotropical bats, genus *Noctilio*. J. Mamm., 49:310–312.

Hopkins, D. M. 1959. Cenozoic history of the Bering land bridge. Science, 129:1519–1528.

Hornocker, M. G. 1970a. The American lion. Nat. Hist., 79:40–49, 68–71.

_____. 1970b. An analysis of mountain lion predation upon mule deer and elk in the Idaho Primitive Area. Wildl. Monogr. No. 21. 39 pp.

Horst, R. 1969. Observations on the structure and function of the kidney of the vampire bat (*Desmodus rotundus murinus*), pp. 73–83. *In* C. C. Hoff and M. L. Riedesel (eds.), Physiological systems in semiarid environments. Univ. New Mexico Press, Albuquerque.

Howell, A. B. 1930. Aquatic mammals. Charles C Thomas, Springfield, Illinois.

_____. 1944. Speed in animals. Univ. Chicago Press, Chicago. 270 pp.

Hudson, J. W. 1962. The role of water in the biology of the antelope ground squirrel. Univ. California Publ. Zool., 64:1–56.

_____. 1965. Temperature regulation and torpidity in the pigmy mouse, *Baiomys taylori*. Physiol. Zool., 38:243–254.

Humbolt, A. von, and A. Bonpland. 1852–53. Personal narrative of travels to the equinoctial regions of America during the years 1799–1804. Henry G. Bohn, London, 3 vols.

Hurley, P. M. 1968. The confirmation of continental drift. Sci. Amer., 218(4):52–64.

Hutchinson, G. E. 1957. Concluding remarks. Cold Spr. Harb. Symp. Quant. Biol., 22:415–427.

Ingles, L. G. 1949. Ground water and snow as factors affecting the seasonal distribution of pocket gophers, *Thomomys monticola*. J. Mamm., 30:343–350.

Irving, L. 1966. Adaptations to cold. Sci. Amer., 214(1):94–101.

_____. 1969. Temperature regulation in marine mammals, pp. 147–173. *In* H. T. Andersen (ed.), The biology of marine mammals. Academic Press, New York.

_____., and J. S. Hart. 1957. The metabolism and insulation of seals as bare-skinned mammals in cold water. Can. J. Zool., 35:497–511.

_____., H. Krog, and M. Monson. 1955. The metabolism of some Alaskan animals in winter and summer. Physiol. Zool., 28:173–185.

Jaeger, E. C. 1950. The coyote as a seed distributor. J. Mamm., 31:452–453.

Jameson, E. W., Jr. 1952. Food of deer mice, *Peromyscus maniculatus* and *P. boylei*, in the northern Sierra Nevada, California. J. Mamm., 33:50–60.

Jay, P. C. (ed.). 1968. Primates, studies in adaptation and variability. Holt, Rinehart and Winston, New York. 529 pp.

Jenkins, F. A., Jr. 1970. Limb movement in a monotreme (*Tachyglossus aculeatus*): a cineradiographic analysis. Science, 168:1473–1475.

Jenkins, H. O. 1948. A population study of the meadow mice (*Microtus*) in three Sierra Nevada meadows. Proc. California Acad. Sci., ser. 4, 26:43–67.

Jepsen, G. L. 1966. Early Eocene bat from Wyoming. Science, 154:1333–1339.

_____. 1970. Bat origins and evolution, pp. 1–64. *In* W. A. Wimsatt (ed.), Biology of bats. Academic Press, New York.

_____., and M. O. Woodburne. 1969. Paleocene hyracothere from Polecat Bench Formation, Wyoming. Science, 164:543–547.

Johannessen, C. L., and J. A. Harder. 1960. Sustained swimming speeds of dolphins. Science, 132:1550–1551.

Johnson, D. R. 1961. The food habits of rodents on rangelands of southern Idaho. Ecology, 42:407–410.
———. 1964. Effects of range treatment with 2,4-D on food habits of rodents. Ecology, 45:241–249.
Jolly, A. 1966. Lemur behavior: a Madagascar field study. Univ. Chicago Press, Chicago.
Jones, C. 1967. Growth, development, and wing loading in the evening bat, *Nycticeius humeralis* (Rafinesque). J. Mamm., 48:1–19.
Jones, F. W. 1923. The mammals of South Australia. Part I: the monotremes and carnivorous marsupials. Government Printer, Adelaide. pp. 1–131.
Jones, F. W. 1924. The mammals of South Australia. Part II: the bandicoots and the herbivorous marsupials. Government Printer, Adelaide. pp. 133–270.
Jones, J. K., Jr. 1964. Distribution and taxonomy of mammals of Nebraska. Univ. Kansas Publ., Mus. Nat. Hist., 16:1–356.
———, and R. R. Johnson. 1967. Sirenians, pp. 367–373. *In* S. Anderson and J. K. Jones, Jr. (eds.), Recent mammals of the world. The Ronald Press Co., New York.

Kahmann, H., and K. Ostermann. 1951. Wahrnehmen und Hervorbringen hoher Tone bei kleiner Saugetieren. Experientia, 7:268–269.
Kalela, O. 1957. Regulation of reproduction rate in subarctic populations of the vole *Clethrionomys rufocanus* (Sund.). Ann. Acad. Sci. Fennicae, Ser. A. Iv. Biol., 34:1–60.
———. 1962. On the fluctuations in the numbers of arctic and boreal small rodents as a problem of production biology. Ann. Acad. Sci. Fennicae, Ser. A, 4(66):1–38.
Kangas, E. 1949. On the damage to the forests caused by the moose and its significance in the economy of the forest. Eripainos: Suomen Riista, vol. 4, pp. 62–90. (English summary, pp. 88–90.)
Kanwisher, J., and H. Leivestad. 1957. Thermal regulation in whales. Norsk Hvalfangst-tid., 1:1–5.
———, and G. Sundnes. 1966. Thermal regulation in cetaceans, pp. 397–409. *In* K. S. Norris (ed.), Whales, dolphins and porpoises. Univ. California Press, Berkeley.
Kellogg, W. N. 1961. Porpoises and sonar. Univ. Chicago Press, Chicago. 177 pp.
———, R. Kohler, and H. N. Morris. 1953. Porpoise sounds as sonar signals. Science, 117:239–243.
Kelsall, J. P. 1970. Migration of the barren-ground caribou. Nat. Hist., 79:98–106.
Kendeigh, S. C. 1961. Animal ecology. Prentice-Hall, Inc., Englewood Cliffs, New Jersey. 468 pp.
Kennerly, T. E., Jr. 1964. Microenvironmental conditions of the pocket gopher burrow. Texas J. Sci., 14(4):397–441.
Kennerly, T. R. 1971. Personal communication.
Kermack, K. A. 1963. The cranial structure of triconodonts. Philos. Trans. Roy. Soc. London, Ser. B, 246:83–103.
———, and F. Musset. 1958. The jaw articulation of the docodonta and the classification of Mesozoic mammals. Proc. Roy. Soc. (London), B, 149:204–215.
King, J. A. 1955. Social behavior, social organization and population dynamics in a black-tail prairie dog town in the Black Hills of South Dakota. Contrib. Lab. Vert. Biol. Univ. Michigan, No. 67:1–123.
———. 1959. The social behavior of prairie dogs. Sci. Amer., 201(4):128–140.
——— (ed.). 1968. Biology of *Peromyscus* (Rodentia). Spec. Publ. No. 2, Amer. Soc. Mammal. 593 pp.
Kinnear, J. E., K. G. Purohit, and A. R. Main. 1968. The ability of the tammar wallaby (*Macropus eugenii*, Marsupialia) to drink sea water. Comp. Biochem. Physiol., 25:761–782.
Klingel, H. 1967. Soziale Organisation und Verhalten freilebender Steppenzebras. Z. Tierpsychol., 24:580–624.
Klingener, D. 1964. The comparative myology of four dipodoid rodents (Genera *Zapus*, *Napaeozapus*, *Sicista* and *Jaculus*). Misc. Publ. Mus. Zool., Univ. Michigan, 124:1–100.
Knappe, H. 1964. Zur Funktion des Jacobsonschen Organs (*Organon vomeronasale Jacobsoni*). Zool. Gart., Lpz., 28:188–194.
Knudsen, V. O. 1931. The effect of humidity upon the absorption of sound in a room, and determination of the coefficients of absorption of sound in air. J. Acoustical Soc. Am., 3:126–138.
———. 1935. Atmospheric acoustics and the weather. Scientific Monthly, 40:485–486.
Komarek, E. V. 1932. Notes on mammals of Menominee Indian Reservation, Wisconsin. J. Mamm., 13:203–209.
Kooyman, G. L., and H. T. Andersen. 1969. Deep diving, pp. 65–94. *In* H. T. Andersen (ed.), The biology of marine mammals. Academic Press, New York.
Kramer, M. O. 1960. J. Am. Soc. Naval Engrs. 25–33.
Krebs, C. 1963. Lemming cycle at Baker Lake, Canada during 1959-62. Science, 146:1559–1560.
———. 1964. The lemming cycle at Baker Lake, Northwest Territories, during 1959-62. Arctic Inst. N. Amer. Tech. Paper No. 15.
———. 1966. Demographic changes in fluctuating populations of *Microtus californicus*. Ecol. Monogr., 36:239–273.
———, and K. T. DeLong. 1965. A *Microtus* population with supplemental food. J. Mamm., 46:566–573.

Krebs, H. A. 1950. Body size and tissue metabolism. Biochem. Biophys. Acta, 4:249–269.

Krishnan, R. S., and J. C. Daniel. 1967. "Blastokinin"—an inducer and regulator of blastocyst development in the rabbit uterus. Science, 158:490–492.

Kruger, L. 1966. Specialized features of the cetacean brain, pp. 232–254. In K. S. Norris (ed.), Whales, dolphins and porpoises. Univ. California Press, Berkeley, California.

Krumrey, W. A., and I. O. Buss. 1968. Age estimation, growth, and relationships between body dimensions of the female African elephant. J. Mamm., 49:22–31.

Kruuk, H. 1966. Clan-system and feeding habits of spotted hyaenas (Crocuta crocuta Erxleben). Nat. Lond., 209:1257–1258.

———. 1970. Interactions between populations of spotted hyaenas (Crocuta crocuta) and their prey species, pp. 359–374. In Adam Watson (ed.), Animal populations in relation to their food resources. Blackwell Scientific Publ., Oxford.

———, and H. Van Lawick. 1968. Hyaenas, the hunters nobody knows. National Geographic, 134(1):44–57.

Krzanowski, A. 1960. Investigations of flights of Polish bats, mainly Myotis myotis. Acta Theriol., 4:175–184.

———. 1961. Weight dynamics of bats wintering in the cave at Pulway (Poland). Acta Theriol., 4:249–264.

Kühme, W. 1965. Freilandstudien zur Soziologie des Hyäenenhundes (Lycaon pictus lupinus Thomas 1902). Z. Tierpsychol., 225:495–541.

———. 1966. Beobachtungen zur Soziologie des Löwens in der Serengeti-Steppe Ostafrikas. Z. Saugetierk., 31:205–213.

Kulzer, E. 1956. Flughunde erzeugen Oreintierung durch Zungenschlag. Naturwissenschaften, 43:117–118.

———. 1958. Untersuchungen über die Biologie von Flughunden der Gattung Rousettus Gray. Zeitschr. Morph. Ökol Biere., 47:374–402.

———. 1960. Physiologische und morphologische Untersuchungen über die Erzeugung der Orientierungslaute von Flughunden der Gattung Rousettus. Z. Vergl. Physiol., 43:231–268.

———. 1961. Über die Biologie der Nil-Flughunde (Rousettus aegyptiacus). Natur U. Volk., 91:219–228.

———. 1963. Temperaturregulation bei Flughunden der Gattung Rousettus Gray. Z. Vergl. Physiol., 46:595–618.

———. 1965. Temperaturregulation bei Fledermäusen (Chiroptera) aus berschiedenen Klimazonen. Z. Vergl. Physiol., 50:1–34.

Kummer, H. 1968. Two variations in the social organization of baboons, pp. 293–312. In P. C. Jay (ed.), Primates, studies in adaptation and variability. Holt, Rinehart and Winston, New York.

Kunz, T. H. 1970. Reproductive patterns and development of Myotis velifer in Kansas. Symposium on bat research in the Southwest. (Mimeo.)

Kurten, B. 1969. Continental drift and evolution. Sci. Amer., 220(3):54–64.

Lack, D. 1948. The significance of litter-size. J. Anim. Ecol., 17:45–50.

———. 1954a. The natural regulation of animal numbers. Oxford Univ. Press, London. 343 pp.

———. 1954b. Cyclic mortality. J. Wildl. Manage., 18:25–37.

———. 1966. Population studies of birds. Clarendon Press, Oxford. 341 pp.

Lackey, J. A. 1967. Growth and development of Dipodomys stephensi. J. Mamm., 48:624–632.

Landry, S. O. 1957. The interrelationships of the New World and Old World histricomorph rodents. Univ. California Publ. Zool., 56:1–118.

Lang, T. G. 1966. Hydrodynamic analysis of cetacean performance, pp. 410–432. In K. S. Norris (ed.), Whales, dolphins and porpoises. Univ. California Press, Berkeley, California.

Langworthy, M., and R. Horst. 1971. Reproductive behavior in a captive colony of Molossus ater. In manuscript.

Laurie, A. H. 1933. Discovery. Rept., 363–406.

Lawrence, B., and A. Novick. 1963. Behavior as a taxonomic clue: relationships of Lissonycteris (Chiroptera). Mus. Comp. Zool., 184:1–16.

Laws, R. M. 1953. The elephant seal (Mirounga leonina, Linn.). I. Growth and age. Falkland Is. Depend. Surv. Sci. Repts., 8:1–62.

Layne, J. N. 1958. Observations on freshwater dolphins in the upper Amazon. J. Mamm., 39:1–22.

———. 1965. Observations on marine mammals in Florida waters. Bull. Florida State Mus., 9:131–181.

———. 1968. Ontogeny, pp. 148–253. In J. A. King (ed.), Biology of Peromyscus (Rodentia). Spec. Publ. No. 2, Amer. Soc. Mamm.

Lear, J. 1970. The bones on Coalsack Bluff: a story of drifting continents. Sat. Rev., 53(6):46–51.

Lechleitner, R. R. 1958a. Certain aspects of behavior of the black-tailed jackrabbit. Amer. Midl. Nat., 60:145–155.

———. 1958b. Movements, density and mortality in a black-tailed jackrabbit population. J. Wildl. Mgmt., 22:371–384.

_____. 1959. Sex ratio, age classes and reproduction of the black-tailed jackrabbit. J. Mamm., 40:63–81.

_____., J. V. Tileston, and L. Kartman. 1962. Die-off of a Gunnison's prairie dog colony in central Colorado. I. Ecological observations and description of the epizootic. Zoonoses Res., 1:185–199.

Lee, A. K. 1963. The adaptations to arid environments in wood rats of the genus *Neotoma*. Univ. California Publ. Zool., 64:57–96.

Leitner, P. 1966. Body temperature, oxygen consumption, heart rate and shivering in the California mastiff bat, *Eumops perotis*. Comp. Biochem. Physiol., 19:431–443.

_____., and J. E. Nelson, 1967. Body temperature, oxygen consumption and heart rate in the Australian false vampire bat, *Macroderma gigas*. Comp. Biochem. Physiol., 21:65–74.

Lenfant, C. 1969. Physiological properties of blood of marine mammals, pp. 95–115. *In* H. T. Andersen (ed.), The biology of marine mammals. Academic Press, New York.

Leopold, A. S., T. Riney, R. McCain, and L. Tevis, Jr. 1951. The jawbone deer herd. California Div. Fish and Game, Game Bull., 4:1–139.

_____., L. K. Sowls, and D. L. Spencer. 1947. A survey of overpopulated deer ranges in the United States. J. Wildl. Manage., 11:162–177.

Leyhausen, P. 1956. Verhaltensstudien an Katzen. Z. Tierpsychol., Beiheft 2:1–120.

_____. 1964. The communal organization of solitary animals. Symp. Zool. Soc. London, 14:249–263.

Lidicker, W. Z., Jr. 1968. A phylogeny of New Guinea rodent genera based on phallic morphology. J. Mamm., 49:609–643.

_____., and P. K. Anderson. 1962. Colonization of an island by *Microtus californicus*, analyzed on the basis of runway transects. J. Anim. Ecol., 31:503–517.

Lillegraven, J. A. 1969. Latest Cretaceous mammals of upper part of Edmonton Formation of Alberta, Canada, and review of marsupial-placental dichotomy in mammalian evolution. Univ. Kansas Paleontol. Contrib., 50:1–122.

Lilly, J. C. 1962. Vocal behavior of the bottle-nosed dolphin. Proc. Amer. Philos. Soc., 106:520–529.

_____. 1963. Distress call of the bottle-nosed dolphin: stimuli and evoked behavioral responses. Science, 139:116–118.

Linsdale, J. M. 1946. The California ground squirrel. Univ. California Press, Berkeley, California. 475 pp.

Linzey, D. W., and A. V. Linzey. 1967. Maturational and seasonal molts in the golden mouse, *Ochrotomys nuttalli*. J. Mamm., 48:236–241.

Lorenz, K. 1950. The comparative method of studying innate behavior patterns. Symp. Soc. Exp. Biol., 4:229–269.

_____. 1963. Das sogenannte Böse. G. Borotha-Schoeler, Vienna. (English version, 1966, On aggression. Methuen, London.)

Louch, C. D. 1958. Adrenocortical activity in two meadow vole populations. J. Mamm., 39:109–116.

Luckens, M. M., and W. H. Davis. 1964. Bats: sensitivity to DDT. Science, 146:948.

Lund, R. D., and J. S. Lund. 1965. The visual system of the mole, *Talpa europaea*. Exp. Neurology, 13:302–316.

Lyman, C. P. 1954. Activity, food consumption and hoarding in hibernators. J. Mamm., 35:545–552.

_____. 1970. Thermoregulation and metabolism in bats, pp. 301–330. *In* W. A. Wimsatt (ed.), Biology of bats, vol. 1. Academic Press, New York.

_____., and W. A. Wimsatt. 1966. Temperature regulation in the vampire bat, *Desmodus rotundus*. Physiol. Zool., 39:101–109.

MacLulich, D. A. 1937. Fluctuations in the numbers of the varying hare *(Lepus americanus)*. Univ. Toronto Studies, Biol. Ser., No. 43.

MacMillen, R. E. 1964a. Population ecology, water relations, and social behavior of a southern California semidesert rodent fauna. Univ. California Publ. Zool., 71:1–66.

_____. 1964b. Water economy and salt balance in the western harvest mouse, *Reithrodontomys megalotis*. Physiol. Zool., 37(1):45–56.

_____. 1965. Aestivation in the cactus mouse, *Peromyscus eremicus*. Comp. Biochem. Physiol., 16:227–248.

_____. 1972. Water economy of nocturnal desert rodents. *In* G. M. O. Maloiy (ed.), Comparative physiology of desert animals. Symp. Zool. Soc. London. Academic Press, New York. In press.

_____., and A. K. Lee. 1967. Australian desert mice: independence of exogenous water. Science, 158(3799):383–385.

_____., _____. 1969. Water metabolism of Australian hopping mice. Comp. Biochem. Physiol., 28:493–514.

_____., _____. 1970. Energy metabolism and pulmocutaneous water loss of Australian hopping mice. Comp. Biochem. Physiol., 35:355–369.

_____., and J. E. Nelson. 1969. Bioenergetics and body size in dasyurid marsupials. Amer. J. Physiol., 217:1246–1251.

Maher, W. J. 1967. Predation by weasels on a winter population of lemmings, Banks Island, Northwest Territories. Canadian Field Natur., 81:248–250.

_____. 1970. The pomarine jaeger as a brown lemming predator in northern Alaska. Wilson Bull., 82:130–157.

Marler, P. R. 1965. Communication in monkeys and apes, pp. 544–584. *In* I. DeVore (ed.), Primate behavior. Holt, Rinehart and Winston, New York.

_____, and W. J. Hamilton III. 1966. Mechanisms of animal behavior. John Wiley & Sons, Inc., New York. 771 pp.

Marshall, L. G. 1972. A study of the peramelid tarsus. Roy. Soc., Victoria. In press.

_____, and G. J. Weisenberger. 1971. A new dwarf shrew locality for Arizona. Plateau, 43:132–137.

Martin, E. P. 1956. A population study of the prairie vole *(Microtus ochrogaster)* in northeastern Kansas. Univ. Kansas Publ., Mus. Nat. Hist., 8:361–416.

Martinsen, D. L. 1968. Temporal patterns in the home ranges of chipmunks. J. Mamm., 49:83–91.

Mayr, E. 1942. Systematics and the origin of species. Columbia Univ. Press, New York. 334 pp.

_____. 1963. Animal species and evolution. Harvard Univ. Press, Cambridge, Massachusetts. 797 pp.

McCabe, T. T., and B. D. Blanchard. 1950. Three species of *Peromyscus*. Rood Associates, Santa Barbara, California. 136 pp.

McCarley, H. 1959. The effect of flooding on a marked population of *Peromyscus*. J. Mamm., 40:57–63.

McCullough, D. R. 1969. The tule elk, its history, behavior, and ecology. Univ. California Publ. Zool., 88:1–209.

McLaren, A. 1970. The fate of the zona pellucida in mice. J. Embryol. Exp. Morph., 23:1–19.

MeLean, D. C. 1944. The prong-horned antelope in California. Bureau of Game Cons., California Div. of Fish and Game, San Francisco, 30(4):221–241.

McNab, B. K. 1966. The metabolism of fossorial rodents; a study of convergence. Ecology, 47:712–733.

Mead, R. A. 1968. Reproduction in western forms of the spotted skunk (genus *Spilogale*). J. Mamm., 49:373–390.

Mech, L. D. 1966. The wolves of Isle Royale. U.S. Nat. Park Serv., Fauna ser. 7. 210 pp.

Menaker, M. 1961. The free-running period of the bat clock; seasonal variations at low body temperature. J. Cell Comp. Physiol., 57:81–86.

Menard, H. W. 1969. The deep-ocean floor. Sci. Amer., 221:126–142.

Merriam, C. H. 1894. Laws of temperature control of the geographic distribution of terrestrial animals and plants. Natl. Geogr. Mag., 6:229–238.

_____. 1899. Life zones and crop zones of the United States. Bull. U.S. Biol. Surv., 10:1–79.

Michael, R. P., E. B. Keverne, and R. W. Bonsall. 1971. Pheromones: isolation of male sex attractants from a female primate. Science, 172:964–966.

Miller, G. S., Jr. 1907. The families and genera of bats. Bull. U.S. Nat. Mus., 57. 282 pp.

Misonne, X. 1959. Analyse zoogéographique des mammiferes de l'Iran, Bruxelles, Inst. Royal des Sci. Nat. de Belgique, Mémoires, 2me sér. 59. 157 pp.

Mizuhara, H. 1957. The Japanese monkey, its social structure. Kyoto: San-ichi-syobo (in Japanese).

Mobius, K. 1877. Die Auster und die Austernwirtschaft. Berlin. pp. 22, 35, 436, 508. (Transl., 1880, The oyster and oyster culture, Rept. U.S. Fish. Comm., 1880:683–751.)

Mohr, E. 1941. Schwanzverlust und Schwanzregeneration bei Nagetieren. Zool. Anzeiger, 135:49–65.

Mohres, F. P. 1953. Über die Ultraschallorientierung der Hufeisennasen (Chiroptera – Rhinolophidae). Z. Vergl. Physiol., 34:547–588.

_____. 1966. Communicative characters of sonar signals in bats, pp. 939–945. *In* R. G. Busnel (ed.), Animal sonar systems, biology and bionics, Tome II. Laboratoire de Physiologie Acoustique, Paris.

_____, and E. Kulzer. 1956. Über die Orientierung der Flughunde (Chiroptera – Pteropodidae). Z. Vergl. Physiol., 38:1–29.

_____, and G. Neuweiler. 1966. Ultrasonic orientation in megadermid bats, pp. 115–128. *In* R. G. Busnel (ed.), Animal sonar systems, biology and bionics, Tome I. Laboratoire de Physiologie Acoustique, Paris.

Morrison, P. 1959. Body temperatures in some Australian mammals. I. Chiroptera. Biol. Bull., 116:484–497.

_____. 1962. Body temperatures in some Australian mammals. III. Cetacea (Megaptera). Biol. Bull., 123:154–169.

_____, and B. K. McNab. 1962. Daily torpor in a Brazilian murine opossum *(Marmosa)*. Comp. Biochem. Physiol., 6:57–68.

_____, _____. 1967. Temperature regulation in some Brazilian phyllostomid bats. Comp. Biochem. Physiol., 21:207–221.

_____, and F. A. Ryser. 1952. Weight and body temperature in mammals. Science, 116:231–232.

_____, _____, and A. R. Dawe. 1959. Studies on the physiology of the masked shrew *Sorex cinereus*. Physiol. Zool., 32:256–271.

Murie, A. 1940. Ecology of the coyote in the Yellowstone. U.S. Dept. Int., Natl. Park Serv., Fauna ser. 4. 206 pp.

_____. 1944. The wolves of Mount McKinley. U.S. Dept. Int., Natl. Park Serv., Fauna ser. 5. 238 pp.

Myers, G. T., and T. A. Vaughan. 1964. Food habits of the plains pocket gopher in eastern Colorado. J. Mamm., 45:588–598.

Negus, N. C., and A. J. Pinter. 1965. Litter sizes of *Microtus montanus* in the laboratory. J. Mamm., *46(3)*:434–437.

———, ———. 1966. Reproductive responses of *Microtus montanus* to plants and plant extracts in the diet. J. Mamm., *47(4)*:596–601.

Neumann, C. A. 1965/1966. Geo-marine Technol., 2:1; as cited by Ridgway (1966).

Noirot, E. 1969. Sound analysis of ultrasonic distress calls of mouse pups as a function of their age. Anim. Behav., *17*:340–349.

Norris, K. S. 1964. Some problems in echolocation in cetaceans, pp. 317–336. *In* W. N. Tavolga (ed.), Marine bio-acoustics. Pergamon Press, Oxford.

———. (ed.). 1966. Whales, dolphins and porpoises. Univ. California Press, Berkeley. 789 pp.

———. 1968. The evolution of acoustic mechanisms in odontocete cetaceans. Peabody Museum Centenary Celebration Volume, Yale Univ.

———. 1969. The echolocation of marine mammals, pp. 391–423. *In* H. T. Andersen (ed.), The biology of marine mammals. Academic Press, New York.

———, H. A. Baldwin, and D. J. Samson. 1965. Deep Sea Res., *12*:505–509.

———, and J. H. Prescott. 1961. Observations on Pacific cetaceans of California and Mexican waters. Univ. California Publ. Zool., *63*:291–402.

———, A. Prescott, D. V. Asa-Doran, and P. Perkins. 1961. An experimental demonstration of echolocation behavior in the porpoise, *Tursiops truncatus* (Montagu). Biol. Bull., *120*:163–176.

Novick, A. 1955. Laryngeal muscles of the bat and production of ultrasonic sounds. Am. J. Physiol., *183*:648.

———. 1958a. Orientation in paleotropical bats. II Megachiroptera. J. Exp. Zool., *137*:443–462.

———. 1958b. Orientation in paleotropical bats. I Microchiroptera. J. Exp. Zool., *138*:81–254.

———. 1962. Orientation in neotropical bats. I Natalidae and Emballonuridae. J. Mamm., *43*:449–455.

———. 1963a. Orientation in neotropical bats. II Phyllostomatidae and Desmodontidae. J. Mamm., *44*:44–56.

———. 1963b. Pulse duration in the echolocation of insects by the bat, *Pteronotus*. Ergebnisse Biol., *26*:1–26.

———. 1965. Echolocation of flying insects by the bat, *Chilonycteris psilotis*. Biol. Bull., *128*:297–314.

———. 1970. Echolocation in bats. Natur. Hist., *79(3)*:32–41.

———. 1971. Echolocation in bats: some aspects of pulse design. Amer. Sci., *59(2)*:198–209.

———, and D. R. Griffin. 1961. Laryngeal mechanisms in bats for production of orientation sounds. J. Exp. Zool., *148*:125–146.

———, and J. R. Vaisnys. 1964. Echolocation of flying insects by the bat *Chilonycteris parnellii*. Biol. Bull., *127*:478–488.

Odum, E. P. 1971. Fundamentals of ecology, 3rd ed. W. B. Saunders Co., Philadelphia. 574 pp.

O'Farrell, M. J., and E. H. Studier. 1970. Fall metabolism in relation to ambient temperatures in three species of *Myotis*. Comp. Biochem. Physiol., *35*:697–703.

O'Farrell, T. P. 1965. Home range and ecology of snowshoe hares in interior Alaska. J. Mamm., *46*:406–418.

O'Gara, B. W., R. F. Moy, and G. D. Bear. 1971. The annual testicular cycle and horn casting in the pronghorn *(Antilocapra americana)*. J. Mamm., *52*:537–544.

Orr, R. T. 1940. The rabbits of California. Occ. Papers California Acad. Sci., *19*:1–207.

———. 1971. Vertebrate biology, 3rd, ed. W. B. Saunders Co., Philadelphia. 544 pp.

Osborn, H. F. 1936-1942. Proboscidea: A monograph of the discovery, evolution, migration, and extinction of the mastodonts and elephants of the world. Vol. 1. Moeritheroidea, Deinotheroidea, Mastodontoidea. Vol. 2. Stegodontoidea, Elephantoidea. Amer. Mus. Nat. Hist., New York. 1675 pp. (Although the taxonomic schemes and the phylogenetic patterns presented in this paper have been seriously questioned, the figures and discussions of structure are useful.)

Parrington, F. R. 1971. On the Upper Triassic mammals. Phil. Trans. Roy. Soc. London, B, *261*:231–272.

Parry, D. A. 1949. The structure of whale blubber and its thermal properties. Quart. J. Microbiol. Sci., *90*:13–26.

Payne, R. S. 1961. The acoustical location of prey by the barn owl *(Tyto alba)*. Am. Zoologist, *1*:379.

———. 1970. Songs of the humpback whale. An LP Record by CRM Records, Del Mar, California.

———, and S. McVay. 1971. Songs of the humpback whales. Science, *173*:585–597.

Pearson, O. P. 1942. On the cause and nature of a poisonous action produced by the bite of a shrew *(Blarina brevicauda)*. J. Mamm., *23*:159–166.

———. 1948. Metabolism of small mammals with remarks on the lower limit of mammalian size. Science, *108*:44.

_____. "1959" 1960. Biology of the subterranean rodents, *Ctenomys*, in Peru. Mem. del Museo de Hist. Nat. "Javier Prado," 9:1–56.

_____. 1963. History of two local outbreaks of feral house mice. Ecology, 44:540–549.

_____. 1964. Carnivore-mouse predation: an example of its intensity and bioenergetics. J. Mamm., 45:177–188.

_____. 1966. The prey of carnivores during one cycle of mouse abundance. J. Anim. Ecol., 35:217–233.

_____. 1971. Additional measurements of the impact of carnivores on California voles (*Microtus californicus*). J. Mamm., 52:41–49.

_____., M. R. Koford, and A. K. Pearson. 1952. Reproduction of the lump-nosed bat (*Corynorhinus rafinesquei*) in California. J. Mamm., 33:273–320.

Perkins, J. 1945. Biology at Little America III, the west base of the United States Antarctic service expedition 1939-1941. Proc. Amer. Phil. Soc., 89:270–284.

Perry, J. S. 1954. Some observations on growth and tusk weight in male and female African elephants. Proc. Zool. Soc. London, 124:97–104.

Peterson, Randolph S. 1955. North American moose. Univ. Toronto Press, Toronto. 280 pp.

Peterson, Richard S. 1965. Behavior of the northern fur seal. Dr. Sci. Thesis. Johns Hopkins Univ., Baltimore. 214 pp.

_____., and G. A. Bartholomew. 1967. The natural history and behavior of the California sea lion. Amer. Soc. Mamm., Spec. Publ. No. 1. 79 pp.

Petter, J. J. 1965. The lemurs of Madagascar, pp. 292–319. *In* I. DeVore (ed.), Primate behavior. Holt, Rinehart and Winston, New York.

Pitelka, F. A. 1957a. Some characteristics of microtine cycles in the arctic. Eighteenth Ann. Biol. Coll., Oregon State College. pp. 73–88.

_____. 1957b. Some aspects of population structure in the short-term cycle of the brown lemming in northern Alaska. Cold Spr. Harb. Symp. Quant. Biol., 22:237–251.

_____., 1964. The nutrient-recovery hypothesis for Arctic microtine cycles. I. Introduction, pp. 55–56. *In* P. J. Crisp (ed.), Grazing in terrestrial and marine environments. Brit. Ecol. Soc. Symp. No. 4. Blackwell, Oxford.

_____., P. Q. Tomich, and G. W. Treichel. 1955. Ecological relations of jaegers and owls as lemming predators near Barrow, Alaska. Ecol. Monogr., 25:85–117.

Poulter, T. C. 1963. Sonar signals of the sea lion. Science, 139:753–755.

Purves, P. E. 1966. Anatomy and physiology of the outer and middle-ear in cetaceans, pp. 320–380. *In* K. S. Norris (ed.), Whales, dolphins and porpoises. Univ. California Press, Berkeley.

Quilliam, T. A. 1966. The problem of vision in ecology of *Talpa europaea*. Exp. Eye Res., 5:63–78.

Ralls, K. 1971. Mammalian scent marking. Science, 171:443–449.

Ratcliff, H. M. 1941. Winter range conditions in Rocky Mountain National Park. Trans. Sixth Amer. Wildl. Conf.: 132–139.

Ratcliffe, F. N. 1932. Notes on the fruit bat of Australia. J. Anim. Ecol., 1:32–37.

Rausch, R. 1950. Observations on a cyclic decline of lemmings (*Lemmus*) on the Arctic coast of Alaska during the spring of 1949. Arctic, 3:166–177.

Ray, C., and W. E. Schevill. 1965. The noisy underwater world of the Weddell seal. Animal Kingdom, New York Zool. Soc., 68:34–39.

Redman, J. P., and J. A. Sealander. 1958. Home ranges of deer mice in southern Arkansas. J. Mamm., 39:390–395.

Reed, C. A. 1944. Behavior of a shrew mole in captivity. J. Mamm., 25:196–198.

_____. 1951. Locomotion and appendicular anatomy in three soricoid insectivores. Amer. Midl. Nat., 45:513–671.

Reeder, W. G., and R. B. Cowles. 1951. Aspects of thermoregulation in bats. J. Mamm., 32:389–403.

Reig, O. A. 1970. Ecological note on the fossorial octodont rodent *Spalacopus cyanus* (Molina). J. Mamm., 51:592–601.

Reysenback de Haan, F. W. 1966. Listening underwater: thoughts on sound and cetacean hearing, pp. 583–596. *In* K. S. Norris (ed.), Whales, dolphins and porpoises. Univ. California Press, Berkeley.

Rice, D. W. 1967. Cetaceans, pp. 291–324. *In* S. Anderson and J. K. Jones, Jr. (eds.), Recent mammals of the world. The Ronald Press Co., New York.

_____., and A. A. Wolman. 1971. The life history and ecology of the gray whale (*Eschrichtius robustus*). Spec. Publ. No. 3, Amer. Soc. Mammal.

Ride, W. D. L. 1970. A guide to the mammals of Australia. Oxford Univ. Press, New York and London. 249 pp.

Ridgway, S. H. 1966. Proc., Third Ann. Conf. Biol. Sonar Diving Mammals, pp. 151–158. Stanford Res. Inst., Menlo Park, California.

Rinker, G. C. 1954. The comparative myology of the mammalian genera *Sigmodon, Oryzomys, Neotoma,* and *Peromyscus* (Cricetinae), with remarks on their intergeneric relationships. Miscl. Publ. Mus. Zool., Univ. Michigan, *83*:1–124.

Robinson, K., and D. H. K. Lee. 1941. Reactions of the cat to hot atmospheres. Proc. Roy Soc. Queensland, *53*:159–170.

Roeder, K. D. 1965. Moths and ultrasound. Sci. Amer., *212(4)*:94–102.

————., and A. E. Treat. 1961. The detection and evasion of bats by moths. Amer. Sci., *49(2)*:135–148.

Romer, A. S. 1966. Vertebrate paleontology, 3rd ed. Univ. Chicago Press, Chicago. 468 pp.

————. 1968. Notes and comments on vertebrate paleontology. Univ. Chicago Press, Chicago. 304 pp.

————. 1969. Cynodont reptile with incipient mammalian jaw articulation. Science, *166*:881–882.

————. 1970. The vertebrate body. W. B. Saunders Co., Philadelphia. 601 pp.

Rood, J. P. 1970. Notes on the behavior of the pygmy armadillo. J. Mamm., *51*:179.

Rosenzweig, M. R., D. A. Riley, and K. Krech. 1955. Evidence for echolocation in the rat. Science, *121*:600.

Rowan, W. 1950. Winter habits and numbers of timber wolves. J. Mamm., *31*:167–169.

Rowell, T. E. 1962. Agonistic noises of the rhesus monkey *(Macaca mulatta).* Symp. Zool Soc. London, 8:91–96.

Ruff, F. J. 1938. Trapping deer on the Pisgah National Game Preserve, North Carolina. J. Wildl. Manage., 2:151–161.

Rutherford, W. H. 1953. Effects of a summer flash flood upon a beaver population. J. Mamm., *34*:261–262.

Saunders, J. K., Jr. 1963. Movements and activities of the lynx in Newfoundland. J. Wildl. Manage., 27:390–400.

Schaeffer, B. 1947. Notes on the origin and function of the artiodactyl tarsus. Amer. Mus. Novitates, *1356*:1–24.

Schaller, G. B. 1963. The mountain gorilla: ecology and behavior. Univ. Chicago Press, Chicago.

————. 1964. The year of the gorilla. Univ. Chicago Press, Chicago.

————. 1967. The deer and the tiger. Univ. Chicago Press, Chicago.

Scheffer, V. B. 1958. Seals, sea lions and walruses. Stanford Univ. Press, Stanford, California. 179 pp.

————., and J. W. Slipp. 1944. The harbor seal in Washington state. Amer. Midl. Nat., *32*:373–416.

Schenkel, R. 1966a. Play, exploration and territoriality in the wild lion. Symp. Zool. Soc. London, No. 18:11–22.

————. 1966b. On sociology and behaviour in impala *(Aepyceros melampus suara* Matschie). Z. Säugetierk., *31*:177–205.

————. 1967. Submission: its features and functions in the wolf and dog. Amer. Zool., 7:319–329.

Schevill, W. E., and B. Lawrence. 1949. Underwater listening to the white porpoise, *Delphinapterus leucas.* Science, *109*:143–144.

————., ————. 1953. Auditory response of a bottle nosed porpoise, *Tursiops truncatus,* to frequencies above 100 kc. J. Exp. Zool., *124*:147–165.

————., and C. Ray. 1965. The Weddell seal at home. Animal Kingdom, New York Zool. Soc., 68:151–154.

————., and W. A. Watkins. 1965. Underwater calls of *Leptonychotes* (Weddell seal). Zoologica, New York Zool. Soc., *50*:45–47.

————., ————. 1966. Sound structure and directionality in *Orcinus* (killer whale). Zoologica, New York Zool. Soc., *51(2)*:71–76.

————., ————., and C. Ray. 1963. Underwater sounds of pinnipeds. Science, *141*:50–53.

————., ————., ————. 1966. Analysis of underwater *Odobenus* calls with remarks on the development and function of the pharyngeal pouches. Zoologica, New York Zool. Soc., *51(3)*:103–106.

Schmidt-Nielsen, B., and K. Schmidt-Nielsen. 1950a. Do kangaroo rats thrive when drinking sea water? Amer. J. Physiol., *160*:291–294.

————., ————. 1950b. Evaporative water loss in desert rodents in their natural habitat. Ecology, *31*:75–85.

————., ————. 1951. A complete account of water metabolism in kangaroo rats and experimental verification. J. Cell Comp. Physiol., *38*:165–182.

————., ————., T. R. Houpt, and S. A. Jarnum. 1956. Water balance of the camel. Amer. J. Physiol., *185*:185–194.

————., ————., ————., ————. 1957. Urea excretion in the camel. Amer. J. Physiol., *188*:477–484.

Schmidt-Nielsen, K. 1959. The physiology of the camel. Sci. Amer., *201*:140-151.

————. 1964. Desert animals, physiological problems of heat and water. Oxford Univ. Press, New York and Oxford. 277 pp.

————., W. L. Bretz, and C. R. Taylor. 1970. Panting in dogs: unidirectional air flow over evaporative surfaces. Science, *169*:1102–1104.

————., F. R. Hainsworth, and D. E. Murrish. 1970. Counter-current heat exchange in the respiratory passages: effects on water and heat balance. Resp. Physiol., *9*:263–276.

_____., and A. E. Newsome. 1962. Water balance in the mulgara *(Dasycercus cristicauda),* a carnivorous desert marsupial. Aust. J. Biol. Sci., *15*:683–689.

_____., and B. Schmidt-Nielsen. 1952. Water metabolism of desert mammals. Physiol. Rev., *32*:135–166.

_____., _____. 1953. The desert rat. Sci. Amer., *189(1)*:73–78.

_____., _____. 1954. Heat regulation in small and large desert animals, pp. 182–187. *In* J. L. Cloudsley-Thompson (ed.), Biology of deserts. Inst. Biol., London. 182 pp.

_____., _____., T. A. Houpt, and S. A. Jarnum. 1956. The question of water storage in the stomach of the camel. Mammalia, *20*:1–15.

_____., _____., _____., _____. 1957. Body temperature of the camel and its relation to water economy. Amer. J. Physiol., *188*:103–112.

Schnitzler, H. U. 1968. Echoortung bie der Ortungslaute der Hufeisen-Fledermäuse (Chiroptera-Rhinolophidae) in verschiedenen Orientierungssituationen. Z. Vergl. Physiol., *57*:376–408.

Scholander, P. F. 1940. Hvalradets Skrifter Norske Videnskaps-Akad. Oslo, *22*:1–131.

_____. 1955. Evolution of climatic adaptation in homeotherms. Evol., *9*:15–26.

_____., R. Hock, V. Walters, and L. Irving. 1950. Adaptation to cold in arctic and tropical mammals and birds in relation to body temperature, insulation, and basal metabolic rate. Biol. Bull., *99*:259–271.

_____., R. Hock, V. Walters, F. Johnson, and L. Irving. 1950. Heat regulation in some arctic and tropical mammals and birds. Biol. Bull., *99*:237–258.

_____., and J. Krog. 1957. Countercurrent and vascular heat exchange; sloths. J. Appl. Physiol., *10*:404–411.

_____., and W. E. Schevill. 1955. Countercurrent and vascular heat exchange; whales. J. Appl. Physiol., *8*:279–282.

_____., V. Walters, R. Hock, and L. Irving. 1950. Body insulation of some arctic and tropical mammals and birds. Biol. Bull., *99*:225–236.

Schubert, G. H. 1953. Ponderosa pine cone cutting by squirrels. J. Forestry, *51*:202.

Schultz, A. M. 1964. The nutrient-recovery hypothesis for Arctic microtine cycles. II Ecosystem variables in relation to Arctic microtine cycles, pp. 57–58. *In* P. J. Crisp (ed.), Grazing in terrestrial and marine environments. Brit. Ecol. Soc. Symp. No. 4 Blackwell, Oxford.

Schultze-Westrum, T. 1964. Nochweis differenzierter Duftstoffe beim Gleitbeulter *Petaurus breviceps papuanus* Thomas (Marsupialia, Phalangeridae). Naturwissenschaften, *51(9)*:226–227.

Schusterman, T. J., and S. N. Feinstein. 1965. Shaping and discriminative control of underwater click vocalization in a California sea lion. Science, *150*:1743–1744.

Sclater, W. L., and P. L. Sclater. 1899. The geography of mammals. Kegan, Paul, Trench, Trübner, London.

Selander, R. K. 1966. Sexual dimorphism and differential niche utilization in birds. Condor, *68*:113–151.

Selye, H. 1955. Stress and disease. Science, *122*:625–631.

_____., and J. B. Collip. 1936. Fundamental factors in the interpretation of stimuli influencing the endocrine glands. Endocrin., *20*:667–672.

Sewell, G. D. 1968. Ultrasound in rodents. Nature, *217*:682–683.

Sharman, G. B. 1970. Reproductive physiology of marsupials. Science, *167*:1221–1228.

Shaw, W. T. 1934. The ability of the giant kangaroo rat as a harvester and storer of seeds. J. Mamm., *15*:275–286.

_____. 1936. Moisture and its relation to the cone-storing habit of the western pine squirrel. J. Mamm., *17*:337–349.

Shkolnik, A., and Borut, A. 1969. Temperature and water relations in two species of spiny mice *(Acomys).* J. Mamm., *50*:245–255.

Shortridge, G. C. 1934. The mammals of Southwest Africa. 2 vols. Heinemann, London.

Siivonen, L. 1954. Features of short-term fluctuations. J. Wildl. Manage., *18*:38–45.

Simmons, J. H. 1971. Echolocation in bats: signal processing of echoes for target range. Science, *171*:925–928.

Simpson, G. G. 1940. Mammals and land bridges. J. Wash. Acad. Sci., *30*:137–163.

_____. 1943. Mammals and the nature of continents. Amer. J. Sci., *24*:1–31.

_____. 1944. Tempo and mode in evolution. Columbia Univ. Press, New York. 237 pp.

_____. 1945. The principles of classification and a classification of mammals. Bull. Amer. Mus. Nat. Hist., *85*:1–350.

_____. 1947. Evolution, interchange, and resemblance of North American and Eurasian Cenozoic mammalian faunas. Evolution, *1*:218–220.

_____. 1950. History of the fauna of Latin America. Amer. Sci., *38*:361–389.

_____. 1951. Horses. Oxford Univ. Press, New York. 247 pp.

_____. 1952. Probabilities of dispersal in geologic time. Bull. Amer. Mus. Nat. Hist., *99*:163–176.

_____. 1953. The major features of evolution. Columbia Univ. Press, New York. 434 pp.

_____. 1959. Mesozoic mammals and the polyphyletic origin of mammals. Evolution, *13*:405–414.

_____. 1961. Historic zoogeography of Australian mammals. Evolution, *15*:431–446.

_____. 1965a. Attending marvels. A Patagonian journal. Time Inc., New York. 289 pp.

_____. 1965b. The geography of evolution. Chilton Books, Philadelphia. 249 pp.

Slaughter, B. H. 1968. Earliest known marsupials. Science, *162(3850)*:254–255.

———., and D. W. Walton (eds.). 1970. About bats. Southern Methodist Univ. Press, Dallas. 339 pp.

Slijper, E. J. 1936. Die Cetacean, vergleichend-anatomisch und systematisch. Capita Zool., 7:1–590.

———. 1962. Whales. Basic Books, Inc., New York. 475 pp.

Smith, J. D. 1972. Systematics of the chiropteran family Mormoopidae. Univ. Kansas Publ., Mus. Nat. Hist. In press.

Smith, H. M. 1960. Evolution of chordate structure. Holt, Rinehart, and Winston, New York. 529 pp.

Smith, P. W. 1965. Recent adjustments in animals' ranges, pp. 633–642. *In* H. E. Wright, Jr. and D. G. Frey (eds.), The Quaternary of the United States. Princeton Univ. Press, Princeton, New Jersey.

Smith, R. B. 1971. Seasonal activities and ecology of terrestrial vertebrates in a Neotropical monsoon environment. M. S. Thesis, Northern Arizona University, Flagstaff, Arizona.

Sorenson, M. W., and C. H. Conaway. 1968. The social and reproductive behavior of *Tupaia montana* in captivity. J. Mamm., 49:502–512.

Southern, H. N. 1964. Handbook of British mammals. Blackwell, Oxford.

Spencer, A. W., and H. W. Steinhoff. 1968. An explanation of geographic variation in litter size. J. Mamm., 49:281–286.

Spencer, D. A. 1958a. Preliminary investigations on the northwestern *Microtus* irruption. U. S. Fish and Wildl. Serv., Denver Wildl. Res. Lab. Spec. Report.

———. 1958b. Biological and control aspects, pp. 15–25. *In* The Oregon meadow mouse irruption of 1957–1958. Federal Cooperative Extension Service, Oregon State College, Corvallis, Oregon.

Spitz, F. 1963. Estude des densities de population de *Microtus arvalis*. Pall. A Saint-Michal-en L'Hern (Vendu). Mammalia, 27:497–531.

Sprankel, H. 1965. Untersuchungen an *Tarsus*. I Morphologie des Schwanzes nebst ethologischen Bemerkungen. Folia Primat., 3:153–188.

Steiniger, F. 1950. Beiträge zur Soziologie und sonstigen Biologie der Wanerratte. Z. Tierpsychol., 7:356–379.

Stenlund, M. H. 1955. A field study of the timber wolf *(Canis lupus)* on the Superior National Forest, Minnesota. Minn. Dept. Cons. Tech. Bull., 4. 55 pp.

Stephenson, A. B. 1969. Temperatures within a beaver lodge in winter. J. Mamm., 50:134–136.

Stevenson-Hamilton, J. 1947. Wild life in South Africa. London, Cassell. 364 pp.

Stirton, R. A., R. H. Tedford, and M. O. Woodburne. 1967. A new Tertiary formation and fauna from the Tirari Desert, South Australia. Records of the South Australian Mus., 15:427–462.

Stock, C. 1949. Rancho La Brea: a record of Pleistocene life in California. L. A. County Mus., Sci. Ser., No. 13:1–81.

Stones, R. C., and J. E. Wiebers. 1965. A review of temperature regulation in bats (Chiroptera). Amer. Midl. Nat., 74:155–167.

———., 1967. Temperature regulation in the little brown bat, *Myotis lucifugus*, pp. 97–109. *In* K. C. Fisher, A. R. Dawe, C. P. Lyman, E. Schonbaum, and F. E. South, Jr. (eds.), Mammalian hibernation, Vol. III. Oliver & Boyd and Amer. Elsevier, New York.

Storer, R. W. 1955. Weight, wing area, and skeletal proportions in three accipiters. Acta 11th Cong. Intern. Ornithol.: 278–290.

Storer, T. I., and R. L. Usinger. 1965. General zoology. McGraw-Hill Book Co., New York. 741 pp.

Struhsaker, T. T. 1967. Behavior of elk (*Cervus canadensis*) during the rut. Z. Tierpsychol., *24(1)*:80–114.

Studier, E. H., and D. J. Howell. 1969. Heart rate of female big brown bats in flight. J. Mamm., 50:842–845.

———., J. W. Proctor, and D. J. Howell. 1970. Diurnal body weight loss and tolerance of weight loss in five species of *Myotis*. J. Mamm., 51:302–309.

Suthers, R. A. 1965. Acoustic orientation by fish-catching bats. J. Exp. Zool., 158:319–348.

———. 1967. Comparative echolocation by fishing bats. J. Mamm., 48:79–87.

Swank, W. G. 1958. The mule deer in Arizona chaparral. State of Arizona Game and Fish Dept., Wildl. Bull. No. 3. 109 pp.

Swinhoe, R. 1870. On the mammals of Hainan. Proc. Zool. Soc., London, 1870:224–239.

Szalay, F. S. 1968. The beginnings of primates. Evolution, 22:19–36.

———. 1969. Origin and evolution of function of the mesonychid condylarth feeding mechanism. Evolution, *23(4)*:703–720.

Taber, R. D., and R. F. Dasmann. 1957. The dynamics of three natural populations of deer *Odocoileus hemionus columbianus*. Ecology, 38:233–246.

Talbot, L. M., and M. H. Talbot. 1963. The wildebeest in western Masailand. Wildl. Monogr., Chestertown, No. 12:1–88.

Talmage, R. V., and G. D. Buchanan. 1954. The armadillo (*Dasypus novemcinctus*): a review of its natural history, ecology, anatomy and reproductive physiology. The Rice Institute Pamphlet, Monograph in Biology, Vol. 41, 135 pp.

Tappen, N. C. 1960. Problems of distributions and adaptations of the African monkeys. Cur. Anthrop., 1:91–120.

Tate, G. H. H. 1933. A systematic revision of the marsupial genus *Marmosa*. Bull. Amer. Mus. Nat. Hist., *66*:1–250.

———. 1947. Mammals of eastern Asia. Macmillan Co., New York. 366 pp.

Tavolga, M. C. 1966. Behavior of the bottlenose dolphin *(Tursiops truncatus)*: social interaction in a captive colony, pp. 718–730. *In* K. S. Norris (ed.), Whales, dolphins and porpoises. Univ. California Press, Berkeley.

Taylor, C. R. 1968. Hygroscopic food: a source of water for desert antelopes? Nature, *219*:181–182.

———. 1969a. The eland and the oryx. Sci. Amer., *220*:89–95.

———. 1969b. Metabolism, respiratory changes, and water balance of an antelope, the eland. Amer. J. Physiol., *217(1)*:317–320.

———, and C. P. Lyman. 1967. A comparative study of the environmental physiology of an East African antelope, the eland, and the Hereford steer. Physiol. Zool., *40(3)*:280–295.

———, C. A. Spinage, and C. P. Lyman. 1969. Water relations of the waterbuck, an East African antelope. Amer. J. Physiol., *217(2)*:630–634.

Taylor, W. T., and R. J. Weber. 1951. Functional mammalian anatomy. D. Van Nostrand Co., New York. 575 pp.

Tevis, L. 1950. Summer behavior of a family of beavers in New York State. J. Mamm., *31*:40–65.

Thomas, R. 1971. Personal communication.

Thompson, D. Q. 1955. The role of feed and cover in population fluctuations of the brown lemming at Point Barrow, Alaska. Trans. N. Amer. Wildl. Conf., *20*:166–176.

Tinbergen, N. 1963. On aims and methods of ethology. Z. Tierpsychol., *20*:410–433.

Tinkle, D. W., and I. G. Patterson. 1965. A study of hibernating populations of *Myotis velifer* in northwest Texas. J. Mamm., *46*:612–633.

Tomich, P. Q. 1962. The annual cycle of the California ground squirrel, *Citellus beecheyi*. Univ. California Publ. Zool., 65:213–282.

Torrey, T. W. 1971. Morphogenesis of the vertebrates. John Wiley and Sons, Inc., New York. 529 pp.

Townsend, M. T. 1935. Studies on some small mammals of central New York. Roosevelt Wildl. Annals, *4*:1–120.

Troughton, E. 1947. Furred animals of Australia. Charles Scribner's Sons, New York. 374 pp.

Trumler, E. 1959. Das "Rossighkeitsgesicht" und ähnliches Ausdrucksverhalten bei Einhufern. Z. Tierpsychol., *16*:478–488.

Tucker, V. A. 1962. Diurnal torpidity in the California pocket mouse. Science, *136*:380–381.

———. 1965. The relation between the torpor cycle and heat exchange in the California pocket mouse *Perognathus californicus*. J. Cell Comp. Physiol., *65*:405–414.

Udvardy, M. D. F. 1969. Dynamic zoogeography. Van Nostrand Reinhold Co., New York. 445 pp.

Van Deusen, H. M. 1967. Personal communication.

———. 1969. Results of the Archbold Expeditions. No. 90. Notes on the echidnas (Mammalia, Tachyglossidae) of New Guinea. Amer. Mus. Novitates, *2383*:1–23.

———. 1971. *Zaglossus*, New Guinea's egg-laying anteater. Fauna, Vol. 1:12–19.

———, and J. K. Jones, Jr. 1967. Marsupials, pp. 61–86. *In* S. Anderson and J. K. Jones, Jr. (eds.), Recent mammals of the world. The Ronald Press Co., New York.

Van Gelder, R. G. 1953. The egg-opening technique of a spotted skunk. J. Mamm., *34*:255–256.

Van Lawick-Goodall, J. 1968. A preliminary report on expressive movements and communication in the Gombe Stream Chimpanzees, pp. 313–374. *In* P. C. Jay (ed.), Primates, studies in adaptation and variability. Holt, Rinehart and Winston, New York.

Van Valen, L. 1965a. The earliest primates. Science, *150*:743–745.

———. 1965b. Treeshrews, primates and fossils. Evolution, *19*:137–151.

———. 1966. Deltatheridia, a new order of mammals. Bull. Amer. Mus. Nat. Hist., *132*:1–126.

Vaughan, T. A. 1954. Mammals of the San Gabriel mountains of California. Univ. Kansas Publ., Mus. Nat. Hist., 7:513–582.

———. 1959. Functional morphology of three bats: *Eumops, Myotis, Macrotus*. Univ. Kansas Publ., Mus. Nat. Hist., *12*:1–153.

———. 1961. Vertebrates inhabiting pocket gopher burrows in Colorado. J. Mamm., *42*:171–174.

———. 1966a. Morphology and flight characteristics of molossid bats. J. Mamm., *47*:249–260.

———. 1966b. Food-handling and grooming behaviors in the plains pocket gopher. J. Mamm., *47*:132–133.

———. 1967. Food habits of the northern pocket gopher on shortgrass prairie. Amer. Midl. Nat., *77*:176–189.

———. 1969. Reproduction and population densities in a montane small mammal fauna, pp. 51–74. *In* J. K. Jones, Jr. (ed.), Contributions in mammalogy. Miscl. Publ., Mus. Nat. Hist., Univ. Kansas, No. 51.

_____. 1970a. Adaptations for flight in bats, pp. 127–143. *In* B. H. Slaughter and D. W. Walton (eds.), About bats. Southern Methodist Univ. Press, Dallas.

_____. 1970b. The skeletal system. The muscular system. Flight patterns and aerodynamics, pp. 97–216. *In* W. A. Wimsatt (ed.), Biology of bats. Academic Press, New York.

_____., and G. C. Bateman. 1970. Functional morphology of the forelimbs of mormoopid bats. J. Mamm., *51*:217–235.

Vessey-FitzGerald, D. F. 1960. Grazing succession among East African game animals. J. Mamm., *41*:161–172.

Villa-R., B. 1966. Los Murciélagos de Mexico. Inst. Biol., UNAM. 491 pp.

_____., and E. L. Cockrum. 1962. Migration in the guano bat, *Tadarida brasiliensis mexicana*. J. Mamm., 43:43–64.

Vogel, V. B., and geb. El-Kareh. 1969. Vergleichende Untersuchungen über den Wasserhaushalt von Fledermäusen (*Rhinopoma, Rhinolophus* und *Myotis*). Z. Vergl. Physiol., *64*:324–345.

Wallace, A. R. 1876. The geographical distribution of animals. 2 vols. Harper, New York. 503 pp. and 553 pp. Reprinted by Hafner Publ., New York.

Walker, E. P. 1968. Mammals of the world, 2nd ed. 2 vols. Johns Hopkins Press, Baltimore.

Walther, F. 1958. Zum Kampf- und Paarungsverhalten einiger Antilopen. Z. Tierpsychol., *15*:340–382.

_____. 1965. Verhaltensstudien an der Grantgazell (*Gazella granti* Brooke, 1872) im Ngorongoro-Krater. Z. Tierpsychol., *22*:167–208.

_____. 1966. Zum Liegeverhalten des Weissschwanzgnus (*Connochaetes gnou* Zimmerman, 1780). Z. Säugetierk., *31*:1–16.

Ward, A. L., and J. O. Keith. 1962. Feeding habits of pocket gophers on mountain grasslands. Ecology, *43*:744–749.

Watkins, W. A., and W. E. Schevill. 1968. Underwater playback of their own sounds to *Leptonychotes* (Weddell seals). J. Mamm., *49*:287–296.

Watson, A. 1958. The behavior, breeding and food-ecology of the snowy owl *Nyctea scandiaca*. Ibis, *99*:419–462.

Weber, N. S., and J. S. Findley. 1970. Warm-season changes in fat content of *Eptesicus fuscus*. J. Mamm., *51*:160–162.

Webster, D. B. 1961. The ear apparatus of the kangaroo rat, *Dipodomys*. Am. J. Anat., *108*:123–148.

_____. 1962. A function of the enlarged middle-ear cavities of the kangaroo rat, *Dipodomys*. Physiol. Zool., *35*:248–255.

_____. 1963. A case of parallel evolution of the ear apparatus. Anat. Rec., *145*:297.

_____. 1966. Ear structure and function in modern mammals. Amer. Zoologist, *6*:451–466.

_____., R. F. Ackermann, and G. C. Longa. 1968. Central auditory system of the kangaroo rat, *Dipodomys merriami*. J. Comp. Neur., *133(4)*:477–494.

_____., and C. R. Stack. 1968. Comparative histochemical investigation of the organ of Corti in the kangaroo rat, gerbil and guinea pig. J. Morph., *129(4)*:413–434.

Webster, F. A., and O. G. Brazier. 1968. Experimental studies on echolocation mechanisms in bats. Aerospace Medical Research Laboratories, Wright-Patterson Air Force Base, Ohio. 156 pp.

_____., and D. R. Griffin. 1962. The role of flight membranes in insect capture by bats. Animal Behav., *10*:332–340.

Wegener, A. 1915. Die Entstehung der Kontinente und Ozeane. Sammlung Vieweg. No. 23. 144 pp. Brunswick. (Translation, 1924. The origin of continents and oceans. Methuen, London. 212 pp.).

Weichert, C. K. 1965. Anatomy of the chordates. McGraw-Hill Book Co., New York. 758 pp.

Weil, W. L. P. 1968. Food habits of the western jumping mouse in north-central Colorado. M.S. Thesis, Colorado State Univ., Fort Collins, 23 pp.

Wells, P. V., and C. D. Jorgensen. 1964. Pleistocene wood rat middens and climatic change in the Mojave Desert: a record of juniper woodlands. Science, *143*:1171–1174.

Wharton, C. H. 1950. Notes on the life history of the flying lemur, *Cynocephalus volans*. J. Mamm., *31*:269–273.

Whitaker, J. O., Jr. 1963. Food, habitat and parasites of the woodland jumping mouse in central New York. J. Mamm., *44*:316–321.

White, A. C. 1948. The call of the Bushveld. Bloemfontein, S. Africa, A. C. White P. & P. Co. 269 pp.

Whittaker, R. H. 1970. Communities and ecosystems. Macmillan Co., New York. 162 pp.

Williams, G. C. 1967. Natural selection, the costs of reproduction, and a refinement of Lack's principle. Amer. Nat., *100*:687–690.

Wilson, R. W. 1960. Early Miocene rodents and insectivores from northeastern Colorado. Univ. Kansas Paleont. Cont., Vertebrata, Art. 7, pp. 1–92.

Wimsatt, W. A. 1944. Further studies on the survival of spermatozoa in the female reproductive tract of the bat. Anat. Rec., *88*:193–204.

_____. 1945. Notes on breeding behavior, pregnancy, and parturition in some vespertilionid bats of eastern United States. J. Mamm., *26*:23–33.

_____. 1969. Transient behavior, nocturnal activity patterns, and feeding efficiency of vampire bats *(Desmodus rotundus)* under natural conditions. J. Mamm., *50*:233–244.

_____. (ed.). 1970. Biology of bats. Academic Press, New York. 406 pp.

_____., and B. Villa-R. 1970. Locomotor adaptations in the disc-winged bat. Amer. J. Anat., *129*:89–120.

Winge, H. 1941. The interrelationships of the mammalian genera. Vol. 1. C. A. Reitzels Forlag, Copenhagen. 418 pp. (Translation from Danish by E. Deichmann and G. M. Allen.)

Wirtz, W. O. II. 1968. Reproduction, growth and development, and juvenile mortality in the Hawaiian monk seal. J. Mamm., *49*:229–238.

_____. 1971. Personal communication.

Wood, A. E. 1955. A revised classification of the rodents. J. Mamm., *36*:165–187.

_____. 1965. Grades and clades among rodents. Evolution, *19*:115–130.

Wood, F. G., Jr. 1959. Underwater sound production and concurrent behavior of captive porpoises, *Tursiops truncatus* and *Stenella plagiodon.* Bull. Mar. Sci. Gulf and Caribbean, *3*:120–133.

Wynne-Edwards, V. C. 1959. The control of population density through social behavior: a hypothesis. Ibis, *101*:436–441.

_____. 1960. The overfishing principle applied to natural populations and their food resources: and a theory of natural conservation. Proc. Int. Orn. Congr., *12*:790–794.

_____. 1962. Animal dispersion in relation to social behavior. Hafner Publ. Co., New York. 653 pp.

Yanagisawa, K., G. Sato, M. Nomoto, Y. Katsuki, E. Ibezono, A. D. Grinnell, and T. H. Bullock. 1966. Federation Proc., Physiol., p. 1539 (abstr.).

Yoakum, J. 1958. Seasonal food habits of the Oregon pronghorn antelope *(Antilocapra americana oregona* Bailey). Inter. Antelope Conf. Trans., *9*:47–59.

Young, J. Z. 1957. The life of mammals. Clarendon Press, Oxford. 820 pp.

Young, S. P., and H. H. T. Jackson. 1951. The clever coyote. Wildl. Manage. Inst., Washington, D.C. 411 pp.

Zimmerman, E. G. 1965. A comparison of habitat and food of two species of *Microtus.* J. Mamm., *46*:605–612.

Zippelius, H., and W. M. Schleidt. 1956. Ultraschalllaute bei jungen Mausen. Naturwissenschaften, *21*:1–2.

INDEX

Page numbers in *italics* indicate illustrations. The symbol *t* indicates a table.